# Public Sector Deficits

# and

# Macroeconomic

# Performance

# Public Sector Deficits and Macroeconomic Performance

Edited by
**William Easterly**
**Carlos Alfredo Rodríguez**
**Klaus Schmidt-Hebbel**

*Published for the World Bank*
Oxford University Press

*Oxford University Press*

OXFORD   NEW YORK   TORONTO
DELHI   BOMBAY   CALCUTTA   MADRAS   KARACHI
KUALA LUMPUR   SINGAPORE   HONG KONG   TOKYO
NAIROBI   DAR ES SALAAM   CAPE TOWN
MELBOURNE   AUCKLAND

*and associated companies in*

BERLIN   IBADAN

© 1994 The International Bank for Reconstruction
and Development / THE WORLD BANK
1818 H Street, N.W.
Washington, D.C. 20433, U.S.A.

Published by Oxford University Press, Inc.
200 Madison Avenue, New York, N.Y. 10016

Oxford is a registered trademark of Oxford University Press.

Manufactured in the United States of America
First printing November 1994

*Library of Congress Cataloging-in-Publication Data*

Easterly, William Russell.
    Public sector deficits and macroeconomic performance / William
Easterly, Carlos Alfredo Rodríguez, Klaus Schmidt-Hebbel.
        p.    cm.
''Published for the World Bank.''
Includes bibliographical references and index.
ISBN 0-19-520988-5
    1. Budget deficits—Developing countries.   2. Fiscal policy—
Developing countries.   I. Rodríguez, Carlos A. (Carlos Alfredo)
II. Schmidt-Hebbel, Klaus.   III. International Bank for
Reconstruction and Development.   IV. Title.
HJ2216.E37     1994
339.5'23'091724—dc20                                           94-11506
                                                                 CIP

# Contents

## Part II. Country Case Studies    99

### 3  Argentina: Fiscal Disequilibria Leading to Hyperinflation    101
*Carlos Alfredo Rodríguez*

### 4  Chile: Fiscal Adjustment and Successful Performance    167
*Jorge Marshall and Klaus Schmidt-Hebbel*

### 5  Colombia: Avoiding Crises through Fiscal Policy    225
*William Easterly*

# Foreword

In setting out the lessons of development experience, the authors of *World Development Report 1991* surely did not hesitate before writing that a "prudent fiscal policy is the foundation of a stable macroeconomy" (p. 110). The consequences of excessive fiscal deficits are well known: inflation, the crowding-out of private investment, debt problems, and balance of payments problems. So why should the World Bank have spent resources on a research project examining the consequences of fiscal deficits?

A sufficient reason is that adjectives such as "prudent" and "excessive" are inadequate guides to action. It is generally agreed that budget deficits which, since 1986, have exceeded 8 percent of gross national product (GNP) bear much of the blame for the current economic disaster in the former Soviet Union. But Pakistan's fiscal deficits have exceeded 8 percent of GNP each year since 1985. If a deficit of 8 percent a year is excessive, why did Pakistan experience steady growth along with single-digit inflation in the 1980s? Are the inevitable consequences merely being delayed? Or are some countries able to sustain large deficits for some time without ill consequences? If so, what determines the size and duration of sustainable deficits? These are among the questions studied in this volume.

Another reason for this study is that the conventional wisdom should always be suspect. The results reported here do contain at least one major surprise: the link between inflation and budget deficits is very weak, at least for the countries examined in this study. The connection between budget deficits and inflation is evidently complicated: inflation of triple and more digits is always associated with large deficits, but large deficits do not necessarily lead to high inflation.

This book shows World Bank research at its best. The country studies draw on detailed knowledge and data from a wide range of countries; they are well constructed, well informed, and valuable in their own right and are the main sources from which conclusions are drawn. The conclusions are presented concisely and convincingly by the editors in the introductory chapters.

The busy policymaker should certainly read the introductory chapters. The serious researcher will benefit from working at and reflecting on the country studies as well.

Stanley Fischer
Professor of Economics
Massachusetts Institute of Technology

*July 1994*

# *Acknowledgments*

The research presented in this volume is the outcome of the World Bank research project on Macroeconomics of Public Sector Deficits. The World Bank Research Committee provided valuable comments at the initiation of the project. We thank the members of the project's advisory committee, Robert Armstrong, Edgardo Barandiarán, Mohsin Khan, Danny Leipziger, Johannes Linn, and Costas Michalopoulos, for their dedication and their contributions to the project's preparation and execution. We are also indebted to Mario Blejer, Vittorio Corbo, and Stanley Fischer for continuing comments and support. Maria Cristina Almero provided valuable research assistance. We are also indebted to Nancy Levine for her excellent editorial support. To all of them and to the participants in the World Bank Conference on the Macroeconomics of the Public Sector Deficit, held June 20–21, 1991, in Washington, D.C., go our thanks.

# Definitions and Data Notes

Statistics are typically for calendar years (Argentina, Chile, Colombia, Côte d'Ivoire, and Morocco) or for fiscal years, denoted by a slash (Pakistan and Zimbabwe). Data for Ghana are typically based on fiscal years through 1982 and on calendar years thereafter.

Variables are defined in each chapter. A dot over a variable represents a time derivative (change per unit of time). A bar denotes an average level of a variable. An asterisk on a variable indicates that the variable is measured in foreign currency. A hat (^) denotes a proportional rate of change. Blanks in tables reporting regression results denote variables omitted from the regressions.

A billion is a thousand million.

# Contributors

| | |
|---|---|
| Christophe Chamley | Professor, Department of Economics, Boston University |
| William Easterly | Principal economist, Macroeconomics and Growth Division, Policy Research Department, The World Bank |
| Riccardo Faini | Professor, Department of Economics, University of Brescia |
| Hafez Ghanem | Senior economist, World Development Report, The World Bank |
| Nadeem Ul Haque | Deputy Division Chief, Research Department, International Monetary Fund |
| Roumeen Islam | Economist, Middle East and North Africa Regional Office, Country Department I, The World Bank |
| Jorge Marshall | Vice President, Central Bank of Chile |
| Peter J. Montiel | Division Chief, Macroeconomics and Growth Division, Policy Research Department, The World Bank |
| Felipe Morandé | Director, ILADES/Georgetown University Economics Program |
| Carlos Alfredo Rodríguez | Director, Centro de Estudios Macroeconómicos de la Argentina |
| Klaus Schmidt-Hebbel | Senior economist, Macroeconomics and Growth Division, Policy Research Department, The World Bank |
| Vito Tanzi | Director, Fiscal Affairs Department, International Monetary Fund |
| Deborah Wetzel | Economist, Office of Development Economics and the Chief Economist, The World Bank |

# Overview

*William Easterly, Carlos Alfredo Rodríguez,*
*and Klaus Schmidt-Hebbel*

The following are the main findings and conclusions of this volume.

1. The correlation of fiscal deficits with any one indicator of macroeconomic imbalance (such as inflation, real exchange rates, or market-determined real interest rates) is close to zero. This is explained by the great variety of ways in which governments finance their deficits.
2. Deficits, however, are unambiguously bad for growth. Reliance on taxation of financial assets to finance deficits—through surges of monetary financing or issuing of domestic debt at controlled nominal interest rates—is bad for private investment and for growth. Reduction of deficits through conventional tax increases is no more contractionary of short-run demand than taxing financial assets and is far preferable for long-run growth.
3. Large fiscal deficits are largely explained by conscious policy choices and not by external shocks or by feedback from domestic economic conditions.
4. We reject the fiscal approach to inflation, whereby inflation is thought to be determined largely by the monetary financing requirements of the government. The long-run association between monetary financing and inflation is indisputable. However, the amount of revenue from monetary financing is small, and some countries are even beyond the level of inflation at which inflation tax revenue is maximized. Thus it is doubtful that revenue motivations are behind inflation. Large temporary accelerations of monetary financing do not consistently result in higher inflation.
5. The conventional wisdom that public investment is good for private investment is contradicted by the evidence in half of the case studies, where public investment has a negative and statistically significant effect on private investment. The negative association in some cases is explained by the likelihood that public investment is replacing rather than complementing private investment. Concentration of public investment on infrastructure and on privatization of other state enterprises would

ensure a complementary relationship between the public and private sectors.

6. Increasing public saving (reducing the deficit) is the only policy measure known to be effective in raising national saving. In only a few cases is there evidence that lower deficits (higher public saving) increase private consumption and thus lower private saving, and even in these cases the effect is far less than one for one.

7. Fiscal deficits lead to current account deficits and overvalued currencies. Conversely, fiscal stabilization is a prerequisite for external adjustment and real depreciation of the currency.

The decade of the 1980s was the decade of fiscal adjustment; so far, the 1990s seem to call for more of the same. The severe erosion of fiscal discipline in the 1970s and the first half of the 1980s has proved difficult to reverse. This widespread fiscal crisis coincided with the worst decade for growth and macroeconomic instability in the post–World War II era. Attaining macroeconomic balance has become the foremost priority of industrial and developing economies alike; often, it is the measure of government success.

Despite the consensus on the need to reduce public sector deficits, developing countries offer a bewildering range of experience with deficits. From country to country, deficits may lead to high and variable inflation, to debt crises, or to low inflation with crowding-out of investment and growth, while in some countries moderately high deficits seem not to generate macroeconomic imbalances at all. What explains these differing outcomes?

This volume presents the findings of a World Bank research project that sought to answer this question through a careful examination of ten country case studies.[1] The case studies and the synthesis chapter use a common framework to assess the macroeconomic consequences of public deficits in a small but representative sample of developing countries.

## Part I: Synthesis and Methodology

Chapter 1 synthesizes and extends the results of the case studies. It finds that simple correlations between fiscal deficits and individual indicators of macroeconomic imbalance—inflation and interest rates—are weak to nonexistent. A simple explanation follows from the theoretical framework of the study: since each of the many ways of financing a high fiscal deficit carries its own macroeconomic outcome, the deficit will not be correlated with any one type of outcome. The chapter, however, finds a strong negative correlation between deficits and growth. No matter how deficits are financed, the consequences are bad news for the long-run health of the economy.

While most of this book treats fiscal deficits as a policy lever chosen by the government, the synthesis chapter examines the extent to which deficits are driven by external shocks (involving, for example, commodity prices and foreign interest rates) or by domestic macroeconomic variables (such as inflation, domestic interest rates, and real exchange rates). Only in Colombia do foreign shocks explain more than 50 percent of the (modest) variation in deficits. External shocks explain very little of the variation in countries with more dramatic changes in deficits, such as Chile, Ghana, and Zimbabwe. The Olivera-Tanzi effect, whereby inflation lowers tax revenue, shows up in only two cases, Ghana and Colombia. A real exchange rate depreciation was found to increase the deficit in Chile and Ghana, lower it in Mexico, and leave it roughly unchanged in Colombia, Thailand, and Zimbabwe. Exogenous policy changes account for most of the variation of deficits in significant episodes of fiscal improvement or deterioration in the case studies.

Although inflation and deficits show no simple correlation, the synthesis confirms other studies that find a long-run association between money creation (as a ratio to gross domestic product—GDP) and inflation. Figure 1.1, in chapter 1, shows that the relationship follows the typical "Laffer curve"; revenues from money creation first rise and then fall as the inflation rate rises. Revenue from money creation is often identified with the "inflation tax" on money holdings (which is slightly inaccurate because revenue from money creation also includes the real growth in money demand). Maximum "revenue" is between 70 and 160 percent inflation. As the maximum "revenue" is approached, the tradeoff between additional monetary financing and inflation becomes more unfavorable: an additional percentage point of GDP of monetary financing induces only 5 additional percentage points of inflation in Thailand but 97 additional percentage points of inflation in Argentina. The case studies find much higher revenue-maximizing inflation rates, compared with our cross-sectional estimate, in the high-inflation countries and much lower rates in the low-inflation countries. The synthesis chapter argues that these findings result from the false assumption that money demand has a constant inflation elasticity. Once a variable inflation elasticity is taken into account, the true revenue-maximizing inflation rate is close to the cross-sectional estimate of between 70 and 160 percent inflation.

With all the sound and fury about money creation, it is easy to forget how small it is as a source of financing. In a sample of fifty-one countries over nearly twenty years, three-quarters of the annual observations of inflation tax revenue are less than 2 percent of GDP. In many countries excise tax revenues on a single product account for as much revenue as the inflation tax does. Considering the steep costs associated with inflation, it is hard to believe the fiscal theories of inflation that ascribe inflation to revenue motivations. The inflation tax tail is just too small to be wagging the inflation dog.

The real significance of money creation is its potential for generating large temporary revenues in times of crisis. Even in those cases, temporary bursts of inflation tax revenue did not always result in accelerations of inflation. The most consistent link between money creation and inflation is the long-run association described above. (Those countries resorting to large temporary use of inflation did have lower growth, however.) Huge one-time taxes on financial assets also lowered the potential for future revenue from money creation. In the most spectacular example, the Ghana study found that an episode of demonetization of the largest-denomination note in 1982 (along with an earlier deposit expropriation) permanently lowered revenue from money creation by 1 to 2 percentage points of GDP.

High fiscal deficits are significantly associated with highly negative real interest rates. Financial repression was a popular means of "financing" fiscal deficits; in our study Ghana, Mexico, and Zimbabwe maintained nominal interest rate controls that resulted in negative real rates. (Mexico liberalized interest rates in 1988.) The control of interest rates was a costly way of mobilizing a small amount of "revenue" (about 1 percent of GDP) through the implicit tax on deposits. The ratio of private credit to GDP in financially repressed economies is only about a third that in unrepressed economies.

The paucity of private credit in high-deficit, financially repressed economies and the increase in interest rates in high-deficit, unrepressed economies both have the effect of driving out private investment. The conventional wisdom that deficits crowd out investment is reaffirmed by these studies. A debt-financed deficit is particularly damaging because rising public debt service crowds out even more investment over time. In Colombia, for example, a debt-financed expansion of 0.9 percentage point of GDP a year in the deficit over three years would lead private investment to fall 0.5 percentage point of GDP in the first year and 0.9 percentage point in the last. Deficits under financial repression had even stronger instantaneous impacts on private investment than those under decontrolled markets.

The conventional wisdom fares less well in the analysis of public investment. The studies of Chile, Colombia, Ghana, and Mexico found *negative* effects of public investment on private investment; the Argentina study found zero effect. These results are consistent with the theoretical ambiguity of the relationship between public and private capital. If public capital substitutes for private capital (a public steel mill, for example), then public investment drives down the rate of return to private capital and lowers private investment. These results do not support the argument that fiscal austerity implemented through cuts in public investment has hurt private investment.

Various theories have raised the hope that private saving would rise when public deficits rose, attenuating the potential crowding-out of private investment. The Ricardian equivalence theory holds that the private sector saves in anticipation of future tax obligations; an alternative theory is that fiscal deficits directly crowd out private consumption through widespread rationing of goods and controls on financial markets. Only three of the case studies—Argentina, Morocco, and Zimbabwe—showed evidence of increased private saving in response to deficits, and the savings were much less than offsetting. The evidence in these cases suggests that direct crowding-out of consumption caused increased saving.

If increased private saving only slightly offsets increases in deficits, it follows that reducing deficits will be effective in raising national saving. Since other research finds little effect of policy variables (such as interest rates or tax incentives) on saving, reducing public expenditure is the only policy measure known to be effective in increasing national saving.

Chapter 1 strongly confirms the prediction of the theoretical framework that fiscal deficits appreciate real exchange rates, as well as the corollary that real devaluation and fiscal adjustment go together. The synthesis also finds a close association between fiscal and external balances in the case studies (figure 1.2 in chapter 1). The remarkable robustness of these relationships across the case studies is the strongest evidence yet in the literature for the ''fiscal approach to the balance of payments''—the idea that external deficits are primarily a result of fiscal deficits. The study also found important evidence for the effects of the *composition* of public spending on the real exchange rate, an effect often predicted and seldom documented. Chapter 2 sets out the framework for interpreting these results and for implementing similar approaches in other countries.

The synthesis points out the key role of economic growth in determining the macroeconomic consequences of public deficits. In the 1980s fast-growing economies like India, Pakistan, and Thailand were able to get away with relatively high deficits because of the accompanying rapid growth of domestic money demand and other means of financing. Countries that had slow or negative growth in the 1980s, such as Argentina, Brazil, and Mexico, found that an exceptionally large dose of fiscal adjustment was required to restore macroeconomic stability.

## Part II: Country Case Studies

The authors of the individual case study chapters have described the particular macroeconomic features of each country and the theoretical tools and econometric tests that were developed to deal with some of those special features. Foremost among the issues discussed are financial repression (Ghana and Zimbabwe), exchange rate controls and black markets for foreign exchange (Ghana), commodity booms and the accommodating fiscal reactions (Colombia, Côte d'Ivoire, and Morocco), financial deepening that allowed for monetary financing of deficits (Pakistan), and the financing of central bank losses—the quasi-fiscal deficit (Argentina and Chile).

There are large differences in the basic macroeconomic backgrounds of the countries described in the case studies. On the one hand, Argentina had a long history of triple-digit annual inflation and finally experienced hyperinflation in 1989 and 1990. Chile, on the other hand, with a past history of high inflation, was able after the mid-1970s to implement serious structural reforms that removed high inflation from the macroeconomic picture. Whereas in Argentina the inflation rate is the overwhelming indicator of macroeconomic imbalance, in Chile discussion centers around export growth, social expenditure, and real interest rates.

The other high-inflation country in the sample of case studies is Ghana. There, however, inflation never reached the triple-digit level. Unlike Argentina, Ghana instituted a wide range of price and exchange rate controls and rationed credit and foreign exchange. During 1972–83 the government ran large fiscal deficits that had to be financed through monetary expansion because of the small size of the local financial market and the reduced availability of external credit. The controls on interest rates generated negative real rates, inducing significant disintermediation and further reducing the base for the inflation tax. A black market for foreign exchange developed because of the negative real interest rates and foreign exchange controls. All these factors worsened the allocation of resources and reduced the chances for implementing structural adjustment.

In Argentina the very high inflation rates and the frequent reliance of authorities on large devaluations induced the public to shift toward holding foreign exchange (mainly U.S. dollars). Because of the lack of a significant degree of indexation in the financial system, the natural inflation hedge was foreign exchange, creating the phenomenon known as "dollarization." Economies with indexed financial markets, such as Brazil or Chile, have not experienced dollarization because the public uses indexed deposits to hedge against inflation. To the extent that dollarization reduces the demand for high-powered money, it magnifies the inflationary impacts of a given degree of monetary deficit financing. The worst case is illustrated by Argentina's experience: dollarization became widespread, the real monetary base shrank to less than 2 percent of GDP, and, as the deficit continued, hyperinflation resulted.

Pakistan is completely different from the high-inflation dollarized economies. Although the average deficit for the public sector was 7 percent of GDP for the period 1973–88, inflation has not exceeded 10 percent since 1976. The financing of deficits was evenly divided between the Central Bank, new issues of domestic debt, and external debt. Low inflation was the result of strong growth in money demand during the past two decades as new activities were incorporated into the monetary economy. (A virtually identical experience has been observed in other countries of the region, including India.) In addition to issuance of base money, the deficits were also financed by imposing on banks liquidity requirements that required the holding of government debt. The remaining fiscal gap has been closed by the steady flow of foreign aid.

From the case study one might think that Pakistan is different because it was able to resort to monetary financing of deficits without inducing high inflation. However, the process of financial deepening cannot continue forever, and the inflationary consequences of deficit financing will eventually be felt. The high fiscal deficits may not have created inflation so far, but they did contribute to a weak external sector. Sustained current account deficits require the continuous inflow of new capital for financing in the context of an economy highly dependent on imports of intermediates.

Zimbabwe represents a combination of the problems faced in Ghana and Pakistan. Although no extraordinary growth in money demand has been observed, the high fiscal deficits have been financed through new issues of domestic debt placed in the private sector. Effective foreign exchange rationing prevents Zimbabweans from converting their portfolios into foreign exchange and prevents imports from booming in response to domestic price controls and the high level of aggregate demand induced by the deficits. The result is a high level of private saving combined with compulsory public debt placements at low, often negative, real interest rates. Zimbabwe is another country that has apparently reached the big spender's paradise of being able to run fiscal deficits without inflation.

The other face of the big spender's paradise is that Zimbabwe's domestic debt rose from 54 percent of GDP in 1980 to 86 percent in 1988, while external debt reached 42 percent of GDP in 1985. At those debt levels, the interest on the debt becomes the most significant item in government spending and starts a dynamic process of its own: high interest requires the issuance of more debt to pay for it, which in turn produces a higher interest rate. Two alternatives are open for a country with levels of indebtedness such as those of Zimbabwe in the mid-1970s: (a) melt the debt through devaluations, as Argentina often did, risking hyperinflation, or (b) generate a level of fiscal deficit that will permit the interest on the debt to be genuinely serviced. According to the authors of chapter 10, levels close to the sustainable fiscal deficit—but still inconsistent with recovery of investment and growth—were attained in Zimbabwe after the partial fiscal adjustment of 1987/88.

Commodity booms appear to help generate public expenditures that are not easily reduced when the boom is reversed. This has been the case in two of the countries studied: Colombia and Morocco. The rise in phosphate prices in the mid-1970s allowed Morocco to increase government spending. As prices fell in the late 1970s, the government did not reduce spending but resorted to external borrowing. Because this foreign borrowing took place at a time of rising interest rates and deteriorating terms of trade, a foreign exchange crisis soon occurred, in 1983. At no point during this difficult period did authorities resort to monetary financing of the deficit; thus inflation never became an issue in Morocco.

In Colombia fiscal deficits increased sharply after the end of the 1975–78 boom in coffee prices, during which the government had helped spend the extra available funds. The increase in government spending had two effects: first, it created permanent spending needs that were not easily reversed when financing was no longer available; and, second, because foreign exchange was used to finance expenditures on domestic goods, the real exchange rate significantly appreciated, damaging the prospects of the nontraditional export sector. Colombia learned from the mismanagement of the 1975–78 coffee boom and did not repeat the mistake of increasing spending during the next boom, in 1986; instead, it properly sterilized the extra foreign exchange proceeds by issuing public debt.

Colombia has had conservative and stable macroeconomic policies and has learned from the few errors it committed in the past. Fiscal deficits have been very low by regional standards. (The highest national government deficit since 1960 has been 5 percent of GDP, in 1982.) The result has been moderate inflation, high growth, and a satisfactory external position.

The Colombia case study precisely quantifies the theoretical relationship between fiscal deficits and inflation. The detailed analysis of money demand presented in the case study shows the limited possibilities for the use of the inflation tax. According to the calculations presented, at 22 percent annual inflation, the revenue from money creation is about 2 percent of GDP. Increasing revenue to 2.65 percent of GDP requires inflation to jump to 100 percent a year. It is clear that the extra 80 points in inflation do not justify an extra revenue collection of barely more than half a point of GDP. Colombians must be aware of this fact because for the past fifteen years inflation has remained in the 15 to 25 percent range.

As in Colombia, Côte d'Ivoire's macroeconomic shocks were mostly external in nature: the boom in cocoa and coffee prices in the mid-1970s allowed for an increase in government spending that was reversed only slowly after prices came back to normal levels. Because of the rules of the CFA zone, the country could not rely on domestic financing of the deficits in any significant amount and instead resorted to external debt. The external financing of deficits contributed to the appreciation of the real exchange rate. The slow fiscal adjustment allowed external debt to grow from 30 percent of GDP in the early 1980s to 100 percent of GDP at the end of the decade. As debt grew and spending was not cut, the government experienced a payments crisis that forced it to make drastic and disorganized spending cuts. We can speculate that as the fiscal adjustment progresses and foreign borrowing ends, the real exchange rate should rise above historical levels in order to generate the resource balance required to service the now-higher external debt.

Argentina is unique among the case studies; the magnitude of its disequilibria makes it a textbook example of the effects of incorrect government policies. In Argentina monetary financing of deficits immediately creates inflation. Fiscal deficits increase domestic absorption and induce trade deficits that also allow for a real appreciation in the exchange rate. Finally, as the deficits are financed with domestic debt, they induce real interest rates much higher than the productivity of capital and thus reduce the ability of the government to repay.

Every year between 1960 and 1988, Argentina ran primary fiscal deficits. The monetary financing of those deficits has implied triple-digit inflation levels since 1975 (except in 1980 and 1987, when inflation was only around 80 percent, and 1989, when it reached 4,927 percent). External debt was also used to finance deficits but was cut as a source of financing after the regional debt crisis of 1982. Since then, external debt has grown because of unpaid interest, reaching about $60 billion in 1990. Domestic debt has been issued in large quantities but has frequently been melted by the huge devaluations to which Argentines became accustomed in the 1970s. The result of the government's melting its own debt was that the market asked for higher real interest rates to continue holding the remaining lower real levels of debt. In consequence, after each devaluation meltdown, interest rates rose and the remaining debt continued rising at rates higher than before; in only a short time the debt reached still higher levels, and the government tried a new devaluation.

The Argentine experience is in sharp contrast to that of Colombia, which kept inflation low by maintaining steady and sound fiscal policies and so grew at satisfactory rates. Argentina presented a picture of steady fiscal imbalance with very high inflation and no growth in income per capita during the twenty-year period 1969–89. Comparison of the experiences of these two countries best exemplifies the high costs of fiscal imbalance and unpredictable intervention on the part of economic authorities.

Chile is the only country studied that has successfully achieved significant structural adjustment during the past two decades. Tax reform, social security reform, and rationalization of the public sector were the main fiscal shocks after the mid-1970s. The economy was opened to international trade to a degree not yet experienced by any other Latin American country. After the 1983 balance of payments crisis, exports did diversify, and they grew more than imports. Chile was able to continue regularly serving its foreign debt and to save substantial amounts of reserves through the copper stabilization fund and thanks to an unusually high world price for copper during 1989–91.

In recent years (1988–90) Chile's nonfinancial public sector has had surpluses with which it has canceled its outstanding debt with the central bank. From this perspective, Chile is completely different from the other countries studied, where the main problem was how to finance the deficits rather than how to invest the surpluses.

Chile's basic economic structure is no different than that of neighboring countries. As in Colombia and Argentina, the Chilean data confirm all the relationships hypothesized at the beginning of the study: monetary financing creates inflation, debt financing raises interest rates, and external financing induces real appreciation and trade deficits. The basic difference lies in the fact that Chile relied on effective adjustment rather than just devising strategies for financing the disequilibria. The reforms implemented before 1982 were an important stabilizing factor in the rapid recovery of the Chilean economy during the 1980s. The three basic secrets of Chile's success with fiscal policy management were its stable policy environment, its consistent policies, and an increased emphasis on human capital and physical infrastructure in public sector investment.

## Conclusion

Financing deficits has been shown to be a significant source of inflation, unsustainable indebtedness, and other types of macroeconomic instability. Price and interest rate controls and foreign exchange rationing often result from a futile attempt to hide the unavoidable costs of deficit spending; they eventually make both macroeconomic instability and growth performance worse. Low and stable fiscal deficits are a necessity for the favorable long-run prospects of a country, as well as for avoiding the short-run macroeconomic ills of high inflation, high real interest rates, and real overvaluation of the currency.

## Note

1. The ten case studies were Argentina, Chile, Colombia, Côte d'Ivoire, Ghana, Mexico, Morocco, Pakistan, Thailand, and Zimbabwe. Because of space constraints, only eight countries are discussed in this volume. (Working papers on Mexico and Thailand are available from the World Bank, Policy Research Department, Macroeconomics and Growth Division, Washington, D.C.) The Statistical Appendix presents complete data for all ten countries and also brings together hard-to-find information, collected in the course of the project, on consolidated nonfinancial public sector deficits, real interest rates, inflation rates, and inflation tax revenues in fifty-nine other countries.

# Part I
# Synthesis and Methodology

# 1

# Fiscal Adjustment and Macroeconomic Performance: A Synthesis

*William Easterly and Klaus Schmidt-Hebbel*

Fiscal deficits were at the forefront of macroeconomic adjustment in the 1980s, in both developing and industrial countries. They were blamed in large part for the assortment of ills that beset developing countries during the decade: overindebtedness, leading to the debt crisis that began in 1982; high inflation; and poor investment and growth performance. In the 1990s fiscal deficits still occupy the center stage in the massive reform programs initiated in Eastern Europe and the former U.S.S.R. and by many developing countries on all continents.

Many issues are raised by the successes and failures of fiscal adjustment. Not the least of these is how to define and measure fiscal adjustment. What are the most meaningful measures of public sector deficits? How should one assess fiscal stance, public sector solvency, and sustainability of deficits? While the analytical literature tends toward a definitional and methodological agreement on this issue (see Blejer and Cheasty 1991 for a comprehensive survey), empirical applications still differ widely.

Once measurement issues are settled—and before analyzing the consequences of deficits—some frequently posed questions are: How important are the macroeconomic causes of fiscal deficits? What role do domestic and foreign shocks play in relation to changes in fiscal policy in the evolution of deficits? What are the most effective policy instruments for fiscal adjustment?

Regarding the macroeconomic impact of deficits, a recurring question is whether larger public deficits are always associated with higher inflation. Sargent and Wallace's (1985) "monetarist arithmetic" answered this question affirmatively. But the relationship is blurred because governments finance deficits by borrowing, as well as by printing money. The relationship is further muddied by other factors, such as unstable money demand, inflationary exchange rate depreciations, widespread indexation practices, and sticky expectations (see Kiguel and Liviatan 1988; Dornbusch and Fischer 1991).

Interest rates are another ambiguous factor. Do deficits raise domestic real interest rates when governments rely heavily on domestic debt financing, or is this relationship also blurred by such factors as financial repression (Easterly 1989; Giovannini and de Melo 1993) and the high degree of substitutability between public debt and other assets held by the private sector?

Looking beyond domestic financial markets, a central issue of fiscal stabilization involves how private consumption and investment react to deficits. Will consumers reduce their spending when taxes are raised and increase it when taxes are lowered? Or will they offset only changes in government consumption—without reacting to changes in government tax or debt financing—as argued by Barro (1974)? This issue is still not empirically settled for industrial countries (see Hayashi 1985; Bernheim 1987; Leiderman and Blejer 1988; and Seater 1993 for surveys of empirical studies on Barro's Ricardian equivalence proposition of one-to-one crowding-out of private consumption by public consumption). There is, however, growing evidence for developing countries against the Ricardian hypothesis (Haque and Montiel 1989; Corbo and Schmidt-Hebbel 1991).

In regard to government investment, does a higher level of public capital spending crowd in or crowd out private investment? Theory predicts that this will depend on the degree of substitutability or complementarity of private and public capital (see Easterly, Rodríguez, and Schmidt-Hebbel 1989), and the limited available evidence for developing countries confirms this ambiguity (see Blejer and Khan 1984; Khan and Reinhart 1990).

Public deficits could also have indirect effects on private consumption and investment if real interest rates rise in response to higher domestic debt financing. Although theory predicts that real interest rates will have an ambiguous effect on private consumption, private investment should decline unambiguously with higher interest rates. A growing body of evidence for developing countries supports the notion that private consumption is insensitive to real interest rates (Giovannini 1983, 1985; Corbo and Schmidt-Hebbel 1991; Schmidt-Hebbel, Webb, and Corsetti 1992). Surprisingly, many studies also show little response of private investment to interest rates in developing countries (see the surveys by Rama 1993 and Servén and Solimano 1993).

Finally, how do fiscal imbalances feed into external deficits? One should expect a strong link between fiscal and current account deficits in financially open economies when either consumers are not Ricardian or the national versus imported composition of public and private sector spending differs. The role that fiscal imbalances played in the overborrowing by developing countries that led to the 1982 debt crisis is widely recognized (see Dornbusch 1985; Sachs 1989). But more systematic evidence linking public deficits with external deficits and real exchange rate appreciations is still lacking.

The underlying theoretical framework of this study is simple. The consequences of deficits depend on how they are financed. As a first approximation, each major type of financing, if used excessively, brings about a macroeconomic imbalance. Money creation to finance the deficit often leads to inflation. Domestic borrowing leads to a credit squeeze—through higher interest rates or, when interest rates are fixed, through credit allocation and ever more stringent financial repression—and the crowding-out of private investment and consumption. External borrowing leads to a current account deficit and real exchange rate appreciation and sometimes to a balance of payments crisis (if foreign reserves are run down) or an external debt crisis (if debt is too high).

In its analysis of the effects of deficits, the method applied here focuses first on the monetary and financial market implications of deficits. Next, the direct and indirect effects of public spending, taxation, and deficits on private consumption and investment are addressed. Finally, the impacts of public deficits on external disequilibria and the real exchange rate are identified.

Each step in this study applies a common framework—for deficit measurement, sustainability, macroeconomic sensitivity, monetary and financial markets, private consumption and investment, the trade deficit, and the real exchange rate—to a set of ten case studies. To put the case study results into broader perspective, selected issues are addressed with the use of a fifty-nine-country sample that includes both developing and industrial countries. The fully specified models based on behavioral relationships can be found in Easterly, Rodríguez, and Schmidt-Hebbel (1989); Marshall and Schmidt-Hebbel (1989); Fischer and Easterly (1990); and Rodríguez, chapter 2 in this volume. This chapter summarizes the empirical evidence—based on econometric estimations and policy simulations for each country—and derives the relevant policy implications.[1] Drawing on a representative set of case studies makes possible inferences on the unsettled issues mentioned above that are more reliable than those based on pooled cross-country studies or individual case studies.

The selection criteria for the ten cases—Argentina, Chile, Colombia, Côte d'Ivoire, Ghana, Morocco, Mexico, Pakistan, Thailand and Zimbabwe—stressed both the diversity of fiscal and macroeconomic regimes and experiences and the sample's ability to represent the developing world at large. As will be clear from the discussion below, the ten countries include countries that underwent fiscal adjustment and those that did not; high- and low-deficit countries; large and small economies; low- and high-inflation countries; countries with and without developed financial markets; and countries with and without access to foreign financing.

This chapter begins with a discussion of alternative measures of fiscal deficits, deficit sustainability, and the interaction between deficits and the macroeconomy. The second section examines the causality from the macroeconomy to the deficits; the contribution of external and domestic shocks in comparison with that of shifts in fiscal policy in changing the deficit; and the most effective policy instruments for fiscal adjustment.

The three subsequent sections address the macroeconomic consequences of deficits. First the focus is on the relationship between the domestic financing of deficits, inflation, and real interest rates; then the relationship of deficits to private consumption and investment is analyzed; and finally the spillover effects of deficits on external imbalances are examined. The chapter closes with the main conclusions and policy implications of the analysis.

## Measurement of Public Deficits

How the public deficit is measured has an important bearing on an accurate analysis of the macroeconomic implications of deficits.[2] Appendix 1.1 discusses briefly two dimensions of deficit measurement: public sector composition and economic relevance. To illustrate how misguided a too narrow measure of public sector deficit can be, box 1.1 compares consolidated nonfinancial public sector (CNFPS) and quasi-fiscal deficits in Argentina and Chile. Box 1.2 illustrates the differences between alternative measures of Morocco's central government deficit.

With regard to deficit measurement, the approach taken by this chapter and by the research project at large is to combine the widest possible sector coverage (subject to data availability) and the most meaningful definitional choice, driven by the issues addressed.[3]

### *Sustainable Public Deficits*

Sustainable public deficit measures are derived by looking at the below-the-line financing constraints of the deficits, following the accounting approach to public sector solvency (see appendix 1.1). Our purpose is to compare actual deficits and sustainable deficit levels. Estimates of sustainable primary deficits are derived by holding constant the ratios of public liabilities to output for feasible values for the macroeconomic variables that determine market demands for public liabilities, as discussed in appendix 1.2.

Figure 1.1 compares sustainable and actual primary surplus levels for six relevant fiscal experiences during the 1980s.[4] Sustainable pri-

## Box 1.1. Quasi-Fiscal Deficits in Argentina and Chile

Quasi-fiscal deficits of the central bank amounted to a cumulative 55 percent of GDP during 1982–85 in Argentina and to a cumulative 41 percent of GDP during the same period in Chile. In Argentina quasi-fiscal deficits were roughly as large as conventional deficits during the period (box figure 1.1); the sum of both exceeded, on average, 25 percent of GDP per year! In Chile quasi-fiscal deficits exceeded, on average, 10 percent of GDP per year, more than double the conventional deficits. Both cases illustrate how misleading nonfinancial public sector deficits can be. For example, while conventional deficits were falling during 1984 in Argentina, the fiscal stance of the overall public sector, including the central bank, deteriorated greatly. In Chile conventional deficits underestimate both the 1981–85 fiscal crisis and the subsequent fiscal adjustment.

Box figure 1.1. Conventional and Quasi-Fiscal Deficits, Argentina and Chile, 1979–89

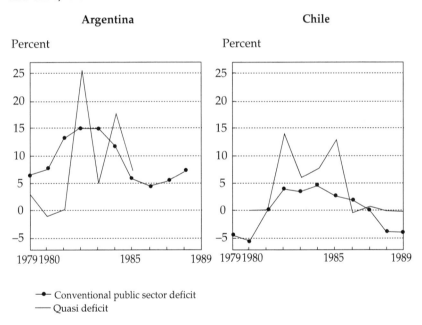

mary surplus levels diverge widely, not only because of the different levels of public liability stocks and macroeconomic variables in each country but also because the calculations were made for different public sector coverages. The levels range from 1.4 percent of GDP for the total (financial and nonfinancial) public sector in Chile to −2.8

## Box 1.2. Alternative Deficit Measures for Morocco

The decline in Morocco's cash-basis deficit gives only a partial picture of the significant fiscal improvement achieved during 1983–88 (see box figure 1.2). The country was able to lower its accruals-basis deficit at an even more rapid pace by reducing accumulation of arrears and starting to repay them in 1986. At the same time, the nominal deficit fell faster than the economically more meaningful operational deficit as a consequence of the decline of the inflation component of domestic interest payments resulting from lower inflation.

**Box figure 1.2. Deficit Measures, Morocco, 1983–88**

Percentage of GDP

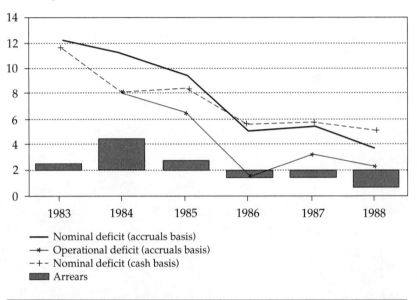

— Nominal deficit (accruals basis)
—*— Operational deficit (accruals basis)
– + – Nominal deficit (cash basis)
▓▓ Arrears

percent of GDP (a sustainable primary deficit) for the central government in Ghana.

How do the calculated sustainable deficits compare with actual levels during the 1980s? In Chile the massive public sector adjustment during the 1980s (comprising both the nonfinancial deficit and the central bank's quasi-fiscal losses) pushed primary surpluses in 1988 and 1989 well beyond the upper bound of sustainable levels. Colombia reached sustainable primary surplus levels in 1987–89 after significantly strengthening its fiscal stance. Morocco also pursued strong fiscal adjustment policies, achieving a primary surplus level in 1988 even higher than that required to reduce public debt as a share of

## Figure 1.1. Actual and Sustainable Public Sector Primary Surplus in Six Countries, 1980s

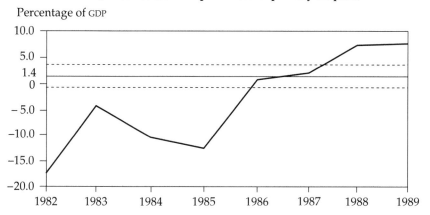

Chile (consolidated total public sector primary surplus)

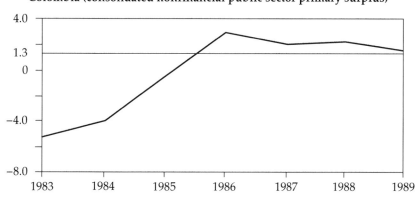

Colombia (consolidated nonfinancial public sector primary surplus)

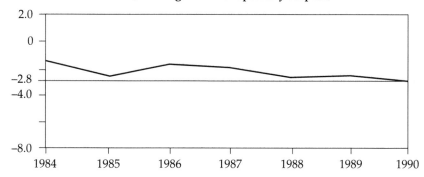

Ghana (central government primary surplus)

*(Figure continues on the following page.)*

**Figure 1.1** (*continued*)

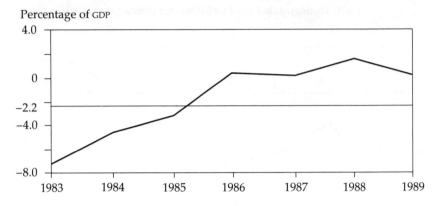

### Morocco (central government primary surplus)

Percentage of GDP

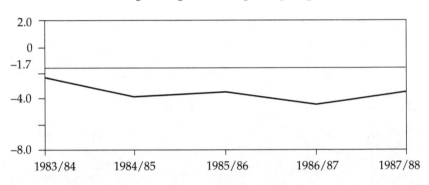

### Pakistan (general government primary surplus)

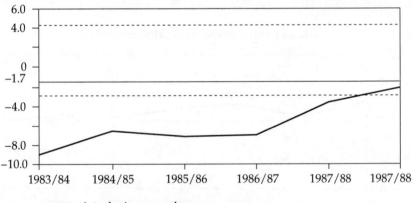

### Zimbabwe (general government primary surplus)

———— Actual primary surplus
———— Sustainable primary surplus (midpoint)
- - - - - Upper and lower bounds of primary surplus

*Source:* Chapters 4, 5, 7, 8, 9, and 10 in this volume.

22

output, although subsequently its fiscal stance deteriorated somewhat. Pakistan's fiscal deterioration raised its primary deficits beyond
the 1.7 percent of GDP level consistent with stable debt-to-output
ratios. Finally, Zimbabwe's modest fiscal adjustment in 1987–89
reduced its primary deficit to within the broad range of values consistent with sustainable levels, but still distant from an upper-bound
level of sustainable primary surplus consistent with an adverse macroeconomic scenario.

Although these calculations are based on simple assumptions, they
provide useful benchmarks for evaluating fiscal stance from a longer-
run perspective. The next section generalizes these results by focusing on the fiscal performance of the ten country cases and distinguishing between the consolidated nonfinancial and quasi-fiscal
deficits.

### Correlations of Deficits with Other Macroeconomic Variables

To obtain an overview of the relationship between fiscal deficits and
relevant macroeconomic variables, we collected data on a large sample of countries, including members of the Organization for Economic
Cooperation and Development (OECD). (For data and sources for the
sample, see the Statistical Appendix in this volume.) The usefulness
of fiscal deficits as an indicator of overall economic performance is
reflected by our calculation of simple correlations between public sector balances (as a percentage of GDP) and other major macroeconomic
variables, as shown in table 1.1.

There are good reasons not to expect very strong correlations. Fiscal deficits are measured in different ways across countries, introducing some measurement error into the sample. In addition, the theoretical relationship between deficits and other macroeconomic
variables depends crucially on the means of financing them. (More
generally, simple correlations may fail to be significant because of the
omission of other variables.) Despite these caveats, we find a significant statistical relationship between the public sector balance and
many, although not all, macroeconomic performance variables. Per
capita growth is significantly and positively related to fiscal surpluses.[5] There is also an interesting negative and significant correlation between per capita growth and the *variance* of fiscal balances.
Low and stable fiscal deficits are associated with high growth.

Fiscal balances are positively related to real interest rates, contrary
to the usual prediction that deficits lead to high interest rates and
surpluses to low rates. Since there are a large number of negative real
interest rates in the sample, this finding is probably explained by an
association between financial repression and fiscal deficits. Fiscal balances are negatively related to money creation (seigniorage), confirm-

**Table 1.1. Cross-Section Correlations of Consolidated Public Sector Balance as a Percentage of GDP with Other Variables**

| Variable | Correlation | t-statistic |
|---|---|---|
| GDP growth | 0.19 | 1.46 |
| GDP per capita growth | 0.37 | 3.02** |
| Real interest rate | 0.31 | 2.34* |
| Money creation (percentage of GDP) | −0.33 | −2.40** |
| Inflation | −0.16 | −1.21 |
| Investment (percentage of GDP) | 0.24 | 1.70* |
| Total consumption (percentage of GDP) | −0.48 | −3.97** |
| Private consumption (percentage of GDP) | −0.38 | −3.00** |
| Real exchange rate | −0.15 | −1.10 |
| Current account (percentage of GDP) | 0.54 | 4.76** |
| Black-market premium | −0.35 | −2.65** |
| Variance of deficits with per capita GDP growth | −0.36 | −2.88** |

\* Significant at 5 percent level (one-tailed).
\*\* Significant at 1 percent level (one-tailed).

*Note:* Public surplus is positive; deficit is negative. Public balances of OECD countries are general government. Sample size varies between fifty and fifty-nine. Period of averages is longest period for which data are available for each pair of concepts for each country. For the real exchange rate, an appreciation is an increase.

*Source:* OECD, *OECD Economic Outlook;* World Bank data. For real exchange rate, Dollar 1990, using purchasing power parity (PPP) comparisons from Summers and Heston 1988.

ing that countries which run high deficits do so in part through greater reliance on seigniorage. (But de Haan and Zelhorst 1990 find that the correlation holds only for high-inflation countries.) However, inflation rates show little correlation with fiscal balances, perhaps reflecting in part the nonlinear relationship between money creation and inflation discussed later in this chapter.

Fiscal balances are positively, although weakly, related to total gross domestic investment, offering at least superficial support to the notion that deficits crowd out investment. The fiscal balance is negatively related to both total and private consumption. This finding is superficially consistent with the hypotheses that taxes crowd out private consumption and that public and private consumption are complementary. It is inconsistent with the Ricardian notion that government spending lowers consumption.

Fiscal balances are correlated with external current account balances across countries. This lends support to the ''fiscal approach to the balance of payments,'' which says that fiscal imbalances are the main sources of external imbalances.[6] A suggestive association is found between fiscal balances and black-market premiums, indicating that countries with high deficits are more likely to control the foreign exchange market tightly and that deficits drive up the premium created by such controls. Real exchange rates, however, show little association with fiscal deficits. This may reflect the extent to which trade intervention differs across countries.

There are two messages to be carried away from this set of generally strong (but occasionally weak) associations between fiscal balances and macroeconomic performance. The first is that despite problems of comparability across countries, the fiscal balance is a useful indicator of macroeconomic health. The second is that in order to trace the relationships between deficits and specific macroeconomic variables, such as inflation and real interest rates, deeper analysis is needed, with careful attention to the bidirectional causality among the variables and the underlying behavioral relationships. The following sections address these issues.

## Causes of and Remedies for Deficits

This section focuses on the determinants of deficits and the components of successful programs for fiscal stabilization. We start by assessing the contribution of foreign and domestic macroeconomic shocks in relation to that of shifts in fiscal policy in the evolution of public deficits. We then identify the main policies used in successful adjustment efforts.

### Sensitivity of Deficits to Foreign Variables

Foreign shocks are a source of fiscal instability in many developing countries. Fluctuating export prices and foreign interest rates mean that commodity exporters and highly indebted countries face an inherent instability which often hinders fiscal adjustment efforts.[7]

Changes in export prices affect the public sector directly, via the profits of the exporting state-owned company or marketing board, or indirectly, through taxes on profits or on exports. (See Gelb and associates 1988 for a detailed study of the macroeconomic and fiscal consequences of commodity price booms in oil-exporting countries.) The quantitative impact of the export price shock on government accounts depends on the tax and property structure, the amount exported, and the magnitude of the price shock. In countries that face high export price volatility and that export through a large state-owned enterprise (as in Chile and Mexico) or a marketing board (as in Côte d'Ivoire and Ghana), fiscal accounts are sensitive to terms of trade shocks. If the foreign trade structure is diversified, the private sector is the main exporter, and export taxes are low or absent (as in Pakistan and Zimbabwe), public sector accounts do not suffer significantly from export price volatility. Import prices affect public expenditure in some countries. In Morocco, for instance, the decline in imported food prices was the main cause of the substantial decline in subsidies to the private sector in the mid-1980s. Changes in foreign interest rates affect highly indebted countries with a high share of variable-interest debt, such as Argentina and Colombia.

In addition to measuring the impact of foreign shocks on public accounts, it is illuminating to assess the contribution of shocks to overall public sector deficits. Both dimensions are presented for six countries in table 1.2. The first column shows the average absolute change in public deficits as a result of different foreign shocks over the relevant sample periods. For instance, foreign shocks contributed, on average, to a variation in the public sector deficit amounting to 2.3 percent of GDP in Chile and 0.3 percent in Zimbabwe. Chile and Thailand are highly sensitive to changes in export prices or terms of trade, while in Colombia and Morocco the average contribution of export price shocks to deficits is only about 1 percent of GDP. In Zimbabwe the influence of terms of trade shocks on government revenue is negligible. In comparison with terms of trade shocks, fluctuations in interest rates have much lower effects on public deficits; they contribute, at most, 0.4 percent of GDP to the variation of public sector deficits in our sample.

The average relative contribution of foreign shocks in the second column measures the degree of correlation between foreign-shock-induced deficits and the overall public deficit. In Chile, Colombia, and Thailand adverse foreign shocks increase deficits, with shares varying between 12 and 50 percent of the total fluctuation of deficits. In Ghana the tiny foreign interest shocks are uncorrelated with deficits. In Morocco and Zimbabwe, however, foreign shocks have the opposite sign from the changes in overall deficits, indicating that domestic macroeconomic shocks and fiscal policy changes more than compensate for the influence of adverse foreign shocks.

Even moderate shocks could explain much variation in deficits. In Colombia, for example, moderate shocks have a huge influence—as much as 50 percent—on the variability of deficits. Because Colombia did not require such substantial fiscal adjustment during the relevant sample period (1984–89), foreign shocks had a more significant role in the evolution of its deficit. In Chile, which experienced the greatest foreign shocks, the relative contribution of these shocks to deficit variability has been a low 12 percent. During 1973–88 Chile embarked on massive fiscal adjustment programs that overshadowed the influence of foreign shocks.

Optimal responses to shocks depend on whether the shocks are temporary or permanent. Purely transitory shocks should be (dis)saved and hence reflected by public deficits, whereas permanent shocks should induce corresponding changes in expenditure or revenue without affecting deficits. In the case of public sectors that own large commodity-exporting companies (as in Chile, Mexico, and

## Table 1.2. Contribution of Foreign Shocks to Public Deficits

| Country and shock | Average absolute variation of public deficits attributable to foreign shocks (percentage of GDP) | Average relative contribution of foreign shocks to variation of public deficits (percentage of variation of deficits) |
|---|---|---|
| *Chile, 1973–88* | | |
| Foreign shocks | 2.3 | 12 |
| Changes in copper price | 2.7 | 15 |
| Changes in foreign interest rates | 0.4 | −3 |
| *Colombia, 1984–89* | | |
| Foreign shocks | 1.0 | 50 |
| Changes in coffee fund | 1.2 | 59 |
| Changes in surplus of oil company | 0.9 | −9 |
| *Ghana, 1972/73 to 1988* | | |
| Changes in foreign interest rates | 0.1 | 0 |
| *Morocco, 1971–88* | | |
| Changes in contributions of phosphate company | 0.8 | −17 |
| *Thailand, 1970–88* | | |
| Changes in terms of trade | 2.2 | 41 |
| *Zimbabwe, 1980/81 to 1988/89* | | |
| Changes in foreign interest rates | 0.3 | −3 |

*Note:* The first column computes the annual average absolute variation of the deficit caused by the corresponding changes in foreign variables. (The exception is Chile; figures are based on period averages for 1973–75, 1975–81, 1981–86, and 1986–88.) If more than one foreign variable is considered, the sum of the average absolute variations for the individual variables differs from the average absolute variation of the combined shocks because of the opposite signs of individual variations. The second column reflects the average relative contribution of foreign shocks to the variation of public deficits, defined as:

$$\left[ \sum_{i=t}^{t+n} dv_i(\text{sign } d_i) \right] \Big/ \sum_{i=t}^{t+n} |d_i|$$

where $d_i$ is the change in the deficit in period $i$, $dv_i$ is the change in the deficit caused by variable $v$ in period $i$, $t$ is the initial period, and $n + 1$ is the total number of periods.

*Source:* Authors' calculations and information from country case studies listed in the references to this chapter.

Morocco) or that collect large revenues from private exporters (Côte d'Ivoire and Ghana), price-stabilization or revenue-stabilization funds, such as those implemented in Chile and Venezuela, or hedging through risk-sharing contracts, are efficient mechanisms for isolating the budget from temporary export price shocks.

### Sensitivity of Deficits to Domestic Macroeconomic Variables

A second group of variables that affect deficits and are also outside the direct control of fiscal policymakers consists of domestic macroeconomic variables. In the following discussion, we concentrate on four variables that often have strong effects on public budgets: inflation, the real interest rate, the real exchange rate, and output.

INFLATION. Inflation affects budget deficits through various channels.[8] Anticipated inflation raises nominal interest payments to domestic debt holders. Inflation also affects the primary deficit (the Olivera-Tanzi effect).[9] Collection lags for taxes that are not fully indexed (for example, nominally fixed excise taxes) lead to declining real revenue when inflation increases. Inflation also tends to lead to public demoralization and hence to lower tax compliance. If, however, income brackets are nonindexed, higher inflation leads to bracket creep and hence to higher direct taxation. Real public current expenditure declines with inflation when public wages or transfers are not indexed. Whereas in many countries the net effect of inflation is to increase primary deficits, the budget structure could conceivably reverse this effect.

Table 1.3 summarizes the effects of inflation on public deficits in seven countries and identifies the channels through which they operate. Results from estimated tax revenue functions allow us to classify countries according to the net influence of inflation on tax revenue. Inflation lowers aggregate tax revenue in Colombia and both direct and indirect taxes in Ghana. The only positive effect of inflation on taxes is found for direct tax revenue in Zimbabwe, where nonindexation of income brackets leads to bracket creep. Short collection lags, indexation of tax revenue, and indexation of income brackets could be behind the nonsignificant effects of inflation on tax revenue in Chile (direct and indirect taxes), Morocco (total taxes), Pakistan (direct, indirect, and trade taxes), and Zimbabwe (indirect and total taxes).

Table 1.3 provides some partial evidence concerning the effects of inflation on expenditure categories. Transfers to the private sector in Chile decline with inflation, presumably as a result of incomplete indexation. No evidence of a significant effect of inflation on aggregate public expenditure could be found in Morocco.

In most countries the net influence of inflation is to raise nominal public sector deficits, as a result of the dominating effect of rising prices on interest payments and tax revenues. A good example is Thailand where, according to econometric results, an increase in inflation of 10 percentage points raises the CNFPS deficit by 0.9 percentage points of GDP.[10]

**Table 1.3. Effect of Inflation on Public Deficits**

| Negative | Zero | Positive[a] |
|---|---|---|
| | *Effect on tax revenue* | |
| Colombia: total taxes (1972–87) Ghana: direct taxes, indirect taxes (1970/71 to 1988) | Chile: direct taxes, indirect taxes (1973–89) Morocco: total taxes Pakistan: direct taxes, indirect taxes, trade taxes (1972/73 to 1987/88) Zimbabwe: indirect taxes, total taxes (1970/71 to 1988/89) | Zimbabwe: direct taxes (1970/71 to 1988/89) |
| | *Effect on public expenditure* | |
| Chile: transfers (1973–89) | Morocco: public expenditure | |
| | *Effect on the public deficit* | |
| | | Thailand (1971–88) |

*Note:* The effect of inflation on deficits via nominal interest payments on the debt is excluded as a separate channel of transmission.

a. Because of nonindexation of income brackets.

*Source:* Country case studies listed in the references to this chapter.

REAL INTEREST RATE. Real interest payments (and hence both the nominal and the operational deficit) obviously increase one-to-one with the real interest rate. Inflation shocks that are unexpected (or, even if expected, are not reflected in higher nominal interest rates because of controls on interest) reduce real interest rates and hence the operational deficit. For instance, in Ghana the one-period rise in inflation from 30 percent in 1982 to 115 percent in 1983 increased the nominal CNFPS deficit only slightly but reduced the operational deficit significantly as a result of the drop in real interest rates to negative levels.

Since the mid-1970s financial liberalization with partial or complete deregulation of interest rates has increased the sensitivity of deficits to the real interest rate. After early and radical financial liberalizations in Chile (1974–75) and Argentina (1977), the 1980s saw partial or complete liberalizations in Mexico, Morocco, and Zimbabwe. Whereas the massive rise in real interest rates in Chile during the 1970s did not impinge on the deficit because of the virtual absence of domestic interest-bearing debt, the increasing domestic debt stocks of the 1980s, in conjunction with moderately high interest rates, added to the burden of the central bank, which holds most of the domestic debt in the public sector. In Morocco partial liberalization of interest rates since 1984 has significantly increased the cost of domestic debt to the treasury. It is estimated that a future increase of rates on government debt to competitive market levels could add 2 percentage points of GDP to the deficit.

Less access to foreign financing after 1982 forced countries to combine deficit reduction with increased reliance on domestic financing. A case in point is Pakistan. Its domestic nonbank borrowing increased from 4.8 percent of GDP in fiscal 1980/81 to 7.4 percent in 1987/88, and its domestic interest payments rose by 1.5 percentage points of GDP as a result of both higher domestic debt stocks and higher interest rates.

REAL EXCHANGE RATE. A real exchange rate (RER) depreciation raises public expenditure (measured in local currency units) by increasing foreign interest payments and the cost of tradable capital and intermediate goods acquired by the public sector. Public sector revenue is boosted by a real depreciation that raises surpluses of firms producing tradable goods, as well as direct and indirect taxation on production or sales of tradable goods. The net effect of the RER on the deficit (in real terms or as a share of GDP) hence depends on the relative weights of traded and nontraded items in public expenditure and revenue.

Table 1.4 summarizes the effects of the RER on tax revenue, profits of state-owned enterprises (SOEs), transfers, and consolidated deficits. In Colombia total tax revenue was reduced by real devaluation—presumably because a more devalued RER reduces quantitative import restrictions or because of a highly elastic import demand. The opposite is true for Ghana and Zimbabwe, where various revenue categories (direct and total taxes in Ghana, direct and indirect taxes in Zimbabwe) are increased by devaluation, presumably because tradable-goods activities (sales and production) are taxed more heavily than nontradable-goods activities. Because the remaining tax categories are shown to be insensitive to the RER, aggregate tax revenue rises with an RER devaluation in Ghana and Zimbabwe.

Real devaluations have positive effects on public budgets in countries in which a significant share of SOEs consists of companies that produce tradable goods. This is especially true when the big commodity exporters are public enterprises, as in Chile, Colombia, Mexico, and Morocco. Devaluations also boost net revenues from profits of agricultural marketing boards; this is clearly the case in Côte d'Ivoire.

A computation of the net effect of the RER on the CNFPS deficit combines the previously mentioned effects on public revenue with

**Table 1.4. Effect of a Real Exchange Rate (RER) Devaluation on Public Deficits**

| Negative | Positive |
|---|---|
| *Effect on tax revenue* | |
| Colombia: total taxes (1972–87) | Ghana: direct taxes, total taxes (1970/71 to 1988) |
| | Zimbabwe: direct taxes, indirect taxes (1970/71 to 1988/89) |

*Effect on profits or transfers from state-owned enterprises (SOEs)*

Chile: surplus of SOEs and copper taxes
Colombia: surplus of coffee fund and state oil company
Côte d'Ivoire: revenue from cocoa and coffee marketing boards
Mexico: surplus of SOE
Morocco: contributions of state phosphate company

*Net effect on the CNFPS deficit*

| Increases deficit | Close to zero | Lowers deficit |
|---|---|---|
| Chile | Colombia | Mexico |
| Ghana | Thailand | |
| | Zimbabwe | |

*Source:* Country case studies listed in the references to this chapter.

the large and positive effect of the RER on foreign interest payments and with any effects on public expenditure. In some countries the interest effect dominates whatever positive effect the RER has on the primary deficit. The opposite is true in Mexico, where the share of oil-related federal revenue in GDP (7.9 percent in 1989) is more than twice as large as interest payments on dollar-denominated debt (3.4 percent in 1989). In Colombia, Thailand, and Zimbabwe a real devaluation has little or no influence on the overall deficit.

OUTPUT. Transitory output shocks affect nonfinancial public deficits because of changing tax bases and transfer payments to the private sector. This anticyclical behavior of public deficits motivated traditional Keynesian prescriptions of using the budget as an automatic stabilizer to counteract "autonomous" demand shocks. In countries with nonindependent central banks or countries under extreme financial crises, the anticyclical behavior of the nonfinancial deficit is reinforced by the anticyclical quasi-fiscal operations of the financial public sector. Cases in point are Argentina and Chile during the financial crisis and recession of the early 1980s (see box 1.1).

Trend growth is sometimes seen as a cure for public deficits; if growth is high enough, it is argued, tax bases expand and countries

### Box 1.3. How Sensitive Are Deficits to Macroeconomic Shocks in Zimbabwe?

Box table 1.3 shows to what extent Zimbabwe's CNFPS deficit is affected by domestic and foreign macroeconomic shocks. (The estimates are based on 1987/88 and 1988/89 CNFPS budgets.) The domestic real interest rate has a significant effect on the deficit as a result of Zimbabwe's high domestic public debt: a 1 percentage point increase in the real interest rate raises the deficit by 0.4 percentage point of GDP, as the ratio of domestic debt to GDP stands at 40 percent. It is interesting that inflation has a lower positive effect on the deficit than does the real interest rate. The reason is that the effect on the deficit via higher nominal interest payments is in part compensated by the positive effect of bracket creep on revenue from income taxes. A devaluation contributes to a slightly lower deficit in Zimbabwe: the higher foreign interest bill is more than compensated by increased tax revenues from import taxes and direct taxes on tradables-producing sectors. Growth seems to have a strong effect on deficits, but its magnitude is overestimated because the calculation considers the influence of GDP only on tax revenue, not on public expenditure.

**Box table 1.3. Macroeconomic Shocks and the CNFPS Deficit**

| *Change in macroeconomic determinants* | *Change in CNFPS deficit (percentage points of GDP)* |
|---|---|
| Increase in domestic inflation (1 percentage point) | 0.31 |
| Increase in domestic real interest rate (1 percentage point) | 0.40 |
| Devaluation of real exchange rate (1 percent) | −0.06 |
| Growth of real GDP (1 percent) | −0.37 |
| Increase in foreign interest rate (1 percentage point) | 0.25 |

*Source:* Chapter 10 in this volume, table 10.5.

can grow out of deficits. This view is flawed for two reasons. First, it neglects the fact that not only tax bases but also successful pressures for higher public expenditure rise with output levels. Second, growth will not materialize if public deficits are high, inflation and real interest rates are high, and private investment is therefore depressed.

Box 1.3 illustrates the preceding analysis by showing how sensitive Zimbabwe's CNFPS deficit is to changes in the four domestic macroeconomic variables and the foreign interest rate.

## Fiscal Policies

In this section we compare the influence of fiscal policy variables and that of foreign and domestic macroeconomic variables in the evolution of public deficits. Using time-series results for the decomposition of public sector deficits according to the three groups of deficit determinants (based on the deficit decomposition methodology of Marshall and Schmidt-Hebbel 1989), we compute the contribution of each group to changes in public deficits. Figure 1.2 presents the average relative contribution of the three groups of variables to changes in CNFPS deficits and the pattern of deficits in Chile, Ghana, and Zimbabwe.[11] The evolution of the public deficit in these three countries reflects the influence both of temporary (or cyclical) shocks and, particularly in the cases of Chile and Ghana, of structural policy shifts that brought about lower trend deficits.

Chile's fiscal experience in 1973–89 reveals four distinct periods: massive fiscal stabilization (1973–76); consolidation of public sector retrenchment (1977–80); crisis and deficit explosion (1981–84); and again, significant fiscal stabilization (1985–89). Fiscal policymakers are the main actors behind this experience, which achieved CNFPS surpluses close to 5 percent of GDP. On average, the relative contribution of fiscal policy variables to changes (and therefore to trend reduction) in the deficit was 142 percent. Changes in fiscal policy variables thus compensated for the strongly negative contribution of domestic macroeconomic variables and the slightly negative contribution of external variables.

Ghana is a case of gradual, but also highly successful, fiscal adjustment. There, too, the contribution of fiscal policy variables to the turnaround was massive, explaining 91 percent of the change in the deficit. Improvements in domestic macroeconomic variables helped to a small extent, contributing 11 percent to the fluctuations and structural correction of the central government deficit in Ghana.

The substantial deterioration in Zimbabwe's CNFPS budget after 1980 was partly reversed when a limited fiscal stabilization began in fiscal 1987/88. Zimbabwean policymakers compensated for the influence of variables beyond their control: fiscal policy variables explain 110 percent of the variation of public deficits, neutralizing the negative contribution of foreign interest shocks to the deficit.

A central conclusion emerges from the cases of Chile, Ghana, and Zimbabwe: fiscal policy variables dominate absolutely these countries' experiences of fiscal adjustment or deterioration. External and domestic macroeconomic shocks play a minor, and often even negative, role in the cyclical variation and the structural changes in public sector budgets. Active fiscal policies are both the main culprits in fiscal crises and effective instruments in bringing about fiscal stabilization and adjustment.

### Figure 1.2. Nominal CNFPS Deficit and Deficit Decomposition by Main Determinants, Chile, Ghana, and Zimbabwe, 1971–89

Deficit (percentage of GDP)

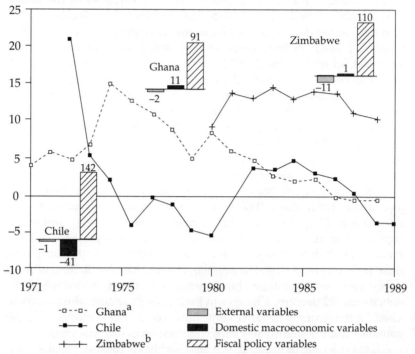

a. Central government deficit; fiscal year data for 1971/72 to 1981/82.
b. Fiscal year.
*Source:* Authors' calculations based on country data in chapters 4, 7, and 10 in this volume.

Table 1.5 identifies the contribution of specific policies in ten relevant country experiences, one for each country in the sample. Three countries (Chile, Mexico, and Thailand) achieved strong and rapid fiscal adjustment. Four (Colombia, Ghana, Morocco, and Zimbabwe) followed a more gradualist approach of fiscal retrenchment. Pakistan experienced moderate deterioration, and Argentina and Côte d'Ivoire experienced massive fiscal deterioration.

Loss of control over public consumption (particularly wage levels and employment levels) is a major cause of a looser fiscal stance. One dramatic example is Argentina, where as a result of the large increase in current expenditure, investment fell by almost 5 percentage points of GDP during the period of extreme fiscal deterioration, 1977–82. The other is Zimbabwe, which, during a period of fiscal retrenchment

(1986/87 to 1988/89), was not able to avoid further increases in its public wage bill by 4.0 percentage points of GDP. Côte d'Ivoire's fiscal deterioration, too, was partly attributable to rising current expenditure. Conversely, the examples of strong austerity policies in Chile (1973–75), Ghana (1975/76 to 1988), Mexico (1986–89), and Thailand (1985–88) illustrate the important role played by reductions in current expenditure and, in particular, by cuts in wages and public employment.

Cutting transfers and subsidies is often an effective way to contribute to both fiscal stabilization and market deregulation. In Ghana and Zimbabwe lower transfers and subsidies contributed greatly to deficit reduction, by 5.4 and 5.0 percentage points of GDP, respectively. On the revenue side, tax reforms are at the heart of efforts to address structural deficits. In Chile reforms of direct taxes and the introduction of the value added tax (VAT) raised revenues by a staggering 10.5 percentage points of GDP, while Zimbabwe's 1988 tax reforms yielded a significant 4.2 percentage points of GDP. Higher tax revenue also helped Colombia, Mexico, and Thailand to reduce their deficits.

Rationalization of public enterprises and reforms of agricultural marketing boards constitute the fourth element of successful stabilization in our ten-country sample. Higher operating surpluses of SOEs contributed significantly to improving structural deficits—in Chile by a dramatic 8.4 percentage points of GDP and in Colombia and Ghana by smaller amounts. Conversely, the drastic deterioration in Côte d'Ivoire was caused by the decline in revenue from the cocoa and coffee revenue stabilization funds as a result of continuing producer price supports during a period of declining world prices.

An encouraging finding from our sample is that successful fiscal retrenchment does not have to rely on lower public investment. In the most dramatic fiscal turnaround (Chile, 1973–75) public capital formation was not reduced. In the three countries in which public investment fell during fiscal adjustment—Colombia, Ghana, and Mexico—the reduction was moderate. Only one case of fiscal retrenchment (Thailand) relied heavily on cutting public investment. Conversely, the two largest declines in public investment occurred in Côte d'Ivoire and Argentina during periods in which public deficits exploded. In the case of Argentina public capital expenditure continued its systematic decline beyond 1982, reaching a thirty-year low of 6.1 percent of GDP in 1987.

We conclude that successful (that is, sustainable) nonfinancial public sector adjustment typically requires simultaneous action on four fronts: reducing an overblown government bureaucracy; cutting transfers and subsidies to the private sector (except for efficient and targeted social programs); enacting tax legislation for increased,

**Table 1.5. Contribution of Policies to Fiscal Adjustment or Deterioration in Ten Countries**
(percentage points of GDP)

| Country experience | Deficit level at start and end of period | Change in deficit and contribution of fiscal policy changes | |
|---|---|---|---|
| Argentina: 1977–82 deterioration | 4.7; 15.1 | Change in deficit | +10.4 |
| | | Higher current expenditure | +15.1 |
| | | Lower capital expenditure | −4.7 |
| Chile: 1973–75 adjustment | 20.6; 2.1 | Change in deficit | −18.5 |
| | | Lower public employment | −4.3 |
| | | Higher revenue from tax reform | −10.5 |
| | | Higher SOE operating surplus | −8.4 |
| Colombia: 1984–89 adjustment | 6.3; 2.2 | Change in deficit | −4.1 |
| | | Lower public wages/salaries | −1.2 |
| | | Lower fixed investment | −2.1 |
| | | Higher tax revenue | −2.1 |
| | | Higher SOE operating surplus | −1.9 |
| Côte d'Ivoire: 1984–89 deterioration | 1.7; 14.4 | Change in deficit | +12.7 |
| | | Higher current expenditure | +3.6 |
| | | Lower capital expenditure | −5.1 |
| | | Lower tax revenue | +2.9 |
| | | Lower revenue commodity fund | +12.7 |
| Ghana: 1975/76–88 adjustment | 15.1; −0.4 | Change in deficit | −15.5 |
| | | Lower wage bill | −1.3 |
| | | Lower expenditure on goods/services | −1.6 |
| | | Lower transfers/subsidies | −5.4 |
| | | Lower public investment | −1.8 |

| Mexico: 1986–89 adjustment | 14.9; 5.1 | Change in deficit | −9.8 |
| | | Lower current expenditure | −2.5 |
| | | Lower other expenditure | −4.6 |
| | | Lower public investment | −0.7 |
| | | Higher direct tax revenue | −3.0 |
| | | Higher VAT revenue | −0.9 |
| Morocco: 1983–88 adjustment | 12.1; 4.1 | Change in deficit | −8.0 |
| | | Lower expenditure on goods/services | −2.9 |
| | | Lower transfers/subsidies | −1.7 |
| | | Lower capital expenditure | −3.3 |
| | | New petroleum levy | −3.4 |
| Pakistan: 1980/81 to 1986/87 deterioration | 4.8; 8.3 | Change in deficit | +3.5 |
| | | Higher noninterest current expenditure | +2.9 |
| | | Lower direct tax revenue | +0.8 |
| | | Lower indirect tax revenue | +1.9 |
| Thailand: 1986–88 adjustment | 8.6; −0.2 | Change in deficit | −8.8 |
| | | Lower public wages/salaries | −1.4 |
| | | Lower public investment | −3.5 |
| | | Higher revenue | −2.2 |
| Zimbabwe: 1986/87 to 1988/89 adjustment | 14.4; 10.0 | Change in deficit | −4.4 |
| | | Higher public wages/salaries | +4.0 |
| | | Lower transfers/subsidies | −5.0 |
| | | 1988 direct tax reform | −2.4 |
| | | 1988 custom duty reform | −1.8 |

*Note:* SOE, state-owned enterprise; VAT, value added tax. Data refer to the central government for Ghana and Morocco and to general government for Pakistan. In all other cases the data refer to the consolidated nonfinancial public sector deficit.

*Source:* Country case studies listed in the references to this chapter.

broadly based direct and indirect taxation; and reforming or privatizing public enterprises and commodity marketing boards. Efficient public investment, particularly in social or physical infrastructure, should not only be exempted from fiscal cuts but should be expanded to encourage economic growth.

## Deficits, Inflation, Real Interest Rates, and Financial Repression

As shown by the cross-country evidence in figure 1.3, the relationships between deficits and inflation and between deficits and interest rates are far from simple. At low to medium rates of inflation, there is no relationship across countries between long-term inflation rates (1980–88) and public deficits. However, the countries with the highest inflation rates—Argentina and Mexico during the 1980s—had significantly higher deficits than countries with lower rates. Similarly, domestic real interest rates show no correlation with public deficits across countries except in the case of high-deficit, high-interest-rate Argentina.

The lack of correlation across countries between deficits and inflation and between deficits and interest rates is primarily attributable to the different ways in which countries finance their public deficits. To account for the effects of these differences, a more detailed understanding is needed of the links between domestic deficit financing and inflation and interest rates (or financial repression).

This section first considers the relationship between seigniorage, inflation, and money demand and compares steady-state seigniorage levels with one-shot seigniorage episodes. It then reviews the empirical evidence on the relationships between specific sources of deficit financing and inflation, real interest rates, and financial repression.

### *Seigniorage Laffer Curve and Misspecified Money Demand*

Any notion that fiscal deficits and inflation display a simple relationship fails for two reasons. The first is that countries make different choices about seigniorage to finance their deficits, partly because they differ in the extent to which other means of finance are available. The second reason is that money creation and inflation are nonlinearly related. The scattergram shown in figure 1.4 suggests a conventional "Laffer-curve" relationship between the inflation rate and revenue from seigniorage, with revenue falling off at some point because of the elastic response of money demand. The exact maximum of the curve is sensitive to the inclusion of the extreme points: with Argentina the maximum is at 160 percent inflation, while without Argentina it is only at 68 percent. (A similar point is made by Fischer and

**Figure 1.3. Fiscal Deficits, Real Interest Rates, and Inflation
Rates in Case Study Countries, 1980–88 Averages**

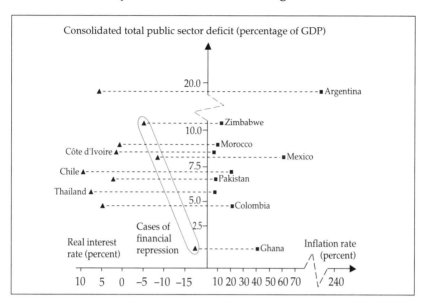

*Note:* The consolidated total public sector deficits are for the CNFPS in all ten
countries, but quasi-fiscal deficits are also included for Argentina, Chile, and Mexico.
*Source:* For the deficit series, the country studies (see References); for the inflation
rates and nominal interest rates (used to compute real interest rates), IMF, *International
Financial Statistics*, various years.

Easterly 1990, who also note that the growth rate affects whether
deficits are inflationary.)

Econometric estimation of a quadratic equation statistically confirms the Laffer curve, as shown by the following cross-country
relationship:

$$T/Y = 0.01 + 0.043\pi - 0.13\pi^2$$
$$(4.9) \quad (4.1) \quad (-2.31)$$

$$R^2 = 0.44$$

where $T/Y$ is average seigniorage revenue as a ratio to GDP in 1970–89
and $\pi$ is average inflation in 1970–89; $t$-statistics are in parentheses.[12]

These cross-section results differ significantly from calculations of
revenue-maximizing inflation from individual time-series results for
the case studies.[13] A regularity is that countries with high-inflation
have very high seigniorage-maximizing inflation rates (in Argentina
the rate is 966 percent, in Chile, 792 percent, and in Ghana, 125
percent); countries with moderate inflation have more moderate maximizing rates (Colombia's is 80 percent); and countries with low infla-

## Figure 1.4. Inflation and Seigniorage

Seigniorage (percentage of GDP)

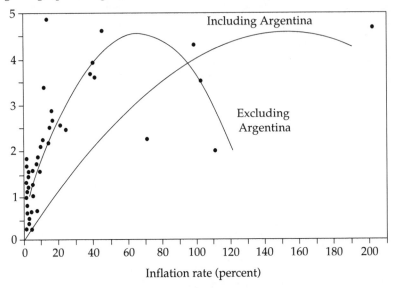

*Source:* Authors' regression and Statistical Appendix, tables A.3 and A.4.

tion have low maximizing inflation rates—Thailand's is only 4 percent!

One hypothesis to explain these huge differences across countries is misspecification of money demand. Conventional estimates of the seigniorage-maximizing inflation rate typically make use of a Cagan money demand, which implies a constant semielasticity of money demand with respect to inflation or interest rates (Cagan 1956). Easterly, Mauro, and Schmidt-Hebbel (forthcoming) show that the elasticity of substitution in transactions between money and bonds determines how the inflation semielasticity of money demand changes as inflation rises. Allowing for a variable semielasticity, the authors report estimates of seigniorage-maximizing inflation—varying between 266 and 303 percent per year—for a panel sample of eleven high-inflation countries. Their results are consistent with a semielasticity that increases with inflation (that is, higher inflation hastens the flight away from money and toward financial assets that provide protection from inflation).[14]

### Steady-State Seigniorage versus One-Shot Seigniorage Episodes

Given the attention devoted to seigniorage in the literature, it is easy to forget how small it is as a source of revenue. Table 1.6 shows the

## Table 1.6. Average Seigniorage in OECD and Developing Countries

| Country | Average seigniorage, 1970–88 (percentage of GDP) | Highest excise tax, 1985 | Product subject to highest excise tax |
|---|---|---|---|
| *OECD countries* | | | |
| Austria | 0.9 | 1.0 | Wine |
| Belgium | 0.5 | 1.1 | Mineral oil |
| Canada | 0.4 | 0.7 | Gasoline |
| Denmark | 0.4 | 1.1 | Cigarettes |
| Finland | 0.6 | 1.3 | Fuel |
| France | 0.6 | 0.4 | Insurance |
| Germany, Fed. Rep. | 0.7 | 1.3 | Mineral oil |
| Greece | 2.8 | 2.2 | Fuel |
| Italy | 2.2 | 1.7 | Mineral oil |
| Japan | 1.0 | 0.6 | Liquor |
| Netherlands | 0.6 | 0.8 | Petroleum |
| Norway | 0.6 | 1.5 | Vehicle transfer |
| Spain | 2.3 | 1.3 | Petroleum |
| Sweden | 0.6 | 1.0 | Petroleum |
| United States | 0.4 | 0.3 | Motor vehicle fuels |
| Average | 1.0 | 1.1 | n.a. |
| *Developing countries* | | | |
| Argentina | 4.2 | 2.5 | Fuel |
| Bangladesh | 1.0 | — | — |
| Bolivia | 2.9 | — | — |
| Brazil | 2.3 | 0.2 | Electricity |
| Burkina Faso | 1.1 | 0.7 | Beverages |
| Chile | 3.7 | — | — |
| Colombia | 2.1 | 0.6 | Gasoline |
| Côte d'Ivoire | 1.3 | 1.1 | Petroleum |
| Dominican Republic | 1.6 | 1.8 | Petroleum |
| Ecuador | 1.8 | 0.3 | Beer |
| Ghana | 3.1 | — | — |
| Honduras | 0.8 | 0.5 | Beer |
| India | 1.5 | 0.7 | Textiles and Jute |
| Indonesia | 1.4 | 0.9 | Tobacco |
| Jamaica | 1.9 | — | — |
| Jordan | 5.0 | — | — |
| Kenya | 1.1 | — | — |
| Korea | 1.6 | 0.8 | Liquor |
| Malawi | 2.0 | — | — |
| Malaysia | 1.3 | 0.7 | Petrol |
| Mexico | 3.1 | 1.4 | Gasoline |
| Morocco | 1.7 | 1.2 | Tobacco |
| Nigeria | 1.1 | — | — |
| Pakistan | 2.0 | — | — |
| Paraguay | 1.9 | 0.9 | Fuel |

*(Table continues on the following page.)*

**Table 1.6** (*continued*)

| Country | Average seigniorage, 1970–88 (percentage of GDP) | Highest excise tax, 1985 | Product subject to highest excise tax |
|---|---|---|---|
| Peru | 3.6 | 4.1 | Gasoline |
| Philippines | 1.0 | — | — |
| Sri Lanka | 1.3 | — | — |
| Thailand | 1.0 | 1.5 | Petroleum products |
| Trinidad and Tobago | 0.9 | — | — |
| Turkey | 3.4 | — | — |
| Venezuela | 1.5 | 0.5 | Liquor |
| Zaire | 4.4 | 0.3 | Tobacco |
| Zambia | 2.0 | 1.9 | Petroleum |
| Zimbabwe | 1.1 | — | — |
| Average | 2.1 | 1.1 | n.a. |

— Not available.
n.a. Not applicable.
*Source:* For excise taxes, IMF 1986; for seigniorage, Statistical Appendix, table A.2.

average seigniorage for a sample of industrial and developing countries for which data are available. Seigniorage is calculated as the ratio
to real GDP of the yearly sum of deflated monthly changes in the
money base. The generally small amount of seigniorage for the ten
case studies is typical of the overall pattern of seigniorage among all
countries. The maximum amount of average seigniorage revenue
over an extended time is less than 5 percent of GDP. Seigniorage is
mainly a phenomenon of developing countries—among industrial
countries, only Greece, Italy, and Spain had seigniorage above 1 percent of GDP. Average seigniorage is more than twice that level in
developing countries.[15]

Seigniorage revenue is of the same order of magnitude as revenue
from individual excise taxes (see table 1.6). Why, then, are macro-
economists so preoccupied with taxes on money, as against taxes on
beer, jute, or cigarettes?

Perhaps one reason is that seigniorage can be a large source of
temporary revenue during times of crisis. The time-series averages
for seigniorage conceal tremendous year-to-year fluctuations. Figure
1.5 shows a frequency distribution of the individual yearly observations for the same sample of countries as in table 1.6. Although nearly
half the sample is concentrated in observations of less than 1 percent
of GDP, there is a significant number of observations of high-
seigniorage revenue, reaching as high as 13 percent of GDP. The average time-series coefficient of variation in the sample is 90 percent.[16]

This suggests that a fruitful approach to seigniorage would be to
study the episodes of high seigniorage to see how they happened and

## Figure 1.5. Frequency Distribution of Annual Seigniorage Observations, Fifty-one Countries, 1970–88

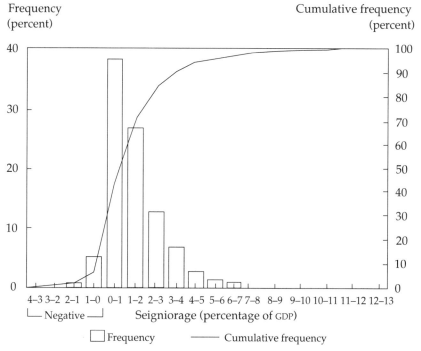

*Source:* Statistical Appendix, tables A.3 and A.4.

what their consequences were. A number of the case studies in this project include such episodes; bursts of seigniorage appear in Argentina in 1975 and 1983, in Chile in 1971, in Ghana in 1978 and 1983, and in Mexico in 1982. A different type of one-shot seigniorage took place in Ghana in 1979, when the government captured 2.5 percent of GDP through a currency conversion and partial expropriation of deposits, and in 1982, when the government again expropriated private wealth through the demonetization of the largest-denomination note. These episodes brought a short-term gain to public finances at considerable long-run cost—the Ghana case study estimates that seigniorage was permanently lowered by 1 to 2 percent of GDP because of the fall in money demand after the 1979 currency expropriation.

Table 1.7 looks at episodes of high (conventional) seigniorage in the broader sample. Of a total of 1,143 observations of forty-nine countries for various years, we identified eighteen instances in which seigniorage was more than 4 percentage points above the average seigniorage-to-GDP ratio in the particular country.[17] We see that

**Table 1.7. Episodes of High Seigniorage Spikes**

| Country | Year of spike | Seigniorage (percentage of GDP) | | Components of spike (percentage of GDP) | | Growth (percent) | | Inflation (percent, December to December) | | |
|---|---|---|---|---|---|---|---|---|---|---|
| | | Seigniorage spike (deviation from average) | Average seigniorage | Change in real money base + average money base | Above-average inflation tax | Growth in year of spike | Average growth rate | Inflation in year of spike | Change in inflation over previous year | Average inflation rate |
| Argentina | 1975 | 9.0 | 4.2 | -4.0 | 6.5 | -0.5 | 2.2 | 336.1 | 296.2 | 105.4 |
| Bolivia | 1982 | 7.5 | 2.9 | -0.8 | 7.9 | -4.4 | 2.7 | 296.6 | 271.4 | 54.5 |
| | 1983 | 5.3 | — | -3.3 | 8.6 | -6.5 | — | 327.8 | 31.3 | — |
| | 1984 | 7.0 | — | -3.9 | 20.8 | 1.0 | — | 2,176.2 | 1,848.4 | — |
| Chile | 1971 | 6.7 | 3.8 | 10.9 | -3.0 | 9.1 | 1.5 | 19.4 | -15.6 | 91.3 |
| Denmark | 1985 | 4.2 | 0.4 | 4.3 | -0.2 | 4.3 | 2.4 | 3.6 | -2.0 | 7.7 |
| Dominican Rep. | 1986 | 4.9 | 1.6 | 5.3 | -0.6 | 3.0 | 5.7 | 6.5 | -21.8 | 13.1 |
| | 1988 | 5.7 | — | 2.2 | 3.0 | 1.3 | — | 57.5 | 32.5 | — |
| Ghana | 1978 | 4.7 | 3.1 | -0.6 | 3.8 | 9.8 | 1.7 | 108.5 | -2.4 | 38.6 |
| Jamaica | 1984 | 5.4 | 1.9 | 4.7 | 1.0 | -1.4 | -0.2 | 31.2 | 14.5 | 17.0 |

| | | | | | | | | | |
|---|---|---|---|---|---|---|---|---|---|
| Mexico | 1982 | 7.9 | 3.1 | 1.4 | 3.7 | -0.6 | 5.0 | 98.9 | 70.2 | 29.8 |
| Peru | 1985 | 6.7 | 3.6 | 2.7 | 2.5 | 2.2 | 2.2 | 158.3 | 46.8 | 82.1 |
| | 1988 | 6.0 | — | -7.3 | 14.8 | -8.0 | — | 1,722.1 | 1,607.5 | — |
| Trinidad/Tobago | 1982 | 4.2 | 0.9 | 4.3 | 0.0 | -4.7 | 1.6 | 10.8 | -0.8 | 10.5 |
| Zaire | 1976 | 4.1 | 4.4 | 0.5 | 2.2 | -5.5 | 3.1 | 78.8 | 42.1 | 47.1 |
| | 1982 | 4.9 | — | 4.4 | -0.6 | -0.4 | — | 41.0 | -12.0 | — |
| | 1987 | 5.7 | — | -0.1 | 4.1 | 0.6 | — | 106.5 | 68.2 | — |
| Zambia | 1986 | 6.0 | 2.0 | 5.4 | 1.1 | 0.2 | 1.8 | 34.6 | -23.7 | 18.9 |
| Average for "spike" episodes or countries | | 8.8 | 2.7 | 1.5 | 4.2 | 0.0 | 2.5 | 311.9 | 236.2 | 43.0 |
| Average for thirty-nine "nonspike" countries in sample | | — | 1.4 | — | — | — | 4.5 | — | — | 11.2 |

— Not available.

*Note:* Spikes are defined as seigniorage more than 4 percentage points of GDP above the average seigniorage-to-GDP ratio for the country. Components do not sum to the "spike" because of the covariance term.

*Source:* Authors' calculations based on data in the Statistical Appendix and in IMF, *International Financial Statistics*, various issues.

"spikes" of high seigniorage are indeed short-lived. The only episode that lasted more than one year was the Bolivian hyperinflation of 1982–84. The episodes are associated with developing countries; among OECD countries only Denmark indulged in a seigniorage spike. One might have thought that these bursts of seigniorage revenue would be associated with accelerations of inflation, but this hypothesis is, surprisingly, not confirmed by the data. Of the sixteen episodes (treating Bolivia, 1982–84, as a single episode), only nine showed rising inflation—roughly the same proportion as in the broader sample. There is no evidence for the supposition that inflation shows a lagged response, as the following year's inflation also shows no tendency to accelerate. Some of the episodes of rising inflation are quite spectacular. Bolivia in 1982–84 and Peru in 1988 experienced classic hyperinflation in which real money demand fell but inflation soared to four digits. Large accelerations of inflation also took place in Argentina in 1975, Mexico in 1982, and Zaire in 1987.

A decomposition of the seigniorage spikes into components associated with the real change in the money base and the inflation tax helps explain the cases in which inflation did not accelerate (see appendix 1.3). Table 1.7 shows that the real change in the money base explains most or all of the above-average seigniorage in seven of the sixteen cases. In six of these cases, inflation declined. The lack of a close association between acceleration of inflation and bursts of seigniorage comes about because in nearly half the cases seigniorage was driven mainly by real money balances. An understanding of this phenomenon would require more careful examination of individual cases, but apparently there was scope for temporary increases in seigniorage revenues through actions such as raising reserve requirements or through exploitation of exogenous increases in demand for money. Price controls were used in Chile in 1971 to generate a "real" change in money demand, but inflation exploded in the following year.[18]

Of course, the classic inflationary method worked just as well as a method for generating bursts of seigniorage. One-time inflation taxes that were more than 8 percentage points of GDP above average were recorded in the hyperinflation in Bolivia and Peru, while less spectacular increases were registered in Argentina, Ghana, and Zaire. The money base fell in all these cases, limiting the potential for further inflation taxes.

The GDP growth rates during the seigniorage spikes were not so high as to make money creation a recommended method of raising revenue. In twelve of the sixteen episodes growth during the episode was below the average growth rate for that country, and eight cases actually registered negative growth of gross—not per capita—output. We must remain skeptical about whether growth was poor because of the unusually high seigniorage or countries resorted to seigniorage because economic recession dried up other revenue sources. But it is interesting that the countries with spikes have a lower average growth than other countries in the sample, perhaps reflecting a tendency toward higher *average* seigniorage and inflation, as well.

In conclusion, seigniorage may be more important as a source of temporary increases in revenue than as a steady-state phenomenon. But the link between these temporary seigniorage surges and inflation is weak. A surprising number of episodes of high seigniorage are attributable to increases in real money balances instead of to accelerated inflation, illustrating the scope for achieving temporary revenue increases through various actions by the monetary authorities other than printing money. But the poor performance of countries that resort to such measures does nothing to further the case for using bursts of seigniorage as an instrument of public finance.

## Fiscal Deficits, Seigniorage, and Inflation

Average long-term (1965–89) monetary financing or seigniorage is 2.3 percent of GDP in the ten sample countries (table 1.8)—close to the average of 2.1 percent for all developing countries, but twice the level of 1.0 percent for the OECD countries (see table 1.6). Seigniorage and inflation show an association across the ten sample countries. The short-run relationship between money financing and inflation is typically blurred by factors such as indexation practices, slowly changing expectations, slow portfolio substitution, and inflationary exchange rate depreciations.

In the long run, however, the tradeoff between inflation and money creation becomes increasingly unfavorable, explaining why seigniorage is generally used only as a last resort. Table 1.8 reports the amount of additional inflation required to achieve another percentage point of GDP in long-run seigniorage revenue. The figures are derived from estimated Cagan-type, constant-inflation or interest-elastic, money demands for the case countries. The tradeoff is still favorable in countries with low inflation (in Thailand the additional inflation is just 5 percentage points), worsens in moderate-inflation countries (15 to 20 percentage points in Colombia and Ghana), and becomes untenable in countries with high inflation (97 percentage points in Argentina), where moneyholders replace most of their local currency holdings with foreign currency and interest-bearing assets.

These results are remarkably similar to the simulation results in table 1.9, which are based on more comprehensive portfolio substitution and deficit-financing models (and, in the case of three countries, general equilibrium macroeconomic frameworks). The long-term effects on price levels of transitory deficits financed by money cre-

**Table 1.8. Seigniorage, Inflation, and Marginal Inflation Revenue, 1965–89**

| Country | Seigniorage (percentage of GDP)[a] | Inflation (percent)[b] | Percentage increase in inflation required to collect additional seigniorage of 1 percentage point of GDP |
|---|---|---|---|
| *Case study countries* | | | |
| Argentina | 4.2 | 115.3 | 97 |
| Chile | 3.7 | 56.6 | 23 |
| Colombia | 2.1 | 17.7 | 15 |
| Côte d'Ivoire | 1.3 | 7.6 | — |
| Ghana | 3.1 | 31.6 | 20 |
| Mexico | 3.1 | 28.9 | — |
| Morocco | 1.7 | 6.1 | 8–26 |
| Pakistan | 2.0 | 8.0 | — |
| Thailand | 1.0 | 5.7 | 5 |
| Zimbabwe | 1.1 | 7.7 | 10 |
| Average | 2.3 | 28.5 | n.a. |
| *Other countries* | | | |
| Average, 35 developing countries | 2.1 | — | n.a. |
| Average, 15 industrial countries | 1.0 | — | n.a. |

— Not available.

n.a. Not applicable.

a. Seigniorage is defined as the nominal change in the money base each month divided by the consumer price index (CPI) for that month. The typical method of calculating the ratio of the nominal change in the money base over the entire year to the annual nominal GDP can seriously overstate seigniorage in high-inflation countries. Although interest paid on reserves should also be subtracted to get a true estimate of seigniorage, the data are generally lacking, and, in any case, few developing countries pay interest on reserves. Where interest is paid, it appears that it is quantitatively unimportant. An important exception is Argentina, where the combination of high inflation and interest paid on reserves makes this adjustment important. We use the Argentine seigniorage series used by Rodríguez in chapter 3 of this volume. Periods covered are generally 1965–89 but vary depending on the availability of data.

b. Average annual rates of change in the CPI between 1964 and 1988.

*Source:* For annual data, Statistical Appendix, table A.2. For average annual rate of change, IMF, *International Financial Statistics*, various years. For Argentina, Colombia, Ghana, and Morocco, data in the last column are from the country case studies listed in the references to this chapter; for Chile, Thailand, and Zimbabwe, these data are calculated from seigniorage and inflation rates in the first two columns and from long-run money demand inflation semielasticities in the country studies.

ation, taking into account feedback effects on inflation from asset substitution (and endogenous output response in the cases of Pakistan and Colombia), are similar to those obtained using the simple Cagan form. For the four reported countries, the additional infla-

**Table 1.9. Simulation Results for Long-Term Effects of Fiscal Deficits on Inflation and Real Interest Rates**
(percent)

| Country | Effect of a 1 percentage point increase in the deficit-to-GDP ratio | |
| --- | --- | --- |
| | On the price level, with money financing | On the real interest rate, with domestic debt financing |
| Chile | 14 | 0.1 |
| Colombia | 14 | 3.0 |
| Morocco | — | 0.2 |
| Pakistan | 18 | 1.1 |
| Zimbabwe | 10 | 2.7 |

— Not available.

*Note:* This table presents the long-term effects of a transitory (one-year) increase in the public deficit, financed by issuing either domestic non-interest-bearing monetary liabilities or domestic interest-paying debt. Both the short-term effects and the cross-effects (of money financing on the real interest rate and of debt financing on inflation) are of less interest and vary from country to country because of differences in model structures. The results for Chile and Zimbabwe are based on portfolio models combined with the public sector budget equation; those for Colombia, Morocco, and Pakistan are based on macroeconomic-portfolio general equilibrium specifications.

*Source:* Country case studies listed in the references to this chapter.

tion required to collect 1 additional percentage point of GDP through seigniorage ranges from 10 percent for Zimbabwe to 18 percent for Pakistan. Considering the unfavorable tradeoff in most cases and the widespread consensus on the undesirability of inflation, it is difficult to believe that revenue motivations alone explain chronic high inflation. (See Blejer and Liviatan 1987 and Kiguel and Liviatan 1988 for similar conclusions.)

*Fiscal Deficits and Interest Rates or Financial Repression*

There are two ways in which fiscal deficits can affect domestic real interest rates and financial markets. First, if interest rates are not controlled, a high fiscal deficit financed through domestic borrowing would be expected to result in high real interest rates. Second, if interest rates can be and are controlled, the implicit tax on financial assets could be a hidden source of revenue for the government.

By liberalizing interest rates, financial reform has shifted deficit financing from implicit financial repression revenue to explicit debt issuing in many developing countries. Argentina, Chile, Colombia, Morocco, Pakistan, and Thailand introduced financial reforms in the 1970s, and their real interest rates reached positive levels in the 1980s (table 1.10). Ghana, Mexico, and Zimbabwe, however, maintained varying degrees of domestic interest controls during most of the 1980s

**Table 1.10.  Real Interest Rates under Financial Reform or Financial Repression in the 1980s**

| Country | Real interest rates on deposits (percent)[a] | | Tax revenue on deposits attributable to financial repression, 1980–88 (percentage of GDP)[b] |
|---|---|---|---|
| | 1970–79 | 1980–88 | |
| Argentina | −17.2 | 4.8 | n.a. |
| Chile | −15.9 | 8.1 | n.a. |
| Colombia | −6.3 | 0.7 | n.a. |
| Ghana | −18.8 | −18.3 | 0.5 |
| Mexico | −4.6 | −8.4 | 1.6 |
| Morocco | −3.1 | 1.8 | n.a. |
| Pakistan | −3.4 | 2.1 | n.a. |
| Thailand | −0.5 | 6.5 | n.a. |
| Zimbabwe | −3.7 | −4.3 | 0.8[c] |

n.a. Not applicable.

a. Average annual real interest rates on time deposits at banking system, calculated using the consumer price index.

b. Average annual revenue calculated as the difference between domestic real interest rates and the average real interest rate in OECD countries (0.9 percent) multiplied by deposits outstanding as a percentage of GDP.

c. 1980–87.

Source: Country case studies listed in the references to this chapter; Statistical Appendix, tables A.4 and A.5.

(Mexico liberalized its rates in 1988), as reflected by negative average real interest rates.

The implicit tax from financial repression is normally computed as the product of the interest tax (typically, the difference between the foreign and domestic interest rates) and the outstanding stock of the relevant public liability or the time deposits in the financial system. In the latter case the financial repression revenue collected by the financial system is often transferred back to the public sector via compulsory placements of government debt in financial institutions or through unpaid reserves held at the central bank.

Average annual revenue for the three countries from financial repression of deposit interest rates during 1980–88 ranged from 0.5 percent of GDP for Ghana to 1.6 percent of GDP for Mexico (see table 1.10). Holding down nominal interest rates under high inflation was a quick and easy way of obtaining revenue to compensate for the loss of external financing after 1982. Table A.5 of the Statistical Appendix presents estimates from other studies of revenue from financial repression. Although calculations differ widely because of different methodologies, there is a consistent finding that Ghana, Mexico, and Zimbabwe reaped significant amounts of revenue from controls on domestic interest rates during the 1980s.[19]

Revenue from financial repression in these three countries is comparable to the 1.0 to 2.1 percent of GDP of average seigniorage collected in OECD and developing countries (see table 1.6). Although the implicit tax from financial repression is less visible than seigniorage and its inflation tax component, its deleterious effects on financial intermediation—and hence on the quantity and quality of private investment—is probably as strong as that of inflation. In fact, controlling interest rates was a costly strategy for private credit and investment, which remained depressed throughout the 1980s.[20] In reaction, partial decontrol of interest rates in Ghana and Zimbabwe and complete interest liberalization in Mexico since the mid-1980s have reduced or abolished taxes on financial intermediation.

Figure 1.6 shows the evolution of domestic private credit in the case studies. There are large differences in domestic private credit stocks between countries with deregulated financial markets—where private credit reaches an average 30 percent of GDP—and those with stringent financial controls, where the corresponding average ratio hovers around 10 percent. Mexico's experience well illustrates the effects of financial repression. Financial controls intensified after 1981 as inflation soared, and the ratio of private credit to GDP dropped below already low levels. Following financial liberalization in 1988, the ratio doubled within two years. In Ghana private credit was at a dismally low level, reflecting years of financial repression, including two episodes of outright expropriation of financial assets. By contrast, countries that abstained from repressive interest rate controls, such as Chile and Thailand, had very high levels of private credit, which may partly explain their superior investment and growth performance in the late 1980s.

The alternative to financial repression is government borrowing at market interest rates. Table 1.9, above, reports simulation results (based on the portfolio and general equilibrium frameworks referred to earlier) for the long-term effects on the real interest rate of a transitory 1 percentage point increase in the deficit (in relation to GDP), financed by floating domestic debt, in five country cases. The effects vary widely in the five countries, reflecting differences in the willingness of asset holders to shift from alternative forms of saving. Flat demands for domestic debt imply that real interest rates increased by a modest 0.1 to 0.2 percentage point in Chile and Morocco. Low asset substitutability between domestic debt and alternative private sector asset holdings in Colombia, Pakistan, and (after interest decontrol) Zimbabwe—partly because of high domestic debt levels—explains increases in real interest rates that range from 1.1 to 2.7 percentage points. The implication for those three countries is that when domestic borrowing is high and costly—which could lead to a domestic debt spiral such as the one in Argentina described in box 1.4—there is no

## Figure 1.6. Private Credit under Financial Liberalization and Repression in Nine Countries, 1980–90

**Without interest rate controls**

Percentage of GDP

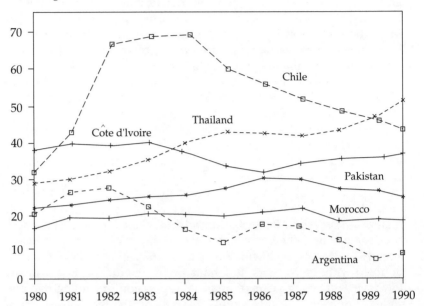

**With interest rate controls**

Percentage of GDP

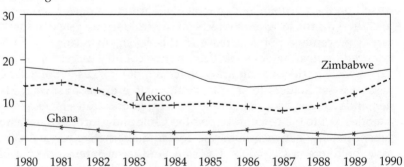

*Source:* IMF, *International Financial Statistics,* various issues.

## Box 1.4. Argentina's Unusual Financial Behavior

The massive decline in Argentina's ratio of private credit to GDP, shown in figure 1.6, reflects an unusual kind of financial behavior. The government, far from controlling interest rates, oscillated between paying high interest rates and ''melting down'' domestic liabilities through surprise devaluations and other methods (including a forced conversion of time deposits into government bonds of questionable value in 1990). This tactic was necessary because the high interest rates themselves fueled the accumulation of more debt, in a classic example of a debt spiral, or Ponzi game. Borrowing was increased in order to pay the interest on the debt, which implied yet higher interest and borrowing in the next period, and so on. Although the government managed to keep persuading the public to buy domestic debt, increasingly high interest rates were required. The Argentina case study in this book chronicles the rise in nominal interest rates at the outset of successive economic plans, each of which opened with a devaluation.

| Stabilization plan and date | Initial devaluation (percent) | Nominal interest rate per month (percent) |
|---|---|---|
| Austral (June 1985) | 40 | 7 |
| Primavera (August 1988) | 24 | 10 |
| Bunge-Born I (July 1989) | 200 | 17 |
| Bunge-Born II (December 1989) | 54 | 60 |
| Erman (January 1990) | 220 | 100 |

choice except to continue financial repression (as Zimbabwe has done up to now) or to pursue the more desirable course of fiscal adjustment (as in Morocco or Colombia throughout the 1980s and to the present).

These results for domestic debt financing and real interest rates (or financial repression) and those for seigniorage and inflation indicate strong correlations in both cases in developing countries. The increasingly unfavorable tradeoffs between these financing sources and the rates of return on government liabilities—leading in the extreme cases to hyperinflation, debt repudiation, or the virtual disappearance of domestic capital markets—imply that there is no alternative to fiscal adjustment for ensuring monetary and financial stability.

## Private Sector Response to Public Deficits

The macroeconomic effects of deficits are determined to a large extent by the direct response of private spending—consumption and investment—to changes in the deficit and in its composition. The way in which governments adjusted their fiscal imbalances during the 1980s (frequently by cutting public investment) was often costly for private sector investment. In the ten sample countries private investment declined sharply from an average of 13 percent of GDP in 1981 to 9 percent in 1986. Meanwhile, consumption, both public and private, was relatively insulated. Not even the sharp increases in public consumption of the 1970s—expansions that had much to do with the subsequent fiscal crises—were moderated during the adjustments of the 1980s.

To provide some insight into how the private sector responds to fiscal policies, we first identify the channels of transmission between fiscal policies and private spending and then assess their empirical relevance. The empirical inferences are derived from econometric estimations (based on annual time series for each country) for both private consumption and investment, consistent with optimizing behavior under liquidity constraints. Here we summarize only the qualitative response to variables related to fiscal policy; the chapters on individual countries provide the estimated coefficients of both fiscal and nonfiscal variables.

### Private Consumption and Fiscal Policies

Fiscal policies affect private consumption and saving through two major channels: disposable income and rate of return (real interest rate).

An increase in the deficit brought about by a cut in current taxes boosts private consumption by increasing disposable income, according to the standard Keynesian hypothesis that consumers increase spending when their current income rises. The permanent income hypothesis asserts that only a permanent tax cut significantly affects consumer spending; thus, if the tax cut is temporary, the consumption effect will be minimal.

Both hypotheses are denied by the Ricardian equivalence hypothesis, which claims that consumers react the same whether government spending is financed through debt or taxes because consumers foresee that a tax cut today, paid for by a deficit and borrowing, will lead to a tax increase in the future. In anticipation of that future tax increase, consumers save rather than spend the income from the tax cut. So a tax cut that simply substitutes debt finance for tax finance of unchanged government spending would leave consumer spending unchanged—and would lower it as a share of (now higher) disposable income. However, if government consumption is increased, private consumption should decline one-to-one with each dollar of higher permanent government spending. The argument, first skeptically postulated by Ricardo and affirmed in the recent literature by Barro

(1974), rests on—in addition to many secondary assumptions—two main and rather stringent assumptions (Seater 1993): that consumers are concerned with their own future welfare and that of their descendants, and that consumers can shift consumption over time by borrowing or lending whenever they wish.

There is another reason—unrelated to the Ricardian hypothesis—why a tax cut could cause private saving to rise or a government consumption increase could cause private consumption to fall. If there are strict government controls on domestic credit and external capital flows, with government having the first claim on credit, an increase in the deficit (a fall in government saving) reduces the credit available to the private sector, forcing private saving to rise or consumption to fall. This effect, which may be difficult to distinguish from the Ricardian hypothesis, may be termed the direct-crowding-out hypothesis.

The real interest rate determines how consumers schedule their consumption over time, assuming that they have access to credit. The effect of the interest rate on today's consumption levels is ambiguous because of the offsetting substitution, income, and wealth effects. An increase in interest rates causes consumers to substitute consumption tomorrow for consumption today, but it also induces consumers to feel richer and thus to spend more both today and tomorrow—unless this wealth stems significantly from future income streams discounted by the (higher) interest rate. Credit controls or borrowing constraints would block the effect of the real interest rate on consumption.

Table 1.11 summarizes the qualitative effects of the above-mentioned fiscal policy variables on private consumption.[21] (For brevity, the discussion excludes any reference to other consumption determinants included in the estimations, such as the terms of trade, foreign saving, or money.) For most of the countries, both current (or temporary) and long-run (or permanent) disposable income levels are found to be important determinants of private consumption—and often by magnitudes halfway between those implied by the Keynesian hypothesis and those implied by the permanent income hypothesis.

Does public saving (or the deficit) affect private consumption directly, as implied by the Ricardian hypothesis and the direct-crowding-out hypothesis? For most countries it does not: permanent public saving is not significant in Chile, Mexico, or Pakistan; current public saving or deficits do not affect consumption in Colombia, Côte d'Ivoire, Ghana, or Pakistan. In three cases, however, changes in public saving (or surplus) cause consumption to move in the same direction, which is consistent with either hypothesis. Private consumption rose with permanent public surpluses in Argentina and

**Table 1.11. Qualitative Effects of Fiscal-Policy-Related Variables on Private Consumption**

| Country | Disposable income | | Public saving | | Public surplus | | Real interest rate |
|---|---|---|---|---|---|---|---|
| | Current | Permanent | Current | Permanent | Current | Permanent | |
| Argentina, 1915–84, 1961–84 | — | (+)[a] | — | — | — | (+)[a] | — |
| Chile, 1960–88 | (+) | (+) | (0) | (0) | — | — | (0) |
| Colombia, 1971–86 | (+) | (+) | (0) | — | — | — | (+) |
| Côte d'Ivoire, 1972–87 | (+) | — | — | — | (0) | — | — |
| Ghana, 1969/70 to 1988 | (+) | (+) | — | — | (0) | — | (0) |
| Mexico, 1981.1–89.4 | (+) | (0) | — | (0) | — | — | (—) |
| Morocco, 1972–88 | — | (+) | — | — | — | (+) | (0) |
| Pakistan, 1963–87 | — | (+) | — | (0) | (0) | — | — |
| Thailand, 1971–87 | (+) | — | — | — | — | — | (+) |
| Zimbabwe, 1965–88 | (+) | (+) | — | (+) | — | — | (0) |

— Not available.

*Note:* The positive and negative signs correspond to statistically significant coefficients; (0) denotes a coefficient not significantly different from zero. Specifications and estimation techniques vary by country. The dependent variable (private consumption) enters in levels for Argentina, Ghana, and Pakistan, in log levels for Morocco and Thailand, in both levels and log levels for Colombia, in the ratio to national income for Côte d'Ivoire, and in the ratio to private disposable income for Chile, Mexico, and Zimbabwe.

a. The specification does not permit a clear distinction between current and permanent forms.

*Source:* Country case studies listed in the references to this chapter.

Morocco and with permanent public saving in Zimbabwe. Although the coefficients were significant and positive, they were much lower than those for permanent disposable income, implying—contrary to the Ricardian hypothesis—that tax cuts would affect consumption. It also implies that public saving would still have a positive net effect on total saving.

These three cases could have supported the Ricardian explanation only if the countries had freely operating financial markets so that consumers could shift their consumption over time in anticipation of future tax increases. In fact, however, Argentina did not liberalize its financial markets until 1977 (late in the sample period), while Morocco and Zimbabwe had institutional arrangements that gave the public sector preferential access to domestic credit. These facts suggest that direct crowding-out of private consumption by public deficits is the more likely explanation. Similar results for a different sample of developing countries were found by Corbo and Schmidt-Hebbel (1991).

The ten case studies provide little evidence that real interest rates have a positive effect on private saving—a result consistent with similar findings for other developing country samples (Giovannini 1983, 1985; Corbo and Schmidt-Hebbel 1991; Schmidt-Hebbel, Webb, and Corsetti 1992). The real interest rate showed significant effects in three countries. Rising real interest rates depressed private consumption in Mexico (signaling the dominance of the intertemporal substitution effect) but increased consumption in Colombia and Thailand. The absence of significant results in five other cases suggests either that the substitution, income, and wealth effects tend to cancel each other out or that borrowing constraints prevent consumers from responding to interest rate swings by shifting consumption over time. Borrowing constraints are also behind Haque and Montiel's (1989) rejection of Ricardian equivalence for a set of developing countries.

### Private Investment and Fiscal Policies

Fiscal policies affect private investment through three major channels: public investment, public deficits, and the user cost of capital.

Public capital could be a close substitute for private capital, driving down the rate of return on private investment. Public investment in steel plants is an obvious example. However, governments also invest in activities that do not attract private investment but that raise the rate of return of other private projects, such as infrastructure projects. Thus the higher the complementarity of public and private capital, the more likely it is that public investment will have a net positive effect on private investment.

## Table 1.12. Qualitative Effects of Fiscal-Policy-Related Variables on Private Investment

| Country | Public capital | | Public sector | | | Cost of capital | |
|---|---|---|---|---|---|---|---|
| | Stock | Flow | Deficit | Consumption | Revenue | User cost | Real interest rate |
| Argentina, 1915–84 | — | (0) | — | (−) | (+) | — | — |
| Chile, 1961–88 | — | (−)/(0) | — | — | — | (−)/(0) | — |
| Colombia, 1925–88 | (−) | — | — | — | — | (0) | — |
| Côte d'Ivoire, 1972–87 | — | — | (−) | — | — | — | — |
| Ghana, 1967–88 | — | (−) | — | — | — | — | (0) |
| Mexico, 1970–89 | — | (−)/(0) | — | — | — | (−) | — |
| Morocco, 1972–88 | (+) | (+) | — | — | — | (−) | — |
| Pakistan, 1972/73 to 1987/88 | — | — | (−) | — | — | (−) | — |
| Thailand, 1971–87 | — | (+) | (−) | — | — | — | — |
| Zimbabwe, 1965–88 | — | (+) | — | — | — | (−) | — |

— Not available.

*Note:* The positive and negative signs correspond to statistically significant coefficients; (0) denotes a coefficient not significantly different from zero. Specifications and estimation techniques vary by country. The dependent variable is private investment for all countries except Côte d'Ivoire and Pakistan; it enters in levels for Argentina, in log levels for Thailand, in the ratio to GDP for Chile, Ghana, Mexico, and Zimbabwe, in the log ratio to GDP for Morocco, and in the level, log level, or ratio to GDP for Colombia. In the case of Pakistan the dependent variable is the ratio of private capital stock to GDP. Because of data limitations for Côte d'Ivoire, the dependent variable is the ratio of domestic investment to national income.

*Source:* Country case studies listed in the references to this chapter.

If there is repression of domestic interest rates and the public sector is given preferential access to domestic credit, the public deficit could crowd out private investment. When interest rates are not regulated, deficit financing through domestic borrowing tends to raise real interest rates, diminishing the profitability of investment by raising the user cost of capital. (The user cost of capital is determined by the real interest rate, the price of investment goods, and investment incentives.)

Table 1.12 summarizes the qualitative effects of fiscal policy variables on investment. (For brevity, the discussion excludes any reference to other investment determinants included in the estimations, such as the marginal product value of capital, foreign saving, firm profits, or banking credit to firms.) Consistent with the theoretical ambiguity of the relationship between public capital and private investment, the case studies found sharply different results. For Pakistan each percentage point increase in the ratio of public capital stock to output results in a 2.1 percentage point increase in the ratio of private capital stock to output. A similar relationship is found for Zimbabwe; a higher public capital stock also raises private investment, but the effect is smaller than in Pakistan. By contrast, an increase in public capital stock in Chile and Colombia tends to lower private investment.

Some of the country studies used public investment rather than public capital stock, again finding opposite effects in different countries. For Ghana and Mexico increasing public investment reduces private investment (although the effect was weak for Mexico), while for Thailand private investment rises with public investment. For Argentina no significant relation was found. The Morocco study found that public investment contributes to growth, from which it can plausibly be inferred that private investment rises with public capital formation.

Thus, only three countries provide direct evidence for the widespread presumption that public sector investment is good for private investment. These findings confirm previous studies for developing countries (Blejer and Khan 1984; Khan and Reinhart 1990), with ambiguous results regarding the effect of public on private investment. By way of comparison, Aschauer (1989) finds that higher public capital strongly increases private investment in the United States. It seems reasonable to infer, then, that for countries with a negative relationship (Chile, Colombia, Ghana, and Mexico) or no relationship at all (Argentina), public investment is concentrated in activities which substitute directly for private investment.

Public deficits have a negative effect on private investment in Côte d'Ivoire, where the effect is weak, and in Thailand, where it is strong. For Argentina the study decomposed the deficit into its three main components and found that public investment does not affect private capital formation but that public consumption and public revenue do, in directions consistent with the crowding-out hypothesis. The inference, then, is that deficits tend to crowd out private investment through domestic financial markets in Argentina, Côte d'Ivoire, and Thailand.

Although many studies have found that private investment is insensitive to real interest rates, the results for the sample countries show a surprisingly strong negative effect in five of them, with only two (Colombia and Ghana) showing no relationship. The effect of interest rates on private investment is strongest in Morocco and Pakistan, moderately strong in Zimbabwe, and weakest in Chile and Mexico.

## Public Deficits, Trade Deficits, and Real Exchange Rates

For the 1980s real exchange rates are closely correlated with the behavior of fiscal deficits in many developing countries. The major fiscal adjustment in Côte d'Ivoire in 1982–85 was accompanied by real depreciation; subsequent fiscal backsliding occurred in 1985–88 together with real appreciation. The large reduction in the fiscal deficit in Colombia in 1983–88 was accompanied by real depreciation. In the same way Chile's real depreciation of 1984–88 was contemporaneous with a fall in the deficit. Ghana's reform program after 1982 included both a deficit reduction and a real depreciation of the official exchange rate (as well as a depreciation of the real black-market exchange rate). Morocco experienced both a deficit reduction and a real depreciation in 1982–85; in Thailand a deficit reduction and a real depreciation occurred together in 1985–88. These associations support the finding of Edwards (1989) that nominal devaluations last as real devaluations only if accompanied by fiscal adjustment.

In order to provide more systematic evidence on the linkages between the fiscal deficit, the trade deficit, and the real exchange rate, the project case studies tested behavioral relations for the two latter variables on the basis of Rodríguez's two-sector dependent-economy model with optimal capital accumulation (see chapter 2 in this volume). This framework permits the derivation of a two-step relationship between the fiscal deficit and the real exchange rate: the fiscal deficit (among other determinants of investment and saving behavior) affects the external deficit, which then determines the real exchange rate consistent with the clearing of the domestic goods market.

The empirical evidence summarized in the following sections is based on econometric estimations (on annual time-series data for each country) for both the trade surplus and the real exchange rate.

Only the qualitative response to fiscal variables is reported here; the chapters on individual countries provide quantitative results on both fiscal and nonfiscal variables.

Table 1.13 summarizes the sensitivity of the trade surplus to three fiscal variables: the deficit, public consumption, and public investment. For eight countries—Argentina, Chile, Colombia, Côte d'Ivoire, Ghana, Mexico, Thailand, and Zimbabwe—there is significant evidence that rising external surpluses are correlated with higher public surpluses. A similar relationship—reducing the fiscal deficit by reducing public investment improves the trade balance—was found for Pakistan on the basis of a comprehensive macroeconomic model. That fiscal adjustment is a major determinant of external adjustment is also implied by the hypothesis that fiscal policy is an effective instrument for raising national saving, as the substantial evidence presented in the preceding section shows.

Table 1.14 summarizes the sensitivity of the aggregate real exchange

### Table 1.13. Qualitative Effects of Fiscal-Policy-Related Variables on the Trade Surplus

| Country | Public surplus | | | Public expenditure | |
|---|---|---|---|---|---|
| | Total | Primary | Operational | Consumption | Investment |
| Argentina, 1963–88 | — | (+) | — | — | — |
| Chile, 1960–88 | — | — | (+) | — | — |
| Colombia, 1970–88 | — | (+) | — | — | — |
| Côte d'Ivoire, 1971–81 | — | (0) | — | — | — |
| Côte d'Ivoire, 1979–89 | — | (+) | — | — | — |
| Ghana, 1970–88 | — | — | — | (−) | — |
| Mexico, 1970–89 | — | — | (+) | — | — |
| Morocco, 1974–88 | — | — | — | (−) | — |
| Pakistan, 1983/84 to 1987/88 | — | — | — | — | (−) |
| Thailand, 1972–89 | (+) | — | — | — | — |
| Zimbabwe, 1965–88 | — | — | (+) | — | — |

— Not available.

Note: The positive and negative signs correspond to statistically significant coefficients;(0) denotes a coefficient not significantly different from zero. Specifications and estimation techniques vary by country. The dependent variable (current account or trade balance) enters as a ratio to GDP for Argentina, Chile, Colombia, Côte d'Ivoire, Mexico, and Thailand; in levels for Ghana, Morocco, and Pakistan; and as a log ratio to GDP for Zimbabwe. The coefficient for Ghana is for aggregate private expenditure. The effects for Morocco and Pakistan are not the coefficients for one structural equation but represent the general equilibrium effect of a change in the exogenous variable on the current account surplus (in Morocco) or the trade surplus (in Pakistan). For Morocco the sign reflects the deterioration in the current account as a result of a foreign-financed increase in government consumption. For Pakistan the sign reflects the improvement in the trade surplus based on the impact of deficit reduction through lower public investment.

Source: Country case studies listed in the references to this chapter.

**Table 1.14. Qualitative Effects of the Trade Surplus and Fiscal-Policy-Related Variables on the Real Exchange Rate**

| Country | Trade surplus | Public expenditure | Public deficit |
|---|---|---|---|
| Argentina, 1964–87 | (−) | (+) | — |
| Chile, 1960–88 | (−) | (−) | — |
| Colombia, 1967–87 | (−) | (−) | — |
| Côte d'Ivoire, 1972–87 | — | (+) | — |
| Côte d'Ivoire, 1972–89 | (−) | (0) | — |
| Ghana, 1970–88 | — | — | (+) |
| Mexico, 1970–89 | (−) | (−) | — |
| Morocco, 1974–88 | (−) | (+) | — |
| Pakistan, 1983/84 to 1987/88 | (+) | — | — |
| Thailand, 1972–89 | (−) | — | — |
| Zimbabwe, 1965–88 | (−) | (+) | — |

— Not available.

*Note:* The increase in the real exchange rate equals appreciation. The positive and negative signs correspond to statistically significant coefficients; (0) denotes a coefficient not significantly different from zero. Specifications and estimation techniques vary by country. The dependent variable (real exchange rate) enters as levels for Côte d'Ivoire, Ghana, and Thailand; as levels distinguished between the relative export price and the relative import price for Chile, Mexico, and Zimbabwe; as natural logarithms of the import price for Argentina; and as natural logarithms of the real exchange rate for Colombia. The effects for Morocco, Pakistan, and Thailand are not the coefficients for one structural equation but represent the general equilibrium effect of a change in the exogenous variable on the corresponding endogenous variable. For Morocco the reported effects combine the simulation results of a domestic-debt-financed increase in public expenditure and a foreign-financed increase in public expenditure. For Pakistan the effect of an appreciation of the real exchange rate is brought about by a 10 percent reduction of the public deficit through lower public investment, which causes domestic prices to rise with a fixed nominal exchange rate. For Thailand the reported effect summarizes the simulation results of domestically financed deficits, which cause a trade deficit and a real exchange rate depreciation.

*Source:* Country case studies listed in the references to this chapter.

rate to the trade surplus and to fiscal variables. For eight countries (Argentina, Chile, Colombia, Côte d'Ivoire, Mexico, Morocco, Thailand, and Zimbabwe) higher trade surpluses lead to depreciation of the real exchange rate. For Ghana a higher public deficit leads directly to appreciation of the real official exchange rate, taking into account the existence of a black market in foreign exchange. The only contrary result was for Pakistan, where deficit reduction through reduced public investment leads to appreciation of the real exchange rate because of the depressing effect of lower public investment on domestic output. These findings, together with those on the positive relation between trade deficits and fiscal deficits, strongly support the central hypothesis of this section: a lower fiscal deficit leads to a lower trade deficit, which in turn leads to a real exchange rate depreciation.

The studies also examined Rodríguez's hypothesis (see chapter 2 in this volume) that, for a given trade deficit, an increase in public spending affects the real exchange rate. This effect occurs because an increase in public spending for a given trade deficit implies a corresponding decline in private spending. If the public sector has a higher propensity than the private sector to spend on imports rather than on domestic goods, a shift toward more public and less private spending implies an increased demand for imports and a corresponding depreciation of the real exchange rate. Tests of this hypothesis show split results for the sample countries. Higher government spending leads to an appreciation of the real exchange rate for Argentina, Côte d'Ivoire, Morocco, and Zimbabwe and to a depreciation for Chile, Colombia, and Mexico.

These empirical results support the notion that the real exchange rate is sensitive to both policy and external variables, including, prominently, the fiscal deficit. The strong contribution of fiscal adjustment to external adjustment and to a corresponding depreciation of the real exchange rate, as found in the ten-country sample, is reflected in figure 1.7. The figure confirms that the dominant macroeconomic policy trend of the 1980s in these countries was fiscal and external adjustment. However, this average trend of steady improvement from 1982 to 1988 was not confined to the sample countries. Other developing countries showed similar, although less pronounced, reductions of the public deficit, and industrial countries also cut their deficits in half during the same period. A major consequence of fiscal adjustment was sharp reductions in current account deficits, supported by massive real exchange rate depreciations.

## Conclusions and Policy Implications

This chapter has summarized empirical evidence on various controversial issues that occupy the center stage in the discussion of the macroeconomics of fiscal adjustment or deterioration. The evidence is drawn mostly from a sample of ten case studies that are highly representative of the structural diversity of developing countries. This feature strengthens the conclusions reached and shows them to be relevant for the developing world at large.

The use of adequate measures of actual and sustainable public deficits was addressed first. Wide public sector coverage (including public enterprises and the central bank) and exclusion of economically irrelevant categories such as the inflation component of domestic interest payments permit the derivation of more meaningful deficit measures. These measures can be compared with estimates for sustainable deficit levels to evaluate the need for fiscal adjustment.

Public budgets are very sensitive to foreign and domestic macroeconomic shocks in the short run. But the empirical findings show

**Figure 1.7. Fiscal and External Balances and Real
Exchange Rate: 1980–88 Averages for Ten Developing Countries**

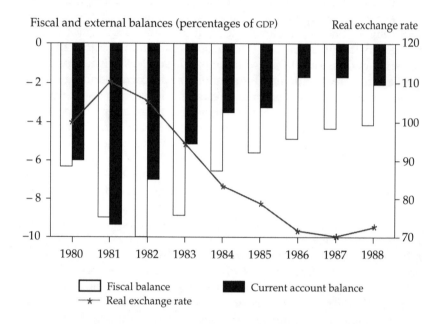

Note: An appreciation of the real exchange rate is shown as an increase.
Sources: Chapters 4, 5, 7, 8, 9, and 10 and Statistical Appendix in this volume;
IMF, *International Financial Statistics*, various issues.

that shocks explain only a minor part of the medium-term variation of public deficits. The major factor explaining change is fiscal policy. Policymakers are to be blamed for fiscal crises and praised for fiscal improvements—luck is a very minor determinant of fiscal stance.

The first set of issues concerning the macroeconomic consequences of deficits that this chapter addresses relates to the linkages between public deficits, inflation, and real interest rates. Although cross-country correlations between deficits and inflation and between deficits and real interest rates were found to be weak at best, the sample countries offer strong evidence that, in the medium term, money financing leads to higher inflation and debt financing leads to higher real interest rates or increased financial repression. As deficit financing mounts, the terms become increasingly unfavorable to the extraction of these unconventional taxes from the private sector.

The evidence soundly refutes the Barro-Ricardian proposition that consumers react the same to conventional taxes, unconventional taxes (inflation or financial repression), and debt financing. The

notion that private saving can be mobilized through higher real interest rates (brought about by increased debt financing or financial liberalization) was also rejected. Both findings are in line with the recent
empirical evidence on private saving behavior in developing
countries.

Higher interest rates, however, have a negative effect on private
investment. This finding is consistent with investment theory, but it
contradicts some of the empirical evidence showing that investment
is insensitive to interest rates in developing countries. Public investment was found to crowd out private investment in some countries
and, in others, to crowd it in. This result confirms previous studies
showing that the net effect of public investment on private investment depends on the composition of the former—on whether it is a
complement to or a substitute for private investment.

Finally, strong and systematic evidence was also found to support
the hypothesis that fiscal deficits spill over into current account deficits, leading, in turn, to a real exchange rate appreciation.

The main policy implications derived from these findings can be
summarized as follows.

• *Fiscal adjustment.* Estimations for six countries suggest that sustainable primary surpluses vary between 2 and −2 percent of GDP
(equivalent to nominal deficits of between 2 and 6 percent of GDP).
These figures, which depend on the combination of macroeconomic
conditions and outstanding public liabilities, are a far cry from actual
deficits in countries such as Côte d'Ivoire and Pakistan.

Successful fiscal adjustment can be seen as proceeding in two
stages. First, deficits are reduced to sustainable levels, consistent with
stable debt-output ratios and normalized financial markets, as, for
instance, in Morocco. A second phase of deeper fiscal adjustment
supporting a strong private sector response—as in Chile or Morocco—
involves reaching nominal public surpluses or only slight deficits,
thus allowing the public sector to reduce its indebtedness in relation
to the domestic private or external sectors.

The ten-country sample suggests that sustainable adjustment typically requires action on four fronts: reducing an overblown government bureaucracy; cutting transfers and subsidies (other than efficient, targeted social programs); enacting tax legislation for increased,
broadly based direct and indirect taxation; and reforming or privatizing public enterprises and commodity marketing boards.

• *Fiscal deficits and inflation.* To the extent that deficits are financed
by money creation, the relationship between fiscal deficits and
inflation is indisputable. However, the tradeoff between additional
inflation and revenue is increasingly unfavorable to the latter, as
documented by the Laffer-curve behavior of money demands.

Governments desperately lacking other income sources and with short planning horizons often start bursts of accelerated money printing and inflation that yield, for a brief period, abnormally high inflation tax revenue. This revenue falls as soon as people reduce their money holdings in response to higher inflation; that reaction leads to growing macroeconomic instability and relative price variability. Since the inflation tax (as well as financial repression, discussed below) *is* a tax, there is no reason to expect conventional fiscal adjustment to be any more contractionary than adjustment through inflation (or financial repression).

• *Fiscal deficits, real interest rates, and financial repression.* If domestic financial markets are not repressed but external financing is not easily available, higher deficits financed by domestic debt raise domestic real interest rates. When domestic financial markets are integrated with world capital markets, higher domestic public borrowing leads to external capital flows and higher foreign debt, without much affecting domestic real interest rates. The story is different when the government represses financial markets through controls on domestic interest rates, compulsory public debt placements, and controls on external capital flows. If the nominal interest rate is fixed, higher fiscal deficits lead to repressed (even negative) real interest rates, implying high taxes on financial intermediation. But the poor economic performance that follows from strong financial repression—depressed private credit and the attendant collapse of private investment—hardly recommends this unconventional form of taxation.

• *Budget deficits and private consumption.* Rejection of the notion that consumers are indifferent between taxes or debt finance carries the policy implication that increasing public saving—or reducing public deficits—is the most effective contribution fiscal policy can make to raising national saving. However, increasing real interest rates through domestic debt financing or financial liberalization will not raise private saving.

• *Budget structure, deficits, and private investment.* Real interest rates and private sector credit do significantly affect private investment. So whether there is financial repression or not, increasing public deficits reduces private investment. The composition of public spending matters as well, since more public investment depresses private investment in some cases—typically, when large public enterprises compete with private firms and enjoy preferential access to domestic financial resources. The implication is that the prospects for higher private investment and growth are improved by three policy measures: restructuring and privatizating public firms and marketing boards; concentrating public investment on public and social infrastructure; and deregulating domestic financial markets—including removal of credit ceilings, compulsory credit allocation, preferential access of the government to credit, and interest controls.

• *Fiscal deficits, trade deficits, and real exchange rates.* The evidence of the strong relationship between public and external deficits complements the policy implication derived from the finding that private saving does not offset changes in public saving: fiscal adjustment is very effective in boosting national saving and therefore in increasing the trade surplus as well. Exchange rates are driven by fundamentals and not the other way around. This should serve as an antidote to the mistaken notion of many policymakers that nominal devaluation alone can restore macroeconomic balance. As Khan and Lizondo (1987) have hypothesized, real exchange rates are also affected by whether the government spends more on tradables than on nontradables. Policymakers should pay attention to the composition of government spending when deciding on an accommodating exchange rate policy.

• *Fiscal deficits and growth.* High deficits are strongly correlated with low growth. Inflation raises uncertainty and distorts relative prices, hurting private investment and resource allocation. The conventional notion that public investment is good for private investment and growth received mixed support. Countries that were forced to shift from external to internal financing of deficits—often because of a debt crisis induced by fiscal mismanagement—showed particularly poor investment and growth performance in the 1980s. Growth itself makes deficits less harmful: countries such as Pakistan and Thailand could sustain larger deficits because of strong growth, while economic collapse worsened the macroeconomic effects of deficits in Argentina, Côte d'Ivoire, and Mexico. The virtuous circle between growth and good fiscal management is one of the strongest arguments for a policy of low and stable fiscal deficits.

## Appendix 1.1. Measurement of Deficits and Evaluation of Public Sector Solvency

Alternative measures for public sector composition stretch from the central government to the consolidated nonfinancial public sector (which consolidates the central government with local government, social security, and nonfinancial public enterprises) and to the consolidated total public sector (adding to the first consolidation the central bank and, possibly, the public commercial banks). Although deficit measures based on the widest public sector coverage are the most accurate measure of fiscal stance and public sector resource transfers, they are often not readily available and are not free of controversy. (See the references in the text on the problems of measuring quasi-fiscal deficits.)

An alternative to cash-based deficits is measurement of deficits on an accruals (or payment-order) basis, which reflects income and spending actions measured at the time they take place, even if they do not immediately involve cash flows. Accumulation of arrears on interest, wage, or goods expenditure would cause accruals-basis deficits to be larger than cash-basis deficits. Box 1.2, above, illustrates the differences between alternative deficit measures for Morocco.

A popular alternative to the nominal cash deficit is the operational deficit, obtained by deducting from the nominal cash deficit the part of nominal interest payments on public debt attributable to inflation, which merely compensates the debt holder for the erosion of the real value of public debt caused by inflation and which, under asset-equilibrium conditions, is reinvested by the debt holder. This correction is particularly important in high-inflation, high-domestic-debt countries. The primary deficit deducts net interest payments from the nominal cash deficit.

Use of accruals-based deficits permits a move away from conventional deficits or intratemporal budget constraints and toward deficit measures consistent with public sector net worth or public intertemporal budget constraints. While the latter would constitute the economically most meaningful measures of fiscal stance and public sector solvency, they are, unfortunately, not observable. Three approaches have been adopted in dealing with this issue. The first is the accounting approach to public sector solvency developed by Buiter (1983, 1985, 1990) and van Wijnbergen (1989), which derives estimates of sustainable deficit levels as those that can be financed without raising debt levels (in relation to GDP) under feasible rates of growth, real interest, and inflation. This methodology has been applied to many countries (for instance, by van Wijnbergen and others 1992 to Turkey and by de Melo 1990 to Morocco), and it is used in deriving the sustainable deficit measures in the sample countries, as reported in the section on sustainable public deficits.

A less stringent requirement than the constancy of debt-output ratios imposed by the accounting approach is to test for the no-Ponzi-game condition on public debt followed by the neoclassical solvency approach. This methodology, developed by Hamilton and Flavin (1988), Grilli (1989), Wilcox (1989), and Buiter and Patel (1990), checks for public sector solvency by comparing the rate of growth of the public debt (in relation to GDP) with the real interest rate. If the debt ratio systematically grows faster than the real interest rate, the public sector is considered insolvent. Among the applications of the methodology are the recent studies by Buiter and Patel (1990) for India and by Werner (1991) for Mexico. The third method diverges from the first two methods by focusing on prices instead of quantities, testing for discounts on public debt paper.

## Appendix 1.2. Sustainable Deficits

This appendix derives the equation for the sustainable primary deficit calculations reported in the first section of this chapter. It is based on the accounting approach to public sector solvency developed by Buiter (1983, 1985, 1990) and van Wijnbergen (1989).

We start with the budget constraint of the consolidated total public sector, which consists of the nonfinancial and financial subsectors, the latter including the central bank. The budget constraint equates the above-the-line total nominal deficit (the sum of the primary deficit and total net interest payments) to below-the-line financing sources (the change in monetary and nonmonetary public debt holdings).

$$(1.1) \qquad \frac{PD}{Py} + i\,\frac{D}{Py} + i^*\,\frac{ED^*}{Py} = \frac{\dot{M}}{Py} + \frac{\dot{D}}{Py} + \frac{E\dot{D}^*}{Py}$$

where $PD$ is the consolidated total public sector primary deficit, $P$ is the GDP deflator, $y$ is real GDP, $i$ is the domestic nominal interest rate, $D$ is the stock of domestic public debt, $E$ is the nominal exchange rate (domestic currency units per unit of foreign currency), $i^*$ is the foreign nominal interest rate, $D^*$ is the foreign public debt stock (in current-price foreign currency units), and $M$ is the base money stock. All variables are in current-price domestic currency units unless otherwise noted.

Simple manipulation of equation 1.1 permits derivation of the ratio of the primary deficit to GDP as:

$$(1.2) \quad pd = \dot{m} + (\pi + n)m + \dot{d} + (n - r)d + \dot{d}^* + (n - r^* - \varepsilon)d^*$$

where the lower-case variables $pd$, $d$, $d^*$, and $m$ are defined as the ratios of $PD$, $D$, $E$, $D^*$, and $M$, respectively, to GDP at current prices; $\pi$ is the domestic rate of inflation; $n$ is real GDP growth; $r$ is the domestic real interest rate, $r^*$ is the foreign real interest rate; and $\varepsilon$ is the rate of real exchange rate depreciation.

Equation 1.2 shows that the primary deficit of the consolidated public sector, as a share of GDP, is constrained to not exceed the sum of six financing sources: revenue from the sum of the inflation tax on the monetary base and growth-induced increase in money demand; the excess of domestic growth over the relevant real interest cost of domestic and foreign debt; and increasing demands for monetary and nonmonetary debt. Primary deficits are sustainable if they do not entail ever-increasing shares of debt and money to income. In the absence of explicit demands for public liabilities, the accounting approach to public solvency defines sustainability in the more restrictive sense of constant ratios of debt to output and of money to output, consistent with steady-state (constant) inflation and interest rates.

Therefore the country applications summarized in the first section of the chapter calculate sustainable primary deficits as determined by equation 1.2, after imposing the steady-state condition of constant liability-to-income ratios. In most cases the calculations assume that 1988–90 liability-to-output ratios are the relevant steady-state values. Country applications differ in public sector coverage (central, general, nonfinancial, or total public sector), and equation 1.2 is modified accordingly.

## Appendix 1.3. Decomposition of Seigniorage

The decomposition is based on the following equation for seigniorage, $T$:

$$(1.3) \quad T_t = \frac{P_t M_t - P_{t-1} M_{t-1}}{P_t} = \left(\frac{\pi_t}{1 + \pi_t}\right) M_{t-1} + M_t - M_{t-1}$$

where $P_t$ is the price level at time $t$, $M_t$ is the real money supply at time $t$, and $\pi_t$ is the inflation rate at time $t$. The steady-state value of seigniorage is given by:

$$(1.4) \qquad \overline{T} = \frac{\overline{\pi}}{1 + \overline{\pi}} \frac{\overline{M}}{1 + \overline{g}} + \frac{\overline{g}}{1 + \overline{g}} \overline{M}$$

where a bar denotes an average level of a variable and it is assumed that real money grows in proportion to output, with output growth given as $\overline{g}$ and the trend value of real money as $\overline{M}$. The first term gives the inflation tax component of seigniorage, while the second gives the seigniorage that accrues from an increase in real money balances.

The deviation of seigniorage from the average can then be given as follows:

$$(1.5) \quad T_t - \overline{T} = \left(\frac{\pi_t}{1 + \pi_t} - \frac{\overline{\pi}}{1 + \overline{\pi}}\right)\frac{\overline{M}}{1 + \overline{g}} + \left(\frac{\overline{\pi}}{1 + \overline{\pi}}\right)\left(M_{t-1} - \frac{\overline{M}}{1 + \overline{g}}\right)$$

$$+ \left(M_t - M_{t-1} - \frac{\overline{g}}{1 + \overline{g}}\overline{M}\right)$$

$$+ \left(\frac{\pi_t}{1 + \pi_t} - \frac{\overline{\pi}}{1 + \overline{\pi}}\right)\left(M_{t-1} - \frac{\overline{M}}{1 + \overline{g}}\right).$$

The first term is the above-average seigniorage attributable to the above-average inflation tax rate. The second term gives the above-average revenue attributable to the above-average real money base. The third term gives the real change in the money base minus the amount that would take place as money grows with output. The last term is the covariance of inflation and money.

## Notes

This chapter draws on all the other chapters in the volume. The authors benefited from discussions with and comments by Jorge Baldrich, Nancy

Birdsall, Mario Blejer, Vittorio Corbo, Giancarlo Corsetti, Shantayanan Devarajan, Ricardo Ffrench-Davis, Nicolás Eyzaguirre, Stanley Fischer, Michael Gavin, Ravi Kanbur, Johannes Linn, Carlos Rodríguez, Vito Tanzi, and Martin Werner, as well as the participants at the World Bank Conference on the Macroeconomics of Public Sector Deficits (Washington, D.C.), the Tenth Latin American Meeting of the Econometric Society (Punta del Este), and seminars at Columbia University, the Ministry of Finance of China (Beijing), the Ministry of Finance of Costa Rica (San José), CEMA–Universidad de San Andrés (Buenos Aires), and the Central Bank of Chile (Santiago). They are also grateful for the comments and assistance of Paolo Mauro, for research assistance by Maria Cristina Almero-Siochi, Piyabha Kongsamut, and Raimundo Soto, and for interaction with the other chapter authors.

1. The chapter presents in more detail the project's findings and implications discussed in Easterly and Schmidt-Hebbel (1993a, 1993b).

2. The most complete study to date on the measurement of fiscal deficits is Blejer and Cheasty (1991). Other references on alternative deficit measures include Tanzi (1985); Eisner (1986); Blejer and Chu (1988); Kotlikoff (1988); Fischer and Easterly (1990); and Buiter (1990). IMF (1986) and United Nations (1968) discuss cash and accrual deficits in more detail. Robinson and Stella (1988) and Teijeiro (1989) survey issues concerning quasi-fiscal deficits.

3. The next section presents measures of sustainable primary deficits for sector coverages that range—because of varying degrees of data availability—from the central government to the total consolidated public sector. The questions addressed in subsequent sections require the use of cash-based operational (or nominal) deficit measures for the widest available public sector coverage. The discussion of correlations of deficits with other economic variables presents separate 1978–89 data for nominal consolidated nonfinancial and central bank quasi-fiscal deficits, while the fifty-nine-country correlations rely on nominal consolidated public sector or general government balances (depending on data availability), as reported in table A.1 of the Statistical Appendix in this volume. The section on causes and remedies is based on measures of the consolidated nonfinancial public sector deficit, and the subsequent section is based on consolidated total (nonfinancial plus quasi-fiscal) public sector deficits—the nominal or operational measures are indifferent here. Finally, the last two sections use operational consolidated nonfinancial public deficits because of the lack of long time-series for quasi-fiscal deficits.

4. In two cases (Chile and Zimbabwe) upper and lower bounds, consistent with possible deviations of the relevant macroeconomic variables from base-case levels, are added to the midpoint estimates. Here, as well as in the other four cases, the relevant macroeconomic variables used are those that determine the primary deficit: the rates of output growth, inflation, domestic and foreign real interest, and real exchange rate devaluation; see equation 1.2 in appendix 1.2.

5. An interesting short-run counterpart to this result is the suggestion of Giavazzi and Pagano (1990) and Blanchard (1990) that fiscal austerity can be expansionary.

6. See, for example, Balassa (1988); Reisen and van Trotsenburg (1988); Bartoli (1989); Sachs (1989); and Rodríguez (chapter 2 in this volume). Note,

however, that the link breaks down if the Ricardian hypothesis of offsetting private saving holds (Frenkel and Razin 1987; Leiderman and Blejer 1988).

7. In broad terms, countries face four types of foreign shocks: changes in the price and interest conditions of their foreign trade and their credit flows, and changes in quantity constraints affecting their foreign trade and their credit flows. While quantity constraints are rather uncommon in foreign trade (abstracting from countries affected by global embargoes or partial trade restrictions on certain goods), massive changes in borrowing constraints are a stylized fact in credit markets. The aftermath of the 1982 debt crisis implied, in fact, a massive change in regime in the form of foreign resource constraints suddenly faced by most developing debtor economies. Although borrowing constraints constitute a strong foreign shock affecting below-the-line financing sources, we focus only on changes in foreign terms of trade and interest rates, which impinge directly on above-the-line deficits.

8. The channels mentioned here add bracket creep and transfer effects to the five-item list of Dornbusch, Sturzenegger, and Wolf (1990).

9. This is also called the Keynes-Olivera-Tanzi effect; see Olivera (1967) and Tanzi (1977). Sometimes the Keynes-Olivera-Tanzi effect is used more restrictively to denote the tax erosion effect mentioned below.

10. The calculation is based on a reduced-form equation estimated for the CNFPS deficit in Thailand and on 1988 data for the deficit, inflation, and GDP.

11. The average relative contribution of each group of deficit determinants is calculated on the basis of the equation presented in the note to table 1.2; hence the equation is now used separately for external, domestic macroeconomic, and fiscal policy variables. However, in order to present the relative contribution of each group of variables, $d_i$ is defined here as the explained change in the deficit, not the actual change as in table 1.2. Hence the average relative contribution of external variables to actual deficits in the second column of table 1.2 differs from the average relative contribution of external variables to explained deficits in figure 1.2. The average absolute deviations between actual and explained deficit changes in percentage points of GDP are 0.9 for Chile (1974–88), 2.0 for Ghana (1972/73–1988), and 1.4 for Zimbabwe (1981/82–1988/89).

12. Seigniorage is defined here as the ratio to GDP of the change in high-powered money during the year, averaged over 1970–89. The number of observations is forty-nine countries, including Argentina; the quadratic term is significant whether or not Argentina is included. See tables A.3 and A.4 of the Statistical Appendix for time-series data on seigniorage and inflation rates for forty-nine OECD members and developing countries.

13. Barro (1990) also suggests that the maximum of the Laffer curve is at inflation rates around 100 percent. Edwards and Tabellini (1990) present suggestive evidence for seigniorage Laffer curves in a number of developing countries.

14. Dornbusch, Sturzenegger, and Wolf (1990) describe the progressive substitution of interest-bearing assets for money in high-inflation episodes.

15. Similar magnitudes were found in the study by Fischer (1982).

16. The coefficient of variation is calculated over 1970–89 for a reduced sample of twenty-six countries with data over that period (to standardize the number of observations, which affects the variance). The coefficient of variation is the standard deviation divided by the mean.

17. There were actually twenty-one such observations, but three were found to involve changes in measurement of the money base and were discarded.

18. This is not a general pattern, however; of the nine cases in which seigniorage is explained largely by the change in the real money base, four had rising inflation the following year, two had essentially unchanged inflation, and three had falling inflation the next year. Dornbusch, Sturzenegger, and Wolf (1990) note the rise in real money balances in the early stages of hyperinflation. Our story is a different one: exogenous or policy-induced rises in real money balances allow large temporary seigniorage without hyperinflation.

19. Estimates by Giovannini and de Melo (1993) find much higher revenue from financial repression for Mexico, Morocco, Pakistan, and Zimbabwe than do the other studies because they calculate the tax rate as the ex-post difference between domestic and foreign interest rates, including devaluation. These countries were experiencing steady real devaluations in the early 1980s, which tends to raise the estimate of the tax rate when this method is used.

20. Chamley and Honohan (1990), Easterly (1989), and Giovannini and de Melo (1993) estimate financial repression revenue and discuss the costs of financial repression. Chapter 5 presents evidence from cross-section regressions that financial repression has a negative effect on long-run growth. Dornbusch and Reynoso (1989), however, argue that financial repression is costly only under very high inflation.

21. The consumption specification of the case studies and table 1.11 include (a) disposable income (gross income, including domestic debt interest payments, less tax revenue) and (b) public saving (tax and nontax revenue less current government expenditure, including domestic debt interest payments) or the public surplus (total government revenue less total government expenditure), as consumption determinants. This follows Corbo and Schmidt-Hebbel (1991) in distinguishing between the Keynesian and permanent-income hypotheses (according to which only disposable income matters) and the Ricardian and direct-crowding-out hypotheses (according to which only gross income net of government consumption matters, so that disposable income and public saving should have a high and positive common coefficient in the above-mentioned specification).

## References

### Country Case Studies

Alberro-Semerena, José Alberto. 1991. "The Macroeconomics of the Public Sector Deficit in Mexico during the 1980s." World Bank, Policy Research Department, Washington, D.C.

Chamley, Christophe, and Hafez Ghanem. 1991. "Fiscal Policy with Fixed Nominal Exchange Rates: Côte d'Ivoire." World Bank Policy Research Working Paper 658. Revised as chapter 6 in this volume.

Easterly, William. 1991. "The Macroeconomics of the Public Sector Deficit: The Case of Colombia." World Bank Policy Research Working Paper 626. Revised as chapter 5 in this volume.

Faini, Riccardo. 1991. "The Macroeconomics of the Public Sector Deficit: The Case of Morocco." World Bank Policy Research Working Paper 631. Revised as chapter 8 in this volume.

Haque, Nadeem Ul, and Peter J. Montiel. 1991. "The Macroeconomics of Public Sector Deficits: The Case of Pakistan." World Bank Policy Research Working Paper 673. Revised as chapter 9 in this volume.

Islam, Roumeen, and Deborah Wetzel. 1991. "The Macroeconomics of Public Sector Deficits: The Case of Ghana." World Bank Policy Research Working Paper 672. Revised as chapter 7 in this volume.

Marshall, Jorge, and Klaus Schmidt-Hebbel. 1991. "Macroeconomics of Public Sector Deficits: The Chile Case Study." World Bank Policy Research Working Paper 696. Revised as chapter 4 in this volume.

Morandé, Felipe, and Klaus Schmidt-Hebbel. 1991. "Macroeconomics of Public Sector Deficits: The Case of Zimbabwe." World Bank Policy Research Working Paper 688. Revised as chapter 10 in this volume.

Ramangkura, Virabongse, and Bhanupongse Nidhiprabha. 1991. "The Macroeconomics of the Public Sector Deficit: The Case of Thailand." World Bank Policy Research Working Paper 633. World Bank, Policy Research Department, Washington, D.C.

Rodríguez, Carlos A. 1991. "The Macroeconomics of the Public Sector Deficit: The Case of Argentina." World Bank Policy Research Working Paper 632. Revised as chapter 3 in this volume.

## *General References*

Aschauer, David. 1989. "Is Public Expenditure Productive?" *Journal of Monetary Economics* 23 (March): 177–200.

Balassa, Bela. 1988. "Public Finance and Economic Development." Policy Research Working Paper 31. World Bank, Office of the Vice President, Development Economics, Washington, D.C.

Barro, Robert F. 1974. "Are Government Bonds Net Wealth?" *Journal of Political Economy* 81 (December): 1095–1117.

———. 1990. *Macroeconomics*. New York: Wiley.

Barro, Robert F., and Xavier Sala-i-Martin. 1990. "World Real Interest Rates." In Stanley Fischer, ed., *NBER Macroeconomics Annual*. Cambridge, Mass.: Massachusetts Institute of Technology Press.

Bartoli, G. 1989. "Fiscal Expansion and External Current Account Imbalances." In Mario I. Blejer and Ke-young Chu, eds. *Fiscal Policy, Stabilization and Growth in Developing Countries*. International Monetary Fund: Washington, D.C.

Bernheim, B. Douglas. 1987. "Ricardian Equivalence: An Evaluation of Theory and Evidence." In *NBER Macroeconomics Annual 1987*. Cambridge, Mass.: Massachusetts Institute of Technology Press.

Blanchard, Olivier. 1990. "Comment on 'Can Severe Fiscal Contractions be Expansionary?'" In Stanley Fischer, ed., *NBER Macroeconomics Annual*. Cambridge, Mass.: Massachusetts Institute of Technology Press.

Blejer, Mario I., and Adrienne Cheasty. 1991. "The Measurement of Fiscal Deficits: Analytical and Methodological Issues." *Journal of Economic Literature* 29(4) (December): 1644–78.

Blejer, Mario I., and Ke-young Chu, eds. 1988. *Measurement of Fiscal Impact: Methodological Issues.* IMF Occasional Paper 59. Washington, D.C.: International Monetary Fund.

Blejer, Mario I., and Mohsin S. Khan. 1984. "Government Policy and Private Investment in Developing Countries." *International Monetary Fund Staff Papers* 31 (2): 379–403.

Blejer, Mario I., and Nissan Liviatan. 1987. "Fighting Hyperinflation: Stabilization Strategies in Argentina and Israel, 1985–86." *International Monetary Fund Staff Papers* 34 (September): 409–38.

Buiter, Willem H. 1983. "Measurement of the Public Sector Deficit and Its Implications for Policy Evaluation and Design." *International Monetary Fund Staff Papers* 30 (June): 306–49.

———. 1985. "A Guide to Public Sector Debt and Deficits." *Economic Policy* (November): 13–79.

———. 1990. "The Arithmetic of Solvency." In Willem H. Buiter, *Principles of Budgetary and Financial Policy.* Cambridge, Mass.: Massachusetts Institute of Technology Press.

Buiter, Willem H., and U. Patel. 1990. "Debt, Deficits and Inflation: An Application to the Public Finances of India." NBER Working Paper 3287. National Bureau of Economic Research, Cambridge, Mass.

Cagan, Phillip. 1956. "The Monetary Dynamics of Hyperinflation." In Milton Friedman, ed., *Studies in the Quantity Theory of Money.* Chicago, Ill.: University of Chicago Press.

Chamley, Christophe, and Patrick Honohan. 1990. "Taxation of Financial Intermediation." Policy Research Working Paper 421. World Bank, Country Economics Department, Washington, D.C.

Corbo, Vittorio, and Klaus Schmidt-Hebbel. 1991. "Public Policies and Saving in Developing Countries." *Journal of Development Economics* 36(1): 89–116.

de Haan, J., and D. Zelhorst. 1990. "The Impact of Government Deficits on Money Growth in Developing Countries." *Journal of International Money and Finance* 9: 455–69.

de Melo, Martha. 1990. "Fiscal Aspects of External Debt: A Case Study of Morocco." World Bank, Policy Research Department, Washington, D.C.

Dollar, David. 1990. "Outward-Oriented Developing Economies Really Do Grow More Rapidly: Evidence from 95 LDCs, 1976–85." World Bank, Country Economics Department, Washington, D.C.

Dornbusch, Rudiger. 1985. "Overborrowing: Three Case Studies." In Gordon W. Smith and John T. Cuddington, eds., *International Debt and the Developing Countries.* A World Bank Symposium. Washington, D.C.

Dornbusch, Rudiger, and Stanley Fischer. 1991. "Moderate Inflation." Policy Research Working Paper 807. World Bank, Office of the Vice President, Developmental Economics, Washington, D.C.

Dornbusch, Rudiger, and Alejandro Reynoso. 1989. "Financial Factors in Economic Development." NBER Working Paper 2889. National Bureau of Economic Research, Cambridge, Mass.

Dornbusch, Rudiger, F. Sturzenegger, and H. Wolf. 1990. ''Extreme Inflation: Dynamics and Stabilization.'' *Brookings Papers on Economic Activity* 2: 1–64.

Easterly, William. 1989. ''Fiscal Adjustment and Deficit Financing during the Debt Crisis.'' In Ishrat Husain and Ishac Diwan, eds. *Dealing with the Debt Crisis.* A World Bank Symposium. Washington, D.C.: World Bank.

Easterly, William, and Klaus Schmidt-Hebbel. 1993a. ''Fiscal Adjustment and Macroeconomic Performance.'' *Outreach* 10 (May). World Bank, Policy Research Department, World Bank.

Easterly, William, and Klaus Schmidt-Hebbel. 1993b. ''Fiscal Deficits and Macroeconomic Performance in Developing Countries.'' *World Bank Research Observer* 8 (2) (July): 211–37.

Easterly, William, Paulo Mauro, and Klaus Schmidt-Hebbel. Forthcoming. ''Money Demand and Seignorage-Maximizing Inflation.'' *Journal of Money, Credit, and Banking.*

Easterly, William, Carlos A. Rodríguez, and Klaus Schmidt-Hebbel. 1989. ''Research Proposal: The Macroeconomics of the Public Sector Deficit.'' World Bank, Country Economics Department, Washington, D.C.

Easterly, William, Robert King, Ross Levine, and Sergio Rebelo. 1992. *How Do National Policies Affect Long-Run Growth? A Research Agenda.* World Bank Discussion Paper 164. Washington, D.C.

Edwards, Sebastian. 1989. *Real Exchange Rates, Devaluation and Adjustment.* Cambridge, Mass.: Massachusetts Institute of Technology Press.

Edwards, Sebastian, and Guido Tabellini. 1990. ''Explaining Fiscal Policies and Inflation in Developing Countries.'' NBER Working Paper 3493. National Bureau of Economic Research, Cambridge, Mass.

Eisner, Robert. 1986. *How Real Is the Federal Deficit?* New York: Free Press.

Fischer, Stanley. 1982. ''Seigniorage and the Case for a National Money.'' *Journal of Political Economy* 90 (2) (April): 295–313. Reprinted in Stanley Fischer, *Indexing, Inflation and Economic Policy* (Cambridge, Mass.: Massachusetts Institute of Technology Press, 1987).

———. 1991. ''Growth, Macroeconomics and Development.''In Stanley Fischer, ed., *NBER Macroeconomics Annual.* Cambridge, Mass.: Massachusetts Institute of Technology Press.

Fischer, Stanley, and William Easterly. 1990. ''The Economics of the Government Budget Constraint.'' *World Bank Research Observer* 5 (2): 127–42.

Frenkel, Jacob A., and Assaf Razin. 1987. *Fiscal Policies and the World Economy.* Cambridge, Mass.: Massachusetts Institute of Technology Press.

Gelb, Alan, and associates. 1988. *Oil Windfalls: Blessing or Curse?* New York: Oxford University Press.

Giavazzi, Francesco, and Marco Pagano. 1990. ''Can Severe Fiscal Contractions Be Expansionary?'' In Stanley Fischer, ed., *NBER Macroeconomics Annual.* Cambridge, Mass.: Massachusetts Institute of Technology Press.

Giovannini, Alberto. 1983. ''The Interest-Elasticity of Savings in Developing Countries.'' *World Development* 11 (7): 601–07.

———. 1985. ''Saving and the Real Interest Rate in LDCs.'' *Journal of Development Economics* 18 (August): 197–217.

Giovannini, Alberto, and Martha de Melo. 1993. "Government Revenue from Financial Repression." *American Economic Review* 83 (4): 953–63.

Grilli, Vittorio. 1989. "Seigniorage in Europe." In Marcello De Cecco and Alberto Giovannini, eds., *A European Central Bank? Perspectives on Monetary Unification after Ten Years of the EMS*. Cambridge, U.K.: Cambridge University Press.

Hamilton, J., and Marjorie Flavin. 1988. "On the Limitations of Government Borrowing: A Framework for Empirical Testing." *American Economic Review* 76 (September): 808–19.

Haque, Nadeem Ul, and Peter J. Montiel. 1989. "Consumption in Developing Countries: Tests for Liquidity Constraints and Finite Horizons." *Review of Economics and Statistics* 71 (3): 408–15.

Hayashi, Fumio. 1985. "Tests for Liquidity Constraints: A Critical Survey." NBER Working Paper 1720. National Bureau of Economic Research, Cambridge, Mass.

IMF (International Monetary Fund). 1986. *A Manual on Government Finance Statistics*. Washington, D.C.

———. Various years. *International Financial Statistics*. Washington, D.C.

Khan, Mohsin, and Samuel Lizondo. 1987. "Devaluation, Fiscal Deficits and the Exchange Rate." *World Bank Economic Review* 1 (2): 357–74.

Khan, M. S., and C. M. Reinhart. 1990. "Private Investment and Economic Growth in Developing Countries." *World Development* 18 (1): 19–27.

Kiguel, Miguel, and Nissan Liviatan. 1988. "Inflationary Rigidities and Orthodox Stabilization Policies: Lessons from Latin America." *World Bank Economic Review* 2 (3): 273–98.

Kotlikoff, Larry. 1988. "The Deficit Is Not a Well Defined Measure of Fiscal Policy." *Science* 241 (August): 791–95.

Leiderman, Leonardo, and Mario I. Blejer. 1988. "Modeling and Testing Ricardian Equivalence: A Survey." *International Monetary Fund Staff Papers* 35 (March): 1–35.

Marshall, Jorge, and Klaus Schmidt-Hebbel. 1989. "Economic and Policy Determinants of Public Sector Deficits." Policy Research Working Paper 321. World Bank, Country Economics Department, Washington, D.C.

OECD (Organization for Economic Cooperation and Development). Various issues. *OECD Economic Outlook*. Paris.

Olivera, J. 1967. "Money, Prices and Inflation Lags: A Note on the Dynamics of Inflation." *Banca Nazionale del Lavoro Quarterly Review* 20: 258–67.

Poterba, James. 1988. "Are Consumers Forward-Looking? Evidence from Fiscal Experiments." *American Economic Review* 48 (2) (May): 413–81.

Rama, Martin. 1993. "Empirical Investment Equations for Developing Countries." In Luis Servén and Andrés Solimano, eds., *Striving for Growth after Adjustment: The Role of Capital Formation*. Washington, D.C.: World Bank.

Reisen, Helmut, and Axel van Trotsenburg. 1988. "Developing Country Debt: The Budgetary and Transfer Problem." Development Centre Study. Organization for Economic Cooperation and Development, Paris.

Robinson, David J., and Peter Stella. 1988. "Amalgamating Central Bank and Fiscal Deficits." In Mario I. Blejer and Ke-young Chu, eds., *Measurement of*

*Fiscal Impact: Methodological Issues.* IMF Occasional Paper 59. Washington, D.C.: International Monetary Fund.

Sachs, Jeffrey, ed. 1989. *Developing Country Debt and the World Economy.* Chicago, Ill.: University of Chicago Press.

Sargent, Thomas J., and Neil Wallace. 1985. "Some Unpleasant Monetarist Arithmetic." *Federal Reserve Bank of Minneapolis Quarterly Review* 9 (Winter): 15–31.

Schmidt-Hebbel, Klaus, Steven B. Webb, and Giancarlo Corsetti. 1992. "Household Saving in Developing Countries: First Cross-Country Evidence." *World Bank Economic Review* 6 (3): 529–47.

Seater, John J. 1993. "Ricardian Equivalence." *Journal of Economic Literature* 21 (1): 142–90.

Servén, Luis, and Solimano, Andrés. 1993. "Private Investment and Macroeconomic Adjustment: A Survey." In Luis Servén and Andrés Solimano, eds., *Striving for Growth after Adjustment: The Role of Capital Formation.* Washington, D.C.: World Bank.

Summers, Robert, and Alan Heston. 1988. "A New Set of International Comparisons of Real Product and Price Levels: Estimates for 130 Countries." *Review of Income and Wealth* 34: 1–25.

Tanzi, Vito. 1977. "Inflation, Lags in Collection, and the Real Value of Tax Revenue." *International Monetary Fund Staff Papers* 24: 154–67.

———. 1985. "Fiscal Management and External Debt Problems." In Hassanali Mehran, ed., *External Debt Management.* Washington D.C.: International Monetary Fund.

Tanzi, Vito, Mario I. Blejer, and Mario O. Teijeiro. 1987. "Inflation and the Measurement of Fiscal Deficits." *International Monetary Fund Staff Papers* 34 (December): 711–38.

Teijeiro, Mario O. 1989. "Central Bank Losses: Origins, Conceptual Issues, and Measurement Problems." Policy Research Working Paper 293. World Bank, Country Economics Department, Washington, D.C.

United Nations. 1968. *A System of National Accounts.* New York: United Nations.

van Wijnbergen, Sweder. 1989. "External Debt, Inflation and the Public Sector: Towards Fiscal Policy for Sustainable Growth." *World Bank Economic Review* 3 (3): 297–320.

van Wijnbergen, Sweder, Ritu Anand, Ajay Chhibber, and Roberto Rocha. 1992. *External Debt, Fiscal Policy, and Sustainable Growth in Turkey.* Baltimore, Md.: Johns Hopkins University Press.

Werner, Martin. 1991. "Is Mexico Solvent? Testing the Sustainability of the Government's Fiscal Policy." Presented at the Tenth Latin American Meeting of the Econometric Society, August, Punta del Este, Uruguay.

Wilcox, David. 1989. "The Sustainability of the Government Deficits: Implications of the Present Value Borrowing Constraint." *Journal of Money, Credit and Banking* 21 (August): 291–306.

World Bank. 1988. *World Development Report 1988.* New York: Oxford University Press.

# 2

# The External Effects of Public Sector Deficits

*Carlos Alfredo Rodríguez*

This chapter analyzes the effects that public sector deficits and the means of financing deficits have on a specific set of macroeconomic variables related to the external sector. These variables include the real exchange rate, the trade balance, the current account, and the level of external indebtedness.

The deficit of the public sector, as measured by public sector borrowing requirements, is the result of the difference between government spending and government tax revenues. Therefore, in describing the effects of a given deficit, it is imperative to separate the effects of the financing of the deficit from those derived from the given levels of government spending or taxation. In order to do this we have to design a conceptual experiment. In our case we shall assume that there is available a neutral tax—for example, a value added tax or a consumption tax such that changes in the level of the tax do not affect the relative structure of demand for goods or assets. The deficit is generated by reducing this neutral tax and increasing the level of debt financing accordingly, whether the financing be external or internal. From this perspective, what we will be analyzing are the effects of tax versus debt financing in the context of an open economy. In the case of internal debt financing, the government may resort to issuing interest-bearing debt (bonds) or non-interest-bearing debt (money).

The issue of tax versus debt financing has received a lot of attention in the literature in reference to the well-known Ricardian equivalence proposition. The Ricardian proposition maintains that a tax reduction financed with debt will have no real effects on the economy if the public discounts the future taxes to service the debt and increases savings by the exact amount of the tax reduction. However, the empirical validity of the Ricardian equivalence is inconclusive. (For a general survey on issues related to the Ricardian equivalence, see Leiderman and Blejer 1988).

In the context of an open economy, the real exchange rate is a crucial relative price for the allocation of resources in the external sector. This relative price certainly will be affected by the composition of government spending and may also be affected by the means of financing such spending. In the first section we discuss the Ricardian equivalence proposition in relation to the external effects of tax versus debt financing. In the next two sections we assume that the equivalence proposition does not hold and describe the resulting short-run and dynamic effects of government spending on variables of the external sector.

## Deficit Financing and the Trade Balance

We are concerned with the short-run effects of deficit financing on the levels of the real exchange rate, the trade and current accounts, the levels of domestic and foreign indebtedness, and finally, the inflation rate to the extent that the deficit is financed with money creation.

The following variables are defined:

(2.1)    $Y$ = gross domestic product (GDP).

$F_{pg}$ = net financing from private sector to government: taxes plus acquisition of domestic paper (debt or currency minus interest collected on domestic debt).

(2.2)    $$F_{pg} = T + dC/dt + dD/dt - i \cdot D$$

where $T$ = taxes, $C$ = money, and $D$ = internal government debt.

$F_{ep}$ = net financing from the foreign to the government sector.

(2.3)    $$F_{ep} = E \cdot dD_p^*/dt - i^* \cdot E \cdot D_p^*$$

where $D_p^*$ = external private debt and $E$ = exchange rate. An asterisk (*) means that the variable is measured in foreign currency.

(2.4)    $G$ = government spending on goods.

$F_{eg}$ = net financing from the foreign to the government sector.

(2.5)    $$F_{eg} = E \cdot d(D_g^*)/dt - i^* \cdot E \cdot D_g^*$$

where $D_g^*$ = external government debt.

### Private Sector Budget Constraint

(2.6)    $G_p = Y + F_{ep} - F_{pg}$ = private spending on goods.

### Government Budget Constraint

(2.7)    $G_g = F_{pg} + F_{eg}$ = government spending on goods.

### Total Spending on Goods

(2.8)    $$GT = G_p + G_g = Y + F_{ep} + F_{eg}.$$

Starting from equation 2.8, we can derive a set of propositions that will be the basis for the subsequent analysis.

- *Proposition 1.* Total spending on goods can exceed total output only if it is externally financed (follows from equation 2.8).
- *Proposition 2.* For a given composition of total spending of goods between tradable and nontradable, the real exchange rate depends on the difference between total spending and total output of goods—in other words, on the trade balance deficit that is equal to the amount of external financing (to be proved later).
- *Proposition 3.* Government financing strategies will affect the real exchange rate only if they affect the trade balance (follows from proposition 2).
- *Proposition 4.* Government financing strategies will affect the trade balance only if the Ricardian equivalence proposition does not hold. If this is the case, a tax reduction financed through increased debt (internal or external) will result in some increase in private spending. As a consequence the trade surplus will deteriorate and the real exchange rate should fall. We would therefore observe that a fiscal deficit generates a real appreciation.

Proposition 4 is the starting point for our analysis. The relevant question is whether the government financing strategies can affect the level of private spending—in other words, the issue of crowding-out, in this case referring also to external borrowing. In order to discuss the effects of deficit financing on the real exchange rate, we have to define a neutral experiment through which the deficit increase does not affect the composition of total spending (which would be an obvious way to affect the real exchange rate). The experiment will be a tax reduction coupled with an equivalent increase in government indebtedness (internal or external). In this way, we assume that a deficit is generated without a corresponding increase in the rate of government spending.

There are three ways to finance such a deficit: increase domestic debt, increase external debt, or increase the rate of money creation. We shall discuss each method separately.

### Tax Reduction Financed by External Government Borrowing

Consider a situation in which the government switches from tax financing to external financing. If the private sector reacts by investing the tax savings in foreign assets, there will be no effect on total spending or on the trade surplus. The real exchange rate will not be affected because government borrowing will be unable to affect the trade balance. In terms of equation 2.8, the increase in $F_{eg}$ is matched exactly by a decrease in $F_{ep}$, so that their sum remains unchanged. This conclusion follows from a straight generalization of the Ricardian equivalence theory for foreign borrowing. The issue was analyzed in the context of an optimal model by Frenkel and Razin (1986), Auernheimer (1987), and Leiderman and Blejer (1988) and receives some empirical confirmation from the Argentine experience during 1978–81.

During 1978–81 the Argentine government acquired a substantial external debt that was matched to a great extent by outflows of private capital. These outflows took place later, when it was already perceived that the government's borrowing and exchange rate policies were doomed to failure. There was a transitional period, however, when the government debt was building up, and during this time the trade deficit deteriorated substantially—although part of the deterioration may have been attributable to the trade liberalization that took place, coupled with the policy of the quasi-fixed exchange rate. It is therefore not clear whether the private capital outflow observed was a private compensation for the increased government debt or a simple speculative movement induced by expectations of a large devaluation. This issue will be addressed in chapter 3.

As mentioned in Leiderman and Blejer (1988) there is a wide variety of reasons why the Ricardian equivalence proposition may not hold, even in an open economy. For example, the authors mention the existence of borrowing constraints, distortionary taxation, uncertainty about the imposition of the required future taxes, and differences in planning horizons for the private and public sectors. We might add to this list the risk-induced differentials in rates of interest at home and abroad and differences in the spending propensities of taxpayers and bondholders.

Given all these arguments, we feel that the degree of substitutability of private and government external borrowing is an empirical issue in need of clarification and that a better understanding can be obtained from the specific experiences related in the case studies.

### *Tax Reduction Financed by Internal Borrowing*

When the government deficit is financed with internal debt, a substitutability similar to that of private and government external borrowing will result. If Ricardian equivalence holds, the lower taxes will be used by the private sector to acquire the increased internal issue of debt, and total private spending will not be increased. However, portfolio composition may indirectly affect the mix of spending between consumption and investment goods.

If the private sector purchases the internal debt with increased foreign debt, external financing will increase and the trade balance and the real exchange rate will be affected. In this case the Ricardian proposition would not hold because private spending has increased by the exact amount of the tax reduction. Here again, the issue should be subject to empirical verification: is government borrowing inter-mediated externally by the private sector or not?

## Tax Reduction Financed through Inflation Tax

A tax reduction financed through inflation tax is the most obvious example of neutrality, since it amounts to the substitution of one tax by another. Thus no direct effect on the rate of private spending should be expected. However, a differential tax has been imposed on a single financial asset—money—and this may have short- and long-run effects on the desired rates of acquisition of other assets, in par-ticular external assets. The higher inflation rate may stimulate larger desired holdings of external assets by the private sector. In the short run this implies larger capital outflows and, through the reduced rate of private spending, a larger trade surplus (and a higher real exchange rate). In the long run, as foreign private assets grow larger, the interest income will increase. This means that the trade surplus must be lower, since the interest earned must be spent on foreign goods. Therefore, the long-run effect should be to lower the real exchange rate.

This analysis suggests that the nonneutrality of the deficit in the case of the inflation tax is attributable to the use of a nonneutral tax on one domestic asset—money—and not to the validity or lack of validity of the Ricardian equivalence proposition.

## General Conclusions

A deficit financed with debt, whether domestic or foreign, will affect the trade surplus only if the reduced taxes affect the rate of private spending. If the private sector uses the reduced taxes to acquire the new issues of internal debt (when the deficit is internally financed) or to acquire foreign assets (when the deficit is externally financed), there will be no effects on the rate of private spending, and therefore there will be no relation between the deficit and the trade balance or the real exchange rate. In this case the Ricardian equivalence proposi-tion will be valid, and the choice of tax or debt financing will be neutral. This result also holds true in the case of an open economy.

Inflationary financing of the deficit will affect the external sector through the portfolio-induced effects on desired private holdings of foreign assets. We expect a higher inflation rate to have opposite effects on the trade balance in the short run and in the long run. In the short run higher inflation should improve the trade balance; in the long run higher inflation should prove detrimental to the trade balance.

In the next two sections we describe in detail the relationship between the real exchange rate and the levels and financing methods of government spending, assuming that Ricardian equivalence does not hold—for example, that government deficits *do* have an impact on trade deficits and therefore on the real exchange rate. The analysis will focus on the short-run impact and the dynamic response of endogenous variables. The purpose will be to derive a set of basic structural relationships that can be used for empirical estimation.

## The Short-Run Process of Determining the Real Exchange Rate

Consider an economy producing two types of goods, tradable ($T$) and nontradable ($N$), with prices $PT$ and $PN$. We define the real exchange rate ($e$) as the relative price of tradable versus nontradable goods: $e = PT/PN$. (The analysis in this section draws from and extends the results presented in Rodríguez 1982.)

Nominal spending on goods by the private sector is denoted by $G_p$, and government spending on goods is denoted by $G_g$. Total spending on goods (absorption) is the sum of private and government spending:

$$(2.9) \qquad G = G_p + G_g.$$

Nominal GDP is denoted by $Y$, and the difference between GDP and nominal absorption is the trade surplus ($TS$):

$$(2.10) \qquad TS = Y - G.$$

On the demand side, assume that the private sector spends a fraction $b(e)$ of total private spending on nontradable goods:

$$(2.11) \qquad G_{pn} = b(e) \cdot G_p.$$

In the same way, the government spends a fraction $b_g$ on nontradable goods:

$$(2.12) \qquad G_{gn} = b_g \cdot G_g.$$

Total nominal spending on nontradable goods is therefore:

$$(2.13) \qquad G_n = G_{pn} + G_{gn} = b(e) \cdot G_p + b_g \cdot G_g.$$

The ratio of government spending to GDP is defined as the policy parameter

$$(2.14) \qquad g = G_g/Y.$$

On the supply side, the nominal value of output of nontradable goods is represented as proportional to nominal GDP:

$$(2.15) \qquad Y_n = a(e) \cdot Y.$$

Equilibrium in the market for nontradable goods requires:

(2.16)                        $G_n = Y_n.$

Substituting (2.13) and (2.15) into (2.16), we obtain:

(2.17)                $b(e) \cdot G_p + b_g \cdot G_g = a(e) \cdot Y.$

Substituting $G_p = G - G_g$ and $G_g = g \cdot Y$, we can express (2.17) as:

(2.18)          $b(e)\,(G - g \cdot Y) + b_g \cdot g \cdot Y = a(e) \cdot Y.$

Collecting terms, we can express this equation as the condition for the excess demand for nontradable goods (EDNT) and make it equal to zero:

(2.19)     EDNT $= b(e) \cdot G - [a(e) + g \cdot (b(e) - b_g)] \cdot Y = 0.$

Finally, defining

(2.20)                    $ts = 1 - (G/Y)$

as the ratio of the trade surplus to GDP and substituting into (2.19), we obtain:

(2.21)     EDNT $= b(e) \cdot (1 - ts) - a(e) + g \cdot [b(e) - b_g]$

$$= 0$$

$$= E(e, ts, g, b_g).$$

Walrasian stability requires that $dE/de > 0$. The other derivatives are:

$$dE/dts < 0 \quad \text{and} \quad dE/db_g < 0$$

where $dE/dg \gtrless 0$ depending on $b(e) \gtrless b_g$.

Given these derivatives, we can solve explicitly for the real exchange rate (the market-clearing relative price) as a function of the other determinants:

(2.22)                    $e = F(ts, g, b_g)$
$$\quad\quad + \;\; ? \;\; -$$

where the signs under the variables indicate the expected signs of the partial derivatives.

According to the previous equilibrium condition, the real exchange rate should depreciate as the trade surplus increases. The reason is simple: a larger trade surplus means a reduction in spending in relation to income. Part of the reduction in spending falls on nontradable goods, so their price *must* fall (the real exchange rate rises). An increase in government spending on nontradable goods, $b_g$, should raise their price, so the real exchange rate must fall. An increase in overall government spending for a given trade surplus implies that the government share in total spending has increased so that it has displaced private spending. In this case the demand for nontradable goods will rise or fall depending on who has a larger propensity to spend on this type of goods; this accounts for the ambiguity in the sign of the partial derivative with respect to $g$.

In this analysis we assumed the constancy of the terms of trade and therefore used an aggregate of tradable goods. A more general analysis would account for at least the existence of export and import-competing sectors. In that case the real exchange rate would measure the relative price of some aggregate of both tradable-goods prices. The equilibrium value of the real exchange rate in this context should also depend on the relative price of both tradable goods—in other words, the terms of trade—as well as trade taxes and subsidies that create a differential between the internal and external terms of trade. The interrelation between commercial policy instruments and the equilibrium level of the real exchange rate has been addressed by Dornbusch (1974), Harberger (1988), Rodríguez (1989), and Sjaastad (1980), among others.

Assume that there are two tradable goods—exportables and importables—with domestic prices determined by the following arbitrage conditions:

(2.23) $$Px = P^*x \cdot (1 - Tx)$$

$$Pm = P^*m \cdot (1 + Tm)$$

where the variables with asterisks refer to the (constant) foreign currency prices and $Tx$ and $Tm$ are ad valorem trade taxes.

There are now two relative prices in this economy that we may denominate as the export and the import real exchange rates:

(2.24) $$ex = Px/Pn$$

$$em = Pm/Pn.$$

Since there are now three goods in the economy, the shares of expenditure and output of nontradable goods should now depend on both relative prices:

(2.25) $$a = a(ex, em)$$

$$b = b(ex, em).$$

Substituting (2.25) into (2.21), it is clear that the equilibrium condition in the market for nontradable goods (2.22) is now changed to:

(2.26) $$ex = ex(em, ts, g, b_g).$$

The internal terms of trade are defined as:

(2.27) $$TT = ex/em$$

$$= (P^*x/P^*m) \cdot (1 - Tx)/(1 + Tm)$$

$$= TT^* \cdot (1 - Tx)/(1 + Tm).$$

This expression allows us to replace *em* in (2.26) by its equivalent in terms of *ex,*, *TT\**, and trade taxes, so that we end up with the following reduced-form equation:

(2.28)  $$ex = F(TT^*, Tx, Tm, ts, g, b_g).$$

Since *em* is a function of *ex, TT\**, and trade taxes, we can also represent the market equilibrium of nontradable goods by the equivalent condition

(2.29)  $$em = G(TT^*, Tx, Tm, ts, g, b_g).$$

Finally, assuming that we still want to refer to a single concept of the real exchange rate, we can define it as an average of the two real exchange rates:

(2.30)  $$e = z \cdot ex + (1 - z) \cdot em$$
$$= z \cdot F(\cdot) + (1 - z) \cdot G(\cdot)$$
$$= H(TT^*, Tx, Tm, ts, g, b_g).$$

As shown in Rodríguez (1989), the average real exchange rate will still be positively correlated with the trade surplus, but the relation with the terms of trade will vary depending on the weights used to construct it.

Equation 2.30 or one of the variations allowed by (2.29) or (2.28) is the basis for the empirical studies on the process of determination of the real exchange rate to be conducted in the country case studies. It follows from this section that government actions affect the real exchange rate at three different levels: total expenditures, composition of expenditures, and external financing of the deficit (which is captured by the effects of financing on the trade surplus).

As discussed previously, the contribution of the government to the trade surplus is directly measured by the government's ability to obtain foreign financing of its deficit, this financing is adjusted by whichever compensating capital flows are generated from the private sector. However, we still have to discover the process of determining the trade surplus of the private sector in relation not only to government-determined parameters but also to the desired rate of accumulation of domestic and foreign assets in the private sector. This is done in the next section.

## Short- and Long-Run Interrelations between the Assets Markets, the Trade Surplus, and the Real Exchange Rate

In the previous section we derived the relationship between the real exchange rate, the terms of trade, trade taxes, the trade surplus, and the level and composition of government spending. We also saw that the trade surplus depends on foreign financing (or lending) from the private and public sectors. While the capital flows of the public sector can be considered a policy variable related to the deficit-financing strategy, private capital flows still have to be explained, since they are an endogenous variable (except in the limiting case in which there is no capital mobility).

In this section we extend the general equilibrium model discussed above to incorporate the assets markets and to determine the equilibrium level of the trade surplus. (For further discussion on the interaction between the trade balance, the real exchange rate, and the assets markets, see Dornbusch 1973, Calvo 1981, Frenkel and Rodríguez 1982, and Rodríguez 1982.)

The interrelation between the assets markets, the trade balance, and the real exchange rate becomes evident when the effects of the imposition of the inflation tax are analyzed. As mentioned in the first section, the effects on the external sector and on the real exchange rate of deficit financing through the inflation tax will not be neutral. The reason is that the inflation tax, as a nonneutral tax on a particular domestic asset (money) sets the incentive for a portfolio shift in favor of foreign assets. Before going into the formal derivation of all the general equilibrium relationships, we provide an intuitive explanation of the most basic interrelation, using the example of the inflation tax.

### The Inflation Tax and the Assets Markets

Consider an economy that produces and consumes both tradable and nontradable goods. Individuals hold domestic money and interest-bearing foreign assets. The differential rate of return between both types of assets is the foreign interest rate plus the expected rate of devaluation. In long-run equilibrium the expected rate of devaluation is equal to the rate of inflation. An increased rate of monetary expansion generates the expectation of higher devaluation and inflation, and a process of substitution of foreign assets for domestic money begins. For analytical simplicity we assume that there is a freely floating exchange rate. (A fixed exchange rate would be inconsistent with the use of the inflation tax.)

In the process of running down cash balances and acquiring foreign assets, the nominal exchange rate is expected to rise because both the stock of money and foreign exchange are fixed at a given moment in time. The rise in the nominal exchange rate (the price of tradable goods) also induces, by substitution, some increase in the price of nontradable goods, although in a smaller magnitude, as we shall show later.

The short-run adjustment is therefore obtained through rises in prices and the exchange rate that reduce the real value of the total asset holdings of the private sector. The reduction in real asset holdings reduces the demand for nontradable goods and thus allows for the increase in the real exchange rate and the improvement in the trade surplus.

The improvement in the trade surplus starts a process of accumulation of foreign assets. As holdings of foreign assets accumulate, the pressure on the exchange rate is reduced, and the real exchange rate begins to fall back to its original level. However, since foreign assets are larger than before, the service account shows a larger surplus. As a consequence, in the new long-run equilibrium the trade balance must show a larger deficit since the current account must be balanced. In conclusion, the imposition of the inflation tax raises the real exchange rate during a transitional period and lowers it in the new long-run equilibrium.

This discussion suggests that the real stocks of assets and the inflation rate should be added as variables in the equation to determine the trade balance because they are linked to the desired rate of accumulation of foreign assets. We now proceed to a formal demonstration of these points in the context of a model that also incorporates domestic issues of public debt.

## A Dynamic General Equilibrium Model of Determination of the Real Exchange Rate

The model developed here describes the dynamic effects on the external accounts and the real exchange rate of changes in the inflation tax, the foreign interest rate, and the stock of internal public debt. The context of the model is an economy in which individuals hold three types of assets: domestic money ($M$), a domestic bond denominated in foreign exchange issued by the government ($b$), and a foreign asset ($D$). The three assets are assumed to be imperfect substitutes, and the relative demands for the assets depend on the differential rates of return offered.

Since we shall be analyzing the effects of the inflation tax, derived from the continuous issuance of money, we have assumed that the government bond is indexed. If it were fixed in nominal terms, as money grew the relative amount of this bond would approach zero. The alternative would be for the government to issue money and nominal bonds in order to keep constant the ratio between them. The assumption that bonds are already indexed to the price level or to some of its components such as the exchange rate simplifies the analysis without loss of relevance.

The economy produces and consumes both tradable and nontradable goods. The excess supply of tradable goods is the trade surplus. The trade surplus plus the interest earnings on foreign asset holdings (the current account) equals the change in the stock of these assets.

Demands for goods depend on the two nominal prices ($Pt$ and $Pn$) and nominal expenditure on goods ($G$). Demands are assumed to be homogeneous of degree zero in all nominal variables. The variable $E$ represents the nominal exchange rate, which is assumed to equal the nominal price of tradable goods ($E = Pt$). For simplicity we assume that the revenues of the inflation tax are redistributed neutrally to the public and that the interest on the internal public debt is also financed through a neutral tax. Supplies of both goods depend on the relative price ($e = Pt/Pn = E/Pn$) and on factor endowments, which we assume are fixed (we abstract here from growth considerations). Given those assumptions, the supply and demand functions take the following form:

(2.31)
$$Cn = Cn(E, Pn, G) = Cn(e, G/E)$$
$$\qquad\qquad\qquad\quad +\qquad +$$

$$Ct = Ct(E, Pn, G) = Ct\ (e, G/E)$$
$$\qquad\qquad\qquad\quad -\qquad +$$

(2.32)
$$Qn = Qn(e)$$
$$\qquad\quad -$$

$$Qt = Qt(e)$$
$$\qquad\quad +$$

We define GDP, measured in terms of tradable goods, as:

(2.33)
$$y(e) = Qt(e) + Qn(e)/e = \text{GDP}.$$

The derivative of $y(e)$ with respect to $e$ is defined as:

(2.34)
$$y'(e) = [QT'(e) + (1/e) \cdot Qn'(e)] - Qn(e)/e^2$$
$$= -Qn(e)/e^2 < 0$$

since the term in brackets is identically equal to zero by the envelope theorem.

The trade surplus, measured in foreign exchange, equals the difference between GDP and expenditure:

(2.35)
$$TS = y(e) - G/E.$$

We define $ts = TS/y(e)$ as the ratio of the trade surplus to GDP. Substituting $ts$ into the demand for $Cn$, we can express it as:

(2.36)
$$Cn = Cn[e, (1 - ts) \cdot y(e)] = Cn(e, ts).$$

If the trade surplus were to be zero, the demand for $Cn$ would unambiguously depend positively on $e$ (this follows from the Slutsky expansion of the price effect on the demand for *Cn*). If $ts < 0$, however, an income effect operating in the wrong direction appears. We assume that the substitution effect dominates, so that the demand for nontradable goods depends negatively on its relative price. We therefore assume the following signs for the partial derivatives of *Cn*:

$$(2.37) \qquad d(Cn)/d(ts) < 0$$

$$d(Cn)/d(e) > 0.$$

Equilibrium in the market for nontradable goods requires that the relative price, *e*, adjust to equal supply and demand:

$$(2.38) \qquad Qn(e) = Cn(e, ts).$$
$$\qquad\qquad\quad - \qquad + \ -$$

It is clear from equation 2.38 that an increase in the trade surplus is associated with a lower level of expenditure and therefore with a higher real exchange rate. As expenditure falls, demand for nontradable goods falls, and its relative price is reduced:

$$(2.39) \qquad e = e(ts) \qquad e' > 0.$$

Equation 2.39 determines the real exchange rate that equilibrates the market for nontradable goods as a function of the proportional excess of expenditure over GNP (*ts*). However, *ts* is also an endogenous variable, and we will now turn our attention to determining it.

Since the trade surplus is associated directly with the desired rate of accumulation of foreign assets, we must turn to the description of the assets markets in order to determine the equilibrium level of the trade surplus.

The level of nominal assets, *A*, is defined as:

$$(2.40) \qquad A = M + E \cdot b + E \cdot D = E \cdot (m + b + D).$$

We assume that there is a desired long-run level of real assets ($a^*$) and that people adjust their expenditures in order to reach it gradually. The desired level of real assets could be defined as a proportion of income or in terms of either commodity. To simplify the analysis, it is convenient to define the desired level of real assets as constant in terms of foreign exchange:

$$(2.41) \qquad A^* = a^* \cdot E.$$

The level of nominal expenditures on goods equals the sum of nominal income [$Y = E \cdot y(e)$], foreign interest earned ($r^* \cdot E \cdot D$), a fraction of the excess of actual asset holdings over the desired long-run level, and the amount of foreign transfer needed to finance local government spending, $f_{eg}$:

$$(2.42) \qquad G = Y + E \cdot r^* \cdot D + z \cdot (A - A^*) + f_{eg}.$$

The trade surplus therefore equals $(Y - G)/E$, using (2.42) and (2.40):

(2.43)          $$TS = z \cdot (a^* - m - b - D) - r^* \cdot D - f_{eg}.$$

Equation 2.43 describes the determination of the equilibrium trade surplus. It is directly related to the desired rate of accumulation of assets and to the interest earned on foreign assets. A fiscal deficit financed abroad should be subtracted from (2.43). What we have determined, then, is the structural form for the desired rate of private foreign savings.

Equation 2.43 still has endogenous variables in the explanation of the trade surplus to the extent that $m$ can change at any instant through jumps in the exchange rate. In order to determine the equilibrium value of $m$, we have to describe the conditions of the portfolio balance equilibrium.

The rate of return for holding domestic money is $-I$, where $I$ is the expected inflation rate. The rate of return on the domestic indexed bond is $d + i - I$, where $d$ is the expected rate of devaluation and $i$ is the dollar rate paid by the bond. Finally, the rate of return for holding the foreign asset is $r^* + d - I$. Since there are three assets, there should be two portfolio-preference functions that we assume depend on the difference between the rates of return of the two assets involved:

(2.44)          $$m/D = L(r^* + d) \qquad L' < 0$$

(2.45)          $$b/D = H(i - r^*) \qquad H' > 0.$$

The stock of the domestic indexed bond is a variable subject to government control. It is clear that the government cannot resort to bond financing as a permanent source of revenue in the absence of growth. We therefore consider $b$ a policy parameter that takes a fixed value and proceed to analyze the effects of changes in its level.

For the moment, we assume that the expected rate of devaluation is a constant parameter. Substituting (2.44) into (2.43), we obtain:

(2.46)          $$TS = z \cdot [a^* - b - (1 + L)D] - r^* \cdot D.$$

According to (2.46), the trade surplus depends on the stocks of domestic and foreign assets held (which are constant at a point in time), the foreign interest rate, and inflationary expectations. We can now normalize $TS$ by $y(e)$ to obtain the variable $ts$:

(2.47)          $$ts = z \cdot [a^* - b - (1 + L) \cdot D - r^* \cdot D/z]/y(e)$$
                $$L = L(r^* + d).$$

Note that in equation 2.47 the real exchange rate enters into the determination of the ratio of the trade surplus to GDP not because it affects the trade surplus but because real GDP depends on it.

Short-run equilibrium is achieved when the home goods market is in equilibrium (2.39) and when *ts* equals the desired rate of assets accumulation (2.47).

Around the steady-state equilibrium, assets equal the desired level, so that $a^* = b + (1 + L) \cdot D$. We now proceed to evaluate the short-run response of the trade surplus to changes in the different parameters when the changes take place in the vicinity of the steady-state equilibrium. These changes are obtained from differentiation of the short-run equilibrium conditions (2.39) and (2.47). After differentiation, the changes in the ratio of the trade surplus to GDP become:

$$d(ts)/d(D)sr = -z \cdot [(1 + L) + r^*]/[y \cdot (1 - J)] < 0$$

$$d(ts)/d(b)sr = -z/[y \cdot (1 - J)] < 0$$

$$d(ts)/d(d)sr = -z \cdot L' \cdot D/[y \cdot (1 - J)] > 0$$

(2.48)

$$d(ts)/d(r^*)sr = -(z \cdot L' \cdot D + D)/[y \cdot (1 - J)] \gtreqless 0.$$

$$d(ts)/d(a^*)sr = z/[y \cdot (1 - J)] > 0$$

$$d(ts)/d(f_{eg})sr = -1/[y \cdot (1 - J)] < 0$$

where

$$J = e' \cdot r^* \cdot D \cdot y'/y < 0.$$

By (2.39), the real exchange rate depends (positively) only on the trade surplus; the effect of other parameters such as trade taxes (*TT*) and the level or composition of government spending have already been analyzed in the previous section and are assumed constant in this section. The partial derivatives in (2.48) therefore also give the sign of the short-run response of the real exchange rate to changes in the different parameters or state variable (*D*).

In particular, it follows that an instantaneous depreciation of the real exchange rate takes place whenever the inflation tax or desired assets are raised, while an appreciation follows from increases in the stocks of domestic or foreign assets held by the private sector (*b* or *D*). Algebraically, these short-run derivatives are:

$$d(e)/d(D)sr = -e' \cdot [z(1 + L) + r^*]/[y \cdot (1 - J)] < 0$$

$$d(e)/d(b)sr = -z \cdot e'/[y \cdot (1 - J)] < 0$$

$$d(e)/d(d)sr = -z \cdot L'e' \cdot D/[y \cdot (1 - J)] > 0$$

(2.49)

$$d(e)/d(r^*)sr = -D \cdot e'(z \cdot L' + 1)/[y \cdot (1 - J)] \gtreqless 0$$

$$d(e)/d(a^*)sr = z \cdot e'/[y(1 - J)] > 0$$

$$d(e)/d(f_{eg})sr = -e'/y \cdot (1 - J)] < 0.$$

In order to close the model, we have to describe the process of formation of the expected rate of devaluation. The model described here is similar in reduced form to that of Calvo and Rodríguez (1977). In that study we closed the model using rational expectations and further showed that a quasi-rational rule of assuming that $d$ equals the rate of monetary expansion yields identical qualitative results.

For simplicity, we assume that expectations of devaluation are equal to the constant rate of monetary expansion, $mu$: $d = mu = (1/M) \cdot (dM/dt)$.[1] At any instant, $ts$ and $e$ are jointly determined by the values of the state variable $D$ and the parameters $r^*$, $d = mu$, and $b$.

Characterizing the dynamic behavior of foreign assets requires specification of the current account, $CA$, which equals the trade surplus plus foreign interest earnings:

(2.50)     $$CA = d(D)/dt = z \cdot [a^* - b - (1 + L) \cdot D].$$

The differential equation 2.50 describing the trajectory of foreign assets is stable. The stock of foreign assets converges asymptotically to the desired long-run level ($Dss$):

(2.51)     $$Dss = (a^* - b)/(1 + L).$$

According to (2.51), the long-run stock of foreign assets depends on $a^*$, the stock of the indexed government bond ($b$), the foreign interest rate ($r^*$), and the inflation tax rate ($d = mu$). Algebraically, these changes are:

(2.52)
$$d(Dss)/d(a^*) = 1/(1 + L) > 0$$
$$d(Dss)/d(b) = -1/(1 + L) < 0$$
$$d(Dss)/d(mu) = -(a^* - b) \cdot L'/(1 + L)^2 > 0$$
$$d(Dss)/d(r^*) = -(a^* - b) \cdot L'/(1 + L)^2 > 0.$$

We can now compute the long-run effects on the real exchange rate of changes in the different parameters. The difference between the short-run effects presented in (2.49) and the long-run effects is that we must take account of the adjustment of $D$ to its long-run value $Dss$. For example, the long-run change in $e$ in response to a change in $mu$ is computed as:

(2.53)   $d(e)/d(mu)ss = d(e)/d(mu)sr + d(e)/d(D)sr \cdot d(D)/d(mu)ss.$

Equations 2.54 summarize the long-run effects of parameter changes on $e$:

(2.54)
$$d(e)/d(mu)ss = e' \cdot r^* \cdot D \cdot L'/[(1 + L) \cdot y \cdot (1 - J)] < 0$$
$$d(e)/d(a^*)ss = -e' \cdot r^*/[(1 + L) \cdot y \cdot (1 - J)] < 0$$
$$d(e)/d(b)ss = e' \cdot r^*/[(1 + L) \cdot y \cdot (1 - J)] > 0$$
$$d(e)/d(r^*)ss = e'.D \cdot [r^* \cdot L' - (1 + L)]/[(1 + L) \cdot y \cdot (1 - J)] < 0.$$

It is interesting to compare the difference between the short- and long-run responses of the real exchange rate to changes in the different parameters. The table below presents the qualitative effects on the real exchange rate of changes in four variables:

|           | $mu$ | $a^*$ | $b$ | $r^*$ |
|-----------|------|-------|-----|-------|
| Short run | +    | +     | −   | ?     |
| Long run  | −    | −     | +   | +     |

The most striking feature of the table is that in all cases the direction of the short-run impact of a parameter change on the real exchange rate is the opposite of the direction of the long-run change (except for $r^*$, which has an ambiguous short-run effect). An increase in the inflation tax rate depreciates $e$ in the short run and appreciates it in the long run. The same qualitative effects take place when the desired level of assets is increased. An increase in the stock of government debt appreciates $e$ in the short run and depreciates it in the long run. The short-run impact of a higher foreign interest rate is ambiguous in the short run but unambiguously induces an exchange rate depreciation in the long run.

## Conclusions

In this chapter we have developed a methodology for analyzing the effects of government spending and the ways that its financing is related to variables in the external sector.

The level and composition of government spending affect the real exchange rate because of existing differences in propensities to spend between the government and private sectors. The fiscal deficit may or may not affect the external sector, depending on the empirical validity of the Ricardian equivalence proposition. If such equivalence does not hold, we expect government deficits to have direct effects on the economy's overall rate of spending and therefore also on the trade balance. Changes in the trade balance are bound to have both impact and dynamic effects on the real exchange rate. The impact effects are derived from the required expenditure switching, which is necessary to convalidate the new level of the trade balance that is compatible with the spending increase. The dynamic effects are the result of induced changes in the desired rate of accumulation of foreign assets.

It follows from our dynamic analysis that it will not be possible to find a stable static relationship between the real exchange rate and the structural parameters without allowing for the fact that the level of foreign assets changes through time. It is therefore necessary to estimate a two-equation model: one equation relating the real exchange rate to the level of the trade balance and another determining the trade balance as a function of a set of parameters that includes the fiscal deficit and the stock of foreign assets held. Finally, the trade balance plus foreign interest earned (the current account) will determine the evolution over time of the stock of foreign assets.

## Note

1. If $d$ is not a constant, the derivation should proceed from the differentiation of the portfolio-balance relation (2.44):

$$mu - \hat{E} = -(L'/L) \cdot (d\hat{E}/dt)^e + (1/D) \cdot (dD/dt).$$

In the above expression, the hat (^) over a variable indicates the proportional rate of change. If there are rational expectations, the expected change in $\hat{E}$ should equal change (abstracting from uncertainty or random shocks). Otherwise it can also be assumed that the expected rate of devaluation is formed according to a process of adaptive expectations. In any event, this expression is the basis for the endogenous determination of the expected rate of devaluation.

## References

Auernheimer, Leonardo. 1987. "On the Outcome of Inconsistent Programs under Exchange Rate and Monetary Rules, or 'Allowing Markets to Compensate for Government Mistakes.' " *Journal of Monetary Economics* 19 (March): 279–305.

Calvo, Guillermo A. 1981. "Devaluation, Levels Versus Rates." *Journal of International Economics* 11 (2): 165–72.

Calvo, Guillermo A., and Carlos A. Rodríguez. 1977. "A Model of Exchange Rate Determination Under Currency Substitution and Rational Expectations." *Journal of Political Economy* 85: 617–25.

Dornbusch, Rudiger. 1973. "Currency Depreciation, Hoarding and Relative Prices." *Journal of Political Economy* 81 (August): 893–915.

———. 1974. "Tariffs and Nontraded Goods." *Journal of International Economics* 4 (May): 177–85

Frenkel, Jacob A., and Assaf Razin. 1986. "Fiscal Policies in the World Economy." *Journal of Political Economy* 94, pt.1 (June): 564–94.

Frenkel, Jacob A., and Carlos A. Rodríguez. 1982. "Exchange Rate Dynamics and the Overshooting Hypothesis." *International Monetary Fund Staff Paper* 29 (March): 1–30.

Harberger, A. C. 1988. "Trade Policy and the Real Exchange Rate." World Bank, Economic Development Institute, Washington, D.C.

Leiderman, Leonardo, and Mario I. Blejer. 1988. "Modeling and Testing Ricardian Equivalence: A Survey." *International Monetary Fund Staff Paper* 35 (March): 1–35.

Rodríguez, Carlos A. 1978. "A Stylized Model of the Devaluation-Inflation Spiral." *International Monetary Fund Staff Paper* 25 (March): 76–89.

———. 1982. "Gasto Público, Déficit y Tipo Real de Cambio: Un Análisis de sus Interrelaciones de Largo Plazo." *Cuadernos de Economia* (Chile) 19 (August): 203–16.

————. 1989. "Macroeconomic Policies for Structural Adjustment." Policy Research Working Paper 247. World Bank, Country Economics Department, Washington, D.C.

Sjaastad, Larry A. 1980. "Commercial Policy, True Tariffs and Relative Prices." In John Black and Brian Hindley, eds., *Current Issues in Commercial Policy and Diplomacy*. New York: St. Martin's Press.

# Part II
## Country Case Studies

# 3

# Argentina: Fiscal Disequilibria Leading to Hyperinflation

*Carlos Alfredo Rodríguez*

This chapter was written during the first months of 1989, as hyperinflation started to develop. The main developments during 1989–93 are not included in the analysis but are briefly described here.

## Recent Developments: Hyperinflation and Political Crisis

On February 6, 1989, Argentina's Central Bank was forced to suspend the convertibility of the austral because of the intense demand for foreign exchange in the daily auctions. The black-market premium skyrocketed, and hyperinflation began to develop. In the presidential elections held in April 1989 the Peronist party defeated the incumbent Radical party; President-elect Menem was scheduled to assume the presidency in October. Quasi-fiscal imbalances and a rush against the austral combined to induce a price explosion. Inflation was 33 percent in April, 78 percent in May, and 114 percent in June. The outgoing administration lost control of the situation, and riots broke out. Three economics ministers were put on trial over a period of three months. Finally President Alfonsin resigned early, transferring power to Menem on July 9, 1989. During that month inflation reached 197 percent.

Contrary to expectations, Menem sided with conservative economic policies and named members of the Bunge and Born group to the Ministry of Economics. Inflation initially subsided, but the money supply continued to be fueled by the significant debt of the Central Bank. In December 1989 another run against currency occurred, the minister of economics resigned, and hyperinflation reappeared, with inflation rates of 40 percent in December, 79 percent in January 1990, 61 percent in February, and 95 percent in March.

In January 1990 the new economics minister, Erman Gonzalez, launched the "BONEX plan," which virtually converted most government debt and bank deposits into a ten-year dollar-denominated bond called BONEX 89. Real $M_2$ (non-interest-bearing money plus time deposits) was reduced to a mere 3.1 percent of GDP, drastically down from 18 percent in 1986. The refinancing of the public debt made possible the elimination of the quasi-fiscal deficit and the adoption of measures to permanently stabilize the economy. The administration announced an ambitious plan of privatization and deregulation.

In February 1991 there was a new run against the currency, which resulted in significant devaluation and an upsurge in inflation. Gonzalez resigned and was replaced by Domingo Cavallo, the foreign minister, who had long been preparing for the job of economics minister. In April Cavallo announced the convertibility plan that became the basis for the stabilization enjoyed in 1993, when this chapter was written. At the insistence of Cavallo, Congress passed a law that effectively converted the Central Bank into a currency board. The law set the exchange rate at $1 to the peso, and the Central Bank was allowed to issue monetary base only for the purpose of purchasing dollars at the established rate.

Since the passage of this law, impressive results in the areas of stabilization and structural adjustment have been achieved. External debt has been renegotiated under the Brady agreement; a significant portion of public enterprises has been privatized (including the telephone company, the national airline, trains, and the electricity, gas, and oil monopolies); and moves have been made toward a more open economy and toward deregulation in the social security system, the tax system, and labor markets. Real GDP grew by about 8 percent annually during 1991–92. Public finances have improved dramatically, and the country has been running primary surpluses since 1991.

The most significant shadow on the economic front is the relative overvaluation of the currency, which is being sustained by the large capital inflows that have taken place in the past three years. These capital inflows in part represent the repatriation of the funds that had left during the 1980s. Dollar deposits at commercial banks grew by about $11 billion after the hyperinflation. In addition, the government used about $5 billion from the proceeds of privatization to finance local spending. Finally, the low U.S. interest rates induced capital flows toward areas with higher rates, including Argentina.

### Background

Argentina has had a sad economic history in recent decades, as illustrated by the indicators shown in table 3.1. In 1989 real GDP per capita was at about the level of the early 1960s (figure 3.1). This meant a quarter century without growth at a time when most of the rest of the world enjoyed a period of economic achievement. GDP per capita reached a historical high in 1974; Argentina matched this level in 1977 and 1979 but has never surpassed it. In 1980 GDP per capita began declining, which resulted in an accumulated decrease of 23.5 percent in the ten-year period 1979–88. During the same period, the con-

**Table 3.1. Key Macroeconomic Indicators, Argentina, 1980–92**

| Year | Inflation (percent)[a] | GDP (millions of australes)[b] | Real exchange rate[c] | Gross investment as share of GDP (percent) | Imports (millions of U.S. dollars) |
|------|------|------|------|------|------|
| 1980 | 87 | 1,129 | 100 | 25.3 | 10,541 |
| 1981 | 131 | 1,054 | 258 | 22.7 | 9,430 |
| 1982 | 209 | 1,002 | 517 | 21.8 | 5,337 |
| 1983 | 433 | 1,032 | 381 | 20.9 | 4,504 |
| 1984 | 687 | 1,060 | 368 | 20.0 | 4,585 |
| 1985 | 385 | 1,014 | 373 | 17.6 | 3,814 |
| 1986 | 81 | 1,072 | 379 | 17.5 | 4,724 |
| 1987 | 174 | 1,095 | 421 | 19.6 | 5,818 |
| 1988 | 387 | 1,068 | 312 | 18.6 | 5,322 |
| 1989 | 4,923 | 1,018 | 556 | 15.5 | 4,200 |
| 1990 | 1,343 | 1,022 | 155 | 14.0 | 4,079 |
| 1991 | 84 | 1,114 | 169 | 14.6 | 8,275 |
| 1992 | 17 | 1,210 | 148 | 16.7 | 14,872 |

a. Annual rate, December to December.
b. Real GDP at 1970 market prices.
c. CPI ratio in relation to U.S. free-market exchange rate.
*Source:* Central Bank of Argentina.

**Figure 3.1. GDP Per Capita (1960 Prices), Argentina, 1960–90**

1960 pesos (thousands)

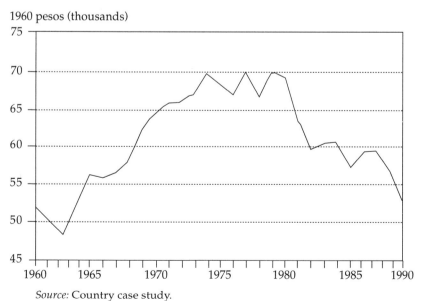

*Source:* Country case study.

sumer price level increased by 3.36 million, equivalent to an annual compound rate of inflation of 349 percent.

We do not intend to explain the reasons for Argentina's economic stagnation; instead we will concentrate on the influence of public sector behavior on that process. In particular, we will be concerned with the effects of government spending, taxation, and deficit financing on the rest of the economy. We will not deal with the effects of government regulation because that topic deserves a separate volume.

Government has played an important role in Argentine society. It taxes, spends, produces a wide variety of goods and services, regulates financial markets, supplies financial services, systematically resorts to incomes policies, and regulates foreign trade. Although the regulatory aspects of government action defy any possible quantitative measure, an impression of the size of government involvement in economic activity can be obtained by looking at the relative share of government spending in GDP. Figure 3.2 shows that government spending systematically tended to grow faster than GDP until the last crisis of the Argentine economy, which began to develop in 1982. After that, spending began to fall, more as a consequence of limited resources than because of deliberate political action.

The fall in the relative share of government spending, however, came too late to avoid a crisis that was already well in the making and that brought the country to a state of hyperinflation in 1989, when the inflation rate approached 5,000 percent and GDP fell by about 7 percent. The December-to-December values of the rate of inflation in the consumer price index (CPI) are shown in table 3.2. It is clear that inflation has been an ever-present phenomenon in recent decades and that it has been explosive. The hyperinflation of 1989 does not appear to be the end of the story. After the monthly inflation rate reached a peak of 195 percent in July 1989, inflation returned to the single-digit level. But in December 1989 a new hyperinflation started. It peaked at 95 percent in March 1990 and apparently ended in April 1990, when the monthly rate was "only" 11.5 percent.

A quick characterization of government action in Argentina would be that the government has generally tended to increase spending and run deficits. Governments do not need to run fiscal surpluses all the time, especially when the economy is growing. But the Argentine economy has been stagnant over the past two decades, and the government has run fiscal deficits every year since at least 1961 (figure 3.3). Indeed, the government has run a primary deficit (excluding interest payments) every year in our sample dating back to 1961. This means that in each of the past twenty-eight years, after paying for current and capital spending, the government has not had any resources left for servicing interest on its internal or external debt.

**Figure 3.2. Total Expenditures of the CNFPS, Argentina, 1961–87**

Percentage of GDP

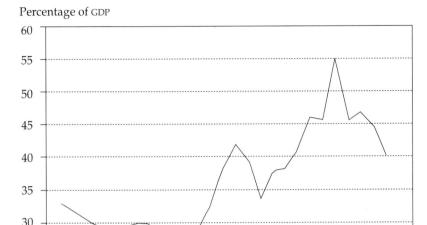

*Note:* CNFPS, consolidated nonfinancial public sector.
*Source:* Country case study.

**Table 3.2.  Annual Inflation Rates in the Consumer Price Index, Argentina, 1961–92**
(percent)

| Year | Rate | Year | Rate |
|------|------|------|------|
| 1961 | 16.5 | 1977 | 160.4 |
| 1962 | 30.4 | 1978 | 169.8 |
| 1963 | 23.9 | 1979 | 139.7 |
| 1964 | 18.2 | 1980 | 87.6 |
| 1965 | 38.2 | 1981 | 131.3 |
| 1966 | 30.0 | 1982 | 209.7 |
| 1967 | 27.3 | 1983 | 433.7 |
| 1968 | 9.6 | 1984 | 688.0 |
| 1969 | 6.6 | 1985 | 385.4 |
| 1970 | 21.7 | 1986 | 81.9 |
| 1971 | 39.1 | 1987 | 174.8 |
| 1972 | 64.2 | 1988 | 387.7 |
| 1973 | 43.8 | 1989 | 4,923.7 |
| 1974 | 40.1 | 1990 | 1,343.0 |
| 1975 | 335.1 | 1991 | 84.0 |
| 1976 | 347.6 | 1992 | 17.5 |

*Source:* National Statistics Institute (INDEC).

**Figure 3.3. Primary Deficit of the CNFPS, Argentina, 1961–88**

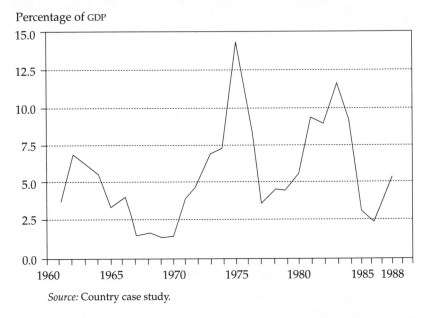

*Source:* Country case study.

(Dornbusch and de Pablo 1987 and Rodríguez 1989 have discussed the fiscal nature of Argentina's apparent inability to serve its mostly public foreign debt.) As a consequence the government has resorted systematically to issuing money and interest-bearing debt. The outcome of this deficit policy has been a systematic tendency of the economy to run high inflation and high real interest rates as a consequence of the pressure on the credit markets from the incremental government borrowing.

For an economy that is not growing and that faces a positive real rate of interest, running a positive primary deficit implies an ever-growing stock of public debt in relation to GDP. Of course, real government debt did not grow continuously because occasionally the existing stock of debt was melted down by bursts of inflation—fueled by large devaluations in the face of a foreign exchange crisis—that exceeded nominal interest rates. (For more on the relationship between deficits and devaluations, captured by increases in the level or in the rate of change of the exchange rate, see Calvo 1981a, Rodríguez 1978, and Fernandez 1989.)

The tendency to melt down the existing stock of debt by implementing unexpected devaluations was eventually discounted by the market; in recent years the market has demanded and has obtained an increasing degree of indexation of the public debt by either the price level or the price of foreign exchange. This meant that the government could no longer melt down the stock of real debt and had to face the critical problem of a growing real debt in the context of a persistent tendency to run primary deficits.

The primary deficit has been falling since 1983 as a result of reduced spending and higher revenues. This trend, however, is not enough to reverse the increasing reluctance of the public to hold the internal government debt or domestic currency. The government has increasingly had to resort to the use of forced investments by the banking system (*depositos indisponibles*). The fall in demand for assets denominated in domestic currency induces real interest rates that are inconsistent with the real equilibrium of the economy. The shift out of domestic currency leads to an increasing degree of dollarization of the economy and a tendency for recurrent currency runs that result in frequent macroeconomic devaluations. The government was forced continuously to raise interest rates so that it could induce the public to keep holding the domestic currency and could roll over the public debt.

The higher interest rates were paid back by issuing more debt and money, and the service of the debt became the major source of money creation. The system finally exploded when the country entered hyperinflation in May 1989. The hyperinflation, however, was not able to melt down the stock of interest-earning government debt because much of the debt reached maturity between one and seven days, and interest rates actually had a tendency to anticipate devaluations. In January 1990 the government replaced all interest-bearing obligations in the financial system—that is, government debt plus interest-earning deposits—with an issue of dollar-denominated government paper that paid the London interbank offered rate (LIBOR) and matured at ten years.

This chapter is concerned with how public sector deficits and the methods of financing them affect inflation, interest rates, the real exchange rate, private savings and investments, and the external balance. The next section takes up measurement issues.

## Measurement of the Public Sector Deficit

In the past fifteen years many important changes have occurred in the distribution of revenues and expenditures among different levels of government in Argentina. These changes were a result of (a) the transfer of important expenditure items, such as elementary and secondary education and local transport systems (including Buenos Aires' subway), from the central to the provincial and city governments; and (b) changes in the tax laws and the rules under which tax revenues are distributed among the central and the provincial governments (the federal coparticipation law). Provincial and local governments have no well-defined budget constraints in Argentina, and the distribution of resources (which are collected mainly by the central government) is highly unpredictable. This makes it especially important in the Argentine case to work on the basis of *consolidated* nonfinancial public sector (CNFPS) data. Studies dealing with the behavior of the consolidated nonfinancial public sector in Argentina include Fundación de Investigaciones Económicas Latinoamericanas—FIEL (1987) and Schenone (1987).

It is also important to account for the operations of the Central Bank as a source of substantial amounts of revenue (because of inflationary money creation) and no less substantial losses. The losses are a result of (a) the purchase of international reserves (including those required to service the foreign debt); (b) the offering of swaps and other "exchange insurance" mechanisms, which have frequently been used to attract short-term foreign financing; and (c) losing operations of the Central Bank with the domestic financial system (including the bailout of failing financial institutions). In a later section we present estimates made by FIEL of losses incurred by the Central Bank of Argentina, which should be considered components of public sector expenditures.

### Consolidated Nonfinancial Public Sector

Table 3.3 presents official data on "above-the-line" CNFPS expenditure, revenue, and financing according to the "international methodology." (The international methodology computes only the operating surpluses or deficits of public enterprises as components of public sector revenue and expenditure; it does not consider the current revenue and expenditure of these enterprises.) Figure 3.4 shows the evolution of the total expenditures and revenues of the CNFPS. Note that revenues historically have shown a growth trend similar to but below the growth of expenditures. Total revenues rose from a low of about 23 percent in 1962–65 to a high of 41 percent in 1985. (The revenue measure does not incorporate current revenues from public enterprises.) This high growth in fiscal revenues in the face of a stagnant economy should be enough to invalidate the commonly voiced claim that Argentina's basic problem is that the public does not pay taxes. The fact is that fiscal pressure is very high and has grown at a much faster rate than has GDP in the past twenty-five years.

Even though fiscal pressure has been high, fiscal spending has also grown and has exceeded revenues. Expenditures of the CNFPS reached a level of 55 percent of GDP in 1983 (up from 30 percent in the 1960s). A serious attempt was then made to reduce government spending. Spending was reduced to 11 percent of GDP, but the decrease was wiped out by the increased quasi-fiscal spending by the

**Figure 3.4. Total Expenditures and Revenues of the CNFPS, Argentina, 1961–87**

Percentage of GDP

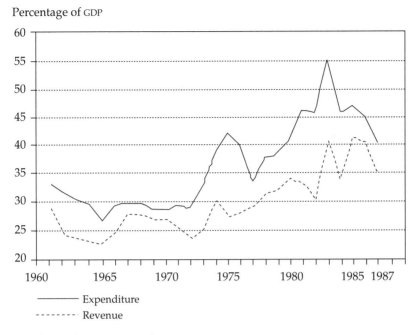

——— Expenditure

- - - - - - - Revenue

*Source:* Country case study.

Central Bank in servicing its accumulated financial liabilities. The last column of table 3.3 represents the difference between total expenditures and resources. A deficit has to be financed through increases in short- and long-run debt (domestic or foreign), advances from suppliers, or money creation in the form of credit from the Central Bank and increases in short-term financial liabilities.

Table 3.4 presents the "below-the-line" items that describe the composition of the financing of the fiscal deficit. In every year from 1961 to 1988 the financial result has been negative; the government has always had to resort to printing money or issuing short-term financial liabilities in order to balance the budget. According to the official data there has been no financing of the CNFPS from the Central Bank since 1986. Until mid-1985 Central Bank financing of the CNFPS was recorded in an account labeled "advances to the treasury." Since the stabilization attempt launched in June 1985, no more drawings have been made on that account, according to official statistics. Actually, since 1985 the CNFPS has borrowed from the Central Bank by obtaining rediscounts for public banks and other public enterprises—

*(Text continues on page 114.)*

**Table 3.3. Consolidated Nonfinancial Public Sector "Above-the-Line" Data, Argentina, 1961–88**
(percentage of GDP)

| Total | Current revenue (1) | Current expenditure (2) | Capital revenue (3) | Capital expenditure (4) | Total revenue[a] (5) | Total expenditure[b] (6) | Deficit or financing needs (6–5) |
|---|---|---|---|---|---|---|---|
| 1961 | 23.17 | 17.41 | 0.55 | 10.68 | 28.67 | 33.03 | 4.36 |
| 1962 | 18.63 | 17.84 | 0.76 | 9.32 | 23.85 | 31.61 | 7.76 |
| 1963 | 18.47 | 17.93 | 0.68 | 8.22 | 23.37 | 30.37 | 7.00 |
| 1964 | 18.00 | 18.11 | 0.37 | 6.89 | 22.81 | 29.43 | 6.62 |
| 1965 | 18.86 | 16.88 | 0.35 | 6.41 | 22.31 | 26.39 | 4.08 |
| 1966 | 20.13 | 18.98 | 0.36 | 6.45 | 24.52 | 29.45 | 4.93 |
| 1967 | 23.35 | 18.26 | 0.48 | 7.65 | 27.78 | 29.86 | 2.08 |
| 1968 | 22.80 | 17.24 | 0.54 | 8.33 | 27.58 | 29.81 | 2.23 |
| 1969 | 22.53 | 16.53 | 0.45 | 8.21 | 26.67 | 28.42 | 1.75 |
| 1970 | 22.83 | 16.65 | 0.43 | 8.48 | 26.68 | 28.55 | 1.87 |
| 1971 | 20.51 | 17.04 | 0.34 | 8.20 | 24.72 | 29.12 | 4.40 |
| 1972 | 18.35 | 15.66 | 0.32 | 8.81 | 23.00 | 28.80 | 5.80 |
| 1973 | 19.01 | 19.11 | 0.11 | 7.50 | 25.07 | 32.55 | 7.48 |
| 1974 | 22.12 | 21.87 | 0.49 | 8.81 | 30.09 | 38.17 | 8.08 |

| Year | | | | | | |
|---|---|---|---|---|---|---|
| 1975 | 15.87 | 22.16 | 0.15 | 8.94 | 27.10 | 42.18 | 15.08 |
| 1976 | 18.14 | 16.92 | 0.15 | 13.06 | 27.78 | 39.47 | 11.69 |
| 1977 | 22.93 | 14.78 | 0.40 | 13.26 | 28.85 | 33.56 | 4.71 |
| 1978 | 25.44 | 19.75 | 0.35 | 12.54 | 31.29 | 37.79 | 6.50 |
| 1979 | 24.44 | 21.10 | 0.25 | 10.51 | 31.84 | 38.35 | 6.51 |
| 1980 | 26.89 | 25.75 | 0.31 | 9.51 | 33.75 | 41.22 | 7.47 |
| 1981 | 24.39 | 29.11 | 0.25 | 9.65 | 32.95 | 46.21 | 13.26 |
| 1982 | 22.53 | 29.90 | 0.49 | 8.56 | 30.51 | 45.63 | 15.12 |
| 1983 | 23.08 | 29.03 | 0.23 | 9.68 | 40.24 | 55.39 | 15.15 |
| 1984 | 22.49 | 27.03 | 0.24 | 7.82 | 33.65 | 45.57 | 11.92 |
| 1985 | 26.96 | 26.90 | 0.25 | 7.06 | 40.92 | 46.94 | 6.02 |
| 1986 | 26.17 | 24.54 | 0.21 | 7.48 | 40.10 | 44.83 | 4.73 |
| 1987 | 23.33 | 23.23 | 0.25 | 6.13 | 34.59 | 40.13 | 5.54 |
| 1988c | — | — | — | — | — | — | 7.39 |

— Not available.

a. Includes carryover from previous fiscal years and contributions.

b. Includes contributions.

c. The data available for 1988 do not include the deficits of provinces and the city of Buenos Aires. For the remainder of the consolidated public sector the deficit is 6.93 percent of GDP. We estimate a deficit for the provinces and Buenos Aires equal to the 1987 deficit, 0.46 percent of GDP.

*Source:* Ministry of Economics, Argentina.

**Table 3.4. Consolidated Nonfinancial Public Sector "Below-the-Line" Data, Argentina, 1961–88**
(percentage of GDP)

| Year | Deficit or financing needs (1) | Total (2) | Net financing[a] | | | | Liabilities | | |
|---|---|---|---|---|---|---|---|---|---|
| | | | Total | Domestic | Credit foreign | Advances from suppliers | Total (1)—(2) | Central Bank | Change |
| 1961 | 4.36 | 2.86 | 2.86 | 2.38 | 0.48 | 0 | −1.50 | 0.19 | 1.32 |
| 1962 | 7.76 | 4.52 | 4.52 | 3.77 | 0.75 | 0 | −3.24 | 0.71 | 2.53 |
| 1963 | 7.00 | 3.15 | 3.15 | 2.26 | 0.89 | 0 | −3.85 | 1.37 | 2.49 |
| 1964 | 6.62 | 1.09 | 1.09 | 0.71 | 0.38 | 0 | −5.53 | 0.49 | 5.04 |
| 1965 | 4.08 | 1.20 | 1.20 | 0.50 | 0.70 | 0 | −2.88 | 0.83 | 2.04 |
| 1966 | 4.93 | 0.31 | 0.31 | 0.01 | 0.30 | 0 | −4.62 | 0.28 | 4.35 |
| 1967 | 2.08 | −0.17 | −0.17 | 0.12 | −0.29 | 0 | −2.25 | 1.00 | 1.26 |
| 1968 | 2.23 | 0.42 | 0.42 | 0.47 | −0.05 | 0 | −1.81 | 0.71 | 1.10 |
| 1969 | 1.75 | 0.94 | 0.94 | 0.57 | 0.37 | 0 | −0.81 | 0.72 | 0.10 |
| 1970 | 1.87 | 0.86 | 0.86 | 0.05 | 0.81 | 0 | −1.01 | 3.38 | −2.36 |
| 1971 | 4.40 | 1.92 | 1.92 | 0.73 | 1.19 | 0 | −2.48 | 2.43 | 0.06 |
| 1972 | 5.80 | 1.93 | 1.93 | 0.89 | 1.04 | 0 | −3.87 | 1.81 | 2.06 |
| 1973 | 7.48 | 0.91 | 0.91 | 0.79 | 0.12 | 0 | −6.57 | 4.47 | 2.10 |

| | | | | | | | | | |
|---|---|---|---|---|---|---|---|---|---|
| 1974 | 8.08 | 1.87 | 1.87 | 1.00 | 0.87 | 0 | −6.21 | 5.60 | 0.62 |
| 1975 | 15.08 | 1.88 | 1.88 | 1.77 | 0.11 | 0 | −13.20 | 9.74 | 3.46 |
| 1976 | 11.69 | 4.14 | 4.14 | 3.03 | 1.11 | 0 | −7.55 | 3.41 | 4.13 |
| 1977 | 4.71 | 2.83 | 2.83 | 1.54 | 1.29 | 0 | −1.88 | 2.44 | −0.56 |
| 1978 | 6.50 | 5.49 | 5.49 | 3.54 | 1.95 | 0 | −1.01 | 0.86 | 0.16 |
| 1979 | 6.51 | 5.53 | 5.53 | 3.77 | 1.76 | 0 | −0.98 | −0.30 | 1.27 |
| 1980 | 7.47 | 2.99 | 3.41 | 1.53 | 1.88 | −0.42 | −4.48 | 3.59 | 0.89 |
| 1981 | 13.26 | 8.03 | 8.27 | 3.97 | 4.30 | −0.24 | −5.23 | 5.32 | −0.09 |
| 1982 | 15.12 | 6.30 | 6.39 | 5.07 | 1.32 | −0.09 | −8.82 | 7.29 | 1.51 |
| 1983 | 15.15 | −1.63 | −1.34 | −1.89 | 0.55 | −0.29 | −16.78 | 16.60 | 0.19 |
| 1984 | 11.92 | −1.49 | −1.48 | −0.59 | −0.89 | −0.01 | −13.41 | 6.19 | 7.22 |
| 1985 | 6.02 | 0.58 | 0.63 | −0.30 | 0.93 | −0.05 | −5.44 | 2.33 | 3.11 |
| 1986 | 4.73 | −0.02 | 0.04 | −1.01 | 1.05 | −0.06 | −4.75 | 0 | 4.74 |
| 1987 | 5.54 | 4.17 | 4.22 | 1.03 | 3.19 | −0.05 | −1.37 | 0 | 1.37 |
| 1988 | 7.39 | — | — | — | — | — | — | — | — |

— Not available.
a. All data are net.
*Source:* Ministry of Economics, Argentina.

and these loans have not been paid back. The CNFPS has also resorted to placing dollar-denominated treasury bills with the Central Bank in exchange for local currency, a procedure that has been labeled as external financing but that is actually equivalent to printing money, since these dollar-denominated treasury bills will very likely never be paid back. The Central Bank also has become the recipient of a large fraction of the service of the foreign debt.

*Quasi-Fiscal Expenditures*

In countries like Argentina central banks often suffer substantial losses—losses that can never be recovered—on loans to the private financial system or from the bailout of failing financial institutions. In 1977 Argentina's Central Bank started to collect interest on the fraction of reserve requirements that corresponded to nonremunerated bank deposits (demand deposits) and to pay interest on the reserves held on account of interest-bearing time deposits. The balance of these operations is kept at the Monetary Regulation Account, which has proved to be a source of additional deficit because the interest paid has exceeded the interest collected. In 1985 the Monetary Regulation Account was modified to incorporate remunerated and nonremunerated reserve requirements. In addition, the Central Bank started to sterilize liquid funds by issuing a variety of short-term liabilities, including short-term certificates of deposit (CDs) and lump-sum mandatory deposits, that absorbed part of the liquidity of the commercial banks.

Table 3.5 presents estimates by FIEL of the quasi-fiscal expenditures of the Central Bank of Argentina for the period 1960–85. It is difficult to determine in advance when the quasi-fiscal deficit will result in additional money creation because much of the interest on the Central Bank's liabilities has been paid by creating *more* such liabilities. This mechanism gave rise to a situation in which the Central Bank gradually absorbed a growing fraction of the lending capability of commercial banks. As of 1989, it is reported, more than 80 percent of the assets of the commercial banks were placed in liabilities of the Central Bank. This way of managing liabilities generated a situation in which the Central Bank, instead of being the "lender of last resort," became the "borrower of first resort." The implications for monetary policy and the eventual development of hyperinflation will be discussed in the section dealing with monetary policy. Table 3.6 summarizes the data on the fiscal deficit.

Another important source of quasi-fiscal deficit was the losses from swaps and other exchange rate insurance procedures. These operations, which were concentrated in the 1982–85 period, led to the Central Bank's absorbing most of the outstanding external debt of the private sector. (The process is described in detail in Rodríguez 1989.)

**Table 3.5. Quasi-Fiscal Expenditures of the Central Bank, Argentina, 1960–85**
(percentage of GDP)

| Year | Net loans[a] | IEA[b] | Swaps[c] | Accumulation of international reserves | Total |
|------|------|------|------|------|------|
| 1960 | −0.32 | | | 2.09 | 1.77 |
| 1961 | 1.82 | | | −2.26 | −0.44 |
| 1962 | 0.34 | | | −3.08 | −2.74 |
| 1963 | −0.85 | | | 0.92 | 0.07 |
| 1964 | −0.52 | | | −0.81 | −1.33 |
| 1965 | 0.11 | | | 0.28 | 0.39 |
| 1966 | −0.39 | | | −0.16 | −0.55 |
| 1967 | −1.24 | | | 2.66 | 1.42 |
| 1968 | −0.03 | | | −0.14 | −0.17 |
| 1969 | 0.94 | | | −1.33 | −0.39 |
| 1970 | −0.54 | | | 0.48 | −0.06 |
| 1971 | 0.66 | | | −1.75 | −1.09 |
| 1972 | −0.51 | | | 0.69 | 0.18 |
| 1973 | −1.83 | | | 0.77 | −1.06 |
| 1974 | −2.79 | | | −1.14 | −3.93 |
| 1975 | −2.27 | | 0.16 | −1.30 | −3.41 |
| 1976 | −6.21 | | 5.00 | 3.46 | 2.25 |
| 1977 | −5.86 | 1.28 | 0.57 | 3.59 | −0.42 |
| 1978 | −2.36 | 2.72 | | 1.98 | 2.34 |
| 1979 | −0.82 | 0.86 | | 2.80 | 2.83 |
| 1980 | 3.17 | −0.61 | | −3.43 | −0.87 |
| 1981 | 2.94 | −1.38 | 3.99 | −5.33 | 0.22 |
| 1982 | 7.57 | 4.98 | 16.63 | −3.74 | 25.45 |
| 1983 | −3.69 | 1.04 | 8.59 | −1.06 | 4.87 |
| 1984 | 3.80 | 4.06 | 10.01 | −0.30 | 17.56 |
| 1985 | −1.78 | 1.56 | 4.63 | 2.99 | 7.00 |

*Note:* Blanks signify not applicable.

a. Annual changes in loans to the financial system minus annual changes in reserve requirements.

b. The Interest Equalization Account (Cuenta de Regulación Monetaria) of 1977 and the cost of remunerated reserve requirements and other remunerated liabilities introduced in its place in 1985.

c. Losses for swaps and other "exchange insurance" mechanisms.

*Source:* FIEL 1987.

The FIEL figures for this period (see the third column from the right in table 3.5) overestimate the quasi-fiscal expenditure impact of these policies because the exchange losses they implied were not presently realized but were simply documented as public external debt. In addition, these figures include interest accrued, although not actually paid, on that debt. To the extent that the external debt was not fully

**Table 3.6. Summary Results for Fiscal Deficit, Argentina, 1961–88**
(percentage of GDP)

| Year | Interest, CPS | Primary deficit of CPS | Total CPS deficit | Quasi-fiscal CPS deficit | Total deficit |
|------|------|------|------|------|------|
| 1961 | 0.67 | 3.69 | 4.36 | −0.44 | 3.92 |
| 1962 | 0.83 | 6.93 | 7.76 | −2.74 | 5.02 |
| 1963 | 0.88 | 6.12 | 7.00 | 0.07 | 7.07 |
| 1964 | 1.11 | 5.51 | 6.62 | −1.33 | 5.29 |
| 1965 | 0.75 | 3.33 | 4.08 | 0.39 | 4.47 |
| 1966 | 0.75 | 4.18 | 4.93 | −0.55 | 4.38 |
| 1967 | 0.59 | 1.49 | 2.08 | 1.42 | 3.50 |
| 1968 | 0.53 | 1.70 | 2.23 | −0.17 | 2.06 |
| 1969 | 0.39 | 1.36 | 1.75 | −0.39 | 1.36 |
| 1970 | 0.41 | 1.46 | 1.87 | −0.06 | 1.81 |
| 1971 | 0.48 | 3.92 | 4.40 | −1.09 | 3.31 |
| 1972 | 0.57 | 5.23 | 5.80 | 0.18 | 5.98 |
| 1973 | 0.53 | 6.95 | 7.48 | −1.06 | 6.42 |
| 1974 | 0.81 | 7.27 | 8.08 | −3.93 | 4.15 |
| 1975 | 0.69 | 14.39 | 15.08 | −3.41 | 11.67 |
| 1976 | 1.47 | 10.22 | 11.69 | 2.25 | 13.94 |
| 1977 | 1.14 | 3.57 | 4.71 | −0.42 | 4.29 |
| 1978 | 1.92 | 4.58 | 6.50 | 2.34 | 8.84 |
| 1979 | 2.03 | 4.48 | 6.51 | 2.83 | 9.34 |
| 1980 | 1.85 | 5.62 | 7.47 | −0.87 | 6.60 |
| 1981 | 3.88 | 9.38 | 13.26 | 0.22 | 13.48 |
| 1982 | 6.28 | 8.84 | 15.12 | 25.45 | 40.57 |
| 1983 | 3.50 | 11.65 | 15.15 | 4.87 | 20.02 |
| 1984 | 2.93 | 8.99 | 11.92 | 17.56 | 29.48 |
| 1985 | 2.93 | 3.09 | 6.02 | 7.40 | 13.42 |
| 1986 | 2.37 | 2.36 | 4.73 | n.a. | n.a. |
| 1987 | 1.88 | 3.66 | 5.54 | n.a. | n.a. |
| 1988 | 2.09 | 5.30 | 7.39 | n.a. | n.a. |

n.a. Not applicable.

*Source:* Tables 3.3 and 3.4.

serviced, this new liability did not result in money creation. Some money creation took place, however, through the implementation of a variety of debt conversion mechanisms, including debt-equity swaps and onlending, which in effect implied the repurchase of external debt with newly printed money or short-term financial liabilities issued in local currency.

## Inflationary Financing of the Deficit

As was mentioned above, we have doubts about the relevance of the official information presented in table 3.4 on the actual financing of the deficit. In particular, that information shows no Central Bank financing of the CNFPS during the years 1986–87. But direct treasury borrowing from the Central Bank was in fact carried out in such a way as not to show openly in the accounts. For example, rediscounts were granted to public enterprises, and treasury paper was denominated in dollars to avoid direct borrowing in australes. Public construction programs were implemented by the Banco Hipotecario (the state mortgage bank) and were fully financed through Central Bank rediscounts that were never returned. Banco Hipotecario is alleged to have lost about $1 billion in 1987–88 because of such operations, and all this was financed through money creation.

*Measuring the Inflation Tax*

Because of these questions, we resorted to measuring directly from the accounts of the monetary sector the fraction of the deficit that was financed by the printing of money. We measured the revenue from money creation as the absolute monthly change in $M_1$ (non-interest-bearing money) divided by the exchange rate for the U.S. dollar in the free market. This provided a monthly series of U.S. dollar revenue from money creation. We then added up the series for each year to obtain annual revenue. Finally, we divided nominal GDP for each year by the average exchange rate in the free market to estimate dollar GDP. Dividing the revenue series by GDP, we obtained revenue from money creation as a percentage of GDP.

This series has a serious problem that makes it not comparable with our series for the CNFPS deficit. Since the Central Bank intervenes in the foreign exchange market by buying or selling foreign exchange, in many instances the changes in $M_1$ are a consequence of increases in money demand that are satisfied through purchases of international reserves. Conversely, when the reserves are lost, we observe a significant fall in the revenue from money creation, as measured. Since reserve purchases or sales are not considered a public expenditure in the accounts of the CNFPS, we have resorted to subtracting those reserve changes from our series of revenue from inflation.

With this correction the series of revenue from money creation obtained from the monetary data shows a clear correlation with the series of the deficit of the CNFPS obtained from the fiscal data. Both series are shown in figure 3.5. Almost systematically, the deficit of the consolidated public sector (the last column in table 3.6) exceeds the revenue from money creation. The difference between the two series can be taken as an approximation of the part of the total deficit that was financed by issuing interest-earning debt. The generation of public debt should have been larger than this amount because the Central Bank had its own quasi-fiscal deficit that also had to be financed. We

**Figure 3.5. Revenue from Money Creation and Total Deficit of the CNFPS, Argentina, 1961–87**

Percentage of GDP

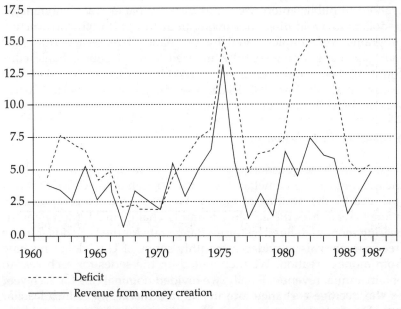

-------- Deficit

———— Revenue from money creation

*Source:* Country case study.

do not believe that the FIEL data in table 3.5 are comparable with our series because part of what FIEL calls quasi-fiscal expenditures may actually be indirect financing of the CNFPS deficit through the mechanisms discussed in the preceding section. The FIEL data also consider the changes in reserves as part of the quasi-fiscal expenditures.

With these caveats, the data in figures 3.5 and 3.6 describe the events quite closely in terms of the actual methods of deficit financing used in recent years. From 1964 through 1975 the deficit of the CNFPS did not exceed significantly the revenue from money creation. This means that the fiscal deficit was mainly financed by creating money rather than by issuing debt. The situation changed drastically in 1976, when debt financing apparently became a significant means of financing the deficit. This change coincided with the fall of the Peronist government and the initiation of the military regime.

In 1977 there was a significant financial liberalization that opened the domestic financial market to foreign investors. The period 1977–79 was marked by foreign borrowing, and revenue from money creation fell well below historical levels. The banking crisis of early 1980 put an end to this stage of foreign financing of the deficit and cleared

**Figure 3.6. Estimated Use of Debt Financing, Argentina, 1961–87**

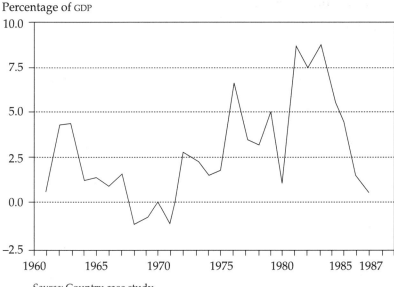

Percentage of GDP

*Source:* Country case study.

the way for the next stage of internal debt financing, which lasted until 1985. Beginning in 1985, a serious effort was made to reduce the total deficit of the CNFPS. The use of the inflation tax, however, did not fall proportionately to the deficit of the CNFPS because of the increasing financial pressures of the service of the internal debt, which was held mostly in the Central Bank.

*The Fiscal Deficit and the Inflation Tax: Statistical Evidence*

In the preceding section we saw that the revenue from money creation appeared to be closely associated with the deficit of the consolidated public sector. This should not surprise us because printing money is one of the only two ways of financing the deficit of the CNFPS—the other being issuance of new interest-earning debt. In this section we identify more closely the nature of this relationship, using regression analysis.

Our methodology consists of considering the revenue from money creation as an endogenous variable that is explained, among other things, by the need of the CNFPS to finance its deficit. We therefore consider the deficit of the CNFPS as the exogenous variable that gives rise to the need to resort to an inflation tax. Because of the possibility of financing the deficit with internal or external debt, we should not expect a stable one-to-one relation between the inflation tax and the deficit. In fact, as the previous discussion indicates, debt financing was more widely used in some periods than in others. Nevertheless, one should consider that debt generation in excess of real growth will eventually have to be canceled or be financed through inflation, as the market for placing new debt eventually dries up.

Since it was also obvious from the inspection of the data that, starting in 1976, more debt financing was used in the financing modality, we have included a dummy (*D76* equals 0 through 1975 and equals 1 from 1976 on). Regression 3.1 reports the results concerning the relation between the inflation tax (ITAX) and the total deficit of the consolidated nonfinancial public sector (DEFT).

(3.1)            ITAX $= 1.248 - 2.138D76 + 0.564$DEFT.
                 $(1.82)\ (-2.85)\qquad (6.19)$

Adjusted $R^2 = 0.6$
Durbin-Watson statistic $= 2.31$
Sample period $= 1963$–$87$

The regression was also run with correction for first-order autocorrelation of residuals, *AR*(1), but the *AR*(1) coefficient did not turn out to be significant. The coefficient of the DEFT variable is 0.56, meaning that 56 percent of the CNFPS fiscal deficit is financed through money creation. The coefficient of DEFT on ITAX is highly significant and justifies the presumption that deficit financing is a significant factor explaining the need for the inflation tax. The negative coefficient on the *D76* variable indicates that since 1976 there has been a tendency to use less inflationary financing for any amount of DEFT and that as a consequence there has been a shift toward more debt financing.

We also tried to use the primary deficit (*PD*) of the CNFPS, which includes no interest service, as the explanatory variable for ITAX and found the surprising result that it works much better than the total deficit (DEFT). The results are reported in regression 3.2.

(3.2)            ITAX $= 0.936 - 1.21D76 + 0.70PD$.
                 $(1.5)\ \ (-1.98)\qquad (7.47)$

$MA(1) = -1.12\ (t\text{-value} = -2.4)$
Adjusted $R^2 = 0.69$
Durbin-Watson statistic $= 1.59$
Sample period $= 1963$–$87$

The results of regressions 3.1 and 3.2 suggest the possibility of a tendency to finance the primary deficit with money and to roll over the interest expenditures in the form of new debt. To check for that possibility, we included the interest expenditure of the CNFPS as an additional explanatory variable, together with the primary deficit. If all deficits, independent of their sources, received the same treatment, we should expect the coefficient on interest expenditure (*IE*) to be the same as the coefficient on the primary deficit. This does not turn out to be the case, as the results of regression 3.2 show. The primary deficit retains the same coefficient of 0.7, while the interest expenditure turns out to be insignificant in explaining the inflation tax. We conclude that over the sample period authorities have used the inflation tax to finance primary expenditures, while interest expenditures that result from the existing stock of government debt have tended to be refinanced by issuing more debt.

(3.3)      ITAX $= 0.929 - 1.26\ D76 + 0.697PD + 0.027IE.$
           (1.4)  $(-1.4)$      (6.78)       (0.08)

$$MA(1) = -0.62\ (t\text{-value} = -2.36)$$
Adjusted $R^2 = 0.68$
Durbin-Watson statistic $= 1.59$
Sample period $= 1963\text{--}87$

To determine the effects of the public sector deficit on inflation, the natural next step is to ascertain the relation between the inflation rate and the revenue from money creation. Such linkage is provided by the demand for real cash balances, which has the inflation rate, taken as a measure of the opportunity cost of holding money, as a determining variable. Precise estimates of money demand for Argentina will be derived in the next section, but we can already obtain some preliminary estimates by running a regression of the inflation rate on the series of revenue from money creation.

The response of the inflation rate to the printing of money in order to finance the deficit need not be instantaneous, since there may be lags in the adjustment of prices to changes in the money supply. We therefore include inflation that lags one year as an explanatory variable for current inflation in addition to current revenue from money creation. The results are presented in regression 3.4, which indicates a clear association between the revenue from money creation and the resulting inflation rate. Here we have assumed a linear relation between ITAX and the inflation rate, a fact that may not be valid for high inflation rates because as inflation raises the base of the tax, real cash balances decrease, and actual revenue may in fact fall. In the next section we derive the precise nonlinear relationship using an estimate of the demand for money. Our results here are therefore an approximation valid for inflation rates to the left of the maximum of the revenue curve.

The results of regression 3.4 indicate that the long-run effect of raising 1 percentage point of GDP from money creation is associated with an additional 97 percent inflation rate. Regressions 3.2 and 3.4 provide a structural framework for the relation between the public sector deficit and inflation (INF). An additional 1 percentage point of primary deficit is financed with 0.7 percent of revenue from money creation (the rest is financed with debt), and the effect of collecting this revenue from money creation is about 67.9 percent of additional inflation (97 × 0.7).

$$(3.4) \qquad \text{INF} = -84.98 + 0.751 \text{ INF } (-1) + 29.34 \text{ ITAX.}$$
$$(-1.76) \quad (5.89) \qquad\qquad (3.60)$$

Adjusted $R^2$ = 0.63
Durbin-Watson statistic = 1.898
Sample period = 1964-87

The implied long-run relation is:

$$\text{INF} = -340 + 97.2 PD.$$

## Inflation, Revenue from Money Creation, and the Structure of Financial Markets

As might be expected in an economy subject to frequent shocks, Argentina's financial markets are highly volatile and cannot be easily described by a simple set of instruments. The financial reform of 1977 freed interest rates and allowed banks to offer interest-bearing deposits (*plazos fijos*).[1] Before the 1977 reform, interest rates were set by the Central Bank, and credit was normally rationed, as real interest rates tended to be negative.

The financial wealth of Argentines can tentatively be divided into five main groups of assets:

- Currency plus demand deposits (aggregate $M_1$)
- Time deposits (*plazos fijos*) denominated in local currency
- Dollar-denominated bonds of the government (BONEX)
- Foreign financial assets, mostly denominated in U.S. currency
- Government paper denominated in local currency, including, at times, treasury bills and Central Bank CDs.

This list roughly describes the alternatives open to the public since the reform of 1977. Unfortunately, we do not have reliable data describing asset holdings prior to that reform. Loosely speaking, however, we can say that the Argentine economy has experienced a sustained process of demonetization and dollarization in recent decades. The real amount of $M_1$ (non-interest-bearing money) has systematically decreased since 1970, as shown in figure 3.7. (The apparent real increase in $M_1$ during 1973–75 was actually a result of a price freeze that in mid-1975 brought about the inflationary explosion known as *Rodrigazo*.) The time path of real $M_2$ ($M_1$ plus time deposits) has been quite dependent on the evolution of interest rates paid on time deposits (figure 3.8).

**Figure 3.7. Real Value of $M_1$, Argentina, 1960–90**

Index (1960 = 100)

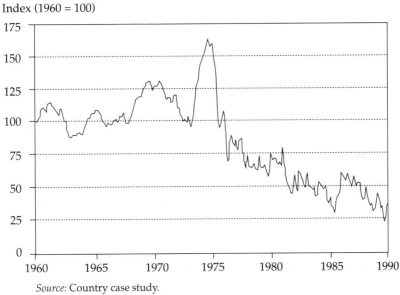

*Source:* Country case study.

**Figure 3.8. Real Value of $M_2$, Argentina, 1960–90**

Index (1960 = 100)

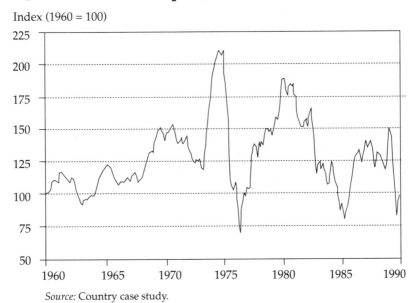

*Source:* Country case study.

The debt policy of the government has much to do with the performance of the financial portfolio of the private sector because the government has gradually become the economy's "borrower of first resort," and as a consequence most of the financial assets of the private sector (except for holdings in foreign exchange) are either directly or indirectly the result of loans to the public sector. Tables 3.7 and 3.8 provide estimates of the domestic interest-earning debt of the treasury and the Central Bank. Practically all of the Central Bank debt is directly held by commercial banks in the form of compulsory reserve requirements (*depositos indisponibles*) or, at times, as voluntary holdings of the Central Bank's CDs. The commercial banks, in turn, obtain their funds by raising interest-earning deposits from the public. What we therefore observe in practice is a system in which most of the public's deposits in commercial banks are lent to the Central Bank and are used to finance the fiscal deficit. Part of the deposits of the public may be lent to the private sector, but that amount has been gradually displaced in favor of lending to the Central Bank (see table 3.10, below). It follows that most of the lending capability generated by the public's demand for $M_2$ is absorbed in the form of domestic liabilities of the Central Bank.

The pressure put on the financial markets by the government debt is best appreciated by evaluating this debt at the commercial exchange rate. Normally authorities try to stabilize the economy by fixing the exchange rate at the level given by the commercial rate. As the credibility of the stabilization plan decreases, interest rates rise, and the stock of government debt tends to rise in terms of dollars. When the stock of debt, particularly the short-term debt of the Central Bank, gets out of line with the available reserves, pressures mount against the currency, and a devaluation eventually follows. Normally devaluations are successful in reducing the dollar value of the government debt denominated in australes but are not as successful in reducing the interest rates in dollar equivalent paid on the remaining stock. As a consequence, immediately after the devaluation the remaining stock of debt continues rising at rates far above the level consistent with a fixed exchange rate, and a new crisis starts to develop.

Table 3.9 shows the evolution of the total domestic liabilities of the Central Bank and the total value of $M_2$ (currency plus demand and time deposits). The ratio between both concepts has oscillated between 59 and 86 percent, depending on how much of the available credit of the commercial banks is absorbed by the Central Bank.

A new stabilization attempt, the Primavera plan, began in August 1988 with a stock of internal debt of $8.6 billion and ended with $18 billion in March 1989, eight months later. Since GDP stood at about $70 billion, it is clear that a rate of debt accumulation on the part of the

**Table 3.7.  Interest-Bearing Internal Debt of the Public Sector, in Australes, Argentina, December 1985–November 1989**

(millions of current australes)

| Year and month | Treasury debt | Central Bank debt | Total |
|---|---|---|---|
| *1985* | | | |
| December | 312 | 3,427 | 3,739 |
| *1986* | | | |
| January | 315 | 3,383 | 3,698 |
| February | 316 | 3,829 | 4,145 |
| March | 307 | 4,293 | 4,600 |
| April | 313 | 4,079 | 4,392 |
| May | 302 | 4,782 | 5,084 |
| June | 300 | 5,039 | 5,339 |
| July | 308 | 5,316 | 5,624 |
| August | 319 | 5,765 | 6,084 |
| September | 322 | 6,250 | 6,572 |
| October | 334 | 5,616 | 5,950 |
| November | 330 | 6,742 | 7,072 |
| December | 1,297 | 7,072 | 8,369 |
| *1987* | | | |
| January | 1,666 | 7,505 | 9,171 |
| February | 1,798 | 8,035 | 9,833 |
| March | 2,409 | 8,399 | 10,808 |
| April | 2,724 | 9,226 | 11,950 |
| May | 2,898 | 9,139 | 12,037 |
| June | 3,311 | 9,796 | 13,107 |
| July | 3,630 | 10,523 | 14,153 |
| August | 5,277 | 11,853 | 17,130 |
| September | 5,743 | 11,859 | 17,602 |
| October | 8,040 | 13,016 | 21,056 |
| November | 8,086 | 14,146 | 22,232 |
| December | 9,117 | 16,755 | 25,872 |
| *1988* | | | |
| January | 11,103 | 17,488 | 28,591 |
| February | 13,152 | 20,160 | 33,312 |
| March | 12,200 | 27,318 | 39,518 |
| April | 13,997 | 31,358 | 45,355 |
| May | 15,282 | 38,226 | 53,508 |
| June | 18,572 | 47,956 | 66,528 |
| July | 22,666 | 60,905 | 83,571 |
| August | 24,503 | 80,441 | 104,944 |
| September | 21,400 | 97,737 | 119,137 |
| October | 22,629 | 107,146 | 129,775 |
| November | 22,635 | 118,341 | 140,975 |
| December | 23,657 | 130,517 | 154,174 |

*(Table continues on the following page.)*

**Table 3.7** *(continued)*

| Year and month | Treasury debt | Central Bank debt | Total |
|---|---|---|---|
| *1989* | | | |
| January | 28,434 | 152,794 | 181,228 |
| February | 45,542 | 170,508 | 216,050 |
| March | 76,457 | 202,606 | 279,063 |
| April | 143,719 | 307,604 | 451,323 |
| May | 207,980 | 707,948 | 915,928 |
| June | 500,913 | 1,558,682 | 2,059,595 |
| July | 877,854 | 2,363,255 | 3,241,109 |
| August | 912,544 | 3,216,690 | 4,129,234 |
| September | 989,721 | 3,767,581 | 4,757,302 |
| October | 995,725 | 4,155,349 | 5,151,074 |
| November | 1,432,058 | 3,743,756 | 5,175,814 |

*Source:* World Bank, *World Debt Tables,* various years.

public sector of 13 percent of GDP in only eight months was unsustainable, and a foreign exchange crisis was inevitable. The collapse of the Primavera plan in February set off a series of devaluations that melted down the debt to $5.7 billion by July 1989. The Bunge and Born plan was then launched by the newly elected authorities with a big devaluation and the announcement of a fixed exchange rate. In the next four months domestic debt rose to $7.9 billion (evaluated at the new exchange rate), mostly because of the interest service of the inherited debt. This time the market was aware of the final effects of rapid rates of debt accumulation and did not wait for debt to reach levels similar to those of the prior stabilization attempt. A new crisis in December forced the abandonment of the Bunge and Born plan and the conversion, on January 1, 1990, of all time deposits (and the reserve requirements that backed them) into a ten-year BONEX. As a counterpart of the conversion of the liabilities of commercial banks into BONEX, all government debt with the banks was also turned into BONEX, as was most of the government paper in the hands of the public. (The exceptions were cash, demand deposits, time deposits up to about $300, and the existing stock of BONEX from previous years.)

## Demand for Money and the Limits on the Inflation Tax

Households have a pure transactions demand for real cash balances of local currency, which we can expect to be positively related to real income and negatively related to the cost of holding those balances (the cost being the expected rate of inflation). In this section we

**Table 3.8. Interest-Bearing Internal Debt of the Public Sector, in U.S. Dollars, Argentina, December 1985–November 1989**

(millions of U.S. dollars at the commercial exchange rate)

| Year and month | Treasury debt | Central Bank debt | Total |
|---|---|---|---|
| *1985* | | | |
| December | 389 | 4,278 | 4,668 |
| *1986* | | | |
| January | 393 | 4,223 | 4,616 |
| February | 395 | 4,780 | 5,175 |
| March | 383 | 5,360 | 5,742 |
| April | 378 | 4,927 | 5,306 |
| May | 356 | 5,629 | 5,985 |
| June | 343 | 5,768 | 6,111 |
| July | 340 | 5,882 | 6,223 |
| August | 331 | 5,974 | 6,304 |
| September | 307 | 5,950 | 6,257 |
| October | 305 | 5,135 | 5,441 |
| November | 287 | 5,858 | 6,145 |
| December | 1,069 | 5,831 | 6,900 |
| *1987* | | | |
| January | 1,289 | 5,805 | 7,094 |
| February | 1,300 | 5,809 | 7,109 |
| March | 1,563 | 5,450 | 7,014 |
| April | 1,768 | 5,987 | 7,755 |
| May | 1,822 | 5,745 | 7,566 |
| June | 1,941 | 5,741 | 7,682 |
| July | 1,916 | 5,555 | 7,472 |
| August | 2,495 | 5,604 | 8,099 |
| September | 2,337 | 4,826 | 7,163 |
| October | 2,479 | 4,013 | 6,492 |
| November | 2,304 | 4,030 | 6,334 |
| December | 2,579 | 4,740 | 7,319 |
| *1988* | | | |
| January | 2,872 | 4,524 | 7,396 |
| February | 3,034 | 4,651 | 7,686 |
| March | 2,510 | 5,619 | 8,129 |
| April | 2,437 | 5,459 | 7,895 |
| May | 2,269 | 5,674 | 7,943 |
| June | 2,301 | 5,941 | 8,242 |
| July | 2,348 | 6,309 | 8,657 |
| August | 2,042 | 6,703 | 8,745 |
| September | 1,783 | 8,145 | 9,928 |
| October | 1,851 | 8,766 | 10,617 |
| November | 1,786 | 9,337 | 11,123 |
| December | 1,802 | 9,943 | 11,745 |

*(Table continues on the following page.)*

**Table 3.8** *(continued)*

| Year and month | Treasury debt | Central Bank debt | Total |
|---|---|---|---|
| *1989* | | | |
| January | 2,081 | 11,181 | 13,262 |
| February | 3,143 | 11,766 | 14,909 |
| March | 4,999 | 13,248 | 18,247 |
| April | 2,811 | 6,017 | 8,829 |
| May | 1,671 | 5,686 | 7,357 |
| June | 2,382 | 7,411 | 9,792 |
| July | 1,546 | 4,163 | 5,710 |
| August | 1,393 | 4,911 | 6,304 |
| September | 1,511 | 5,752 | 7,263 |
| October | 1,520 | 6,344 | 7,864 |
| November | 2,186 | 5,716 | 7,902 |

*Source:* World Bank, *World Debt Tables,* various years.

assume that the expected rate of inflation equals the actual rate in the current period. Knowledge of the demand for real cash balances is critically important for estimating the effect of deficit-induced money creation on inflation, as well as for estimating the limits of money creation as a means of financing.

Given the level and volatility of the Argentine rates of inflation, as well as the unending series of radical changes in monetary policy, one would not expect to find stable demand-for-money parameters over extended periods of time. Rather, we should expect to see the revenue-maximizing rate of inflation, and the maximum revenue itself, changing over time. For this reason, we have chosen to estimate the demand for real cash balances through two different approaches.

1. The first approach makes use of monthly data for the period January 1984–June 1988, a period that covers the time from the return to democracy to the onset of the Primavera stabilization plan, ending with the first hyperinflationary episode of 1989. The most important monetary event of this period was the Austral stabilization plan, launched in June 1985.
2. The second approach uses long-term series of annual data (1960–88) and incorporates a time-dependent dummy variable to capture the possibility of structural change. We found that an additive dummy variable ($D77$) which takes the value of 1 for the period from 1977 on and 0 otherwise best captures the structural change that took place as a consequence of the financial liberalization of 1977.

**Table 3.9.  Domestic Liabilities of the Central Bank and Money Stock, Argentina, December 1985–March 1989**

(millions of current australes; end-of-month data)

| Year and month | Central Bank domestic liabilities (1) | Total stock of domestic money ($M_2$) (2) | Ratio, (1)/(2) |
|---|---|---|---|
| *1985* | | | |
| December | 8,581 | 10,033 | 0.86 |
| *1986* | | | |
| January | 8,365 | 10,538 | 0.79 |
| February | 8,366 | 10,971 | 0.76 |
| March | 9,443 | 11,517 | 0.82 |
| April | 9,319 | 12,079 | 0.77 |
| May | 10,214 | 13,044 | 0.78 |
| June | 10,863 | 13,840 | 0.78 |
| July | 11,447 | 14,675 | 0.78 |
| August | 12,351 | 15,336 | 0.81 |
| September | 12,532 | 15,876 | 0.79 |
| October | 11,615 | 17,288 | 0.67 |
| November | 13,184 | 18,176 | 0.73 |
| December | 14,015 | 20,120 | 0.70 |
| *1987* | | | |
| January | 14,867 | 21,126 | 0.70 |
| February | 14,978 | 21,964 | 0.68 |
| March | 15,593 | 23,011 | 0.68 |
| April | 16,619 | 24,097 | 0.69 |
| May | 17,286 | 25,765 | 0.67 |
| June | 18,082 | 27,679 | 0.65 |
| July | 19,717 | 29,898 | 0.66 |
| August | 21,095 | 31,530 | 0.67 |
| September | 21,568 | 34,314 | 0.63 |
| October | 24,881 | 38,811 | 0.64 |
| November | 26,139 | 42,602 | 0.61 |
| December | 30,274 | 48,075 | 0.63 |
| *1988* | | | |
| January | 31,864 | 52,343 | 0.61 |
| February | 33,752 | 56,984 | 0.59 |
| March | 42,924 | 65,494 | 0.66 |
| April | 48,575 | 74,191 | 0.65 |
| May | 58,660 | 85,827 | 0.68 |
| June | 71,448 | 102,730 | 0.70 |
| July | 92,428 | 120,617 | 0.77 |
| August | 117,793 | 150,250 | 0.78 |
| September | 143,378 | 175,167 | 0.82 |
| October | 155,276 | 193,011 | 0.80 |

*(Table continues on the following page.)*

**Table 3.9** (continued)

| Year and month | Central Bank domestic liabilities (1) | Total stock of domestic money ($M_2$) (2) | Ratio, (1)/(2) |
|---|---|---|---|
| November | 174,917 | 215,552 | 0.81 |
| December | 200,514 | 259,270 | 0.77 |
| *1989* | | | |
| January | 221,780 | 287,405 | 0.77 |
| February | 243,199 | 315,141 | 0.77 |
| March | 277,949 | 368,864 | 0.75 |

Source: For $M_2$ data, FIEL 1987; for remunerated Central Bank liabilities, World Bank data; for nonremunerated monetary base, IMF, *International Financial Statistics*, various years.

The estimation on the basis of monthly data precludes the use of income series to estimate a velocity function. As a consequence, we estimate a demand for real cash balances with the inflation rate as the only explanatory variable in our regressions. To avoid the simultaneous determination problem (between current-period real cash balances and current-period inflation), we have used a two-stage least squares (TSLS) estimation procedure with inflation lagged up to three periods. TSLS is also used to estimate annual data.

The following equations report the regression results for the annual data. Table 3.10 reports the results for the monthly data.

(3.5)  $$LV = 1.9090 + 4.5484 \, \text{INF} + 0.67087D77$$
$$(49.49) \quad (5.83) \quad\quad (9.89)$$

Instruments: INF($-1$), INF($-2$), $LV(-1)$, D77
$$MA(1) = 0.925 \ (t\text{-value} = 3.98)$$
Adjusted $R_2 = 0.96$
Durbin-Watson statistic $= 1.85$
$F$-statistic $= 206$

where

| | | |
|---|---|---|
| $LV$ | = | $A0 + A1 \cdot \text{INF}$ |
| $LV$ | = | $\ln(\text{GDP}/M_1)$ |
| $V$ | = | annual velocity of circulation of $M_1$ |
| GDP | = | annual nominal GDP |
| $M_1$ | = | average of monthly holdings of nominal $M_1$ |
| INF | = | equivalent monthly inflation rate for the year |
| D77 | = | financial reform dummy (1 for 1977–88, 0 otherwise). |

Looking at figure 3.9, we can appreciate the significant changes that have occurred over time in the demand for real $M_1$ as measured

**Table 3.10. Regressions for Monthly Money Demand**

| Variable | (a) | (b) | (c) |
|---|---|---|---|
| C | 13.213 | 13.185 | 8.202 |
|   | (195) | (346) | (5.4) |
| $LM1(-1)$ | — | — | 0.378 |
|   |   |   | (3.29) |
| INF | −3.01 | −2.85 | −1.763 |
|   | (−6.04) | (−10.2) | (−4.29) |
| $MA(1)$ | — | 0.50 | 0.45 |
|   |   | (3.4) | (2.8) |
| $AR(1)$ | 0.308 | — | — |
|   | (2.56) |   |   |
| Estimated long-run $A1$ coefficient | −3.01 | −2.85 | −2.83 |
| Adjusted $R^2$ | 0.61 | 0.67 | 0.81 |
| Durbin-Watson | 1.31 | 1.92 | 1.89 |
| F-statistic | 42 | 56 | 75 |

*Note:* Method of estimation, two-stage least squares; instruments, INF($-1$), INF($-2$), INF($-3$), and (only in regression c) $LM1(-1)$; sample period, February 1984–June 1988. Regression (a): $LM1 = C + A0\ LM1(-1) + A1$ INF, where $LM1 = \ln(M1/\text{CPI})$. Regression (b): INF $= $ CPI/CPI($-1$) $-1$.

by its velocity of circulation. During 1960–74 velocity remained approximately stable in the range of 6–7. In 1975 velocity began an upward swing that showed no signs of stopping and that took it to a value of 33 in 1988. Unofficial data estimates put velocity around 50 in the second half of 1989, after the hyperinflation of June and July of that year. The rise in velocity in 1975–76 may have been caused by high price instability during those two years; we have reason to believe, however, that a structural change took place in 1977, when interest rates were totally freed for the first time in decades and the public was able to invest in short-term time deposits at market-determined interest rates. This structural change was bound to reduce the real demand for $M_1$, as alternative assets were now available. Our empirical results confirm this assumption; the value for the structural change dummy is 0.67, meaning that there was a 95 percent [exp(0.67) − 1] upward shift in velocity as a result of the financial liberalization.

The estimates from the monthly data yield long-run estimates of the semilog elasticity (coefficient $A1$ corrected for the effect of lagged $M_1$) that range between −2.83 and −3.01, implying a monthly revenue-maximizing inflation rate between 33 and 35 percent. The regression using annual data yields a comparable estimate for $A1$ of −4.54, which is compatible with a maximum revenue rate of 21.9

**Figure 3.9. Annual Velocity of Circulation of $M_1$, Argentina, 1960–90**

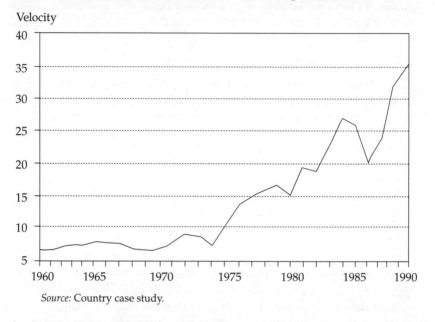

Source: Country case study.

percent per month. All of these estimates of the maximum revenue inflation rate should be corrected by the monthly real growth rate, if any real growth is to be assumed.

Our estimates of the semilog elasticity of demand for $M_1$ fall within the range of other empirical studies. For example, Kiguel and Neumeyer (1989), using monthly data for the period July 1982–March 1985, find values for $A1$ in the range $-2.4$ to $-3.8$. Their estimates of $A1$ for the fixed-exchange-rate period of the *tabla cambiaria* (January 1979–January 1981), when there was a high degree of capital mobility, yields somewhat higher estimates, from $-4.9$ to $-6.0$. In order to estimate the possible range for the revenue from money creation, we have decided to use the results from the regression based on annual data, since this procedure allows for a direct estimate of velocity.

*Computing the Maximum Revenue from Money Creation*

In this section we derive an estimate of the relation between monetary financing of the deficit and the resulting inflation rate. In particular, we derive an estimate of the maximum revenue that can be obtained from the inflation tax before the economy enters into hyperinflation.

Consider a demand for high-power money of the following form:

(3.6) $$M \cdot V(le) = Y = p \cdot Q$$

where $Q$ = GDP, $p$ = price level, $V$ = velocity, and $M$ = stock of high-power money. By issuing high-power money, the government can finance part of its current expenditures, in the amount $(dM/dt)/p$. As a fraction of GDP, the revenue from money creation is:

$$(3.7) \qquad IR = [-(1/V)\,(dV/dt) + (I + g)]/V(Ie).$$

We shall be concerned here with the possibility of sustainable deficit financing through the inflation tax. Leaving aside the transitory effects of the transition from one equilibrium to another, the sustainable steady rate of deficit financing through the inflation tax is derived from (3.7) under the assumption that velocity remains constant at the level determined by the actual inflation rate and that expectations have been adjusted so that actual and expected inflation rates are identical. Under those circumstances, the steady-state sustainable revenue from inflation becomes:

$$(3.8) \qquad IR = (I + g)/V(I).$$

There is a maximum amount of revenue that can be raised with the inflation tax before generating hyperinflation. This amount corresponds to the solution of the maximum value for the equation: maximum inflation tax = $\max(I + g)/V(I)$. Attempts to raise a higher revenue than this maximum will require ever-increasing rates of money creation and inflation.

Assume the following form for the velocity function:

$$(3.9) \qquad \log(V) = V_0 + b \cdot I.$$

In terms of the standard form for velocity shown in (3.9), the steady-state revenue from inflation ($IR$) is:

$$(3.10) \qquad IR = (I + g)/V(I) = (I + g) \cdot \exp(-V_0 - bI).$$

The function $IR$ takes a maximum when

$$(3.11) \quad d(IR)/dI = \exp(-V_0 - bI) - (I + g) \cdot b \cdot \exp(-V_0 - bI) = 0.$$

The solution to this expression yields the maximum revenue inflation tax:

$$(3.12) \qquad \text{IMAX} = (1/b) - g$$

which is the continuous time rate of inflation for the period over which velocity is defined.

The corresponding maximum revenue is derived as

$$(3.13) \qquad \text{IRMAX} = [\exp(-V_0 - 1 + b \cdot g)]/b.$$

It is clear from this analysis that (3.10) is valid provided the expression in parentheses is less than IRMAX. Any deficit in excess of IRMAX cannot be financed through the inflation tax because equation 3.10

will not have a solution—in other words, ever-increasing rates of monetary expansion and inflation will be necessary to finance the deficit, and the system will enter into hyperinflation.

When annual data are used, the estimate of money demand yields the following expression for the sustainable revenue from the inflation tax as a percentage of annual GDP:

$$(3.14) \qquad IR = 12 \cdot (\text{INF} + g) \cdot \exp(-2.5798 - 4.5484 \cdot \text{INF})$$

where INF and $g$ are the corresponding monthly inflation and growth rates and $IR$ is the annual revenue from the inflation tax as a fraction of annual income.

Assuming a real growth rate of 2 percent per year, $g$ takes a value of 0.00165 per month. With this value for $g$, the rate of monthly inflation that maximizes $R/Y$ is equal to:

$$(3.15) \qquad\qquad \text{INF}^* = (1/b) - g = 0.2184.$$

This maximum revenue rate is 21.8 percent per month, or 966 percent per year. The associated maximum revenue from money creation is 7.4 percent of GDP, and velocity at this inflation rate takes the value of 35.5.

To illustrate the workings of the maximum inflation tax, assume a GDP of $70 billion (about the level of Argentina's GDP in 1989). With velocity at about 36, money demand is $1,945 million. At a monthly inflation rate of 21.8 percent, revenue from inflation is $424 million per month (0.218 × 1,945), or $5,088 million per year, which is equivalent to 7.3 percent of annual GDP.

It should be noted that as the inflation rate approaches INF*, the revenue function becomes increasingly elastic with respect to the inflation rate. This means that small changes in revenue require large changes in inflation. As revenue reaches the maximum level, there is no increase in the inflation rate capable of generating any sustainable increase in the rate of revenue. Figure 3.10 shows the relation between revenue and inflation derived from our estimated revenue function, assuming a growth rate of 2 percent per year. The figure shows little gain and much cost from raising inflation above 10 percent per month. At 10 percent per month the revenue is about 6 percent of annual GDP. Raising the maximum amount of extra revenue, 1.3 percent, requires that the inflation rate increase from 10 to 21.8 percent per month. Therefore, at the margin, the extra 1.3 percentage points of GDP in additional revenue requires an increase in the annual inflation rate from 213 to 966 percent. Raising the first 6 percentage points of revenue requires only 213 percent inflation.

In conclusion, our estimate for the money demand equation for Argentina yields an estimated maximum-revenue inflation tax of about 22 percent per month and an estimated maximum revenue of

### Figure 3.10. Annual Revenue from Inflation and the Monthly Inflation Rate, Argentina

Annual inflation revenue (percentage of GDP)

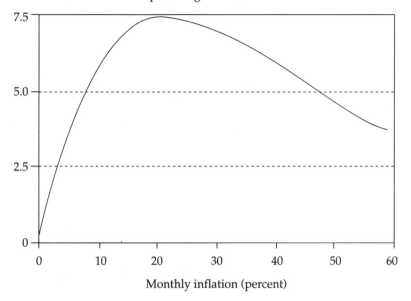

Monthly inflation (percent)

*Source:* Country case study.

7.3 percent of GDP. The revenue estimate should be modified if all of the tax on $M_1$ is not collected by the monetary authorities. As of December 1988 the monetary base was 52 billion australes and $M_1$ was 53 billion australes, suggesting that our estimate of the revenue from inflation is approximately correct, since the $M_1$ multiplier of the monetary base is about unity.

## The Process of Determining Public Debt and the Interest Rate

There are two financial markets in Argentina, the formal market and the informal market. The informal market consists of the *mercado interempresario* (interfirm financial market), in which firms borrow and lend among themselves through a set of institutions called "money-market desks." No record of the volume of these transactions exists, although newspapers do quote the interest rates at which transactions are settled. There is no accepted estimate of the size of this market. Most of the transactions are handled in one to seven days.

The formal financial market is concentrated in the commercial banks and *financieras*. Both are subject to Central Bank regulation, under which deposits are guaranteed and reserve requirements are set. The foremost instrument on the borrowing side of the banks and *financieras* is the interest-earning time deposit. From time to time a variety of other instruments have been offered to the public, including a wide range of indexed deposits (indexed to the dollar, the CPI, or components of the price index). Time deposits are sold to the public and traditionally bear a maturity ranging from seven to thirty days. The average maturity for deposits has rarely exceeded fourteen days and in recent years has been close to seven days. During 1989, following several episodes of long weekends coupled with forced bank holidays, it was not uncommon for 90 percent of the time deposits of the system to come due on the same day.

Banks use their deposits to grant credit to the private sector or to acquire Central Bank assets. In 1989 about 80 percent of all assets of the commercial banks consisted of liabilities of the Central Bank. Regulations on how those Central Bank liabilities are remunerated have changed over time, but on average the liabilities have paid an interest rate equal to the average cost to the commercial banks of raising the funds plus a spread that covers operating costs and allows for a profit. In some instances banks seem to have been interested in having their liquid funds absorbed by the Central Bank through increases in remunerated reserve requirements. In a country like Argentina which is subject to many financial uncertainties, it may actually be safer to lend to the entity that regulates the industry (the Central Bank), particularly since such lending is profitable. (For further information on the dynamics of the financial system that led to the hyperinflation of 1989, see Almansi and Rodríguez 1989 and Fernandez 1989.)

From this description we can conclude that for all practical purposes commercial banks raised deposits in order to lend them to the Central Bank. The result was a grossly distorted financial system in which, as of December 1989, about 2,500 bank branches with 140,000 employees administered $4,700 million worth of deposits, or $33,600 per employee. After the meltdown of January 1990, total bank deposits sank to $1,384 million, but approximately the same number of employees were trying to retain their jobs, so that there was about $9,885 worth of deposits per bank employee! About half the bank deposits and half the employees were in state banks. The obviously needed restructuring of the Argentine financial system has not yet taken place, even though the system is grossly oversized.

In such a system it is very difficult to ascertain how the interest rate on deposits is determined. One might think that the marginal 20 percent of private borrowers would generate a tendency for rates to approximate the productivity of investment, but this is not so. Most private creditors borrow on a daily basis not for investment purposes but to finance temporary financial disequilibria by overdrawing their checking accounts. The interest rate appears to be determined by the short-run liquidity available in the system.

*Short-Run Analysis*

We have found a significant relation between the stock of time deposits outstanding and the real interest rate that these deposits yield. The general rule for banks in determining interest rates is to set the rates at the level required to roll over all deposits (principal plus interest), unless the Central Bank intervenes by providing the required cash so the banks can reduce the outstanding level of deposits.

In terms of the standard money-supply multipliers, the monetary base of the Central Bank (its liabilities) is equal to the sum of reserve requirements:

$$(3.16) \qquad MB = a_1M_1 + a_2D$$

where $a_1$, the average reserve requirement on $M_1$, is very close to unity since the reserve requirement on demand deposits has tended to equal unity. The coefficient $a_2$ of reserve requirements on interest-earning deposits ($D$) is also close to unity as a result of the Central Bank's policy of gradually absorbing most of the lending capability of commercial banks. The demand for $M_1$ is assumed to take the form:

$$(3.17) \qquad M_1 = pL(pie)$$

where *pie* is expected inflation.

The demand for real-interest-earning bank deposits depends on the expected real return to be earned:

$$(3.18) \qquad D/p = d(i - pie).$$

The growth rate of the total monetary base, in turn, equals:

$$(3.19) \qquad d(MB)/dt = (i + s) \cdot a_2 \cdot D + \text{DEF}$$

where $s$ is the spread paid over the cost to the banks of raising the deposits and DEF is the financing needs of the nonfinancial public sector.

Equations 3.16 through 3.18 are not enough to determine the four endogenous variables $M_1$, $D$, $p$, and $i$. The fourth missing equation is the key to determining the interest rate. This equation depends on the structure of the operations of financial markets. If there is still enough of a link with the real sector through bank credit, we can assume that the real interest rate is determined by the marginal productivity of capital. Given inflationary expectations, this condition determines $i$, and equations 3.16–3.18 will determine the remaining variables $M_1$, $D$, and $p$.

If there is perfect mobility of capital, the real interest rate will be determined by the external rate, and the nominal interest rate will therefore also be exogenously determined, given inflationary expectations. In recent years (1982–89), however, the currency in general has not been convertible, and we do not observe any close link between the productivity of capital and the cost of bank credit. For practical purposes, most bank credit goes to the Central Bank, which simply pays the cost of raising the credit. A missing element in this description is the Central Bank's method of determining the level or the cost of its interest-earning debt. The growth rate of this debt depends on the interest rate and the shares of remunerated and nonremunerated liabilities in the total monetary base. We hypothesize that the Central Bank at times has aimed at controlling the ratio of remunerated to nonremunerated liabilities in the monetary base to ensure that its total liabilities grow at the desired rate. This means that at any point in time the total amount of interest-earning deposits in the system is fixed in nominal terms. Since the public can freely shift from $D$ to $M_1$, the constancy of $D$ means that banks must offer whatever interest rate is needed to induce the public to roll over all of its deposits (principal plus interest) at a given point in time. The alternative is to try to keep the nominal interest rate constant by allowing depositors to shift freely between $M_1$ and interest-earning deposits.

Consider the tradeoff faced by the Central Bank when the demand for interest-earning deposits falls. The bank has the option of keeping the interest rate constant by allowing depositors to convert all their excess supply of $D$ into $M_1$. If that is done, however, the price level must jump because the real demand for $M_1$ has not changed. Therefore, if interest-earning deposits fall and the Central Bank feeds the run by substituting a nonremunerated for a remunerated monetary base, the price level increases (or, in other terms, the currency is devalued in the black market). This behavior characterizes the periods of low interest rates and high black-market premiums.

The alternative is to keep $M_1$ and $D$ constant and to allow the nominal interest rate to adjust so that the run is stopped by a higher return on deposits. The higher real interest rate reduces pressures in the parallel foreign exchange market, and the black-market premium initially falls. In this case there is no effect on $M_1$. (We have assumed that $M_1$ depends only on expected inflation; otherwise we would have to make some minor adjustments to the analysis that follows, but the thrust of the analysis would remain the same.) Since the nominal interest rate has increased, the growth rate of the monetary base is now higher (according to equation 3.19). This behavior characterizes periods of high real interest rates and low black-market premiums for the currency. This situation is, however, unsustainable because the monetary base starts growing at rates inconsistent with price stability, and the black-market exchange rate starts to depreciate. Eventually the system breaks down as the Central Bank is forced to devalue the official exchange rate and to allow $M_1$ to increase in order to reduce interest rates and the pressure of the quasi-fiscal deficit.

The inherently unstable system implies wide oscillations in real interest rates and the real exchange rate as depositors try to anticipate sudden devaluations (the preferred instrument for melting down the quasi-fiscal deficits) by exchanging their deposits for cash with which to buy U.S. dollars in the black market. If the government tries to stop the portfolio shift, it must resort to higher interest rates to force the rollover of the deposits. This mechanism reduces the short-run pressures on the currency but increases the rate of growth of the monetary base and therefore generates expectations of even larger devaluations. The run feeds on itself until authorities give in to pressures and devalue.

*Empirical Estimates*

The magnitude of the changes in the interest rate needed to accommodate fluctuations in the demand for deposits is bound to depend on the interest elasticity of such demand. The more inelastic the demand is, the larger should be the increase in the interest rate necessary to induce the rollover of the deposits in the face of an exogenous fall in demand. We have attempted to estimate this elasticity using data for the period following the financial reform of 1977, when a fundamental structural change allowed the free-market determination of interest rates. The result was an unprecedented change in the ratio of $M_2$ to $M_1$ (figure 3.11). The increase in the ratio of interest-earning deposits to $M_1$ began in 1977 and was completed by approximately mid-1980. Since then, this ratio has oscillated in response to the relative returns on both assets.

Regression (a), presented in the first column of table 3.11, shows the estimates of the real demand for interest-earning deposits using the ordinary least squares (OLS) estimation method and monthly data covering the period January 1978 through December 1988. Regression (b), shown in the second column of table 3.11, estimates the same demand for the subperiod January 1984 through December 1988. We assume that the real demand for deposits depends on the nominal interest rate for the current month and on the expected inflation rate. The regressions shown use the current-period inflation rate as the measure of the expected inflation rate. No significant changes were found when actual inflation figures for the next period were used. The high value and significance of the coefficient for the lagged endogenous variable strongly indicates a slow adjustment process. Regres-

**Figure 3.11. Ratio of Interest-Bearing Bank Deposits to $M_1$, Argentina, 1961–90**

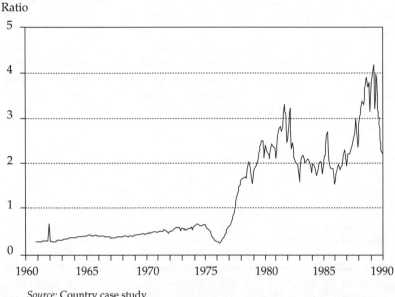

Ratio

*Source:* Country case study.

sion (a) uses the data for longer periods and regression (b) uses the data for shorter periods. Regression (c), shown in the last column, is calculated for the longer period, using instrumental variables, to correct for the problem of simultaneous determination of interest rates and inflation rates.

In all cases the sign of the nominal interest rate and inflation rate is what was theoretically expected: the nominal interest rate is positive and the inflation rate is negative and has an absolute value equal to that of the interest rate. The results confirm the assumption that the demand for real deposits depends on the real interest rate paid on them. The coefficients are significant at the 1 percent confidence level and, when the OLS estimation method is used, do not differ in absolute value between the two periods, indicating that no significant structural change took place in the post-financial-reform years 1978–88.

The estimates in regression (c), measuring the longer period with the use of instrumental variables, remain highly significant, but their absolute values for the interest elasticity are higher than in the OLS estimate. In this case the semilog interest elasticity with respect to the monthly real interest rate is about 1.1 in the short run and 11.0 in the long run. Thus a 10 percent decrease in the demand for deposits would be compensated by an increase in the monthly interest rate of 9 percentage points in the short run and 0.9 percentage points in the

**Table 3.11. Demand Estimates for Interest-Bearing Deposits**

| Variable | (a) | (b) | (c) |
|---|---|---|---|
| C | 0.981 | 0.758 | 0.991 |
| | (4.6) | (2.1) | (4.0) |
| A1 | 0.929 | 0.946 | 0.928 |
| | (60) | (35) | (51) |
| A2 | 0.629 | 0.524 | 1.10 |
| | (6.0) | (3.92) | (3.59) |
| A3 | −0.765 | −0.659 | 1.09 |
| | (−9.6) | (−6.2) | (−4.52) |
| Adjusted $R^2$ | 0.977 | 0.964 | 0.973 |
| Durbin-Watson | 1.72 | 1.87 | 1.83 |
| F-statistic | 1,876 | 530 | 1,601 |
| Number of observations | 132 | 60 | 132 |

*Note:* For regressions (a) and (b): method of estimation, ordinary least square; instruments, $\log(D/P) = C + A1 \cdot \log(D/P)(-1) + A2 \cdot \text{INT} + A3 \cdot \text{INF}$, where $D = M_2 - M_1$ (interest-bearing deposits from FIEL databank), $P$ = consumer price index, and INT = nominal interest on deposits, monthly basis (IMF, *International Financial Statistics*, various years, line 60L); sample period for regression (a), January 1978–December 1988; for regression (b), January 1984–December 1988. For regression (c): method of estimation, two-stage least squares; instruments, $\log(DP(-1))$, INF$(-1)$, INF$(-2)$, INT$(-1)$, INT$(-2)$; sample period, January 1978–December 1988.

long run. Unfortunately, the instrumental variables technique did not yield satisfactory results for the period 1984–88, probably because the instruments used could not capture the sharp fluctuations experienced in both nominal interest rates and inflation rates.

In the system described above, the real interest rate on time deposits of the institutionalized financial system depends on a delicate equilibrium between expectations of meltdowns and the government's need to refinance its debt with the financial system—a debt that is the counterpart of most of the interest-earning deposits at the commercial banks. If we assume that the government controls the rate of devaluation and can always outsmart the market by devaluing by more than was expected, we arrive at the conclusion that the government determines the ex post real rate of interest to a certain extent. The ex post real interest rate in the financial system is basically determined by the only significant creditor of the system: the Central Bank. This determination is not made on a monthly basis, to be sure, but through the meltdowns that systematically take place.

Why do depositors agree to remain in such an uncertain asset situation? One reason is that they are systematically tempted with attractive ex ante real interest rates. Devaluations happen inevitably, but they are usually instituted by a newly appointed economics minister—the previous minister having been just fired because of his now-obvious policy mistakes. Why people have believed in each new minister and have accepted the rollover of the remaining deposits in spite of just having been melted down remains unexplained, but in fact the credibility of new ministers is decreasing over time as the same pattern is repeated over and over again.

Consider the initial nominal interest rates after the last full-fledged stabilization plans, all of which were based on a large devaluation that, it was promised, would be the last ever. The Austral plan (June 1985) devalued by 40 percent; immediately after the devaluation, the interest rate was about 7 percent per month. The Primavera plan (August 1988) devalued by 24 percent; the subsequent interest rate was about 10 percent per month. The first part of the Bunge and Born plan devalued by 200 percent; the initial interest rate was 17 percent per month. The second part of the Bunge and Born plan (December 1989) devalued by 54 percent; the initial interest rate after the devaluation was about 60 percent per month. Finally, the Erman plan (January 1989) made mandatory the conversion of all time deposits into ten-year dollar-denominated government bonds that started trading at about 30 to 40 percent of par value, and the initial interest rate was about 100 percent. In the days prior to the announcement of the Erman plan, monthly interest rates reached levels of 600 percent per month for large depositors because of the expectations of a forthcoming meltdown. After the meltdown, which took place on January 2, 1990, nominal interest rates fell only to 100 percent per month—an indicator of the market's lack of confidence in the success of the new plan.

It is clear that the institutionalized credit market in australes has no significant role in generating new credit. All of the potential credit tends to be absorbed by the government as a result of its need to finance deficits. Since 1985 or earlier, the primary role of the market has been in refinancing the stock of public debt that has its counterpart in the stock of bank deposits. A change may have taken place after March 1990, when the government announced yet another effort at fiscal adjustment and a reform of the Central Bank's charter was undertaken, for the second time in less than a year. It is still too soon to judge the success of these efforts.

An additional source of credit for the government has been the issuance of different types of treasury debt and Central Bank instruments that are indexed either to prices or, most commonly, to the U.S. dollar. One such government paper is the BONEX series of bonds, first issued in 1980. BONEX bonds pay LIBOR rates and have a ten-year maturity. They usually trade below par, in spite of the fact that so far all BONEX bonds have been regularly served. The internal rate of return on BONEX bonds (if held to maturity) is usually taken as a measure of the marginal cost of borrowing in U.S. dollars for Argentina.

BONEX bonds are in high demand by Argentines; they are widely held as collateral for loans in the informal financial market. Since BONEX bonds can be legally traded in secondary markets for either australes or U.S. dollars, they are the mechanism through which firms and individuals can buy or sell U.S. dollars legally. (Holding U.S. dollars is legal; trading them in the free market rate has usually been illegal.) Because of the wide acceptability of BONEX bonds, their internal rate of return (IRR) provides a good measure of the equilibrium U.S. dollar interest rate in Argentina; other U.S. dollar or indexed rates tend to use the BONEX rate as the preferred reference rate. Instruments that are subject to devaluation risk or have dubious possibility of collection pay rates higher than the IRR on the BONEX The BONEX rate puts a floor under other U.S. dollar rates in the informal system; no one in Argentina will lend at less than this rate. The IRR on the BONEX has shown significantly less variability than the ex post realized U.S. dollar rate on austral deposits (figure 3.12).

As of January 1990 the stock of outstanding BONEX (series issued from 1980 to 1987) was stable at about $2,200 million, an amount larger than the U.S. dollar value of $M_1$. The BONEX rate sets the reference rate for an even wider informal market for U.S. dollar–indexed operations. The BONEX rate is not affected by exchange rate expectations (it is already set in U.S. dollars), and it is the last asset that the market expects to be melted down by default. Changes in the BONEX rate can be assumed to be determined by changes in world U.S. dollar rates and in the creditworthiness of the government.

The creditworthiness of the government is not something to be taken for granted. The government stopped serving its external debt with commercial banks in 1987. The debt traded at less than 20 percent of par in 1988 and fell to 12 percent in early 1990. The BONEX trades at parities above 70 percent because it has attained a sort of preferred status at the moment of servicing the debt. Other supposedly preferred instruments have already been subject to forced refinancing and melting down; an example is the LEDOL, a 90-day dollar-denominated bill issued by the Central Bank in August 1989. LEDOL was refinanced compulsorily with a ten-year treasury bond that immediately started trading at 30 percent of par. In January 1990, before the treasury bond was even issued, it was replaced by the new issue of BONEX 1989, which was used to purchase most of the outstanding austral-denominated government debt (and the *plazo fijos*) under the Erman I plan of January 2, 1990.

The best measure of the creditworthiness of the government is its ability to generate a fiscal surplus in relation to its level of indebtedness. As has already been seen, the government has systematically run a primary deficit in recent years. As a result, the government was eventually forced practically to default on its external debt, a process

**Figure 3.12. Ex Post Dollar Rate on Austral Deposits and BONEX Rate, Argentina, 1986–August 1989**

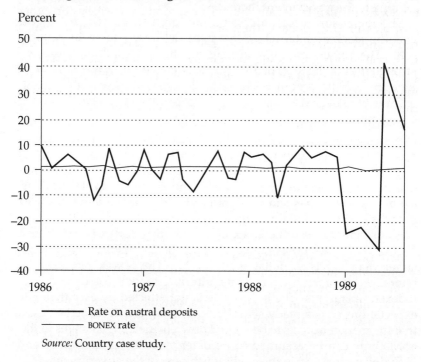

Percent

Rate on austral deposits
BONEX rate

*Source:* Country case study.

that began gradually in 1982. The service of BONEX actually competes with all other instruments of the internal debt. We therefore suggest the existence of a positive tradeoff between the internal rate of return on the BONEX and the stock of internal debt of the government. In addition, we would expect that the BONEX rate is related to the opportunity cost of external funds as measured by a risk-free rate such as prime or LIBOR.

In defining the relative value of the stock of government debt, three deflators come to mind: the price level, the official exchange rate, and the free-market exchange rate. We have found that the official exchange rate is the deflator that yields the best econometric results in the sense that the real stock of debt (measured in U.S. dollars at the official exchange rate) is the one measure best associated with the interest rate on BONEX. One reason may be that much of the stock of government debt was generated by the Central Bank as a result of attempts to maintain constant the nominal value of the official exchange rate. The U.S. dollar value of the stock of debt at the official exchange rate is therefore a measure of the pressures against the sustainability of that rate. The higher is the stock of government debt evaluated at the official exchange rate, the less likely it is that the current set of policies can be maintained, and the risk premium of lending to the government increases.

The regression results assessing the link between the BONEX rate and the stock of internal government debt are shown in equation 3.20. The regression estimates, using the OLS method, indicate a strong effect of the level of government debt on the BONEX rate. In the long run a 10 percent increase in the U.S. dollar value of the government debt results in an increase of 3.2 percentage points in the annual BONEX rate.

$$(3.20) \quad \text{IRRBONEX} = -9.13 + 0.797 \text{IRRBONEX}\,(-1) + 6.53 \log(\text{DGCOM})$$
$$\phantom{(3.20) \quad \text{IRRBONEX} = } (-2.5) \quad (11.7) \phantom{+ 0.797 \text{IRRBONEX}\,(-1) +} (3.37)$$

Adjusted $R_2 = 0.84$
Durbin-Watson statistic $= 1.66$
Durwin $h = 1.23$
Sample period $=$ April 1986–September 1989

where DGCOM is the dollar value of the stock of internal public debt evaluated at the commercial exchange rate and IRRBONEX is the annual equivalent of the internal rate of return on BONEX. Correction for first-order autocorrelation did not yield a statistically significant $AR(1)$. The Durwin $h$ test also shows no significant autocorrelation.

The external interest rate, measured by the monthly U.S. prime rate, was not significant in previous regressions and was not included here. A possible explanation for the lack of significance of the external interest rate might be that since 1982 Argentina has not had access to external credit and therefore a link between domestic and external interest rates should not be expected. The fact is, however, that Argentines do hold an estimated $30 billion or more in foreign assets (as against $2 billion–$3 billion for $M_1$ in local currency). Since Argentines can freely shift the composition of their portfolio between foreign assets and local paper, we should expect some association between the local and external interest rates. We have no doubt that such a relationship exists, but it is difficult to capture statistically for a short period, such as the one analyzed here, especially since the U.S. prime rate did not experience any significant variation in comparison with the sharp oscillations in the BONEX rate that were induced by changes in the market's evaluation of the risk of lending to the government (see figure 3.13)

*Interest Rates and Inflation in the Steady State*

In this section we turn to the long-run steady-state tradeoff between interest rates and government debt policy. In order to focus on the

**Figure 3.13. Stock of Government Debt (Official Dollars) and BONEX Rate, Argentina, 1986–August 1989**

Deviation from sample average

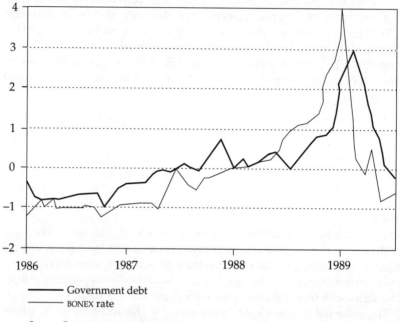

*Source:* Country case study.

essential elements of the process—in particular the fact that interest is being paid on money by printing more money—we will assume that there is a 100 percent reserve requirement on both remunerated ($D$) and nonremunerated ($M_1$) money. Under this assumption, we can write the monetary base, $MB$, as

$$(3.21) \qquad MB = M_1 + D.$$

We also assume the following behavioral relationships between the demand for the two kinds of monies and the inflation rate ($pi$) and nominal interest rate:

$$(3.22) \qquad M_1 = P \cdot L(pi) \qquad L' < 0.$$

$$(3.23) \qquad D = P \cdot F(i - pi) \qquad F' > 0.$$

The change over time of the monetary base is given by:

$$(3.24) \qquad d(MB)/dt = i \cdot D + def$$

where the variable *def* represents the nominal budget deficit and $i \cdot D$ represents the Central Bank's remuneration of interest-bearing deposits $D$.

The real budget deficit, $g$, is given by:

(3.25) $$g = def/P.$$

Finally, we use the Fisher equation to define the real interest rate, $R$, as:

(3.26) $$R = i - pi.$$

In the steady state the following equality must hold:

(3.27) $$[d(MB)/dt] \cdot (1/MB) = pi.$$

Dividing (3.24) by $MB$ and using (3.25), we can write:

(3.28) $$[d(MB)/dt] \cdot (1/MB) = i \cdot D/MB + g \cdot P/MB.$$

From equations 3.21, 3.22, and 3.23 we can express the real monetary base as:

(3.29) $$MB/P = L(pi) + F(R).$$

Hence, using equations 3.23, 3.26, 3.28, and 3.29, we can rewrite the steady-state equilibrium condition, (3.27), as follows:

(3.30) $$L(pi) \cdot pi = F(R) \cdot R + g.$$

In an economy in which the real sector, or the international capital market, determines the real interest rate, the equilibrium condition established by equation 3.30 determines the inflation rate. The equilibrium inflation rate is the one that delivers the inflation-tax revenue required to pay for the real budget deficit, $g$, and the real service of the Central Bank's debt (the real quasi-fiscal deficit), $F(R) \cdot R$.

There are, of course, real interest rates for which equation 3.30 has no solution. This corresponds to the situation in which the amount of resources required by the sum of the two deficits exceeds the maximum stationary inflation tax.

There will normally be many real interest rates for which there are two solutions for equation 3.30. For these cases we assume that the monetary authority chooses the lowest-inflation-rate (or, the efficient) solution. This assumption is important because, as we will see later, it determines the sign of the equilibrium relationship between the real interest rate and the inflation rate. Differentiating equation 3.30, we observe that along the steady-state equilibrium relationship between $r$ and $pi$ we must have:

(3.31) $$d(pi)/d(R) = (F + R \cdot F')/(L + pi \cdot L').$$

As long as the economy stays on the efficient side of the inflation-tax Laffer curve, the right-hand side of equation 3.31 must be positive—

that is, an increase in the real interest rate corresponds to an increase in the equilibrium inflation rate.

It is obvious from our model that in an economy in which the real sector does not determine the real interest rate, the financial sector alone cannot determine both the real interest rate and the inflation rate. The financial sector provides us with one equation, (3.30), but we have two unknowns, $R$ and $pi$. The Argentine economy seems to be precisely such an economy. On the one hand, the supply of loanable funds comes from deposits, both remunerated and nonremunerated, that people hold not as an alternative investment but simply for liquidity reasons. On the other hand, the demand for loanable funds is related merely to the short-term liquidity needs of business firms. Saving and investment in Argentina—the capital market—no longer function in local currency.

As is described in this chapter, the Argentine monetary authorities tried to use the apparent degree of freedom provided by the lack of connection between the financial market and the real sector during the 1982–89 period to manipulate the real interest rate and so control the composition of the Central Bank's liabilities. In particular, the authorities tried to prevent the expansion of the Central Bank's non-remunerated liabilities by raising the real interest rate as much as required by the market to hold remunerated liabilities. This was done in the belief that it is only the expansion of the nonremunerated liabilities that causes prices to rise over time. The result of this monetary policy has been a sort of "unpleasant monetarist arithmetic." By raising the real interest rate the Argentine monetary authorities increased the required inflation-tax revenue, thus making necessary an increase in the equilibrium inflation rate.

By choosing the composition of its liabilities between remunerated and nonremunerated debt, the Central Bank chooses a point in the tradeoff given by equation 3.31. This can be seen by dividing equation 3.23 by 3.22 and denoting by $\sigma$ the ratio of remunerated to non-remunerated Central Bank debt. The relationship between $\sigma$, $pi$, and $R$ is given by:

$$(3.32) \qquad \sigma = F(R)/L(pi).$$

Assuming that the equilibrium is at the efficient side of the Laffer revenue curve, equation 3.30 describes an upward-sloping relationship between $R$ and $pi$. For a given $\sigma$, equation 3.31 describes a downward-sloping relation between $R$ and $pi$. The intersection of both schedules determines the unique steady-state values of $R$ and $pi$. A higher $\sigma$ is associated with a rightward shift in the downward-sloping schedule (equation 3.31) and therefore with a higher $R$ and $pi$ in the new steady state.

The nature of the tradeoff between remunerated debt and inflation is now clear. In the short run, increasing σ (reducing liquidity by issuing interest-earning debt) helps reduce pressures on inflation. In the long run, the rate of nominal monetary expansion must be higher in order to finance not only the previous deficit but also the real interest service on the larger remunerated debt. As a consequence, the inflation rate, as well as the real interest rate, must be higher, since the real stock of debt is higher and depositors in the banks have to be induced to hold the extra deposits with which to finance the extra government debt—whether these deposits be called remunerated reserve requirements, treasury bills held by banks, compulsory bank investments, or something else.

## The External Effects of Public Sector Deficits

The external effects of public sector deficits can be analyzed in a two-step process, taking, first, the effects of the fiscal deficit on the level of aggregate spending and therefore on the trade balance deficit, and, second, the effects on the real exchange rate of the changes in aggregate spending, as measured by the trade balance. Additional side effects are caused by portfolio shifts induced by changes in the rate of inflation that result in changes in the desired rate of accumulation of foreign assets and therefore in the trade balance. Finally, the rate of government spending may affect the real exchange rate if the government's propensity to consume nontradable goods is different from that of the private sector.

### *The Theoretical Framework for the Determination of the Real Exchange Rate*

Consider an economy with three broad aggregates of goods: exportable goods, import-competing goods, and nontradable goods, with nominal prices $Px$, $Pm$, and $Ph$, respectively. The concept of the real exchange rate (RER) is intended to be a measure of some aggregate of the nominal prices of tradable goods ($Px$ and $Pm$) in terms of nontradable goods ($Ph$). In general, however, this economy must have two relevant relative prices: $Px/Ph$ and $Pm/Ph$, which we shall term the export real exchange rate (RERX) and the import real exchange rate (RERM).

As relative prices, RERX and RERM are endogenously determined and so cannot be considered policy variables. For a given (equiproportional) change in the equilibrium values of the RERs, however, an accommodation can be made in the nominal exchange rate so that the RERs get to their new equilibrium values without need for variation in the domestic prices of nontradable goods. In addition, a nominal exchange rate policy that indexes this nominal variable to some aggregate price level may force the RER measures to depart from their equilibrium levels for long periods of time. It is important, therefore, to have available a structural model of determination of the equilibrium values of the RER so that nominal exchange rate policy does not force such a departure. (Earlier studies on real exchange rate determination in Argentina include Cavallo and Peña 1984, Diaz-Alejandro 1981, and Rodríguez and Sjaastad 1979.)

The model used here follows the one presented in Rodríguez (1989a). This model assumes that for internal balance to be achieved, there must be an equilibrium relationship between the three nominal prices and the rate of nominal spending. Such a relationship can be interpreted as the condition for equilibrium in the market for nontradable goods. In functional form, the equilibrium can be expressed as:

(3.33)        $Dh(Ph, Pm, Px) \cdot A - Sh(Ph, Pm, Px) \cdot Y = 0$

where $A$ is the nominal rate of absorption, $Y$ is nominal income, $Dh(\cdot)$ is the fraction of total absorption of goods devoted to the purchase of nontradable goods, and $Sh$ is the fraction of the value of the output of nontradable goods in total nominal GDP. In equation 3.33 we have assumed that supply of and demand for nontradable goods are homogeneous of degree one with respect to the levels of nominal output or absorption, respectively.

Since equation 3.33 must be homogeneous of degree zero in all nominal variables, we can deflate by $Ph$ to obtain:

(3.34)      $Dh[Pm/Ph, (Px/Pm) \cdot (Pm/Ph)]$

$$\cdot (1 - ts) = Sh[Pm/Ph, (Pm/Px) \cdot (Px/Pm)]$$

or

(3.35)                    $\text{RERM} = G[(Pm/Px), ts]$

where $ts = (Y - A)/Y$ is the trade balance surplus normalized by GDP. In logarithmic form, equation 3.35 can be expressed as:

(3.36)        $\log(\text{RERM}) = C_0 + w \cdot \log(Pm/Px) + z \cdot ts.$

Since $\log(\text{RERM}) = \log(Pm/Ph)$, we can interpret (3.36) as the equation that determines the equilibrium value of $Ph$ given the exogenous values of $Px$, $Pm$, and $ts$. Both $Px$ and $Pm$ are determined by foreign prices and commercial policy, whereas $ts$ is determined by macroeconomic variables related to the equilibrium rate of foreign savings (to be analyzed later). From this perspective, we expect $w$ to be positive and between 0 and 1 because it is the elasticity of $Ph$ with respect to an increase in the nominal price of exports when $ts$ and $Pm$ are held constant. (For detailed analyses of these relationships, see Dornbusch 1974, Harberger 1988, and Sjaastad 1980.)

Since RERM = RERX · ($Px/Pm$), it follows that equation 3.36 can also be expressed as:

(3.37)        $\log(\text{RERX}) = C_0 + (1 - w) \cdot [\log(Px/Pm)] + z \cdot ts.$

In general, we expect the parameter $z$ to be positive because it represents the effect on $Ph$ of an increase in absorption in relation to income; as some extra spending falls on nontradable goods, $Ph$ rises, and thus RERX and RERM must fall. In terms of (3.34), as $ts$ rises, absorption falls and so does $Ph$; it follows that $z$ must be positive.

It is usual to refer to the real exchange rate as the relative price of some average price of tradable goods. Assume that this average is formed in the following way:

(3.38)        $\log(\text{PTA}) = a \cdot \log(Px) + (1 - a) \cdot \log(Pm).$

The average real exchange rate would then be:

(3.39)   $\log(\text{AVRER}) = a \cdot \log(Px) + (1 - a) \cdot \log(Pm) - \log(Ph).$

Since $\log(Ph) = -C_0 + w \cdot \log(Px) + (1 - w) \cdot \log(Pm) - z \cdot ts$, we can substitute this expression into equation 3.39 to obtain:

(3.40)        $\log(\text{AVRER}) = C_0 + z \cdot ts + (a - w) \cdot \log(Px/Pm).$

It follows from equation 3.40 that if the aggregation parameter, $a$, is chosen to be identical to the structural parameter, $w$, the AVRER will not depend on the terms of trade or commercial policy. In general, depending on the aggregation weights used, an average RER could depend on the terms of trade in any conceivable way.

Assume now that the government also demands nontradable goods according to $Gh \cdot Ag$, where $Gh$ is the share of government expenditure that consists of nontradable goods and $Ag$ is total spending by the government. Market equilibrium is now given by:

(3.41)                $Dh \cdot Ap + Gh \cdot Ag = Sh$

where $Ap$ is private sector absorption.

Defining total absorption as $A = Ap + Ag$, it follows that all of the prior analysis is still valid if $Gh$ is identically equal to $Dh$—for example, if the demand of the government is identical to the demand of the private sector. If $Gh$ is greater than $Dh$, an increase in government spending, for a constant total absorption, implies that the demand for nontradable goods will rise and that therefore the RER must fall (there has been a shift in the composition of absorption toward the sector with the higher propensity to consume nontradable goods). Conversely, if $Gh$ is less than $Dh$, an increase in government spending, for constant total absorption, will mean a higher RER at equilibrium. As a consequence, the relationship between the RER and the rate of government spending is subject to empirical determination.

The final expression to be tested empirically—one that incorporates the possibility of government spending at a rate different from that of the private sector—is therefore:

$$(3.42) \qquad \log(\text{RERM}) = C_0 + w \cdot \log(Pm/Px) + z \cdot ts + \epsilon \cdot Ag$$

where the sign of $\epsilon$ is the same as the sign of the difference between the propensities of the government and the private sector to consume nontradable goods. If the government has a higher demand for non-tradable goods than the private sector, an increase in government spending for a given level of total demand implies a shift in demand toward nontradable goods and, therefore, a fall in the import real exchange rate.

Unfortunately, monthly or quarterly series of government spending are not available, and we were therefore restricted to the use of annual data for the period 1964–87. The results presented below show the estimation of the structural relationship for the import real exchange rate. This variable is constructed as the ratio of the imported component of the wholesale price index (WPI) to the consumer price index (CPI):

$$\text{LRERM} = \log(\text{price of imports in WPI/CPI}).$$

The explanatory variables are:

- TSGDP, the ratio of the trade account balance to GDP. (Since the trade account is denominated in U.S. dollars, the nominal GDP was converted into U.S. dollars using the official exchange rate for commercial transactions from the FIEL data bank.) The expected sign of the effect of this variable on RERM is positive.
- LPXM, the internal terms of trade—in logarithmic form, $\log(Px/Pm)$, equal to the ratio of the agricultural component of the WPI to the imported component of the same price index. This variable incorporates the substitution effects attributable to the external terms of trade and of taxes and subsidies on foreign trade. The expected sign of the effect of this variable on RERM = $Pm/Ph$ is negative (a rise in $Px$ increases $Ph$ by a smaller proportion and thus reduces $Pm/Ph$).
- Government spending, captured by the ratio of nominal government spending to nominal GDP. The expected sign of this variable is negative, under the reasonable assumption that the government has a larger propensity to consume nontradable goods than does the private sector.

(3.43)   LRERM $= 5.16 + 0.07$TSGDP $- 0.49$LPXM $- 0.02$TEGDP.
         (17.3)   (2.5)          $(-2.8)$        $(-2.9)$

Instruments: TSGDP$(-1)$, TSGDP$(-2)$, $PD$, $PD(-1)$,
LPXM, LPXM$(-1)$, TEGDP, TEGDP$(-1)$
$AR(1) = 0.34$ (1.56)
Adjusted $R^2 = 0.53$
Durbin-Watson statistic $= 1.63$
Sample period $= 1964$–$87$

The empirical results show that all three variables have the expected signs and are highly significant in the determination of Argentina's import real exchange rate. The trade surplus coefficient equals 0.07, indicating that a 1 percentage point increase in the ratio of the trade surplus to GDP is associated with a 7 percent increase in the import real exchange rate. The coefficient of the internal terms of trade is approximately equal to 0.5. This means that a 10 percent rise in the price of imports (or of exports) results in a 5 percent rise in the CPI, which is our measure of the nontradable goods price index. Finally, the coefficient of the ratio of government spending to GDP is equal to $-0.02$, meaning that a 1 percent increase in this ratio is associated with a 2 percent decrease in the real exchange rate. This result implies that the government has a higher propensity to spend on nontradable goods than does the private sector.

To correct for simultaneous determination bias, the regression was estimated with two-stage least squares. Instrumental variables were used for the ratio of the trade surplus to GDP (instruments were the current and lagged primary deficits of the public sector, lagged ratios of trade balance to GDP, and current and lagged remaining explanatory variables). All the coefficients are significantly different from 0 at the 2 percent confidence level or less. Figure 3.14 shows the actual relationship between the ratio of the trade balance to GDP and the import real exchange rate. (Both variables are normalized by their means.)

Since the late 1970s Argentina's foreign debt has risen from a negligible level to almost 100 percent of GDP. (In 1990 the debt—including the arrears accumulated since 1988—reached \$66 billion.) Servicing this debt would require a trade surplus of about 10 percent of GDP just for the nominal interest (assuming an interest rate of 10 percent per year and also assuming that the government, which is the main debtor, has a fiscal surplus large enough to purchase the trade surplus required to pay its debt).

According to our regression results, generating a ratio of trade balance surplus to GDP (TSGDP) of 10 percent, as required to fully service the nominal interest on the Argentine foreign debt, would imply a real exchange rate 70 percent higher than if there were no

**Figure 3.14. Real Exchange Rate and Trade Balance, Argentina, 1964–87**

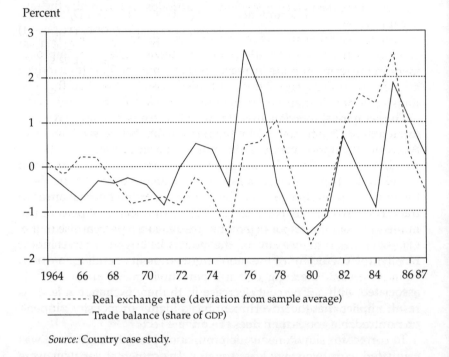

Percent

------- Real exchange rate (deviation from sample average)
———— Trade balance (share of GDP)

*Source:* Country case study.

need to service external debt. Actually, however, the real exchange rate in 1989 was at levels similar to those that prevailed in the early 1970s, when the debt was nonexistent. This is in part a reflection of the fact that Argentina has not serviced its external debt since 1988.

Other factors also have worked toward the maintenance of a relatively low real exchange rate. In particular, the increase in real government spending, according to our results, has been biased toward nontradable goods and has therefore tended to lower the equilibrium real exchange rate. Government spending as a fraction of GDP increased from about 27 percent in the early 1970s to 56 percent in 1983, falling back to 43 percent in 1987. According to the empirical results in equation 3.43, each additional percentage point of GDP in government spending is associated with a 2.1 percentage point fall in the real exchange rate.

### The Trade Balance and the Fiscal Deficit

As explained in chapter 2 in this volume, the trade deficit may or may not be influenced by the fiscal deficit, depending on the validity of the

Ricardian equivalence hypothesis. Other variables that affect the trade deficit are debt service, debt levels, and inflation rates, which may affect the desired composition of portfolios and therefore the rate of capital flow.

Another variable that theoretically belongs in the trade balance equation is the terms of trade (Laursen-Metzler) effect: an improvement in the terms of trade increases real income and may thus induce an increase in the desired stock of foreign assets. However, the relationship is not clear and unambiguous, since a debtor country facing an increase in real income may decide it can support a larger stock of foreign debt. The final word on the relationship between the terms of trade and the trade balance, therefore, will be empirical.

The deficit variable chosen consists of the deficit of the consolidated public sector before any interest service (primary deficit), normalized by GDP. The terms of trade series are from the Economic Commission for Latin America and the Caribbean (ECLAC). No series on the stock of foreign assets held by Argentines are available. We tried two variables as proxies: (a) the balance on the service account of the balance of payments and (b) a measure of foreign assets held, constructed from the accumulated sum of current account surpluses.

None of the variables tested to capture the effects of the level of foreign indebtedness turned out to be significant in explaining the trade balance. One would expect that the effect of the level of foreign debt (or its service) on the trade balance surplus would have a positive sign, indicating that the economy makes some adjustment in order to service its debt. In the regression presented in equation 3.44, however, the debt service variable has the correct sign but is not significantly different from 0, as indicated by the $t$-value of only 0.83. In the regression presented in equation 3.45 we test the series generated for the stock of foreign debt by accumulating the past current account deficits from 1960 onward (arbitrarily setting the initial level of debt equal to 0). This variable (lagged one period in order to represent initial debt during the current period) also has the correct sign, but again the coefficient is not statistically significant. These results simply verify the obvious observation that the Argentine economy has not paid its external debt and therefore has not faced up to the need to adjust the level of the external trade surplus.

It turns out that only two of the variables, inflation and the fiscal deficit, are significant in explaining the trade surplus. In both cases the coefficients of the variables have the theoretically expected signs and are significantly different from 0 at the 2 percent or less confidence level. Higher current inflation improves the trade balance surplus in accordance with what is theoretically expected from the portfolio model discussed earlier. (It is not possible to illustrate the opposite long-run effect because of the insignificance of the coefficient of the foreign assets held.) The primary deficit of the CNFPS is shown to deteriorate the trade balance surplus, as expected in an economy in which the Ricardian equivalence proposition is not fully valid. The coefficient value of 0.32 indicates that one-third of the value of the primary fiscal deficit results in a trade deficit. The trade deficit in turn requires a lower real exchange rate (a real appreciation) for the home goods market to clear. We therefore find a negative relationship between the real exchange rate and the primary deficit of the CNFPS.

The next three regressions use the following variables:

TSGDP  = ratio of trade surplus to GDP
INF    = December-to-December inflation rate of the CPI
*PD*     = ratio of primary deficit in the CNFPS to GDP
FASGDP = ratio of foreign assets to GDP (constructed by accumulating current account surpluses)
SERGDP = ratio of the service account surplus to GDP
*MU*     = rate of expansion of $M_1$ (instrumental variable for inflation).

All regressions were performed using the two-stage least squares method in order to correct for the simultaneous determination of TSGDP and inflation (*PD*, FASGDP, and SERGDP were considered exogenous variables).

(3.44)    TSGDP = 1.63 + 0.14SERGDP + 0.009INF − 0.29*PD*.
             (2.3)    (0.83)          (2.5)       (−2.2)

Instruments: SERGDP, INF(−1), *PD*, *PD*(−1), *MU*, *MU*(−1)
Adjusted $R^2$ = 0.42
Durbin-Watson statistic = 1.60
Sample period = 1963–88

(3.45)  TSGDP = 1.90 − 0.005FASGDP (−1) + 0 .01INF − 0.33*PD*.
            (2.5) (−0.09)                    (3.2)     (−2.2)

Instruments: FASGDP (−1), INF (−1), *PD*, *PD* (−1), *MU*, *MU* (−1)
Adjusted $R^2$ = 0.36
Durbin-Watson statistic = 1.65
Sample period = 1963–88

(3.46)            TSGDP = 1.94 + 0.01 INF − 0.32*PD*.
                    (3.17)  (4.5)     (−2.7)

Instruments: INF (−1), *PD*, *PD* (−1), *MU*, *MU* (−1)
Adjusted $R^2$ = 0.39
Durbin-Watson statistic = 1.65
Sample period = 1963–88

## General Summary and Conclusions

Argentina has had a very troubled economy in recent decades, and the behavior of its public sector may have been instrumental in that process. Government spending grew systematically faster than GDP until the last crisis of the Argentine economy began to develop in 1982. Since then, spending has started to fall, more because of resource constraints than as a result of deliberate political action. The decrease in the relative size of government spending came too late to prevent the financial crisis that brought the country into a state of hyperinflation in 1989.

The government ran a primary deficit (not including any interest payments) every year from 1961 to 1989. As a consequence, it had to resort to issuing money and interest-bearing debt. This in turn led to a systematic tendency of the economy to experience high real interest rates and inflation.

A permanent positive primary deficit, coupled with high real interest rates and a stagnant economy, would be expected to lead to an ever-growing stock of public debt in relation to GDP. Real government debt did not grow continuously, however, because occasionally the existing stock of debt was melted down by outbursts of inflation driven by the large devaluations accompanying a foreign exchange crisis.

Revenues of the CNFPS have historically shown a growth trend similar to that of expenditures, although at systematically lower levels. This high growth in fiscal revenues in the face of a stagnant economy should be enough to invalidate the commonly voiced claim that Argentina's basic problem is that the private sector does not pay taxes. In fact, not only is fiscal pressure very high, but it has also grown at a much faster rate than that of GDP in the past twenty-five years. Even though fiscal pressure has been high, fiscal spending has also grown and has systematically exceeded revenues during the three decades covered by this study.

The significant changes experienced by Argentina after 1990 were a result of a genuine desire on the part of authorities to promote structural adjustment, an attitude partly induced by the debacle caused by hyperinflation. It is clear that after the adjustment that has taken place, it is very unlikely that the economy will revert to its earlier condition. The significant reduction in the size and role of the public sector is not likely to be reversed, either.

## Appendix

This appendix complements the discussion of asset markets by assessing the impact of fiscal policy variables on private saving and investment.

*Private Consumption*

In this section we examine how Argentine private consumption reacts to an increase in the public sector financing needs as a result of either an increase in public expenditure or a decrease in public revenue. A critical issue in this context is whether domestic public debt can or cannot be considered net private wealth in Argentina. Assuming rational economic behavior, the question becomes whether capital market imperfections are strong enough to produce net private wealth effects from changes in public financing strategies (debt versus taxes).

Because of data limitations, what we present here is a simplified version of the framework proposed in the project. We estimate a private consumption function that depends on fiscal expenditures and revenues. In contrast to the usual practice, we do not include the real interest rate as an argument of the consumption function. Between the 1940s and the early 1970s Argentina experienced nearly permanent financial repression, rendering the government-imposed interest rate (charged in the formal financial market for rationed credit) meaningless from the point of view of resource allocation. This situation changed after the financial reform of 1977, which eliminated the former direct regulation of credit. However, as we argue in "The Process of Determining Public Debt and the Interest Rate," above, the interest rate in local currency, which was determined in a financial market working with a horizon of seven days or less, basically reflected the state of short-run liquidity rather than the intertemporal tradeoffs faced by economic agents in Argentina.

*Data*

In the regressions reported below, we used the annual database of the Fundación Mediterranea for the years 1913–84. The original variables taken from that database were:

    PBIPM = real GNP at market prices
CONSUMP = real private consumption
     GGN = nominal national government expenditure
     IGN = nominal national government revenue
  PPBICF = implicit prices of GNP at factor cost.

Using the GNP implicit prices (PPBICF) series as a deflator, we constructed the series for *real* national government expenditure (RGGN) and revenue (RIGN).

*Regressions*

To account for structural change, all the regressions were conducted both for the entire database sample, 1914–84, and for the 1960–84 period. The private consumption functions we estimated are:

(3.47)   CONSUMP($t$) = 50.51 + 0.79PBIPM($t$ − 1)
             (4.21)  (17.80)

             − 61.08RGGN($t$ − 1) − 22.51RIGN($t$ − 1).
             (−2.79)                      (−0.66)

$$R^2 = 0.98$$
Adjusted $R^2 = 0.98$
Durbin-Watson statistic = 1.36
Sample period = 1914–84

(3.48)   CONSUMP($t$) = 176.81 + 0.70PBIPM($t$ − 1)
             (3.11)  (9.88)

             − 46.18RGGN($t$ − 1) − 38.70RIGN($t$ − 1).
             (−1.62)                      (−0.83)

$$R^2 = 0.927$$
Adjusted $R^2 = 0.916$
Durbin-Watson statistic = 1.98
Sample period = 1960–84

In both regressions only public expenditure has a coefficient significantly different from zero. Furthermore, the coefficient is negative, as we would expect either from a simple wealth effect or from combined wealth and substitution effects. Government revenue has no significant effect, consistent with the Ricardian equivalence proposition.

Correcting for first-order autocorrelation does not change our results. Only public expenditure has a significant negative impact on private consumption. Given that both the consumption and the fiscal series are autocorrelated, we also tried the alternative procedure of normalizing the private consumption, public expenditure, and revenue data with GNP. The normalized variables are:

NCON 1913:1 1984:1 = CONSUMP($T$)/PBIPM($T$)
NGGN 1913:1 1984:1 = 100*[GGN($T$)/PPBICF($T$)]/PBIPM($T$)
NIGN 1913:1 1984:1 = 100*[IGN($T$)/PPBICF($T$)]/PBIPM($T$).

In this case we have included the one-period lagged value of normalized private consumption and both past and current values of normalized government expenditure and revenue as explanatory variables. The ordinary least squares regression estimates for both sample periods are:

(3.49)          $\text{NCON}(t) = 0.49 + 0.46\text{NCON}(t - 1)$
                  (5.44)  (4.5)

$+\ 0.03\text{NGGN}(t) - 0.72\text{NGGN}(t - 1)$
(0.12)              (−2.30)

$-\ 0.10\text{NIGN}(t) + 0.08\text{NIGN}(t - 1).$
(−0.24)              (0.21)

$R^2 = 0.58$
Adjusted $R^2 = 0.55$
Durbin-Watson statistic = 1.85
Sample period = 1914–84

(3.50)          $\text{NCON}(t) = 0.36 + 0.53\text{NCON}(t - 1)$
                  (2.89)  (3.82)

$+\ 0.41\text{NGGN}(t) - 0.86\text{NGGN}(t - 1)$
(1.74)              (−3.76)

$-\ 0.25\text{NIGN}(t) + 0.59\text{NIGN}(t - 1).$
(−0.88)              (1.64)

$R^2 = 0.673$
Adjusted $R^2 = 0.587$
Durbin-Watson statistic = 2.078
Sample period = 1960–84

Again, the lagged value of public expenditure has a negative impact on private consumption, while public revenue shows no statistically significant effect. On the basis of this evidence, it would be tempting to declare Argentina a Ricardian economy. That is not possible, however, because in Argentina changes in (conventional) tax revenues have normally been associated with changes in inflation-tax revenue. To illustrate, we have regressed the revenue from money creation, computed as $\text{INFT}(t) = [M_1(t) - M_1(t - 1)]/\text{PPBICF}(t)$, against the government deficit ($\text{DEF} = \text{RGGN} - \text{RIGN}$), obtaining the following results:

(3.51)              $\text{INFT}(t) = 0.12 + 0.78\text{DEF}(t).$
                      (2.11)  (9.55)

$AR(1) = 0.56$
$R^2 = 0.866$
Adjusted $R^2 = 0.864$
Durbin-Watson statistic = 2.07
Sample period = 1914–84

As equation 3.52 shows, a substantial fraction of the government deficit, as conventionally measured, has been financed by money creation. Actually, only in the 1970s did it become common practice to finance the deficit in the capital markets. As equation 3.53 shows, about 90 percent of the deficit seems to have been financed with money creation between 1914 and 1970.

$$(3.52) \qquad \text{INFT}(t) = 0.03 + 0.90\text{DEF}(t).$$
$$(0.79) \quad (7.00)$$

$$AR(1) = 0.53$$
$$R^2 = 0.762$$
Adjusted $R^2 = 0.758$
Durbin-Watson statistic $= 2.08$
Sample period $= 1914\text{–}70$

Clearly, for most of this century the traditional choice in Argentine public finances was between conventional taxes and the inflation tax, not between conventional taxes and debt, as a test of the Ricardian hypothesis would require.

Equation 3.53 shows very neatly the shift to debt financing in the 1970s. As we have discussed elsewhere, the shift can be dated to 1977. Unfortunately, lack of private consumption data with a quarterly frequency prevents us from testing the Ricardian hypothesis for the only period (the past twelve years) for which a test would really make sense.

$$(3.53) \qquad \text{INFT}(t) = 0.49 + 0.62\text{DEF}(t).$$
$$(4.07) \quad (5.55)$$

$$AR(1) = 0.05$$
$$R^2 = 0.707$$
Adjusted $R^2 = 0.682$
Durbin-Watson statistic $= 2.09$
Sample period $= 1970\text{–}84$

Taking into account the importance of inflationary finance in Argentina, we have also estimated equations 3.47 and 3.48 by computing the revenue from money creation as part of the total tax revenues of the Argentine government. In this form we make sure that any change in "revenues," given public expenditure, must mean a change in public debt. In equations 3.54 and 3.55, the variable (TAX) is the sum of conventional public revenues (RIGN) and the revenue from money creation (INFT).

$$(3.54) \quad \text{CONSUMP}(t) = 76.55 + 0.69\text{PBIPM}(t-1)$$
$$(4.35) \quad (15.54)$$

$$- 110.39\text{RGGN}(t-1) + 81.63\text{TAX}(t-1).$$
$$(-3.13) \qquad\qquad (2.26)$$

$$AR(1) = 0.37$$
$$R^2 = 0.986$$
Adjusted $R^2 = 0.986$
Durbin-Watson statistic $= 2.10$
Sample period $= 1915\text{–}84$

(3.55)    $\text{CONSUMP}(t) = 259.42 + 0.54\text{PBIPM}(t-1)$
                      (2.74)  (5.40)

$$- 116.07\text{RGGN}(t-1) + 109.70\text{TAX}(t-1).$$
$$(-2.08) \qquad\qquad (1.61)$$

$$AR(1) = 0.26$$
$$R^2 = 0.922$$
$$\text{Adjusted } R^2 = 0.911$$
$$\text{Durbin-Watson statistic} = 1.998$$
$$\text{Sample period} = 1961\text{--}84$$

We again observe the negative impact of public spending. The revenue variable, TAX, is significantly different from 0, but it is positive, which probably captures the expansionary effect of money creation rather than the contractionary effect that the inflation tax would produce if the Ricardian hypothesis were not valid.

### Private Investment

In this section we expect to find a negative impact of deficit financing on investment as a result of rising domestic interest rates or of stricter credit rationing (in the case of financial repression).

DATA. We use the same database described in the section on consumption behavior. The additional original variables taken from that database were:

INVEST = aggregate investment
INVSTGOB = government investment
PUBCON = public consumption.

From aggregate (INVEST) and government (INVSTGOB) investment we calculated private investment (PINV) and normalized private investment (NPINV) as follows:

PINV = INVEST − INVSTGOB
NPINV = PINV/PBIPM.

REGRESSIONS. Following a procedure similar to that used with consumption, we regressed normalized private investment against the normalized fiscal variables. The results of regressing normalized private investment against *current* normalized public expenditure and revenue, for the 1914–84 sample, are:

(3.56)      $\text{NPINV}(t) = 0.78\text{NPINV}(t - 1) - 0.10\text{NGGN}(t)$
         $(10.49)$           $(-0.73)$

        $+\ 0.49\text{NIGN}(t).$
        $(2.45)$

$R^2 = 0.67$
Adjusted $R^2 = 0.66$
Durbin-Watson statistic $= 1.58$
Sample period $= 1914\text{–}84$

As expected, there is a significant positive impact of public revenue on private investment. Given public expenditure, an increase in public revenue reduces the deficit and this has a positive impact on investment. Since we saw in the consumption module that changes in public revenue do not affect private consumption and, hence, private saving, it is likely that the positive impact on investment comes from a freer or perhaps merely a smoother working of the capital market. The introduction of lagged values of public revenue and expenditure does not change the conclusion in any fundamental way. Equation 3.57 shows the results of regressing normalized private investment against the current values of normalized public expenditure and revenue for the 1960–84 sample period.

(3.57)     $\text{NPTNV}(t) = 0.70\text{NPINV}(t - 1) + 0.02\text{NGGN}(t) + 0.48\text{NIGN}(t).$
            $(3.71)$           $(0.11)$         $(1.45)$

$R^2 = 0.238$
Adjusted $R^2 = 0.169$
Durbin-Watson statistic $= 1.77$
Sample period $= 1960\text{–}84$

We can see that the results are considerably weaker than for the larger sample. The coefficient of revenue still has the correct sign, and it has a substantially larger *t*-statistic value. It is not, however, significantly different from 0 at a 5 percent significance level.

In both equations 3.56 and 3.57 the coefficient of public spending turns out to be insignificant. To see if aggregation made a difference, we regressed private investment against public consumption and government investment. Equations 3.58 and 3.59 show the results. Again, as we observed with normalized variables, public revenue has a significant positive impact on private investment. The difference comes from the expenditure side: public consumption has, for both sample periods, a significant negative impact on private investment. Contrary to what one would expect, if we were to assume that government investment is bound to produce positive externalities on private economic activity (as it definitely did in the nineteenth century with the railroads), government investment has no significant effect on private investment.

(3.58)     PINV($t$) = $-14.87 + 0.24$PBIPM($t - 1$)
                    $(-1.26)$  $(5.28)$

           $- 1.35$PUBCON($t - 1$) $+ 39.07$RIGN($t - 1$)
           $(-2.72)$                  $(1.77)$

           $+ 0.24$INVSTGOB($t - 1$).
           $(0.52)$

$$AR(1) = 0.42$$
$$R^2 = 0.947$$
Adjusted $R^2$ = 0.944
Durbin-Watson statistic = 1.85
Sample period = 1915–84

(3.59)   PINV($t$) = $61.75 + 0.23$PBIPM($t - 1$) $- 1.99$PUBCON($t - 1$)
                  $(1.16)$  $(3.29)$                  $(-3.15)$

         $+ 45.85$RIGN($t - 1$) $+ 0.49$INVSTGOB($t - 1$).
         $(1.37)$                  $(0.68)$

$$AR(1) = 0.20$$
$$R^2 = 0.789$$
Adjusted $R^2$ = 0.745
Durbin-Watson statistic = 1.93
Sample period = 1961–84

## Notes

The author is grateful for very useful comments and suggestions by Aquiles Almansi, who provided invaluable help for the econometric analysis presented in the appendix.

1. The financial liberalization was but one of the many policy instruments of what came to be known as the Stabilization Plan of December 1978. The plan began to be put into effect after the military coup of March 1976; the financial opening of 1977 was followed by the *tabla cambiaria* (prefixed exchange rate path) and the trade reform of 1978. The plan was abandoned after March 1981 in the middle of a set of serious disadjustments, among them currency overvaluation, persistence of inflation, and the external debt problem. The lack of fiscal adjustment has been cited as the main reason for the failure of this stabilization attempt that aimed at making structural adjustment the centerpiece of the policies being followed. Literature covering developments during this period include Calvo (1981b, 1986); de Pablo (1983); Fernandez (1982); Rodríguez (1982a, 1982b, 1983); and Sjaastad (1982). The sequential order of financial and trade liberalization has also been mentioned as a factor contributing to the failure of the plan, an issue that is analyzed in Edwards (1984).

# References

Almansi, Aquiles, and Carlos Rodríguez. 1989. ''Reforma Monetaria y Financiera en Hiperinflación.'' Working Paper 67. Centro de Estudios Macroeconómicos de Argentina (CEMA), Buenos Aires.

Auernheimer, Leonardo. 1987. ''On the Outcome of Inconsistent Programs under Exchange Rate and Monetary Rules, or 'Allowing Markets to Compensate for Government Mistakes.' '' *Journal of Monetary Economics* 19 (March): 279–305.

Calvo, Guillermo A. 1981a. ''Devaluation, Levels versus Rates.'' *Journal of International Economics* 11: 165–73.

———. 1981b. ''Reflecciones Teoricas Sobre el Problema de Estabilización en Argentina.'' Working Paper 29. Centro de Estudios Macroeconómicos de Argentina (CEMA), Buenos Aires.

———. 1985. ''Currency Substitution and the Real Exchange Rate: The Utility Maximization Approach.'' *Journal of International Money and Finance* 4 (June): 175–88.

———. 1986. ''Fractured Liberalism: Argentina under Martinez de Hoz.'' *Economic Development and Cultural Change* 34: 511–33.

Calvo, Guillermo A., and Carlos Rodríguez. 1977. ''A Model of Exchange Rate Determination under Currency Substitution and Rational Expectations.'' *Journal of Political Economy* 3: 617–25.

Cavallo, Domingo, and Angel Peña. 1983. ''Deficit, Endeudamiento del Gobierno y Tasa de Inflación: Argentina 1940–1982, Estudios.'' Fundación Mediterranea, Córdoba, Argentina.

———. 1984. ''Gasto Público y Tipo Real de Cambio.'' Instituto de Estudios Económicos sobre la Realidad Argentina y Latinoamericana (IEERAL), Córdoba, Argentina.

de Pablo, Juan Carlos. 1983. ''El Enfoque Monetario de la Balanza de Pagos en la Argentina: Analisis del Programa del 20 de Diciembre de 1978.'' *El Trimestre Económico* (Mexico) 50 (April–June): 6450–69.

Diaz-Alejandro, Carlos. 1981. ''Tipo de Cambio y Términos de Intercambio en la Republica Argentina.'' Working Paper 22. Centro de Estudios Macroeconómicos de Argentina (CEMA), Buenos Aires.

Dornbusch, Rudiger. 1974. ''Tariffs and Nontraded Goods.'' *Journal of International Economics* 4 (May): 177–83.

Dornbusch, Rudiger, and Juan Carlos de Pablo. 1987. ''Argentina: Debt and Macroeconomic Instability.'' NBER Working Paper Series 2378. National Bureau of Economic Research, Cambridge, Mass.

Edwards, Sebastian. 1984. ''The Order of Liberalization of the External Sector in Developing Countries.'' *Princeton Essays in International Finance* 156. Princeton, N.J.

Fernandez, Roque. 1982. ''La Crisis Financiera Argentina.'' Working Paper 35. Centro de Estudios Macroeconómicos de Argentina (CEMA), Buenos Aires.

———. 1989. ''Hiperinflacion, Repudio y Confiscacion: Los Limites del Financiamiento Inflacionario.'' Working Paper 65. Centro de Estudios Macroeconómicos de Argentina (CEMA), Buenos Aires.

FIEL (Fundación de Investigaciones Económicas Latinoamericanas). 1987. *El Gasto Público en Argentina 1960–1985*. Buenos Aires.

Frenkel, Jacob A., and Assaf Razin. 1986. "Fiscal Policies in the World Economy." *Journal of Political Economy* 94, pt. 1 (June): 564–94.

Frenkel, Jacob A., and Carlos Rodríguez. 1982. "Exchange Rate Dynamics and the Overshooting Hypothesis." *International Monetary Fund Papers* 29 (March): 1–30.

Harberger, A. C. 1988. "Trade Policy and the Real Exchange Rate." World Bank, Economic Development Institute, Washington, D.C.

IMF (International Monetary Fund). Various years. *International Financial Statistics*. Washington, D.C.

Kiguel, Miguel A., and Pablo Andrés Neumeyer. 1989. "Inflation and Seignorage in Argentina." Policy Research Working Paper 289. World Bank, Country Economics Department, Washington, D.C.

Leiderman, Leonardo, and Mario I. Blejer. 1988. "Modeling and Testing Ricardian Equivalence, A Survey." *International Monetary Fund Staff Papers* 35 (March): 1–35.

Rodríguez, Carlos A. 1978. "A Stylized Model of the Devaluation-Inflation Spiral." *International Monetary Fund Staff Papers* 25 (March): 76–89.

———. 1982a. "The Argentine Stabilization Plan of December 20th." *World Development* 10 (September): 801–11.

———. 1982b. "Gasto Público, Deficit y Tipo Real de Cambio: Un Análisis de sus Interrelaciones de Largo Plazo." *Cuadernos de Economía* (Chile) 19 (August): 203–16.

———. 1983. "Politicas de Estabilización en la Economía Argentina." *Cuadernos de Economía* (Chile) 20 (April): 21–42.

———. 1989. "Managing Argentina's External Debt: The Contribution of Debt Swaps." World Bank, Latin America and the Caribbean Regional Office, Washington, D.C.

Rodríguez, Carlos A., and Larry A. Sjaastad. 1979. "El Atraso Cambiario en Argentina: Mito o Realidad." Working Paper 2. Centro de Estudios Macroeconómicos de Argentina (CEMA), Buenos Aires.

Schenone, Osvaldo H. 1987. "El Comportamiento del Sector Público en Argentina; 1970–1985." Working Paper 60. Centro de Estudios Macroeconómicos de Argentina (CEMA), Buenos Aires.

Sjaastad, Larry A. 1980. "Commercial Policy,'True' Tariffs and Relative Prices." In John Black and Brian Hindley, eds., *Current Issues in Commercial Policy and Diplomacy*. New York: St. Martin's Press.

———. 1982. "The Failure of Economic Liberalism in the Southern Cone." 1982 Bateman Lecture. University of Western Australia, Nedlands.

World Bank. Various years. *World Debt Tables*. Washington, D.C.

# 4

# Chile: Fiscal Adjustment and Successful Performance

## Jorge Marshall and Klaus Schmidt-Hebbel

After almost two decades characterized by extreme volatility in economic performance, Chile has been showing remarkable results since the mid-1980s. Gross domestic product (GDP) has grown at an average 6.7 percent since 1986, unemployment fell from 28 percent in the early 1980s to 5 percent in 1992, and inflation has been cut in half, to 13 percent. Gross domestic investment increased from 12 percent of GDP in 1984 to 21 percent in 1992, while the current account deficit shrank from 10.7 percent of GDP in 1984 to 1.5 percent in 1992. External debt has been serviced regularly, and its total amount has been reduced through debt conversion. The reasons behind this success are to be found in the deepening of structural reforms and the adoption of coherent macroeconomic policies, among which fiscal policy played a key role.

This chapter analyzes the macroeconomic determinants and effects of public sector deficits in Chile after 1974. The focus is on key issues related to the interaction between public deficits and macroeconomic variables and the channels through which deficits are transmitted to the economy.

We begin with a brief historical overview of the background against which fiscal policy was implemented.[1] The evolution of selected macroeconomic indicators during the past two decades is summarized in table 4.1. Table 4.2 reports the structure of the consolidated nonfinancial public sector (CNFPS) deficit and its financing during 1974–88.[2] Table 4.3 presents an estimate of quasi-fiscal Central Bank losses during the 1980s.

The first half of the 1970s was dominated by serious macroeconomic imbalances inherited from the Allende government. The last year of the Allende administration, 1973, was characterized by a huge CNFPS deficit resulting from large losses by public enterprises, low fiscal revenue, and overexpanded public sector employment, wages, and social security payments. The deficit, which exceeded 20 percent of GDP, was largely financed by Central Bank monetized loans.

After the 1973 military takeover, Chile followed a strategy of market deregulation, privatization of public enterprises, and opening of the

**Table 4.1. Selected Macroeconomic Indicators, Chile, 1970–92**
(percent)

| Year | CNFPS deficit as share of GDP | Central Bank quasi-fiscal losses as share of GDP | GDP growth | Inflation | Current account deficit as share of GDP |
|---|---|---|---|---|---|
| 1970–73 | 23.4 | — | 0.7 | 204.3 | 2.4 |
| 1974 | 5.5 | — | 1.0 | 369.2 | 0.5 |
| 1975 | 2.1 | — | −12.9 | 343.3 | 5.2 |
| 1976 | −4.0 | — | 3.5 | 197.9 | −1.7 |
| 1977 | −0.4 | — | 9.9 | 84.2 | 3.7 |
| 1978 | −1.4 | — | 8.2 | 37.2 | 5.2 |
| 1979 | −4.6 | — | 8.3 | 38.9 | 5.4 |
| 1980 | −5.4 | — | 7.8 | 31.2 | 7.0 |
| 1981 | −0.4 | — | 5.5 | 9.5 | 14.4 |
| 1982 | 3.9 | 14.0 | −14.1 | 20.7 | 9.2 |
| 1983 | 3.5 | 6.1 | −0.7 | 23.1 | 5.4 |
| 1984 | 4.6 | 7.9 | 6.3 | 23.0 | 10.7 |
| 1985 | 2.9 | 13.1 | 2.4 | 26.4 | 8.3 |
| 1986 | 2.0 | −0.3 | 5.6 | 17.4 | 6.9 |
| 1987 | 0.2 | 0.8 | 5.7 | 21.5 | 4.3 |
| 1988 | −3.6 | −0.1 | 7.4 | 12.7 | 0.7 |
| 1989 | −3.8 | −0.2 | 10.0 | 21.4 | 3.1 |
| 1990 | −0.5 | — | 2.1 | 27.3 | 2.1 |
| 1991 | −1.4 | — | 6.0 | 18.7 | −0.5 |
| 1992 | −0.8 | — | 10.4 | 12.7 | 1.5 |

— Not available.

*Note:* The consolidated nonfinancial public sector (CNFPS) deficit is in nominal terms and comprises general government and nonfinancial public enterprises.

*Source:* For Central Bank quasi-fiscal losses, authors' estimates. For other data, Central Bank of Chile and Budget Office, Ministry of Finance, Chile.

economy to foreign trade and capital flows. Along with these structural reforms, the government applied a strict stabilization program through monetary control, fiscal discipline, and general retrenchment by the public sector. Government spending was cut sharply by reducing real wages, employment, and public investment. Tax rates were raised, and many firms that had been transferred to the public sector by the Allende government were returned to the private sector. The remaining public enterprises were forced to maximize profits and had to face hard budget constraints. The 1975 tax reform introduced value added taxation, significantly simplified the tax system, and indexed tax bases to inflation. As a result, the 1975 CNFPS deficit fell drastically, to 2.1 percent of GDP.

From 1976 to 1981 the public sector showed sizable surpluses. Fiscal deficits having been stabilized, public finances were further favored by a recovery of growth, the new, more broadly based tax system,

**Table 4.2. Nonfinancial Public Sector Revenue, Expenditure, Surplus, and Financing, Chile, 1974–88**

(percentage of GDP)

| Item | 1974 | 1975 | 1976 | 1977 | 1978 | 1979 | 1980 | 1981 | 1982 | 1983 | 1984 | 1985 | 1986 | 1987 | 1988 |
|---|---|---|---|---|---|---|---|---|---|---|---|---|---|---|---|
| *General government* | | | | | | | | | | | | | | | |
| Current revenues | 30.3 | 34.9 | 37.4 | 38.6 | 33.2 | 32.5 | 32.9 | 32.1 | 29.9 | 27.7 | 28.7 | 28.6 | 28.2 | 28.4 | 28.7 |
| Direct taxes | 5.7 | 8.3 | 7.0 | 5.4 | 5.3 | 5.2 | 5.4 | 5.5 | 4.8 | 3.1 | 3.4 | 3.1 | 3.2 | 3.1 | 2.9 |
| Indirect taxes | 13.0 | 14.6 | 14.0 | 14.9 | 13.7 | 13.3 | 13.4 | 14.8 | 13.8 | 14.6 | 16.3 | 17.1 | 17.1 | 17.3 | 14.5 |
| Copper taxes | 1.7 | 1.3 | 2.6 | 2.6 | 2.8 | 1.9 | 1.9 | 0.2 | 1.0 | 1.9 | 1.3 | 0.5 | 0.6 | 1.1 | 4.7 |
| Social security | 3.1 | 3.4 | 3.4 | 3.7 | 3.7 | 5.3 | 5.6 | 4.7 | 3.3 | 2.8 | 2.8 | 2.4 | 2.5 | 2.2 | 1.9 |
| Other income | 6.8 | 7.3 | 10.3 | 12.1 | 7.7 | 6.7 | 6.6 | 6.9 | 7.2 | 5.3 | 5.0 | 5.5 | 4.9 | 4.7 | 4.7 |
| Current expenditures | 26.4 | 27.6 | 31.0 | 33.0 | 26.8 | 24.8 | 24.5 | 26.7 | 31.9 | 30.6 | 30.7 | 29.5 | 27.4 | 26.3 | 23.2 |
| Purchases | 4.3 | 4.4 | 3.7 | 4.9 | 4.7 | 3.0 | 3.1 | 2.9 | 3.3 | 3.2 | 3.4 | 3.2 | 3.0 | 3.3 | 2.7 |
| Wages | 10.0 | 9.3 | 8.9 | 11.0 | 10.0 | 9.2 | 9.1 | 8.8 | 10.3 | 8.9 | 8.5 | 7.8 | 7.4 | 6.8 | 6.2 |
| Domestic interests | 0.4 | 0.5 | 0.2 | 0.8 | 0.9 | 0.7 | 0.5 | 0.2 | 0.0 | 1.3 | 1.8 | 2.4 | 1.2 | 1.7 | 1.8 |
| External interests | 0.9 | 2.1 | 2.0 | 0.9 | 0.7 | 0.5 | 0.4 | 0.3 | 0.5 | 0.5 | 0.6 | 0.8 | 1.1 | 1.3 | 1.3 |
| Total transfers | 6.4 | 10.3 | 12.3 | 12.2 | 9.8 | 10.8 | 10.9 | 14.1 | 17.3 | 16.3 | 16.2 | 15.0 | 14.3 | 13.0 | 11.2 |
| Other expenditures | 4.3 | 1.1 | 3.9 | 3.2 | 0.7 | 0.5 | 0.5 | 0.4 | 0.4 | 0.4 | 0.2 | 0.3 | 0.3 | 0.2 | 0.1 |

*(Table continues on the following page.)*

**Table 4.2** (continued)

| Item | 1974 | 1975 | 1976 | 1977 | 1978 | 1979 | 1980 | 1981 | 1982 | 1983 | 1984 | 1985 | 1986 | 1987 | 1988 |
|---|---|---|---|---|---|---|---|---|---|---|---|---|---|---|---|
| Saving | 3.9 | 7.3 | 6.4 | 5.6 | 6.5 | 7.7 | 8.4 | 5.5 | -1.9 | -2.9 | -2.0 | -0.9 | 0.9 | 2.1 | 5.5 |
| Public investment | 8.7 | 5.7 | 3.1 | 4.2 | 3.5 | 3.2 | 2.6 | 2.5 | 2.1 | 2.1 | 2.3 | 3.1 | 3.3 | 3.0 | 2.9 |
| Other expenditures | 1.9 | 1.6 | 0.1 | 0.5 | 0.8 | -0.7 | 0.3 | 0.1 | -1.8 | -1.9 | -0.7 | -0.4 | -0.8 | -0.6 | -0.8 |
| Surplus | -6.6 | 0.0 | 3.2 | 0.9 | 2.1 | 5.1 | 5.5 | 2.9 | -2.3 | -3.1 | -3.5 | -3.6 | -1.7 | -0.3 | 3.4 |
| *Public enterprises* | | | | | | | | | | | | | | | |
| Current revenues | 32.8 | 35.9 | 31.0 | 26.6 | 24.7 | 27.0 | 25.8 | 20.8 | 24.2 | 29.5 | 29.8 | 35.3 | 35.4 | 34.7 | 37.8 |
| Current expenditures | 25.3 | 30.2 | 20.5 | 18.4 | 18.3 | 17.6 | 16.3 | 15.6 | 16.5 | 18.5 | 19.2 | 21.2 | 22.6 | 21.8 | 22.2 |
| Saving | 7.5 | 5.7 | 10.5 | 8.2 | 6.4 | 9.4 | 9.5 | 5.2 | 7.7 | 11.0 | 10.6 | 14.1 | 12.8 | 12.9 | 15.6 |
| Taxes and transfers | 4.8 | 5.4 | 6.9 | 5.2 | 4.4 | 8.3 | 7.5 | 5.6 | 7.1 | 8.6 | 8.2 | 9.6 | 9.1 | 9.7 | 12.6 |
| Investment | 3.9 | 3.5 | 3.0 | 2.7 | 3.2 | 1.9 | 2.6 | 2.6 | 2.6 | 2.6 | 3.7 | 4.0 | 4.7 | 4.0 | 3.3 |
| Other income | 2.3 | 1.1 | 0.3 | -0.8 | 0.4 | 0.2 | 0.5 | 0.4 | 0.3 | -0.1 | 0.3 | 0.3 | 0.5 | 0.9 | 0.6 |
| Surplus | 1.2 | -2.1 | 0.9 | -0.5 | -0.8 | -0.6 | -0.1 | -2.5 | -1.6 | -0.4 | -1.1 | 0.8 | -0.5 | 0.1 | 0.3 |
| *Nonfinancial public sector* | | | | | | | | | | | | | | | |
| Consolidated surplus | -5.5 | -2.1 | 4.0 | 0.4 | 1.3 | 4.6 | 5.4 | 0.4 | -3.9 | -3.5 | -4.6 | -2.9 | -2.1 | -0.2 | 3.7 |
| Financing | 5.5 | 2.1 | -4.0 | -0.4 | -1.3 | -4.6 | -5.4 | -0.4 | 3.9 | 3.5 | 4.6 | 2.9 | 2.1 | 0.2 | -3.7 |
| Internal | 5.2 | 5.0 | -0.9 | -0.1 | -4.0 | -4.6 | -5.3 | -3.2 | 1.7 | 2.4 | 1.9 | -1.2 | -1.0 | -2.2 | -7.3 |
| External | 0.3 | -2.9 | -3.1 | -0.3 | 2.7 | 0.0 | -0.1 | 2.8 | 2.2 | 1.1 | 2.7 | 4.0 | 3.1 | 2.4 | 3.6 |

*Source:* Budget Office, Ministry of Finance, Chile.

**Table 4.3.  Central Bank Losses from Quasi-Fiscal Operations, Chile, 1982–89**

(percentage of GDP)

| Item | 1982 | 1983 | 1984 | 1985 | 1986 | 1987 | 1988 | 1989 |
|---|---|---|---|---|---|---|---|---|
| Loans to bankrupt financial institutions | 8.6 | 0 | 0 | 0 | −0.3 | −0.2 | — | — |
| Purchase of commercial banks' bad loans | 0.6 | 2.4 | −0.2 | 0 | 0.3 | −0.1 | −0.1 | −0.2 |
| Domestic debt rescheduling | 4.8 | 2.9 | 1.5 | 4.6 | 0.9 | 0 | — | — |
| Preferential exchange rate program | 0 | 0.3 | 0.9 | 1.1 | 0.1 | 0.3 | — | — |
| Exchange rate guarantee program | 0 | 0.5 | 2.8 | 3.5 | 0.2 | 0 | — | — |
| Foreign exchange capital losses | 0 | 0 | 2.9 | 3.9 | −1.5 | 0.8 | — | — |
| Total quasi-fiscal losses | 14.0 | 6.1 | 7.9 | 13.1 | −0.3 | 0.8 | −0.1 | −0.2 |

— Not available.

*Note*: This table presents annual losses consistent with the capitalized losses of table 4.5. For losses from purchase of bad loans and rescheduling of domestic debt, the estimates of net losses were obtained by multiplying gross losses (presented in tables 4 and 6 in Eyzaguirre and Larrañaga 1990) by the ratios of net loss to present value of disbursements calculated from table 4.5. These ratios are 0.193 for purchase of bad loans and 0.248 for domestic debt rescheduling.

*Source*: Eyzaguirre and Larrañaga 1990; authors' calculations.

and the profit-maximizing behavior of public enterprises. The liberalization-cum-stabilization process seemed to be very successful. Inflation, which had reached 1,000 percent in 1973, fell to 30 percent in 1980 and almost converged toward international levels in 1981—the stated objective of the fixed-exchange-rate policy pursued since mid-1979. GDP growth averaged 7.2 percent during 1976–81. However, private consumption and investment grew explosively, at rates that subsequently proved unsustainable. Although part of the private sector deficit was financed by public surpluses, the private sector increasingly relied on foreign saving; the current account deficit reached a record 14.4 percent of GDP in 1982.[3]

At the end of 1981 Chile was burdened by a huge foreign debt and an unstable domestic financial system—a result of inconsistent exchange rate and wage policies, a government-supported euphoria that contributed to private overspending, and lax regulation of the financial system. Given this setting, a slight negative shock could have triggered a crisis. What happened was that a number of strong adverse foreign shocks hit the economy simultaneously. Declining terms of trade, higher foreign interest rates, and the end of voluntary foreign lending contributed to the massive 1982–83 financial crash and economic recession. In 1982 alone, real GDP declined by 14.1 percent, and inflation climbed to 20.7 percent. The recession lowered public sector revenues, particularly those from trade-related taxes. The huge unemployment rate forced the government to undertake a series of emergency support programs that raised government spending.

In 1984 a structural adjustment program was initiated to address three pressing problems faced by the Chilean economy: excessive concentration of exports, low levels of national saving, and a precarious financial system. From the perspective of fiscal policy this period is characterized by a major fiscal adjustment program aimed at both stabilization and consolidation of the structural reforms begun in 1974. Six policy reforms affected the nonfinancial public sector. First, current expenditures were curtailed. Second, income taxes were reduced significantly, with the purpose of encouraging private saving and investment. Third, revenues were falling as a result of the 1981 social security reform.[4] Fourth, a new impetus was given to the privatization program initiated in 1974. Fifth, as terms of trade improved in 1988 and 1989, a new tax reduction—this time affecting value added rates—transferred additional resources to the private sector. Sixth, government expenditure was delinked from volatile public sector copper revenue by a copper price stabilization fund, established in 1985. The outcome of these reforms, boosted by a favorable evolution of foreign and domestic macroeconomic conditions, was a gradual but massive improvement in fiscal stance. The CNFPS deficit, which had reached 4.6 percent of GDP in 1984, was almost eliminated by 1987, and a sizable CNFPS surplus was recorded in 1988–89.

Important public policy innovations during the 1980s also took place within the Central Bank. In response to the 1982–83 financial and external crises, the government decided to bail out the private sector by providing Central Bank loans at subsidized rates and extending exchange rate subsidies to debtors in foreign currency.[5] The losses from these operations reached staggering levels: 14.0 percent of GDP in 1982 and an average annual 10.3 percent of GDP throughout 1982–85, the four-year period of financial bailouts and interest and exchange rate subsidies. The Central Bank financed most of its losses by issuing domestic debt, which increased by US$5.7 billion between 1982 and 1989. In addition, its foreign liabilities increased by US$4.0 billion during the same period.

The combination of fiscal reform in the nonfinancial public sector and the lender-of-last-resort emergency financing provided by the Central Bank contributed to a swift recovery of the economy and public finances after 1985. Resource allocation and private spending also responded strongly to the massive real devaluations of 1982–85, which were supported by a nominal crawling-peg exchange rate policy, based on purchasing power parity, that was occasionally interrupted by corrective discrete devaluations. As a result of these policies, and with the help of favorable terms of trade after 1988, the current account deficit fell to an average 1.9 percent in 1988–89. GDP growth resumed strongly after 1985, as a consequence first of capacity recovery and then of a vigorous expansion of capacity as private investment responded to the incentives provided by structural reforms and a stable macroeconomy. In 1988 and 1989 monetary and fiscal policies were relaxed (the latter through tax cuts), contributing to some overheating of the economy. By the end of 1989 annualized inflation had reached almost 30 percent, and GDP growth hit double-digit levels.

In 1990 the newly independent Central Bank pursued a strongly contractionary monetary policy, and the new democratically elected Aylwin government obtained parliamentary approval for a tax increase. Tax revenue was raised by about 3 percentage points of GDP from 1991 onward through a combination of a higher value added tax rate (18 percent, up from 16 percent), a higher profit tax rate (15 percent, up from 10 percent), an expanded tax base, and higher personal income taxes. After monetary policy was eased in 1991, GDP growth—which was fairly low in 1990—increased vigorously, reaching a record 10.4 percent in 1992, while inflation fell to 12.7 percent. The CNFPS showed systematic surpluses that averaged 0.9 percent in 1990–92. The current account showed an average deficit of 1.0 percent of GDP during the same three-year period, implying that the private sector ran an average deficit of 1.9 percent of GDP, evenly financed by the public sector and the rest of the world.

To analyze the interaction between fiscal deficits and macroeconomic variables, this chapter focuses on four key issues.

- The impact of macroeconomic and external variables on public sector deficits. The focus is on how vulnerable public finances are to variables beyond the direct control of fiscal policymakers and how fiscal policy reacts to exogenous shocks. Our hypothesis is that even though the Chilean public sector faces an unstable external environment, active fiscal policies have been able to counteract both external and domestic shocks.
- The effect of the deficit on inflation and interest rates, and the effects of alternative strategies of deficit financing. Fiscal policy was an important factor behind the relatively low inflation and interest rates during the 1984–90 period. A puzzle to be explained is why the huge quasi-fiscal deficit did not destabilize the economy.

- The effect of fiscal policies on private sector consumption and spending. This chapter focuses on the relative contributions of foreign saving and fiscal policy to the private consumption boom in the late 1970s and early 1980s and to the decline in private consumption during the adjustment period. Public policies also contributed to the strong expansion of private investment in 1984–92. Perhaps the most important issue regarding the relation between fiscal variables and private spending in the Chilean case is the explanation for the opposite trajectories of private consumption and investment since the mid-1980s.
- The key role of the exchange rate in Chile's adjustment. The strategy of trade deficit correction and export diversification required a sustained real exchange rate depreciation.[6] Our main hypothesis is that fiscal adjustment has contributed to a higher trade surplus and hence to a more depreciated real exchange rate.

We begin with a review of the evolution of fiscal policy in Chile during the 1974–88 period, emphasizing the structure and decomposition of CNFPS deficits, the scope of quasi-fiscal operations, and the sustainability of public sector deficits. A three-asset portfolio model is then developed in order to estimate the impact of domestic deficit financing on interest and inflation rates and to assess the implications of alternative strategies of deficit financing. The direct and indirect effects of fiscal policy on private consumption and investment are analyzed, and the impact of public sector deficits on real exchange rates and the current account is estimated. The final section summarizes the main findings and their policy implications.

## Fiscal Policy, Decomposition of Nonfinancial Deficits, Quasi-Fiscal Deficits, and Sustainability

Fiscal deficits are affected by fiscal policies and exogenous variables; the former are under the direct control of policymakers and the latter are not. An assessment of the relative contribution of these two types of variables to the evolution of fiscal deficits helps in understanding the sign and the net effect of fiscal policy actions. Since 1974 Chile has faced severe shocks, and fiscal policy has been systematically oriented so as to compensate for the effects of these shocks on the deficit. For example, when falling copper prices increased the CNFPS deficit by 3.1 percentage points of GDP in 1973–75, tax and expenditure policies were used to reduce the deficit. Similarly, lower tax revenue was allowed to increase the deficit during 1987–89, when the rise in copper prices reduced the deficit by almost 4 percentage points of GDP.

*Decomposition of Nonfinancial Public Sector Deficits*

The decomposition of the CNFPS attempts to measure the contribution of exogenous and endogenous fiscal policy variables to the deficit, following closely the methodology presented elsewhere.[7] Endogenous variables are defined as those under the control of fiscal policymakers, such as tax rates, public investment, expenditure, and public sector wages. Exogenous determinants of the deficit include domestic variables (such as output, the real exchange rate, and inflation) and external variables (for example, copper prices and foreign interest rates) that are beyond the direct influence of fiscal policymakers. Appendix 4.1 presents estimated functions for tax revenue, public enterprise surplus, and transfers to the private sector and identifies the role of fiscal policy and exogenous variables in their behavior. The decomposition of the CNFPS deficit makes use of these estimated relations.

Table 4.4 summarizes the net impact of fiscal policy and of domestic macroeconomic and external variables on the nonfinancial public sector deficit for 1973–88. The data are shown for four subperiods that are relevant from the point of view of both the fiscal stance and the state of the overall economy: sharp stabilization (1974–75); reform, recovery and euphoria (1976–81); crisis and stabilization-cum-adjustment (1982–86); and recovery under continuing reform (1987–88).

The sharp decrease in the 1973–75 public sector deficit is largely explained by fiscal policy variables. The main policy shifts that were conducive to the massive fiscal adjustment were tax reforms (which raised indirect and direct effective tax rates), the decline in public employment, and the surge of the public enterprise surplus. Fiscal balance was restored two and a half years after the experience of huge public sector disequilibria at the end of the Allende administration.

The 1976–81 period was characterized by relatively stable surpluses. The data in table 4.4, however, suggest that this stability was the result of two opposing forces. On one side, fiscal policy variables—effective tax rates, public sector real wages, and the external public debt—in conjunction with a declining inflation rate and a backward indexation scheme, raised the deficit. On the other side, large increases in income and the recovery of copper prices lowered the deficit.

During 1982–86 the deficit increased as a result of fiscal policy actions, including reduced direct tax rates, a declining number of affiliates to the state-run social security system, and the rising public debt. These changes reflected, respectively, the tax reform of 1982, the social security reform of 1981, and increasing debt-financing of deficits swollen by government intervention in the economy during the crisis.

**Table 4.4. Consolidated Nonfinancial Public Sector (CNFPS) Deficit Decomposition, Chile, 1973–88**
(change in public sector deficit, in percentage points of GDP)

| Item | 1974–75 | 1976–81 | 1982–86 | 1987–88 |
|---|---|---|---|---|
| Fiscal policy variables | −14.1 | 13.2 | 5.7 | 5.2 |
| Indirect tax rate | −5.4 | 5.5 | −2.4 | 1.8 |
| Direct tax rate | −3.3 | 3.5 | 2.2 | 0.7 |
| Copper tax rate | −1.8 | 0.0 | 0.0 | −0.8 |
| Public employment | −4.3 | −2.1 | −0.1 | 0.0 |
| Affiliates to social security | −0.1 | 1.8 | 1.1 | 2.4 |
| Public investment | −0.8 | −1.8 | 0.8 | 0.0 |
| Public sector real wage | 0.2 | 3.3 | −1.5 | 0.4 |
| Foreign public debt | 1.7 | 2.8 | 2.9 | 0.5 |
| Domestic public debt | 0.0 | 0.0 | 2.7 | 0.1 |
| Real purchases of goods and services | −0.3 | 0.2 | 0.0 | 0.1 |
| Domestic macroeconomic variables | 4.7 | −7.5 | −2.8 | −2.7 |
| Real GDP growth | 3.4 | −10.8 | −1.3 | −2.9 |
| Real exchange rate | −0.7 | 0.3 | −1.2 | 0.0 |
| Inflation rate | 2.1 | 4.6 | −0.1 | 0.1 |
| Real wage growth | −0.1 | −1.6 | −0.4 | −0.1 |
| Domestic interest rate | 0.0 | 0.0 | 0.2 | 0.2 |
| External variables | 2.9 | −1.8 | 0.8 | −3.6 |
| Copper price | 3.1 | −2.3 | 1.3 | −3.9 |
| Foreign interest rate | −0.2 | 0.5 | −0.5 | 0.3 |
| Nondecomposed accounts | | | | |
| Public enterprise surplus (after transfers) | −8.4 | 0.4 | −2.1 | −0.7 |
| Other public sector income | −4.2 | 0.6 | 2.0 | 0.1 |
| Other public sector expenditure | −3.5 | −2.3 | −0.9 | −0.2 |
| Explained change in deficit | −22.6 | 2.3 | 2.7 | −1.9 |
| Nonexplained change in deficit | 4.1 | −4.7 | −0.9 | −3.9 |
| Total change in deficit | −18.5 | −2.4 | 1.8 | −5.8 |

*Note:* For instance, the CNFPS deficit fell by 2.4 percentage points of GDP between 1975 and 1981 (compare the corresponding levels in table 4.1), as reflected by the last line of the second column. The change in the deficit explained by our variables is an increase of 2.3 percentage points of GDP (third line from bottom), which is overexplained by the combined effect of fiscal policy variables, causing a deficit rise of 13.2 percentage points of GDP (first line).

*Source:* Authors' calculations based on the methodology of Marshall and Schmidt-Hebbel 1989a, 1989b.

The 1987–88 recovery of the nonfinancial public sector was associated with favorable changes in domestic and external sector variables—in particular, the surge in real income and copper prices. However, fiscal policy variables such as a decrease in the value added tax rate and the continuing reduction of affiliates to the social security system had a negative impact on public finances.

To compare the role of fiscal policies with the influence of exogenous variables in the evolution of the CNFPS deficit in Chile, we computed the average contribution to the explained change in the deficit over the 1974–88 period of three factors: fiscal policy (142 percent), domestic macroeconomic variables (−41 percent), and external variables (−1 percent).[8] These figures confirm the massive predominance of fiscal policy changes in both the cyclical and trend evolution of the CNFPS deficit. Although external and domestic shocks beyond the control of fiscal policymakers are important in shaping deficits—as reflected in table 4.4—they have been counteracted by massive policy changes. Hence, even in an economy such as Chile's in which the budget is subject to strong shocks, policymakers are to blame for fiscal deterioration and are to be praised for successful stabilization.

### Quasi-Fiscal Operations and Domestic Debt

An essential feature of the Chilean adjustment process after 1982 has been the quasi-fiscal operations of the Central Bank in support of a bankrupt financial system—in effect subsidizing financial institutions and private debtors. During the early 1980s the Central Bank, at that time under the aegis of the Ministry of Finance, responded to an acute balance of payments and domestic financial crisis (caused in part by a lax banking law) with a program of massive bailouts, credit, and exchange rate subsidies. These operations, however, were limited to the 1982–85 emergency period.

The main quasi-fiscal programs were:

- *Loans to bankrupt financial institutions.* In 1981 the Central Bank provided emergency loans to eight financial institutions that faced insolvency. All eight institutions went bankrupt, and the outstanding Central Bank loans turned into losses.
- *Purchase of bad loans.* The Central Bank purchased bad loans from private banks, which made future repurchase commitments. The loans were extended at subsidized interest rates.
- *Domestic debt rescheduling.* The Central Bank rescheduled domestic debts and financed this operation through loans at negative spreads.
- *Preferential exchange rate.* After the massive 1982–83 devaluations, which raised the domestic-currency cost of foreign debt service, private debtors with foreign-currency-denominated liabilities were allowed to purchase, at a subsidized exchange rate, foreign exchange for servicing their debt.
- *Exchange rate guarantees.* The Central Bank purchased foreign exchange with the commitment to sell it back after one year at the

purchase exchange rate adjusted by domestic inflation. The massive 1982–85 real exchange rate depreciation caused corresponding losses to the Central Bank.
* *Foreign exchange capital losses.* The acquisition of private external debt pushed the Central Bank into a net foreign debtor position, which implied capital losses as a result of the large real devaluations.

Table 4.5 reports estimated capitalized losses incurred by the Central Bank as a consequence of quasi-fiscal operations. Losses were estimated as the present value—as of December 1989—of Central Bank disbursements net of the value of assets acquired through quasi-fiscal operations. The total estimated loss is about US$9.0 billion, almost 40 percent of 1989 GDP. The equivalent cash flow is about US$540 million, taking into account an average 6 percent interest rate on Central Bank liabilities.

The Central Bank financed most quasi-fiscal operations by issuing domestic debt. As a result, its outstanding domestic debt increased by US$5.7 billion during the 1982–89 period, while its external debt increased by US$4.0 billion. To compensate for the increase in Central Bank liabilities, during 1983–86 the general government transferred to the Central Bank approximately US$7.2 billion in low-interest treasury bonds. In this way the government recognized that the Central Bank was its financial agent during the financial crisis. The Central Bank still exhibits a significant cash-flow deficit (table 4.6) as a result of its holdings of low-yield treasury bonds, which pay a minimum real return of 2 percent per year, and its high-interest liabilities, which cost an average real 6 percent interest per year. The deficit is financed predominantly by the Bank's seigniorage revenue.

**Table 4.5. Capitalized Value of Central Bank Losses from Quasi-Fiscal Operations as of December 1989, Chile**
(millions of U.S. dollars)

| Quasi-fiscal operation | Present value of disbursements | Estimated value of assets | Net loss |
|---|---|---|---|
| Credit to bankrupt financial institutions | 1,930 | 0 | 1,930 |
| Purchase of bad loans | 3,114 | 2,513 | 601 |
| Domestic debt rescheduling | 1,570 | 1,180 | 390 |
| Preferential exchange rate | 3,320 | 0 | 3,320 |
| Exchange rate guarantees | 1,585 | 0 | 1,585 |
| Foreign exchange capital losses | 1,227 | 0 | 1,227 |
| Total | 12,746 | 3,693 | 9,053 |

*Source:* Eyzaguirre and Larrañaga 1990.

**Table 4.6 Estimated Central Bank Deficit, Chile, 1991**
(percent)

| Item | Share of GDP |
|---|---|
| Expenditure | 2.95 |
| Domestic liabilities | 1.66 |
| External liabilities | 0.83 |
| Dollar-denominated domestic liabilities | 0.48 |
| Revenue | 1.05 |
| International reserves | 0.58 |
| Loans to the financial sector | 0.47 |
| Treasury bonds | 0.43 |
| Deficit | 1.47 |

*Source*: Eyzaguirre and Larrañaga 1990.

## How Sustainable Is the Deficit?

Whereas the nonfinancial public sector showed an average annual surplus of 0.9 percent of GDP over 1990–92, the Central Bank of Chile reported in 1991 a deficit close to 1.5 percent of GDP (table 4.6). A 0.6 percent total public sector deficit—approximately equivalent to a primary deficit of 3.5 percent of GDP for the total public sector—seems manageable. However, public finances could be placed under stress by the large total public debt, which increased from US$7.9 billion in 1981 to more than US$22 billion in 1989. This figure includes both domestic and external debt, held by the financial and nonfinancial public sectors.

To illustrate the implications of public debt for public finances, we perform a simple test of deficit sustainability, based on a condition of a nonincreasing debt-to-GDP ratio imposed on the consolidated total public sector budget constraint.[9] If the deficit is financed through issuing money, domestic debt, or foreign debt, financing of the primary deficit as a fraction of GDP ($d$) can be written as:

$$(4.1) \quad d = \dot{m} + m(\pi + n) + \dot{b} - b(r - n) + \dot{b}^* - b^*(r^* - n + \mu)$$

where $\pi$ is the inflation rate, $n$ is the GDP growth rate, $r$ is the real interest rate paid on domestic debt, $r^*$ is the real interest paid on foreign debt, and $\mu$ is the rate of real devaluation.[10] The outstanding stocks of base money, domestic public debt, and external public debt as shares of GDP are denoted by $m$, $b$, and $b^*$, respectively. A dot over a variable denotes the time derivative.

Table 4.7 reports simulation results for the primary public surplus levels required for (a) maintaining a constant debt-to-GDP ratio and (b) decreasing this ratio by 5 percentage points of GDP per year. Two macroeconomic scenarios—a base scenario and an adverse case—are

**Table 4.7. Consolidated Total Public Sector Primary Surpluses under Alternative Public Debt Paths**
(percent)

| Item | Real interest rate | Real GDP growth | Primary surplus as share of GDP |
|---|---|---|---|
| *Constant ratio of public debt to GDP* | | | |
| Base scenario | 6.0 | 5.0 | −0.8 |
| Adverse scenario | 10.0 | 2.0 | 3.6 |
| *Decrease in ratio of public debt to GDP by 5 percent per year* | | | |
| Base scenario | 6.0 | 5.0 | 4.2 |
| Adverse scenario | 10.0 | 2.0 | 8.6 |

*Note:* Simulations assume 15 percent inflation, a constant real exchange rate, and domestic real interest rates equal to foreign real interest rates.

considered for each alternative. Under the base scenario, with output growing at 5 percent and with a real interest rate of 6 percent, the public sector can maintain its debt ratio by running a primary deficit of 0.8 percent of GDP, equivalent to a 4.8 percent total deficit. Alternatively, the public sector is able to reduce significantly its debt-to-GDP ratio—by 5 percentage points of GDP per year—when running a primary surplus of 4.2 percent of GDP. Under an adverse macroeconomic scenario, with 2 percent growth and real interest rates at 10 percent, the public sector would have to generate a primary surplus of 3.6 percent of GDP to maintain a constant debt-to-GDP ratio and would need a primary surplus of 8.6 percent of GDP to reduce the debt-to-GDP ratio.

The implication is that the current deficit levels of the total consolidated public sector (0.6 percent total, 3.5 percent primary) are sustainable under current macroeconomic conditions. Ratios of public debt to GDP could be maintained at their current levels even under more adverse conditions.

## Deficit Financing, Interest Rates, and Inflation

This section analyzes the effect of the public deficit on inflation and interest rates in Chile under alternative financing strategies. The performance of the key rates and financial aggregates is depicted in figures 4.1 and 4.2. Interest and inflation rates have followed a rather similar pattern since the mid-1970s. Their relative stability—by Latin American standards—contrasts with the erratic behavior of money and particularly with the upward trend of domestic public debt. The latter aggregate, starting at a stable and low level of about 10 percent of quarterly GDP before the 1982 crisis, rose sharply, to 80 percent of

## Figure 4.1. Interest and Inflation Rates, Chile

**Real interest rate, 1975:1–1988:4**

Percentage per month

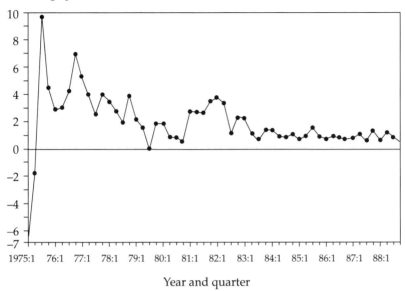

Year and quarter

**Inflation rate, 1975:2–1988:4**

Percentage per month

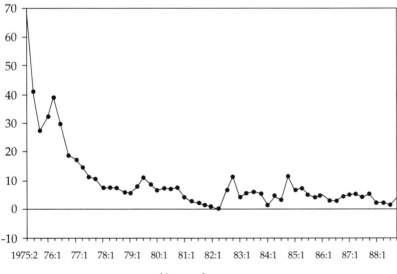

Year and quarter

*Source:* See appendix 4.4.

**Figure 4.2. Money and Public Debt, Chile**

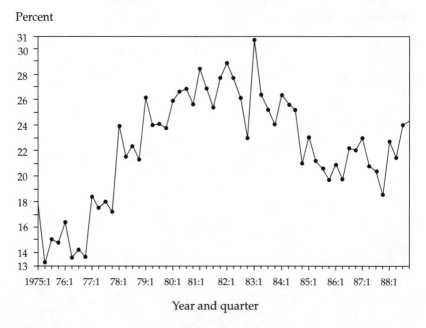

**Money as a share of GDP, 1975:1–1988:4**

**Public debt as a share of GDP, 1981:4–1988:4**

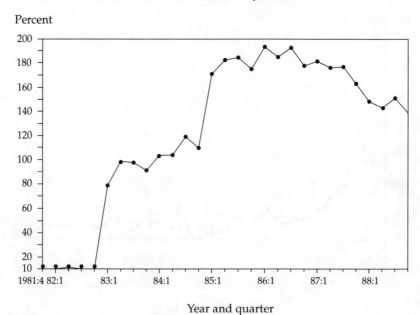

*Source:* See appendix 4.4.

GDP in the first quarter of 1983, and reached a peak of 190 percent of GDP in the first quarter of 1986, after which it started to decline.

As was true for Latin America at large, Chile overborrowed during 1978–81 and faced an adverse external environment in the early 1980s. This triggered a crisis of vast proportions, forcing major adjustments in macroeconomic policy in general and fiscal policy in particular. What is unique to the Chilean experience is that the 1982 crisis and the decline in external financing hit a stabilized public sector that had shown significant surpluses before 1982. This favorable initial condition allowed the government (including the Central Bank) to finance the deterioration in the public sector budget after the debt crisis without destabilizing the economy. However, both the 1981 pension reform and the quasi-fiscal operations of the Central Bank in 1982–85 raised measured total public sector deficits. To a large extent these deficits were financed by increased domestic public debt, a liability demanded by the private social security system put in place by the pension reform. The empirical model presented in this section emphasizes these specific features of the post-1982 fiscal adjustment undertaken in Chile, introducing modifications to the framework proposed by Easterly (1989) on the basis of a model developed by Mujica (1990).

### Public Deficit Financing and Private Sector Portfolios

A growing literature on fiscal deficits deals with the relation between fiscal deficits and economic growth, the sensitivity of macroeconomic variables to changes in the public sector budget, and the channels through which fiscal deficits impinge on the economy. This literature emphasizes the implications of an increased reliance on domestic financing of public deficits as a result of the sharp reduction in external financing in most highly indebted developing countries (see, for instance, Buiter 1988; Dornbusch 1985a, 1985b, 1988; and van Wijnbergen 1989).

The point of departure of the model is the basic government financing identity, which may be stated by rewriting equation 4.1 for the nominal total consolidated public sector deficit:

$$(4.2) \qquad G_t + i_{t-1}D_t + i^*_{t-1}e_tD^*_t = \dot{H}_t + \dot{D}_t + e_t\dot{D}^*_t$$

where $G$ is the primary deficit, $D$ is the stock of domestic public debt, $D^*$ is the stock of foreign public debt, $H$ is the stock of high-powered money, $i$ is the domestic nominal interest rate, $i^*$ is the foreign nominal interest rate, and $e$ is the nominal exchange rate. The identity states that the total public deficit, comprising the primary deficit and interest payments, is financed by changes in foreign and domestic public liability holdings. However, since for most developing countries after 1982 external borrowing has been quasi-exogenous, an increase in the primary deficit must be financed basically by issuing bonds or by creating high-powered money. Therefore the model emphasizes the tradeoff associated with the choice between money and domestic debt financing for given levels of the primary deficit and foreign financing. The choice has implications for both real and nominal variables. This section focuses on inflation and interest rates.

The model uses a standard portfolio framework for the private sector's holdings of money ($M_1$), public debt, and foreign interest-bearing assets. Focusing on the demands for the first two assets and assuming short-run stock adjustment costs, the short-run equilibrium values of real balances and real public bonds holdings can be expressed as follows (see appendix 4.2 for a formal derivation):

(4.3)
$$m_t - p_t = \phi_0 + \phi_1 y_t + \phi_2 i_t + \phi_3(i_t^* + \delta_t)$$
$$+ \phi_4(m_{t-1} - p_{t-1}) + \phi_5 S_t + \phi_6 w_t$$

(4.4)
$$d_t - p_t = \Phi_0 + \Phi_1 y_t + \Phi_2 i_t + \Phi_3(i_t^* + \delta_t)$$
$$+ \Phi_4(d_{t-1} - p_{t-1}) + \Phi_5 \dot{d}_t + \Phi_6 w_t$$

where (in addition to the variables defined earlier) $m$ is the nominal money stock, $w$ is real financial wealth, $y$ is real GDP, $p$ is the domestic price level, $d$ is the nominal public debt stock, $S$ is a nominal supply shock (defined as the difference between the actual and expected rates of change of money), $e$ is the nominal exchange rate, and $\delta$ is the expected change in the exchange rate. All variables except $i$, $i^*$, and $\delta$ are expressed in natural logarithms. Subscripts $t$ and $t-1$ denote time. The $\phi_i$ and $\Phi_j$ coefficients are nonlinear combinations of the structural coefficients, as shown in appendix 4.2.

The money shock variable $S$ captures the shock-absorber hypothesis proposed by Carr and Darby (1981), whereby unexpected monetary supply shocks cause unplanned transitory monetary hoarding. A similar term is included in the specification of the public debt market equilibrium condition: the actual rate of change in the stock of nominal public debt ($\dot{d}$) is hypothesized to affect the level of short-term public bond holdings. This variable may reflect the demand for public debt by the private social security corporations since 1981, which accommodated—at least in part—the financing of deficits originating in the social security reform and the quasi-fiscal operations of the Central Bank. This partial demand accommodation to the large supply of public debt implies that the interest rate would not rise by as much as in the absence of demand accommodation.

Equations 4.3 and 4.4 are quite general because they do not merely incorporate features of open and closed economies but also allow for the possibility of distinguishing between the short- and long-run effects of shocks in the money and public bond markets by assuming partial short-term adjustment in both markets. Since the nominal stocks of money and domestic public debt are determined by the government budget equation (4.2) and real wealth is exogenous, equations 4.3 and 4.4 describe an adjustment mechanism for domestic prices and interest rates. Inverting them allows the derivation of the following reduced-form equations for the two latter variables (where the $\Omega$ and $\pi$ coefficients are defined as in appendix 4.2):

(4.5)     $i_t = \Omega_0 + \Omega_1 y_t + \Omega_2(i_t^* + \delta_t) + \Omega_3(m_{t-1} - p_{t-1})$
$$+ \Omega_4(d_{t-1} - p_{t-1}) + \Omega_5 m_t + \Omega_6 d_t + \Omega_7 d_{t-1}$$
$$+ \Omega_8 S_t + \Omega_9 w_t$$

(4.6)     $p_t = \pi_0 + \pi_1 y_t + \pi_2(i_t^* + \delta_t) + \pi_3(m_{t-1} - p_{t-1})$
$$+ \pi_4(d_{t-1} - p_{t-1}) + \pi_5 m_t + \pi_6 d_t + \pi_7 d_{t-1}$$
$$+ \pi_8 S_t + \pi_9 w_t.$$

## Empirical Results

This section presents the results obtained from the estimation of the short-run demands for real balances and real public debt, based on quarterly data for the Chilean economy.[11] We also show results for the reduced-form equations for the interest rate and the price level in order to assess the ability of the model to describe inflation and interest rate paths directly.

The estimated structural asset demands correspond to equations 4.3 and 4.4, amended by including exponential time trends, as reflected by the first lines of tables 4.8 and 4.9 The results for short-run money demand in table 4.8 report coefficients with the expected signs, and almost all are significant at conventional levels. In particular, the significance levels of the unexpected nominal monetary shock ($S_t$) and the lagged dependent variable ($m_{t-1} - p_{t-1}$) suggest that real balances have been highly sensitive to money supply shocks and that the effect of any variable on holdings of real balances is spread over a significant period. Finally, the negative coefficient of the time variable is related to the strong financial innovations that took place between the mid-1970s and the late 1980s as a consequence of financial liberalization. Taken as a whole, these results are consistent with the hypotheses of stock adjustment and unexpected monetary shocks implied by the model and suggest that conventional money demand models cannot sufficiently explain the behavior of real balances during this period in Chile.

The results for the short-run demand for public debt in table 4.9 are consistent with some of the hypotheses advanced earlier. For exam-

**Table 4.8. Estimation Results for Money Demand, Chile, 1976:1–1988:4**

| Variable | OLS | AR1 | NLS |
|---|---|---|---|
| | | *Method* | |
| $\phi_0$ | −1.582 | −1.959 | |
| | (1.3) | (1.5) | |
| $\phi_1$ | 0.415 | 0.486 | |
| | (3.5) | (3.9) | |
| $\phi_2$ | −3.101 | −3.595 | |
| | (4.1) | (4.6) | |
| $\phi_3$ | −0.165 | −0.102 | |
| | (0.9) | (0.6) | |
| $\phi_4$ | 0.702 | 0.660 | |
| | (15.2) | (12.6) | |
| $\phi_5$ | 0.148 | 0.151 | |
| | (3.5) | (4.1) | |
| $\Gamma_1$ | −0.004 | −0.004 | |
| | (4.1) | (4.0) | |
| $\sigma$ | 0.298 | 0.340 | |
| $\alpha_1$ | 1.390 | 1.430 | 1.394 |
| $\alpha_2$ | −10.410 | −10.590 | −10.414 |
| $\alpha_3$ | −0.550 | −0.300 | −0.555 |
| Adjusted $R^2$ | 0.972 | 0.979 | 0.972 |
| Durbin's $h$ | 1.947 | 0.570 | 1.947 |
| $\rho$ | | 0.285 | |

*Note:* The equation is:

$$m_t - p_t = \phi_0 + \phi_1 y_t + \phi_2 i_t + \phi_3(i_t^* + \delta_t) + \phi_4(m_{t-1} - p_{t-1}) + \phi_5 S_t + \Gamma_1\text{TIME}.$$

The estimation methods were ordinary least squares (OLS), maximum likelihood estimation with first-order autoregressive residuals (AR1), and nonlinear least squares (NLS). Coefficients $\alpha_1$, $\alpha_2$, and $\alpha_3$ are the long-run asset demand coefficients, and parameter $\sigma$ denotes the short-run asset adjustment parameter (see appendix 4.2). Numbers in parentheses are $t$-values, and $\rho$ is the residual first-order correlation coefficient. A blank denotes the omission of the specific variable from the regression.

ple, slow stock adjustment is reflected by the significant coefficient of lagged public debt holdings. However, the actual rate of change in the stock of nominal public debt was not significant in preliminary regressions; hence table 4.9 reports the estimates of equation 4.4 without this variable. Output has a negative effect on debt demand—a result that reflects more the peculiarity of the sample period than any specific feature of the Chilean economy. Another interesting result refers to the relative size of the interest rate semielasticities of money and public bond holdings and the speed of adjustment in the money and

**Table 4.9. Estimation Results for the Demand for Public Debt, Chile, 1983:1–1988:4**

| Variable | *Method* | | |
|---|---|---|---|
| | OLS | AR1 | NLS |
| $\Phi_0$ | 55.005 | 51.932 | |
| | (3.8) | (3.7) | |
| $\Phi_1$ | −4.249 | −3.956 | |
| | (3.3) | (3.2) | |
| $\Phi_2$ | 19.024 | 19.175 | |
| | (2.6) | (2.7) | |
| $\Phi_3$ | −2.816 | −2.820 | |
| | (2.1) | (2.1) | |
| $\Phi_4$ | 0.305 | 0.281 | |
| | (4.6) | (4.2) | |
| $\Phi_5$ | | | |
| $\Gamma_2$ | 0.081 | −0.082 | |
| | (4.5) | (4.4) | |
| $\mu$ | 0.695 | 0.719 | |
| $\beta_1$ | −6.115 | −5.502 | −6.117 |
| $\beta_2$ | 27.380 | 26.668 | 27.381 |
| $\beta_3$ | −4.052 | −3.922 | −4.054 |
| Adjusted $R^2$ | 0.872 | 0.816 | 0.872 |
| Durbin's $h$ | 0.375 | 0.570 | |
| $\rho$ | | 0.285 | |

*Note:* The equation is:

$$d_t - p_t = \Phi_0 + \Phi_1 y_t + \Phi_2 i_t + \Phi_3(i_t^* + \delta_t) + \Phi_4(d_{t-1} - p_{t-1}) + \Phi_5 \dot{d}_t + \Gamma_2 \text{TIME}.$$

For methods and definitions, see table 4.8. A blank denotes the omission of the specific variable from the regression.

public bond markets. The estimates confirm a larger impact on inflation when fiscal deficits are money financed than when they are debt financed.

Table 4.10 reports the reduced-form equation results for the interest rate and price level equations (4.5 and 4.6), on the basis of the 1983.1–1988.4 sample. All variables with an unambiguous effect on the interest rate and the price level present the expected signs. The only exception is lagged real public bond holdings, which has the wrong sign in both the interest rate and the price level equations but is not significant.[12] In addition, for the variables with a priori ambiguous signs in the reduced-form equations, the signs of the estimated coefficients coincide in all cases with the signs implied by the results of the structural equations.

**Table 4.10. Estimation Results for Reduced-Form Equations for the Rates of Interest and Inflation, Ordinary Least Squares Method, Chile, 1983:2–1988:4**

| Independent variable | Interest rate | Price level |
|---|---|---|
| Constant | −1.170 | −8.810 |
| | (2.5) | (1.3) |
| $y_t$ | 0.084 | 0.170 |
| | (1.9) | (0.3) |
| $i_t^* + \delta_t$ | 0.106 | 0.632 |
| | (4.8) | (1.9) |
| $m_{t-1} - p_{t-1}$ | 0.055 | −0.110 |
| | (2.8) | (0.4) |
| $d_{t-1} - p_{t-1}$ | 0.00026 | 0.064 |
| | (0.03) | (0.6) |
| $m_t$ | −0.014 | 0.514 |
| | (−1.9) | (2.0) |
| $d_t$ | 0.004 | 0.182 |
| | (0.5) | (1.6) |
| $S_t$ | 0.0039 | −0.068 |
| | (0.5) | (0.6) |
| Adjusted $R^2$ | 0.757 | 0.977 |
| Durbin-Watson | 1.72 | 1.50 |

*Note:* Figures in parentheses are *t*-values.

Figure 4.3 shows the steady-state relationship between inflation rates and the inflation-tax revenue derived from the estimated money demand coefficients and assuming no growth in exogenous variables. At an inflation rate of 2 percent per month, the inflation tax is about 0.5 percent of GDP. Little tax gain is obtained from higher inflation. Raising the inflation rate from 2 to 20 percent per month—the revenue-maximizing rate—increases seigniorage revenue by only 1.5 percentage points of GDP.

*Simulation Results*

The results reported in the first columns of tables 4.8 and 4.9 were used to simulate the impact on inflation and interest rates of a temporary increase in the fiscal deficit under alternative strategies of domestic financing. To simulate the twelve-quarter effects of a transitory one-quarter public sector deficit, we assume that the economy is at an initial steady-state equilibrium with a money-to-GDP ratio of 0.2 and a debt-to-GDP ratio of 1.0.[13]

Table 4.11 reports the short- and long-term effects of money and debt-financed one-quarter fiscal deficits of 2 percent of GDP. The

**Figure 4.3. Seigniorage Revenue, Chile**

Percentage of GDP

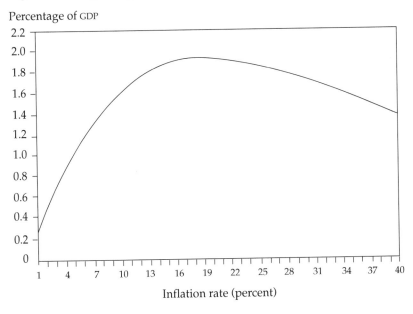

Inflation rate (percent)

*Source:* Authors' calculations.

money-financed fiscal shock causes a fall in the interest rate of 0.39 percentage point in the first period and a long-run decline of about 0.27 percentage point. The effect on inflation is about 7 percent in both the short and the long run. Hence most of the consequences of a money-financed deficit are felt during the first period. In addition, since the increase in the price level never matches the increase in the stock of money required to finance the fiscal shock, real monetary holdings are affected. The dynamic behavior of the interest and inflation rates in response to this money-financed deficit is shown in figure 4.4.

**Table 4.11. Simulation Results of a Temporary Fiscal Deficit, Chile**

| Item | Short run | Long run |
|---|---|---|
| *Effects of a money-financed increase in the fiscal deficit* | | |
| Change in interest rate (percentage points) | −0.39 | −0.27 |
| Change in the price level (100 *d* log *p*) | 7.33 | 7.25 |
| *Effects of a public debt-financed increase in the fiscal deficit* | | |
| Change in interest rate (percentage points) | 0.09 | 0.05 |
| Change in price level (100 *d* log *p*) | 0.28 | 0.55 |

**Figure 4.4. Simulation Effects of a Money-Financed Deficit on Interest and Inflation Rates, Chile**

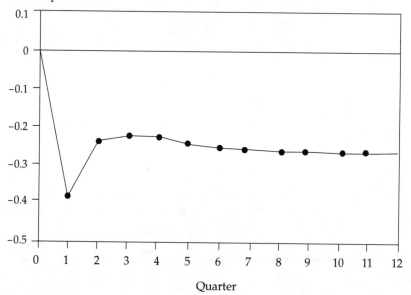

**Interest rate**

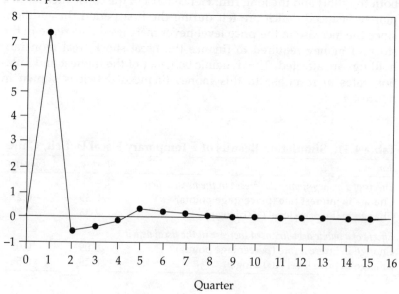

**Inflation rate**

A temporary deficit of 2 percent of GDP financed by public debt raises slightly both the interest rate and the price level (see figure 4.5 for the dynamic response). The increase in the price level required to maintain the portfolio equilibrium is much lower now than under the money-financed deficit because of the higher interest rate.

To summarize, these results suggest that the main impact of a change in the fiscal deficit occurs in the first quarter. The only exception is the gradual rise in prices when the fiscal deficit is financed by domestic borrowing. The results also suggest that debt-financed deficits have a small effect on interest and inflation rates—an outcome that reflects the specific features of the estimation sample period. The net transfer payments to the private sector implied by the social security reform and the quasi-fiscal operations of the Central Bank during the 1980s, which were accommodated by the rising demand for public debt paper by the new private social security corporations, are the main reasons for the small impact of debt-financed fiscal deficits on interest rates and prices.

## Crowding-Out and Crowding-In of Private Consumption and Private Investment by the Public Sector

This section assesses the impact of the public sector on private sector spending in Chile. The focus is on the sensitivity of private consumption and investment to fiscal variables, in addition to the indirect effects of these variables through interest rates, inflation, and private disposable income. How private saving and capital formation are affected by fiscal policies has significant implications for both short-run stabilization issues and long-run growth prospects.

Figure 4.6 shows the evolution of the ratios of private consumption to private disposable income and private investment to GDP during the past three decades. It is not surprising that private consumption has tended to be countercyclical—in other words, consumption does not decline as strongly as does private disposable income during recessions such as those of 1975 and 1982–83. Two high-consumption episodes coincided with the fiscal-monetary expansion of the early 1970s and the foreign-financed private spending boom of ten years later. A distinct structural change took place after 1984: private consumption as a share of disposable income reached the lowest values in three decades and remained at that level throughout the 1985–88 recovery—an important counterpart to the significant current account adjustment that took place in the 1980s.

The private investment rate showed wild swings during the 1960–88 period in Chile. After reaching a historical low of 5 percent of GDP during the early 1970s, private investment began a continuous recovery and eventually boomed in the early 1980s. Private capital forma-

**Figure 4.5.  Simulation Effects of a Debt-Financed Deficit on Interest And Inflation Rates, Chile**

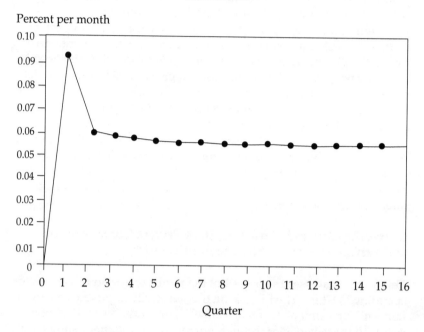

Interest rate

Percent per month

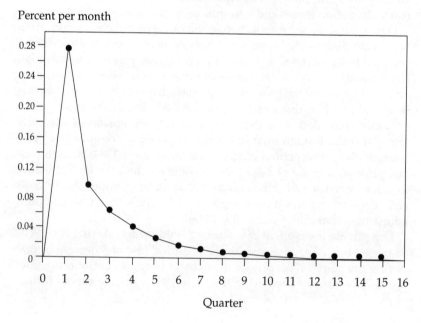

Inflation rate

Percent per month

**Figure 4.6. Private Consumption and Investment, Chile**

**Ratio of private consumption to private disposable income, 1960–88**

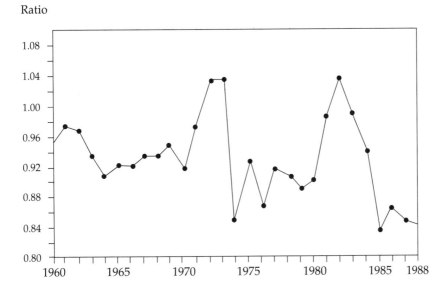

**Ratio of private investment to GDP, 1961–88**

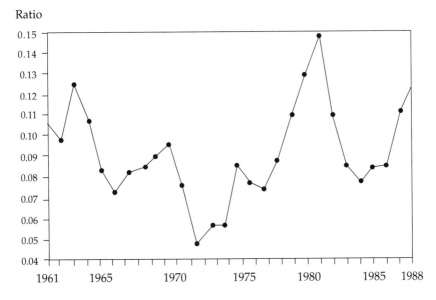

*Source:* See appendix 4.4.

tion then took a dive that coincided with the 1982–83 recession and the subsequent policy uncertainty. Investment remained low in 1984–86 before picking up again in 1987 and 1988 and growing to remarkably high levels—about 20 percent—in the early 1990s.

### Private Consumption

The direct effects of fiscal policies on consumption or saving operate through public saving (or the deficit) and its composition. If the stringent conditions required for Ricardian equivalence are satisfied, a permanent rise in public saving—if it comes about via lower public spending—is exactly offset by an increase in consumption. Because disposable income does not change, the reduction in private saving matches the increase in public saving. A rise in public saving does not affect consumption at all if it is a result of higher taxes. But as disposable income is reduced by the size of the tax, the reduction in private saving also matches the increase in public saving (the latter being equivalent in macroeconomic terms to issuing additional public debt). The opposite results are predicted by the Keynesian hypothesis (and the permanent income hypothesis without Ricardian equivalence): current (permanent) taxes matter for consumption, not current (permanent) public spending levels.[14]

The indirect effects of public (fiscal-monetary) policies on consumption operate through the impact of public deficits and their financing on major relative prices—real interest rates, inflation, and the real exchange rate—that affect private consumption. Real interest rates affect private consumption only if the substitution, income, and human wealth effects do not offset each other. But this seems not to be the case, to judge from the growing evidence showing that consumption is not sensitive to real interest rates.[15] Although the first-order effect of inflation is on the composition of the savings portfolio and not on saving or consumption flows, its second-order effects may reduce saving if inflation is associated with capital flight or raises precautionary saving as a result of higher uncertainty. An expected devaluation reduces the consumption-based real interest rate and leads to capital flight; the first effect could increase actual consumption, while capital flight raises the measured consumption-to-income ratio if it reduces measured income.[16] Finally, substitution effects from fiscal policies could arise if fiscal spending on privately appropriated services such as education, health, and nutrition, and direct transfers to households reduce the need for private consumption expenditure on these items, leading to possible declines in aggregate private consumption (see Easterly, Rodríguez, and Schmidt-Hebbel 1989).

Equation 4.7 specifies the ratio of private consumption to private disposable income, following the previous discussion and a framework developed by Corbo and Schmidt-Hebbel (1991):[17]

$$(4.7) \quad \frac{C_{pt}}{DY_{pt}} = \beta_0 + \beta_1 \frac{PDY_{pt}}{DY_{pt}} + \beta_2 \frac{PS_{gt}}{DY_{pt}} + \beta_3 r_{ct} + \beta_4 \pi_{ct} + \beta_5 \frac{P_{cmt}}{P_{cnt}}$$

$$+ \beta_6 \frac{CPTR_t}{DY_{pt}} + \beta_7 \frac{H_t}{DY_{pt}} + \beta_8 \frac{FS_t}{DY_{pt}}$$

where $C_p$ is private consumption expenditure, $DY_p$ is current private disposable income, $PDY_p$ is permanent private disposable income, $PS_g$ is permanent public saving, $r_c$ is the consumption-based real interest rate, $\pi_c$ is consumption inflation, $P_{cm}$ and $P_{cn}$ are deflators for imported and domestic private consumption goods, respectively, CPTR is the sum of public expenditure on privately appropriated services and direct transfers to consumers, $H$ is base money, and $FS$ is foreign saving.

The coefficients $\beta_0$, $\beta_1$, and $\beta_2$ are associated with the corresponding Keynesian, permanent income, and Ricardian hypotheses.[18] Note, however, that $\beta_2$ could be positive if the public sector has preferential access to bank credit, forcing private consumption (and investment) to adjust residually. This is a form of institutional or direct crowding-out of private saving by public sector saving that is obviously unrelated to the Ricardian notion of unconstrained, forward-looking private consumers who internalize the government's intertemporal budget constraint. Expected signs of the coefficients are: $\beta_0$, $\beta_1$, $\beta_2$, $\beta_7$, $\beta_8 > 0$; $\beta_6 < 0$; $\beta_3$, $\beta_4$, $\beta_5 \gtrless 0$.[19]

Equation 4.7 was tested using annual data for the 1960–88 period, which is the sample period used most frequently in this section and the next. This period is characterized by significant structural breaks and changes in policy regime, which, because of their intensity, necessitate caution in the interpretation of the statistical results presented below.

Table 4.12 presents a selected number of empirical results for the consumption function. In general, the results for the static-expectations version dominate statistically those related to the partial perfect-foresight case.[20] The average Chilean consumer appears to be partly Keynesian, partly permanent-income forward-looking: under the static version marginal propensity to consume out of current private disposable income is approximately 0.70, and marginal propensity to consume out of permanent income is about 0.30. Under the partial perfect-foresight version the corresponding propensities are of a similar magnitude, close to 0.50. It is interesting to note that consumers do not react to permanent public saving; that is, the consolidated public sector deficit (less public investment) has no effect on private consumption. The policy implication is clear: an increase in

**Table 4.12. Private Consumption Estimation Results, Chile, 1960–88**

| Equation | $C$ | $\frac{PDY_p}{DY_p}$ | $\frac{PS_g}{DY_p}$ | $r_c$ | $\pi_c$ | $\frac{P_{cm}}{P_{cn}}$ | $\frac{CPTR}{DY_p}$ | $\frac{H}{DY_p}$ | $\frac{FS}{DY_p}$ | $\rho$ | $R^2A$ | Durbin-Watson |
|---|---|---|---|---|---|---|---|---|---|---|---|---|
| *Static expectations* | | | | | | | | | | | | |
| 1 OLS | 0.71 (10.4) | 0.28 (4.1) | 0.18 (0.8) | 0.05 (1.7) | -0.02 (-0.5) | -0.14 (-3.0) | 0.12 (0.4) | 0.29 (3.2) | 0.24 (1.7) | | 0.79 | 1.77 |
| 2 OLS | 0.74 (9.8) | 0.25 (3.2) | 0.36 (1.6) | | | -0.13 (-5.2) | | 0.23 (2.6) | 0.37 (2.5) | | 0.72 | 1.52 |
| 3 ML | 0.66 (10.0) | 0.26 (3.7) | 0.29 (1.2) | | | -0.08 (-2.2) | | 0.45 (3.8) | 0.41 (2.8) | 0.59 (3.5) | 0.90 | 1.75 |
| 4 OLS | 0.72 (9.4) | 0.27 (3.5) | | | | -0.13 (-5.0) | | 0.24 (2.6) | 0.48 (3.6) | | 0.70 | 1.38 |
| 5 ML | 0.67 (10.0) | 0.27 (3.9) | | | | -0.09 (-2.4) | | 0.44 (3.7) | 0.42 (2.8) | 0.57 (13.4) | 0.89 | 1.50 |
| *Partial perfect foresight* | | | | | | | | | | | | |
| 6 OLS | 0.49 (2.6) | 0.49 (2.8) | 0.002 (0.005) | 0.0004 (0.0008) | 0.08 (1.2) | -0.19 (-3.6) | 0.04 (0.09) | 0.28 (1.5) | 0.87 (4.0) | | 0.63 | 1.32 |
| 7 ML | 0.38 (2.9) | 0.50 (3.6) | 0.38 (0.8) | -0.03 (-0.6) | 0.03 (0.4) | -0.11 (-2.1) | 0.24 (0.5) | 0.52 (3.0) | 0.68 (4.1) | 0.67 (4.1) | 0.89 | 1.39 |
| 8 ML | 0.42 (3.4) | 0.48 (3.7) | 0.14 (0.4) | | | -0.10 (-2.4) | | 0.58 (4.8) | 0.71 (4.6) | 0.62 (3.9) | 0.89 | 1.41 |
| 9 ML | 0.43 (3.5) | 0.48 (3.80) | | | | -0.10 (-2.6) | | 0.57 (4.9) | 0.73 (4.9) | 0.60 (3.7) | 0.90 | 1.43 |

*Note:* The dependent variable is the ratio of private consumption to private disposable income ($C_p/DY_p$). The estimation methods were ordinary least squares (OLS) and maximum likelihood (ML). Figures in parentheses are *t*-statistics; $\rho$ is the first-order residual correlation coefficient. $R^2A$ is adjusted $R^2$. A blank denotes the omission of the specific variable from the regression.

public sector saving has a strong effect on national savings, as it is offset only in part by a decline in private saving.

Of all relevant prices, only the real exchange rate (given by the relative prices of the imported and domestic consumption goods components of aggregate consumption) affects private consumption. A 10 percent real devaluation, which would affect directly this relative price, would reduce aggregate consumption by approximately 1 percentage point of private disposable income. But neither the real interest rate nor the inflation rate has any consistently significant separate effects on private consumption. This outcome corroborates earlier studies on the interest insensitivity of consumption in Chile (Arrau 1989; Schmidt-Hebbel 1987, 1988). In addition, there is no evidence of a substitution effect of public spending on privately appropriate services (such as education and health) or of transfers on aggregate consumption. Monetary holdings and foreign saving flows exert significant positive influences on private consumption. Whereas higher base money holdings tend to relax domestic borrowing constraints faced by consumers, higher foreign saving is associated with weaker foreign resource constraints. The effect of weaker foreign constraints on private consumption is very strong: from these results one can infer that between 42 and 71 percent of the dramatic current account correction that took place after 1981 was borne by private consumers.

The policy implications of our results are clear. First, there is no evidence of a direct effect of the public deficit (taken separately from the influence of taxes through disposable income) on private consumption; consumers in Chile are neither Ricardian nor directly crowded-out by public spending. A transitory deficit reduction financed by a transitory tax hike reduces private disposable income and hence private consumption and saving. If the tax hike is permanent, private consumption will take the brunt of the reduction in (current and permanent) private disposable income, without private saving being affected significantly.

Second, domestic financing of public deficits has no indirect effects on consumption through changes in the real interest and inflation rates. However, if inflation reduces real base money, private consumption will be affected. Foreign adjustments have strong effects on private consumption in Chile. The combination of lower foreign saving and a higher real exchange rate has a strong negative effect on private consumption expenditure, as evidenced during the post-1982 adjustment period.

### Private Investment

Following Easterly, Rodríguez, and Schmidt-Hebbel (1989) and empirical investment studies of Chile by Solimano (1992) and of

**Table 4.13. Private Investment Estimation Results, Chile, 1961–88**

| Equation | C | D60s | D70s | PUCK | $\frac{PK_{p,-1}}{Y}$ | $\frac{K_{g,-1}}{Y}$ | $\frac{PRO}{Y}$ | $\frac{H_{-1}}{Y}$ | $\frac{FS}{Y}$ | VUCK | VY | ρ | $R^2A$ | Durbin-Watson |
|---|---|---|---|---|---|---|---|---|---|---|---|---|---|---|
| *Static expectations* | | | | | | | | | | | | | | |
| 1 OLS | 0.22 (3.5) | 0.06 (0.8) | -0.10 (-1.5) | 0.12 (2.3) | -0.05 (-1.5) | -0.11 (-2.2) | 0.17 (2.0) | -0.11 (-1.4) | -0.07 (-0.5) | -0.01 (-1.8) | -0.01 (-0.1) | | 0.62 | 1.29 |
| 2 ML (AR1) | 0.17 (2.7) | -0.03 (-0.5) | 0.05 (0.1) | 0.04 (1.1) | -0.06 (-1.8) | -0.07 (-1.6) | 0.26 (2.8) | -0.11 (-1.6) | 0.05 (0.4) | -0.002 (-0.7) | -0.27 (-1.3) | 0.79 (5.9) | 0.59 | 1.93 |
| 3 ML (AR1) | 0.10 (1.7) | -0.08 (-1.2) | 0.05 (1.6) | 0.01 (0.5) | -0.07 (-1.9) | | 0.25 (2.5) | | 0.13 (1.4) | | -0.26 (-1.2) | 0.81 (6.2) | 0.51 | 1.72 |
| *Partial perfect foresight* | | | | | | | | | | | | | | |
| 4 OLS | 0.21 (3.4) | -0.27 (-2.4) | -0.09 (-0.8) | 0.04 (0.8) | 0.04 (0.8) | -0.18 (-3.1) | 0.03 (0.3) | 0.22 (0.9) | -0.11 (-0.9) | -0.02 (-1.3) | -0.15 (-0.7) | | 0.71 | 1.19 |
| 5 ML (AR1) | 0.18 (2.7) | -0.21 (-2.1) | -0.15 (-1.7) | 0.16 (3.1) | -0.02 (-0.2) | -0.12 (-2.0) | 0.16 (1.7) | -0.10 (-0.4) | -0.06 (-0.05) | -0.001 (-1.1) | -0.18 (-0.8) | 0.63 (3.2) | 0.67 | 1.81 |
| 6 ML (AR1) | 0.20 (2.5) | -0.09 (-1.0) | -0.001 (-0.002) | 0.08 (2.0) | -0.12 (-2.3) | | 0.17 (1.8) | | 0.17 (2.0) | | -0.46 (-2.2) | 0.80 (5.9) | 0.57 | 1.96 |

*Note:* The dependent variable is the ratio of private investment to GDP $(I_p/Y)$. The estimation methods were ordinary least squares (OLS) and maximum likelihood (ML). Figures in parentheses are $t$-statistics; ρ is the first-order residual correlation coefficient; $R^2A$ is adjusted $R^2$. A blank denotes the omission of the specific variable from the regression.

Morocco by Schmidt-Hebbel and Müller (1992), we specify a behavioral function for private investment that will depend on profit, cost, and tax variables, as well as on liquidity constraints and risk determinants. To reduce the incidence of spurious correlation, we scale all nonstationary variables to GDP. Therefore we specify the following generic equation for the ratio of private investment to GDP:

$$
(4.8) \quad \frac{I_{pt}}{Y_t} = \frac{I}{Y}\left( \underset{(-)}{PUCK_t}, \underset{(?)}{\frac{PK_{pt-1}}{Y_t}}, \underset{(?)}{\frac{P_{ipmt}}{P_{ipmt}}}, \underset{(-)}{\frac{PCOT_t}{Y_t}}, \underset{(+)}{\frac{K_{gt-1}}{Y_t}}, \underset{(+)}{\frac{PRO_t}{Y_t}}, \underset{(+)}{\frac{FC_t}{Y_t}}, \underset{(+)}{\frac{H_{t-1}}{Y_t}}, \underset{(+)}{\frac{FS_t}{Y_t}}, \underset{(-)}{VUCK_t}, \underset{(-)}{VY_t} \right)
$$

where $I_p$ is private fixed-capital investment, $Y$ is GDP, UCK is the user cost of capital, PUCK is the estimated permanent UCK, $PK_{p-1}$ is the permanent estimate of private sector capital, $P_{ipm}/P_{ipn}$ is the price ratio of imported and domestic private investment components, COT is corporate tax revenue, PCOT is the permanent estimate of COT, $K_g$ is public sector capital stock, PRO is corporate profits, FC is banking credit flows to firms, $H$ is base money, FS is foreign saving, VUCK is the coefficient of variation of UCK, and VY is the coefficient of variation of GDP. The expected signs of partial derivatives are denoted below each variable in the equation.[21] The empirical results for the private investment functions are presented in tables 4.13 and 4.14, for both the static and partial perfect-foresight expectation alternatives for all relevant right-hand variables.[22]

In all the equations reported in table 4.13, the permanent user cost of capital (PUCK) has a positive effect, which is also significant in most cases. This seems to be a reflection of the extremely strong structural breaks that occurred during the 1960–88 period with regard to the functioning of the Chilean financial market, the determination and the levels of the real interest rate, and hence the dependence of investment on the cost of domestic financial capital. Two corrections were made to face this problem. First, multiplicative dummies for PUCK were specified, separately for the 1960s (D60s) and the early 1970s (D70s) (see table 4.13). Second, the original specification was tested for the 1976–88 period, which began after the 1974–75 financial interest liberalization (see table 4.14). However, the brevity of this more homogeneous period suggests the need for caution in interpreting the results of the table because of the small number of degrees of freedom.

The coefficients of the period-specific dummies in table 4.13 (for 1961–88) tend to present the correct negative sign and are significant in some of the reported results. For example, regression 5 shows that

**Table 4.14. Private Investment Estimation Results, Chile, 1976–88**

| Equation | C | PUCK | $\frac{PK_{p,-1}}{Y}$ | $\frac{K_{g,-1}}{Y}$ | $\frac{PRO}{Y}$ | $\frac{H_{-1}}{Y}$ | $\frac{FS}{Y}$ | VUCK | VY | ρ | R²A | Durbin-Watson |
|---|---|---|---|---|---|---|---|---|---|---|---|---|
| *Static expectations* | | | | | | | | | | | | |
| 1 OLS | 0.35 (4.1) | 0.05 (0.7) | −0.09 (−1.8) | −0.14 (−2.2) | 0.02 (0.14) | 0.51 (1.7) | −0.02 (−0.1) | 0.003 (0.5) | −0.26 (−1.0) | | 0.88 | 3.69 |
| 2 OLS | 0.36 (0.5) | −0.06 (−2.0) | −0.17 (−7.3) | | | 0.26 (1.1) | | | 0.56 (−2.3) | | 0.82 | 2.95 |
| 3 ML (AR1) | 0.37 (16.9) | −0.07 (−3.1) | −0.18 (−11.7) | | | 0.36 (1.9) | | | −0.65 (−3.3) | −0.51 (−1.9) | 0.95 | 2.19 |
| *Partial perfect foresight* | | | | | | | | | | | | |
| 4 OLS | 0.52 (2.6) | 0.16 (1.3) | −0.19 (−1.0) | −0.07 (−0.5) | −0.24 (−1.2) | −0.32 (−0.5) | −0.05 (−0.2) | 0.001 (0.1) | −0.27 (−0.5) | | 0.86 | 2.91 |
| 5 ML (AR1) | 0.45 (13.8) | 0.10 (3.2) | −0.24 (−11.1) | | | −0.49 (−2.5) | | | −0.55 (−2.9) | −0.37 (−1.1) | 0.95 | 2.09 |

*Note:* The dependent variable is the ratio of private investment to GDP ($I_p/Y$). The estimation methods were ordinary least squares (OLS) and maximum likelihood (ML). Figures in parentheses are *t*-statistics; ρ is the first-order residual correlation coefficient; R²A is adjusted R². A blank denotes the omission of the specific variable from the regression.

while the coefficient of PUCK is still positive for the 1974–88 period (as reflected by its value of 0.16), its net influence is close to zero for the early 1970s (0.16 for PUCK minus 0.15 for D70s gives a net effect of 0.01) and is negative for the 1960s (0.15 for PUCK minus 0.21 for D60s gives a net effect of −0.06). However, for the 1976–88 period and the static-expectations alternative, the user cost of capital is negative and significant—although its coefficient is small—in the best results reported in table 4.14.

The lagged ratio of capital to current output appears to be negative and highly significant in most reported results. This is not surprising: private investment is strongly procyclical. The consistently negative influence of public sector capital stock on private investment suggests that either public investment tends to substitute for private investment or private investment is directly crowded out by public investment because of the public sector's preferential access to domestic saving. Although direct crowding-out could have been present until the mid-1970s, the domestic financial liberalization rules it out for the post-1975 period. Whatever the reason, clearly there is no evidence for a complementarity of public and private capital that leads to crowding-in of private investment by public sector capital spending in Chile. However, to assess possible noise stemming from the inclusion of this variable, we also present results that omit it.

The influence of liquidity constraint variables depends strongly on the period selected. This is very consistent with the change in the role of interest rates and hence of the user cost of capital during the 1960–88 period: quantity constraints were of greater importance before domestic financial liberalization than afterward. Precisely the same is true of firm profits and foreign saving, which had a strong influence on investment during the 1960–88 period as a whole but disappeared from the scene during the 1976–88 period. Base money plays an ambiguous and unstable role: whereas in most subperiods and specification alternatives it exerts a negative, nonsignificant influence, it presents a positive and significant, although minor, effect only in regression 3 (static expectations) for the 1976–88 period (see table 4.14).

The coefficient of variation of the user cost has a negative, although usually insignificant, influence during the complete 1976–88 period, and its effect disappears during the second subperiod. Much more important is the variability of GDP, which affects private investment consistently and negatively during the entire period and particularly strongly in more recent years.

From the empirical results we can draw a number of policy implications. Financing of public deficits through debt would be felt increasingly by private investors in Chile because of its upward pressure on domestic real interest rates in unrepressed financial markets. The other side of financial liberalization is the weakening of both domestic and foreign liquidity constraints: firm profits and foreign saving were important investment determinants before financial liberalization but vanished afterward. However, foreign adjustment has affected private investment negatively through the relative price of investment goods, which increased as a consequence of the real devaluations and so raised the user cost of capital.

No evidence was found on public investment's crowding-in effects on private investment. The relevant variable—public capital stock—tends to reduce private capital formation, suggesting that public investment competes with private capital formation.

Finally, investment is negatively affected by the variability of GDP. More stability in the external environment and in domestic macroeconomic policies influences private capital formation positively.

## Relative External Prices and the Trade Balance

This section focuses on the impact of fiscal policies on the external sector, especially with regard to the trade deficit and the real exchange rate. The framework applied here is derived in two steps. The first involves determination of the relative prices of exportables and importables from the equilibrium condition in the nontradable goods market; the second implies derivation of an equation for the trade balance.

Figure 4.7 depicts the evolution of relative export and import prices and the ratio of the trade surplus to GDP during the past three decades in Chile. The price of exports relative to nontradables is strongly correlated with the international copper price, which is the main component of the country's external export prices and terms of trade. The price of imports relative to nontradables is a more useful measure of the real exchange rate relevant for decisions on spending and production (other than for copper). However, both measures of the real exchange rate tended to move together during 1975–88, a period of smaller terms of trade fluctuations than those experienced during 1960–75.

The two periods of massive real exchange rate appreciations—the early 1970s and the early 1980s—coincide with the policy-induced spending frenzy of the Allende government and the foreign-financed private spending spree of Pinochet's *plata dulce*. Trade deficits, depicted in the lower panel of figure 4.7, reached their highest levels during those years—the most dramatic observation being the 10 percent of GDP trade deficit observed in 1981. The required current-account corrections after 1973 and 1981 led to drastic reversals of the trade deficits as a result of massive exchange rate depreciations and reductions in domestic spending. A distinct feature of the post-1984

## Figure 4.7. Relative Export and Import Prices, Chile

**Export and import prices, 1960–88**

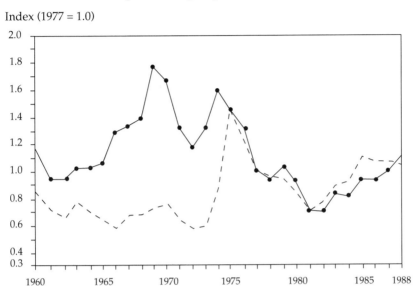

$\bullet$ = $P_X/P_N$, ratio of exportable to nontradable prices

$-$ = $P_M/P_N$, ratio of importable to nontradable prices

*Note:* $P_X/P_N$, export prices; $P_M/P_N$, import prices.

**Ratio of trade surplus to GDP, 1960–88**

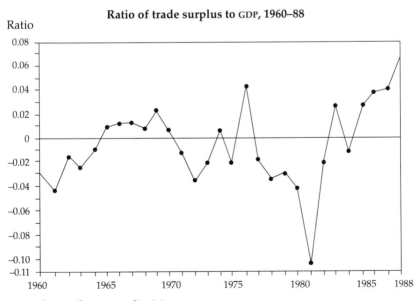

*Source:* See appendix 4.4.

**Table 4.15. Relative Export and Import Prices Estimation Results, Chile, 1960–88**

| Equation | C | TT* | $t_M$ | $\frac{TS}{Y}$ | $\frac{G}{Y}$ | D75 | ρ | R²A | Durbin-Watson |
|---|---|---|---|---|---|---|---|---|---|
| *Relative export price* | | | | | | | | | |
| 1 OLS | -0.21 (-1.2) | 0.63 (8.5) | -0.04 (-0.8) | 2.48 (4.0) | 3.64 (3.0) | 0.51 (4.5) | | 0.86 | 1.29 |
| 2 ML (AR1) | 0.01 (0.04) | 0.59 (6.9) | -0.05 (-0.9) | 2.22 (3.4) | 2.50 (1.7) | 0.38 (4.0) | 0.52 (3.0) | 0.77 | 1.76 |
| 3 ML (AR1) | -0.02 (-0.01) | 0.56 (7.7) | | 2.39 (3.8) | 2.80 (2.0) | 0.37 (4.0) | 0.51 (3.0) | 0.77 | 1.74 |
| *Relative import price* | | | | | | | | | |
| 4 OLS | 1.20 (6.7) | -0.15 (-2.0) | -0.09 (-1.9) | 2.10 (3.4) | -0.62 (-0.5) | 0.61 (5.4) | | 0.75 | 1.03 |
| 5 ML (AR1) | 1.26 (5.4) | -0.14 (-1.8) | -0.09 (-1.7) | 2.00 (3.3) | -1.09 (-0.8) | 0.47 (5.7) | 0.62 (3.9) | 0.72 | 1.65 |
| 6 ML (AR1) | 1.09 (10.7) | -0.13 (-1.7) | -0.09 (-1.6) | 2.04 (3.5) | | 0.47 (5.7) | 0.60 (3.6) | 0.72 | 1.62 |
| 7 ML (AR1) | 1.08 (9.8) | -0.18 (-2.5) | | 2.26 (3.8) | | 0.46 (5.5) | 0.66 (4.6) | 0.70 | 1.58 |

*Note:* The dependent variables are the relative export price (*ex*) and the relative import price (*em*). The estimation methods were ordinary least squares (OLS) and maximum likelihood (ML). Figures in parentheses are *t*-statistics; ρ is the first-order residual correlation coefficient; R²A is adjusted R². A blank denotes the omission of the specific variable from the regression.

external adjustment was the maintenance of massive trade surpluses, exceeding 6 percent of GDP in 1988. These surpluses are the counterpart of the significant decline in private consumption as a share of income (discussed in the preceding section) and a progressive correction in public sector deficits (discussed in the section on fiscal policy, above).

## Relative Prices of Exports and Imports

Following Easterly, Rodríguez, and Schmidt-Hebbel (1989) and Rodríguez, chapter 2 in this volume, the real exchange rate, defined as the relative price of tradable and nontradable goods, is derived from the continuous market-clearing condition for nontradable goods in the Salter-Swan-Corden-Dornbusch tradition for the small open economy. Extension of this paradigm to a three-sector distinction between exportables, importables, and nontradables permits rewriting of the market-clearing condition as either one of the following functions for the relative prices of exportables and importables:[23]

$$(4.9) \qquad ex_t \equiv \frac{P_{Xt}}{P_{Nt}} = ex \left( \underset{(+)}{TT_t^*}, \ \underset{(?)}{t_{Mt}}, \ \underset{(+)}{\frac{TS_t}{Y_t}}, \ \underset{(?)}{\frac{G_t}{Y_t}}, \ \underset{(-)}{\frac{G_{Nt}}{G_t}} \right)$$

$$(4.10) \qquad em_t \equiv \frac{P_{Mt}}{P_{Nt}} = em \left( \underset{(-)}{TT_t^*}, \ \underset{(+)}{t_{Mt}}, \ \underset{(+)}{\frac{TS_t}{Y_t}}, \ \underset{(?)}{\frac{G_t}{Y_t}}, \ \underset{(-)}{\frac{G_{Nt}}{G_t}} \right)$$

where $ex$ and $em$ are the relative prices of exportables and importables, respectively; $P_X$, $P_M$, and $P_N$ are the absolute prices of exportables, importables, and nontradables; $TT^*$ is foreign terms of trade; $t_M$ is the ad valorem tariff rate for imports; $TS$ is the current-price trade surplus; $G$ is current-price (general) government spending (public consumption plus public investment); $G_N$ is current-price (general) government spending on nontradable goods; and $Y$ is current-price GDP.

This framework was applied to Chile using annual data for the 1960–88 period. The results, reported in table 4.15 for both equations, tend to favor the relative export price interpretation as the relevant reduced-form equation for the equilibrium in the nontradable goods market. The external terms of trade are an important determinant of both export and import prices, although their quantitative influence on export prices is much stronger. The average tariff rate has a negative (although not significant) influence on both relative prices—a surprising result in the case of the import price, which should be positively affected by tariffs.

The ratio of the trade surplus to output has an extremely high, consistent, and symmetric effect on both relative prices: an increase in the trade surplus of 1 percent of GDP implies a 2.0 to 2.5 percent devaluation of the real exchange rate. Finally, aggregate government spending has a strongly positive impact on the relative prices of exports and a nonsignificant negative influence on the relative price of imports. In general, its influence could be of either sign.[24]

### The Trade Balance

The trade balance is both the difference between output and absorption and the goods-markets counterpart of the accumulation of net foreign assets. In Rodríguez, chapter 2 in this volume, and Easterly, Rodríguez, and Schmidt-Hebbel (1989), the trade surplus is directly related to the accumulation of net foreign assets (NFA). Private net foreign asset accumulation depends on the difference between desired and actual private NFA holdings, the former being replaced by its main determinants: the covered interest differential between domestic and foreign rates, domestic public debt, the terms of trade, and income. Public NFA accumulation will reflect directly, with a negative sign, the (operational) public sector deficit for given stocks of domestic public debt and base money. Hence, under this foreign asset accumulation version the equation for the trade balance can be written as:[25]

(4.11)

$$\frac{TS_t}{Y_t} = \frac{TS}{Y} \left[ \underset{(+)}{C}, \underset{(-)}{\frac{i_t - (i_t^* + \hat{E}_t^e + i_t^* \hat{E}_t^e)}{1 + (i_t^* + \hat{E}_t^e + i_t^* \hat{E}_t^e)}}, \underset{(-)}{\frac{NFA_{t-1}}{Y_t}}, \underset{(-)}{\frac{B_{t-1}}{Y_t}}, \underset{(+)}{TT_{t}^*}, \underset{(-)}{\frac{OD_{gt}}{Y_t}} \right]$$

where $TS$ is the current-price trade surplus, $Y$ is current-price GDP, $C$ is a constant, $i$ is the average annual domestic nominal interest rate (using the average of active and passive rates), $i^*$ is the average annual nominal external interest rate paid on net foreign assets, $\hat{E}^e$ is the expected rate of nominal devaluation (defined below), $B_{-1}$ is current-price domestic public sector debt (at the end of the preceding period), $NFA_{-1}$ is current-price net foreign assets (total foreign debt less international reserves at the end of the preceding period), and $OD_g$ is the current-price operational public sector deficit. The expected signs of partial derivatives are denoted in parentheses below each variable.

Another way of viewing the trade balance is to derive it from the macroeconomic equilibrium condition (see appendix 4.3 for a detailed derivation):

(4.12)

$$\frac{TS_t}{Y_t} = \frac{TS}{Y} \left[ \underset{(+)}{C}, \underset{(?)}{\frac{i - (i_t^* + \hat{E}_t^e + i_t^* \hat{E}_t^e)}{1 + (i_t^* + \hat{E}_t^e + i_t^* \hat{E}_t^e)}}, \underset{(?)}{\frac{NFA_{t-1}}{Y_t}}, \underset{(?)}{\frac{B_{t-1}}{Y_t}}, \underset{(+)}{TT_{t}^*}, \underset{(-)}{\frac{OD_{gt}}{Y_t}}, \underset{(-)}{\frac{YP_t}{Y_t}} \right]$$

where $YP$ is permanent GDP at current prices.

Two differences arise between the asset-accumulation version of the trade surplus (equation 4.11) and the macroeconomic equilibrium version (equation 4.12). First, the effect of domestic public debt, net foreign assets, and the interest rate differential are ambiguous in equation 4.12. Second, equation 4.12 includes the ratio of permanent to current output (an inverse measure of the business cycle). The subsequent empirical application will compare the relative relevance of these two approaches to the trade surplus for the Chilean economy.

The empirical results are shown in table 4.16. Lines 1 and 3 present results for the complete specification. Of the right-hand determinants, the interest rate differential (IRD), total net foreign asset holdings, and the stock of domestic public debt holdings present positive, mostly significant signs for expectations under both versions. To capture a possible break of the relationship between the trade surplus and the stock of net foreign asset holdings in 1981, the multiplicative dummy $D81$ for NFA was included for 1981. The positive signs of IRD, $NFA_{-1}$, and $B_{-1}$ contradict the asset-accumulation version of the trade balance in equation 4.11 but are consistent with the ambiguous signs postulated by the alternative version of equation 4.12.

The current external terms of trade have a consistently significant but low positive influence on the trade balance. The influence of the business cycle is not significant. A dummy for 1979–82 ($D7982$) signals the particular regime of high access to foreign credit that characterized those years. The trade deficit was 2 to 3 percentage points of GDP higher during the period, unhindered by either domestic or foreign limits on foreign borrowing (see McNelis and Schmidt-Hebbel 1993).

Finally, the operational public sector deficit has a negative, although not significant, effect on the trade surplus under the static-expectations complete specification for the dependent variable (line 1). To test for the influence of this variable when most other variables are omitted, we run the specification of line 2. Keeping in mind the variable exclusion bias, we conclude that an increase in the public deficit of 1 percentage point of GDP tends to reduce the Chilean trade surplus by a maximum 0.29 percentage point of GDP.

## Conclusions and Policy Implications

A number of important findings and lessons emerge from this study of the macroeconomic causes and consequences of budget sector deficits in Chile. The first, unavoidable observation concerns the wild gyrations of public sector deficits during the past two decades. In fact, each decade comprises a complete cycle of fiscal and macroeconomic

**Table 4.16. Trade Surplus Estimation Results , Chile, 1960–88**

| Equation | C | IRD | D81 | $\frac{NFA_{-1}}{Y}$ | $\frac{B_{-1}}{Y}$ | TT* | $\frac{OD_g}{Y}$ | $\frac{YP}{Y}$ | D7982 | $R^2A$ | Durbin-Watson |
|---|---|---|---|---|---|---|---|---|---|---|---|
| *Static expectations* | | | | | | | | | | | |
| 1 OLS | -0.04 (-0.3) | 0.01 (1.5) | -1.76 (-3.1) | 0.50 (1.9) | 0.04 (1.0) | 0.03 (2.0) | -0.12 (-1.0) | -0.03 (-1.0) | -0.03 (-2.1) | 0.68 | 1.84 |
| 2 OLS | 0.006 (1.0) | | | | | | -0.29 (-4.1) | | -0.06 (-2.6) | 0.39 | 1.84 |
| *Partial perfect foresight* | | | | | | | | | | | |
| 3 OLS | -0.08 (-0.5) | 0.03 (1.9) | -1.61 (-2.6) | 0.50 (2.4) | 0.07 (2.0) | 0.04 (3.3) | 0.03 (0.2) | -0.02 (-0.1) | -0.02 (-1.4) | 0.73 | 1.97 |

*Note:* The dependent variable is the ratio of the trade surplus to GDP (*TS/Y*). The estimation method was ordinary least squares (OLS) and maximum likelihood (ML). Figures in parentheses are *t*-statistics. A blank denotes the omission of the specific variable from the regression; $R^2A$ is adjusted $R^2$.

crisis, recovery, and consolidation. Starting with a CNFPS deficit of 23.4 percent of GDP in 1970–73, which, in conjunction with adverse foreign stocks, forced a major stabilization effort that was very successful on the fiscal side, the 1970s culminated in a 5.4 percent surplus in 1980. In the early 1980s the conjunction of negative external developments and misguided wage and exchange rate policies caused a major macroeconomic crisis, which led to deterioration of the public finances of the financial and nonfinancial public sector. Quasi-fiscal rescue operations by the Central Bank directed at domestic debtors and the private financial system, combined with social emergency programs and a recession-induced fall of revenue in the nonfinancial public sector, caused large total public sector deficits in 1982–85.

A strong conclusion can be drawn regarding the relative roles of fiscal policy shifts and of shocks exogenous to the control of the fiscal authority: fiscal policy shifts were responsible for more than 100 percent of the cyclical variations and structural corrections of nonfinancial public sector deficits during 1974–88. Public sector retrenchment and reform were major causes of changes in CNFPS deficits: the tax reforms, public sector rationalizations, and privatizations of the mid-1970s corrected the initial deficits, while the 1981 social security reform and the tax reductions in the mid- to late 1980s contributed to the deficits. The large swings in copper prices were traditionally the most destabilizing factors outside the control of policymakers until the copper stabilization fund introduced a stronger separation between public spending and copper revenue. An implication of the high quasi-fiscal deficits was the huge buildup of total public debt. Nonetheless, our calculations show that the solvency of the total public sector is not jeopardized by the current fiscal stance: a deficit of 0.6 percent of GDP—reflecting a 1.5 percent deficit of the Central Bank and a 0.9 percent surplus of the CNFPS—is perfectly sustainable under normal macroeconomic conditions.

In the section on deficit financing, we explored the implications of domestic financing of public deficits for the interest rate and the price level in Chile. The effects of alternative strategies of domestic financing were discussed and contrasted in the context of a portfolio model with a partial adjustment structure for both the money and the public debt markets. The simulations based on the estimated portfolio model showed relatively conventional results: money-financed public deficits reduce interest rates and are strongly inflationary, while debt-financed deficits raise interest rates and are only weakly inflationary. Less conventional was the result that the moderate fluctuations of interest rates and price levels during the 1980s were attributable in large part to the rising demand for public debt by the new private pension funds. This rise permitted the accommodation of domestic financing of deficits that were largely caused by quasi-fiscal operations and social security reform. Therefore, one should not generalize Chile's experience with fiscal deficits and their financing during the 1980s.

In the section on crowding-out and crowding-in of private consumption and investment, we went a step further in analyzing the macroeconomic implications of public sector deficits. The focus was on the sensitivity of private consumption and investment to fiscal variables, in addition to the indirect effects of fiscal policies through interest rates, inflation, and private disposable income. Clear policy implications can be drawn from our empirical results on private consumption. There is no evidence of a direct effect of the public deficit on private consumption; consumers are neither Ricardian nor directly crowded-out by public spending in Chile. A transitory deficit reduction financed by a transitory tax hike reduces private disposable income and hence private consumption and saving. If the tax hike is permanent, private consumption will take the brunt of the reduction in (current and permanent) private disposable income without significantly affecting private saving. Also, domestic deficit financing through changes in the real interest rate and the real inflation rate exercise no indirect effects on consumption. If, however, inflation were to reduce real base money, private consumption would be affected. Adjustment to foreign shocks has strong effects on private consumption in Chile. The combination of lower foreign saving and a higher real exchange rate has a large negative impact on private consumption expenditure. This was clearly observed during the post-1982 adjustment period, when private consumption fell to historical lows, while private investment reached historical highs.

From the empirical results for private investment behavior, the following implications can be drawn. Financing of the public deficit through issuance of domestic debt is increasingly felt by private investors in Chile, as a consequence of the upward pressure of this form of financing on domestic real interest rates. The other side of the coin is the weakened role of both domestic and foreign liquidity constraints: whereas firm profits and foreign saving were important investment determinants before financial liberalization, they have not been significant investment determinants since the mid-1970s. No evidence was found for public investment crowding-in effects. In fact, the relevant variable (public capital stock) tends to reduce private capital formation, reflecting public-private investment substitution. Foreign adjustment has affected private investment negatively through the relative price of investment goods, which has increased with the real devaluations. However, external adjustment coincided with a strengthening of the development strategy and stable policy rules, reducing significantly policy uncertainty and fluctuations in output. This reduction of systemic risk and macroeconomic uncertainty (which negatively affect irreversible investment decisions) has had a major beneficial impact on private capital formation in Chile, as illustrated by our results and by the current record levels reached by this variable.

External adjustment during the 1980s has implied a large real exchange rate depreciation and a 12 percentage point correction of the ratio of the trade surplus to GDP (from a deficit of 10 percent of GDP in 1981 to a surplus of 2 percent in 1992), which was required to ensure continuous debt servicing. Our results reflect the sensitivity of the trade surplus to the public sector deficit; however, its low parameter is also consistent with the fact that a significant share of the external adjustment was borne by private consumers.

What lessons emerge for fiscal policy management from the Chilean experience? First, the Chilean case suggests that there are no quick-fix remedies for domestic and external imbalances. Adjusting fundamentals—among which public sector imbalances are paramount—requires perseverance by policymakers in order to build up the credibility of private agents that is required for an adequate financial-markets and investment response. Second, balanced public accounts provide the best foundation for adjusting to external shocks. There is little doubt that the basic stabilization-cum-structural-reform strategy implemented in Chile before the 1982 debt crisis (even taking into account some serious policy mistakes) was the major determinant of the rapid recovery and subsequent high-growth path followed by the Chilean economy during the 1980s and early 1990s. Finally, although this chapter has not focused in detail on public investment, we infer from other country experiences that public investment in infrastructure and human capital is a key to economic growth. The public sector should focus increasingly on high-return investment in human capital and physical infrastructure; this would be the government's main contribution toward ensuring that Chile continues on a high-growth path.

## Appendix 4.1. Revenue and Expenditure Functions

Tax revenue functions for three tax categories, the public enterprise surplus, and transfers to the private sector were estimated before performing the CNFPS deficit decomposition in the section on fiscal policy. The estimation results for the revenue functions are reported in table 4.17.

Income is the main exogenous (for the policymaker) determinant of direct and indirect tax revenue. Inflation shows a negative sign consistent with the Olivera-Tanzi effect in the case of indirect taxes but is not significant at the 10 percent level. Tax reforms (or endogenous

**Table 4.17. Estimation Results for Behavioral Public Sector Variables, Chile, 1973–89**

| Dependent variable | Independent variables | | | | | | | $R^2A$ | Durbin-Watson |
|---|---|---|---|---|---|---|---|---|---|
| | Constant | Income | Exchange rate | Inflation rate | Copper price | D1 | D2 | | |
| 1. Direct taxes[a] | 506.3 (0.93) | 3.36 (1.92) | | | | 298 (2.2) | −494 (−3.0) | 0.73 | 1.7 |
| 2. Indirect taxes[b] | 364.9 (0.5) | 10.65 (4.4) | | −0.91 (−1.5) | | 786 (2.1) | 1,833 (2.9) | 0.95 | 1.6 |
| 3. Copper taxes[c] | −84.9 (−0.2) | −6.34 (−1.9) | 7.41 (1.6) | | 27.7 (3.4) | | | 0.54 | 1.4 |
| 4. Surplus of public firms[c] | −2,869 (−4.6) | | 43.0 (10.3) | | 23.5 (3.3) | | | 0.92 | 1.9 |
| 5. Total transfers to private sector[d] | 1,121 (0.5) | 11.2 (1.8) | | −3.8 (−1.6) | | | | 0.65 | 1.6 |

*Note:* The simulation method was ordinary least squares. Figures in parentheses are *t*-statistics; $R^2A$ is adjusted $R^2$. A blank denotes the omission of the specific variable from the regression.

a. Income is lagged by one period; D1 is 1 for 1975–81 and 0 for 1982–89. D2 is 0 for 1973–84 and 1 for 1985–89.

b. D1 is 1 for 1973–83 and 0 for 1984–89. D2 is 0 for 1975–83 and 1 for 1984–89.

c. The sample period is 1975–89.

d. Income is lagged by one period.

fiscal policies) are very significant determinants of revenue, as reflected by the period dummies that signal tax reforms. Revenue from special taxes on public sector copper mines is countercyclical and, not surprisingly, increases with the real exchange rate and the international copper price. The latter two exogenous determinants also contribute to the state-owned enterprise surplus, which is dominated by tradable-goods-producing firms, including the copper industry. Finally, total transfers to the private sector (including social security payments) appear to be procyclical and tend to fall with inflation.

## Appendix 4.2. Derivation of the Asset Portfolio Model

This appendix presents the basic building blocks of the model discussed in the section on deficit financing. The point of departure is the budget constraint of the public sector given by equation 4.2 in the text. In each period individuals allocate their wealth (which is fixed at any point in time) into three broad assets: domestic money, public sector bonds, and interest-bearing foreign assets. The long-run portfolio demands for these three assets are:

(4.13)    $m_t^d - p_t = \alpha_0 + \alpha_1 y_t + \alpha_2 i_t + \alpha_3(i_t^* + \delta_t) + w_t$

(4.14)    $d_t^d - p_t = \beta_0 + \beta_1 y_t + \beta_2 i_t + \beta_3(i_t^* + \delta_t) + w_t$

(4.15)    $e_t + m_t^* = \tau_0 + \tau_1 y_t + \tau_2 i_t + \tau_3(i_t^* + \delta_t) + w_t$

where $w$ is real financial wealth, $y$ is domestic real output, $p$ is the domestic price level, $i$ is the domestic nominal interest rate, $i^*$ is the foreign nominal interest rate, $m$ is the domestic nominal money supply, $m^*$ is foreign nominal asset holdings, $d$ is public domestic debt, $e$ is the nominal exchange rate, and $\delta$ is the expected rate of change of the exchange rate. All variables except $i$, $i^*$, and $\delta$ are expressed in natural logarithms. The subscript $t$ refers to time, and the superscript $d$ denotes demand. The expected signs are denoted below each coefficient. The expected signs of the coefficients are as follows: $\alpha_0, \alpha_1 \lessgtr 0$; $\alpha_2, \alpha_3 < 0$; $\beta_0, \beta_1 \lessgtr 0$; $\beta_2 > 0$; $\beta_3 < 0$; $\tau_0, \tau_1 \lessgtr 0$; $\tau_2 < 0$; $\tau_3 > 0$. Wealth elasticities of long-term asset demands are assumed to be 1.

The wealth adding-up constraint implies that only two of the three demand equations are independent, and we will therefore focus on the equilibrium conditions in the domestic money and public debt markets. Assuming a partial adjustment mechanism as a result of the existence of implicit adjustment costs, the change in the stock of real balances and public debt holdings can be written:

(4.16)    $(\dot{m}_t - \dot{p}_t) = \sigma[(m_t^d - p_t) - (m_{t-1} - p_{t-1})] + \theta S_t$

(4.17)    $(\dot{d}_t - \dot{p}_t) = \mu[(d_t^d - p_t) - (d_{t-1} - p_{t-1})] + \cap \dot{d}_t.$

The expected signs of the coefficients are as follows: $\sigma > 0$; $\mu \leq 1$; $\theta > 0$; $\cap < 1$. Dots over variables denote time derivatives.

In addition to the partial adjustment mechanism, equation 4.16 includes a nominal money supply shock term $(S)$, defined as the difference between actual and expected rates of change in nominal money balances. This term captures the shock-absorber hypothesis proposed by Carr and Darby (1981), which states that when the government changes the rate at which it creates money, the result is a net unplanned increase in money holdings by individuals.

As in the case of the money market, a partial adjustment mechanism is assumed in the market for public bonds. Equation 4.17 also includes the actual rate of change in the stock of nominal public debt as part of the determination of the accumulation of real public bonds. This term stands for the increase since 1981 in the demand for public debt by the private pension funds, which accommodated the financing of deficits implied by the social security reform and the quasi-fiscal operations of the Central Bank in Chile. This partial demand accommodation to the increase in the public debt implies that the interest rate did not rise by as much as if demand accommodation were absent.

After substituting equations 4.13 and 4.14 into 4.16 and 4.17 and solving for the current values of real balances and real public bonds, equations 4.3 and 4.4 are obtained. The short-run coefficients of asset demand equations (4.3–4.4) and the coefficients of the reduced-form equations for the nominal interest rate and the price level in 4.5–4.6 are determined by the structural coefficients of equations 4.13, 4.14, 4.16, and 4.17 as follows:

$$\phi_0 = \sigma\alpha_0 \qquad \Phi_0 = \mu\beta_0$$

$$\phi_1 = \sigma\alpha_1 \qquad \Phi_1 = \mu\beta_1$$

$$\phi_2 = \sigma\alpha_2 \qquad \Phi_2 = \mu\beta_2$$

$$\phi_3 = \sigma\alpha_3 \qquad \Phi_3 = \mu\beta_3$$

$$\phi_4 = 1 - \sigma \qquad \Phi_4 = 1 - \mu$$

$$\phi_5 = \theta \qquad \Phi_5 = \cap$$

$$\phi_6 = \sigma \qquad \Phi_6 = \mu$$

$$\Omega_0 = \frac{\sigma\alpha_0 - \mu\beta_0}{\mu\beta_2 - \sigma\alpha_2} \qquad \pi_0 = \frac{\sigma\mu(\alpha_2\beta_0 - \beta_2\alpha_0)}{\mu\beta_2 - \sigma\alpha_2}$$

$$\Omega_1 = \frac{\sigma\alpha_1 - \mu\beta_1}{\mu\beta_2 - \sigma\alpha_2} \qquad \pi_1 = \frac{\sigma\mu(\alpha_2\beta_1 - \alpha_1\beta_2)}{\mu\beta_2 - \sigma\alpha_2}$$

$$\Omega_2 = \frac{\sigma\alpha_3 - \mu\beta_3}{\mu\beta_2 - \sigma\alpha_2} \qquad \pi_2 = \frac{\sigma\mu(\alpha_2\beta_3 - \alpha_3\beta_2)}{\mu\beta_2 - \sigma\alpha_2}$$

$$\Omega_3 = \frac{1 - \sigma}{\mu\beta_2 - \sigma\alpha_2} \qquad \pi_3 = \frac{(\sigma - 1)\mu\beta_2}{\mu\beta_2 - \sigma\alpha_2}$$

$$\Omega_4 = \frac{\mu - 1}{\mu\beta_2 - \sigma\alpha_2} \qquad \pi_4 = \frac{(1 - \mu)\sigma\alpha_2}{\mu\beta_2 - \sigma\alpha_2}$$

$$\Omega_5 = \frac{-1}{\mu\beta_2 - \sigma\alpha_2} \qquad \pi_5 = \frac{\mu\beta_2}{\mu\beta_2 - \sigma\alpha_2}$$

$$\Omega_6 = \frac{1 - \cap}{\mu\beta_2 - \sigma\alpha_2} \qquad \pi_6 = \frac{\sigma\alpha_2(\cap - 1)}{\mu\beta_2 - \sigma\alpha_2}$$

$$\Omega_7 = \frac{\cap}{\mu\beta_2 - \sigma\alpha_2} \qquad \pi_7 = \frac{-\sigma\cap\alpha_2}{\mu\beta_2 - \sigma\alpha_2}$$

$$\Omega_8 = \frac{\theta}{\mu\beta_2 - \sigma\alpha_2} \qquad \pi_8 = \frac{-\theta\mu\beta_2}{\mu\beta_2 - \sigma\alpha_2}$$

$$\Omega_9 = \frac{\sigma - \mu}{\mu\beta_2 - \sigma\alpha_2} \qquad \pi_9 = \frac{\sigma\mu(\alpha_2 - \beta_2)}{\mu\beta_2 - \sigma\alpha_2}$$

## Appendix 4.3. Derivation of the Trade Surplus Equation

This appendix focuses on the derivation of equation 4.12. The trade surplus in current prices, defined by the excess of exports over imports, is equal to the difference between current-price output and total absorption ($A$):

$$(4.18) \qquad TS = P_x x - P_m m = Y - A$$

$$= Y + (rB_{-1} + i^* NFA_{g-1} - T) - OD_g - A_p$$

where $r$ is the domestic real interest rate paid on government bonds, $T$ is total taxes net of transfers paid by the private sector, $A_p$ is private absorption (private consumption and investment expenditure), $A_g$ is public sector absorption, and $OD_g$ is the operational public sector deficit.

After substituting general functional forms for output and absorption consistent with theory (as, for instance, in McNelis and Schmidt-Hebbel 1993) into equation 4.18, we obtain:

$$(4.19) \quad TS = \underset{(-)(-)}{Y(i^*, r, \dots)} + [rB_{-1} + i^* NFA_{g-1} - T] - OD_g$$

$$\underset{(+)\ \ (+)(0,+)(?,-)(?,-)(+)(+)}{- A_p(NFA_p, \ B, \ NFA_g, \ i^*, \ \ r, \ \ TT^*, YP)}$$

Equation 4.12 is obtained by rewriting equation 4.19 as a functional form for the ratio of the trade surplus to $Y$ after scaling all relevant variables to current-price GDP.

## Appendix 4.4. Data Sources, by Section

*Deficit Financing, Interest Rates, and Inflation*

The definitions of and sources for the variables in this section are:

$i$ = domestic nominal interest rate. Monthly average of the effective interest rate paid on short-term deposits (30–89 days). Source: Banco Central de Chile (1990).

$i^*$ = foreign interest rate. Monthly average LIBOR for 180-day loans in U.S. dollars. Source: IMF, *International Financial Statistics*.

$\delta$ = expected devaluation rate of the nominal exchange rate. Sources: for 1975:1–1984:3, Le Fort and Ross (1985); for 1984:4–1988:4, a univariate autoregressive integrated moving-average process was estimated for the nominal exchange rate of the banking system on the basis of data from Banco Central de Chile (1990).

$M_1$ = narrow money. The original series (sources: Vial and Marín 1986; Banco Central de Chile 1990) was deseasonalized. Anticipated money holdings were estimated from an autoregressive process of order four.

$D$ = quarterly domestic public debt. Source: annual data from Larrañaga (1989), corresponding to the debt of the Central Bank held by the private sector. The quarterly interpolation was performed following the method described in Jadresic (1990).

$Y$ = quarterly real GDP. Sources: Haindl (1986); Banco Central de Chile (1990).

*Crowding-Out and Crowding-In of Private Consumption and Private Investment by the Public Sector*

The definitions of and sources for the variables are:

$C_p$ = private consumption. Source: Banco Central de Chile (1990).
$C_g$ = public consumption. Source: Banco Central de Chile (1990).
$Y$ = annual real GDP. Source: Banco Central de Chile (1990).
$I_p$ = private fixed-capital investment. Sources: Solimano (1992); Zucker (1988).
UCK = user cost of capital. Source: Banco Central de Chile (1990).
MPK = marginal product of capital. Source: Banco Central de Chile (1990).

$K$ = aggregate real capital stock. This series was elaborated on the basis of Solimano (1992) and Zucker (1988) by splicing and assuming an initial capital-to-GDP ratio of 2.5. In addition, the ratio of public to private capital stock was assumed to be equal to the ratio between public and private investment.

$H$ = nominal base money stock. Source: Banco Central de Chile (1990).

The remaining variables were obtained directly from Banco Central de Chile or were derived from the variables defined above according to the definitions provided in the text.

*Relative External Prices and the Trade Balance*

The definitions of and sources for the variables are:

$P_X/P_N$ = relative price of exportable and nontradable goods. Sources: for exportable goods, Banco Central de Chile (1990); for the nontradable goods index, Schmidt-Hebbel, Castro, and Leng (1990).

$P_M/P_N$ = relative price of importable and nontradable goods. Sources: for the price of importable goods, Banco Central de Chile (1990); for the nontradable goods index, Schmidt-Hebbel, Castro, and Leng (1990).

$t_M$ = nominal ad valorem tariff rate for imports. Sources: Ffrench-Davis (1984); Lagos and Aedo (1984).

The remaining variables were obtained from Banco Central de Chile (1990).

## Notes

The authors benefited from extensive discussions with Osvaldo Larrañaga and Patricio Mujica. Bela Balassa, Edgardo Barandiarán, William Easterly, James Hanson, Miguel Kiguel, and Juan Carlos Lerda gave us excellent comments on previous drafts. Francisco Berrasconi and Eduardo Saavedra provided efficient research assistance. We thank all of them, and we retain responsibility for any remaining errors.

1. References on macroeconomic adjustment and reforms in Chile during the past two decades include Corbo (1985, 1990); Edwards and Edwards (1987); Ffrench-Davis and de Gregorio (1987); and Morandé and Schmidt-Hebbel (1988).

2. This chapter uses three public sector deficit measures differentiated by public sector coverage: the consolidated nonfinancial public sector deficit (comprising general government and nonfinancial public enterprises); the financial deficit (comprising quasi-fiscal losses of the Central Bank); and the total public sector deficit (comprising the first two ). In addition, it distinguishes between the total (or nominal) deficit and the primary deficit, the

difference between the two latter being explained by interest payments on domestic public debt.

3. For a discussion of the causality between domestic spending, foreign lending, and the real exchange rate during this period see Morandé and Schmidt-Hebbel (1988), in particular chapter 5.

4. Since May 1981 Chile has gradually replaced the existing pay-as-you-go state pension scheme with a privately managed and fully funded system. The transition deficit caused by this reform started at 3.2 percent of GDP in 1982, reached a peak of 4.8 percent in 1991, and fell gradually thereafter (Arrau 1991, 1992). Although this type of pension reform increases measured public deficits (while private sector surpluses increase by the same amount), it is important to point out that such a reform, if financed by issuing public debt, has no first-order effects on the macroeconomy. The reform provides the means of financing the transition deficit by creating a demand for public debt paper by the new private pension funds (see Arrau 1991 and Arrau and Schmidt-Hebbel 1993 for detailed discussions).

5. For a detailed calculation of Central Bank quasi-fiscal operations and losses, see Eyzaguirre and Larrañaga (1990).

6. The exchange rate definitions used throughout this chapter are, for the nominal exchange rate, the price of foreign currency in units of domestic currency and, for the real exchange rate, the relative price of tradable goods in units of nontradables. Hence a real devaluation means a higher real exchange rate.

7. See Marshall and Schmidt-Hebbel (1989a, 1989b) and a compact presentation in the appendix of Morandé and Schmidt-Hebbel (1991).

8. The average relative contribution of each group of deficit determinants to the explained variation of the deficit (see table 4.4) is computed as:

$$\left[ \sum_{i=1}^{4} dv_i(\text{sign } d_i) \right] \left/ \left| \sum_{i=1}^{4} |d_i| \right. \right.$$

where $d_i$ is the explained change in the deficit, $dv_i$ is the change in the deficit caused by variable category $v$ (the total variations that resulted from fiscal policy, macroeconomic, and external variables in table 4.4, with nondecomposed accounts as part of fiscal policy variables), sign $d_i$ is the sign (positive or negative) of $d_i$, and the subperiods $i$ are the four considered in table 4.4.

9. See Buiter (1988) and van Wijnbergen (1989) for this notion of deficit sustainability and its relation to public sector solvency.

10. The real domestic interest rate $r$ is defined as the difference between the nominal domestic interest rate and domestic inflation. The real foreign interest rate $(r^*)$ is the difference between the nominal foreign interest rate and foreign inflation. The rate of real devaluation is defined as the rate of increase of the nominal exchange rate plus foreign inflation minus domestic inflation.

11. The sample period for the money equation is 1976:1–1988:4, while for the domestic debt equation it covers 1982:1–1988:4.

12. This finding can be explained in terms of the values of the estimated coefficients of the structural equations. When the value of the coefficient related to the speed of adjustment in the public bond market $(\mu)$ is high—which is the case according to our estimations—the model implies small

values for the coefficients of lagged real public bonds in the reduced-form equations (see appendix 4.2).

13. Recall that flow variables, such as GDP, are quarterly figures. Therefore the corresponding annual ratios have to be divided by 4.

14. The Ricardian hypothesis has been widely rejected in empirical studies for industrial countries. Most of these studies identify the existence of pervasive borrowing constraints as the main cause for its rejection. A study for a set of developing countries by Haque and Montiel (1989) tests for two different causes, which could explain a deviation from Ricardian equivalence: higher private than government discount rates (attributable to Blanchard-Yaari infinite-lived households that face a mortality probability) and liquidity constraints. They find significant evidence that the latter causes a deviation from Ricardian equivalence without much support for such a role for the former. Borrowing constraints, proxied by current income or financial asset holdings, are also major determinants in the cross-developing-country studies for private consumption by Rossi (1988) and for household saving by Schmidt-Hebbel, Webb, and Corsetti (1992).

15. Among the empirical studies showing saving to be insensitive to the real interest rate in developing countries, see Corbo and Schmidt-Hebbel (1991), Giovannini (1985), and Schmidt-Hebbel, Webb, and Corsetti (1992). For an alternate view, see Balassa (1989) and Fry (1988).

16. For the theory and the Latin American experience on the role of consumption-based real interest rates in intertemporal consumption allocation, see Dornbusch (1983, 1985a, 1985b, 1988), and for the relation between saving and capital flight see Dornbusch (1989).

17. The scaling of nonstationary variables to private disposable income and of nonstationary variables to GDP in the private investment and trade surplus equations that follow reduces the incidence of problems derived from nonstationary time series.

18. Note that equation 4.7 allows for testing the separate influence of permanent public saving ($PS$), in addition to the influence of permanent private disposable income (PDY). If $PS$ is not significant but PDY is (that is, if Ricardian equivalence is rejected in favor of the permanent income hypothesis), taxes affect consumption, but government spending does not.

19. Permanent private disposable income and permanent public saving are consistent with the following definitions for their corresponding current values: $DY_{Pt} = GDP_t - NFP_{Pt} - T_t + r_t D_t$, and $S_{Gt} = T_t - C_{Gt} - NFP_{Gt} - r_t D_t$, where GDP is gross domestic product, NFP is net foreign payments made by the private sector, $S_G$ is current public saving, $C_G$ is public consumption, and $NFP_G$ is net foreign payments made by the public sector. Note that $D$ now refers only to the domestic public debt. For the expected "permanent" values of any variable (private disposable income and public saving in this section, and other variables in the investment section that follows), we specify two alternatives. The first is to assume expectations of the permanent values that are consistent with partial perfect foresight, defined as the simple average of the contemporaneous variable and two periods into the future, for any variable $x$: $Px_t = (x_t + x_{t+1} + x_{t+2})/3$, where $Px$ defines the expected permanent value of variable $x$. The second alternative is the simple static-

expectations specification, which allocates a 100 percent weight to the contemporaneous value in the preceding average. Similar assumptions are made with respect to expected consumption inflation (and expected investment inflation in the section on private investment). A first alternative takes actual inflation between today and tomorrow as the relevant proxy for rationally expected inflation. The second alternative is static, specifying the expected price change to be equal to the actual price change between yesterday and today.

20. This result mimics the results for the quarterly structural consumption function for Chile in Schmidt-Hebbel (1988), which are stronger for the backward- than for the forward-looking expectations specifications.

21. The definition of permanent variables is the same as for consumption determinants, explained in footnote 19. The current real user cost of capital is defined as:

$$\text{UCK}_t = (P_{it}/P_t)[(i_{Ft} - \hat{P}^e_{It})/(1 + \hat{P}^e_{It}) + \delta]$$

where $P_{it}$ is the private investment deflator, $P_t$ is the GDP deflator, $i_{Ft}$ is the nominal interest rate on banking loans to firms, $\hat{P}^e_{It}$ is the expected rate of change of the private investment deflator, and $\delta$ is the real rate of capital depreciation. The private sector capital-output ratio (the inverse of the average product of capital) stands for both the neoclassical marginal product of capital (which is a linear transformation of the marginal product under a Cobb-Douglas technology) and the Keynesian potential-to-actual-output ratio. Note that private and public capital add to the total domestic capital stock: $K_t = K_{Gt} + K_{pt}$. Expected investment inflation is based on an estimated autoregressive structure. All expected permanent variables are specified according to the two hypotheses mentioned above: the partial perfect-foresight alternative and the static version. Finally, the two coefficients of variation are defined as five-period moving variances for two periods back, the current period, and two periods forward.

22. The final results reported in the tables exclude three variables that appear in equation 4.8: the relative price of the two aggregate investment components, corporate taxes, and firm credit. The exclusion of the first variable from preliminary results was justified by its implausibly high coefficient, which affected many other parameters. The other two variables were not included because of lack of data.

23. For the same reason discussed in the preceding section (to avoid spurious correlation), we scale nonstationary variables (such as the trade surplus, government expenditure, and net foreign assets) to appropriate scale variables in equations 4.9–4.12.

24. It was not possible to include the nontradable component of public spending as an additional determinant because of lack of data. A significant fraction of the massive 1975 devaluation could not be explained by any of the preceding variables and hence was treated as an outlier.

25. Current-price GDP is used as the relevant scale variable. Hence the positive sign of the constant term C reflects the hypothesized positive effect of income on the trade balance, when multiplying equation 4.11 by the latter and abstracting from the presence of multiplicative terms involving income and all nonscaled right-hand variables.

# References

Arrau, Patricio. 1989. "Intertemporal Monetary Economics: Evidence from the Southern Core of Latin America." Ph.D. diss. University of Pennsylvania, Department of Economics, Philadelphia.

———. 1991. "La Reforma Previsional Chilena y su Financiamiento durante la Transición." *Colección Estudios CIEPLAN* (Chile) 32 (June): 5–44.

———. 1992. "El Nuevo Régimen Previsional Chileno." In Fundación Friedrich Ebert de Colombia (FESCOL), *Regímenes Previsionales*. Bogotá.

Arrau, Patricio, and Klaus Schmidt-Hebbel. 1993. "Macroeconomic and Intergenerational Welfare Effects of a Transition from Pay-As-You-Go to Fully-Funded Pension Systems." World Bank, Policy Research Department, Washington, D.C.

Balassa, Bela. 1989. "The Effects of Interest Rates on Saving in Developing Countries." Policy Research Working Paper 56. World Bank, Office of the Vice President, Development Economics, Washington, D.C.

Banco Central de Chile. 1990. *Indicadores Económicos y Sociales 1960–1988.* Santiago.

Buiter, Willem H. 1988. "Some Thoughts on the Role of Fiscal Policy in Stabilization and Structural Adjustment in Developing Countries." NBER Working Paper 2603. National Bureau of Economic Research, Cambridge, Mass.

Carr, Jack, and Michael R. Darby. 1981. "The Role of Money Supply Shocks in the Short-Run Demand for Money." *Journal of Monetary Economics* 8 (2): 183–99.

Corbo, Vittorio. 1985. "Reforms and Macroeconomic Adjustments in Chile during 1974–1984." *World Development* 13 (18): 893–916.

———. 1990. "Public Finance, Trade and Development: The Chilean Experience." In Vito Tanzi, ed., *Public Finance and Economic Development*. Detroit, Mich: Wayne State University Press.

Corbo, Vittorio, and Jaime de Melo. 1989. "External Shocks and Policy Reforms in the Southern Cone: A Reassessment." In Guillermo A. Calvo, ed., *Debt, Stabilization and Development: Essays in Memory of Carlos Diaz-Alejandro*. Oxford, U.K.: Basil Blackwell.

Corbo, Vittorio, and Klaus Schmidt-Hebbel. 1991. "Public Policies and Saving in Developing Countries." *Journal of Development Economics* 36 (1): 89–116.

Cortázar, René, and Jorge Marshall. 1980. "Indice de Precios al Consumidor en Chile: 1970-1978." *Colección Estudios CIEPLAN* (Chile) 4: 159–201.

Dornbusch, Rudiger. 1983. "Real Interest Rates, Home Goods, and Optimal External Borrowing." *Journal of Political Economy* 91 (1): 141–53.

———. 1985a. "External Debt, Budget Deficits and Disequilibrium Exchange Rates." In Gordon W. Smith and John T. Cuddington, eds., *International Debt and the Developing Countries*. A World Bank Symposium. Washington, D.C.

———. 1985b. "Overborrowing: Three Case Studies." In Gordon W. Smith and John T. Cuddington, eds., *International Debt and the Developing Countries*. A World Bank Symposium. Washington, D.C.

————. 1988. "Mexico: Stabilization, Debt and Growth." In Georges de Ménil and Richard Portes, eds., *Economic Policy: A European Forum*. Cambridge, U.K.: Cambridge University Press.

————. 1989. "Capital Flight: Theory, Measurement and Policy Issues." Massachusetts Institute of Technology, Department of Economics, Cambridge, Mass.

Easterly, William. 1989. "Fiscal Adjustment and Deficit Financing during the Debt Crisis." In Ishrat Husain and Ishac Diwan, eds., *Dealing with the Debt Crisis*. A World Bank Symposium. Washington, D.C.

Easterly, William, Carlos A. Rodríguez, and Klaus Schmidt-Hebbel. 1989. "Research Proposal: The Macroeconomics of the Public Sector Deficit." World Bank, Country Economics Department, Washington, D.C.

Edwards, Sebastian, and Alejandra C. Edwards. 1987. *Monetarism and Liberalization: The Chilean Experiment*. Cambridge, Mass.: Ballinger.

Eyzaguirre, Nicolás, and Osvaldo Larrañaga. 1990. "Macroeconomía de los Operaciones Cuasifiscales en Chile." Economic Commission for Latin America and the Caribbean, Santiago.

Ffrench-Davis, Ricardo. 1984. "Indice de Precios Externos." In *Colección Estudios CIEPLAN* (Chile) 13: 87–106.

Ffrench-Davis, Ricardo, and José de Gregorio 1987. "Orígenes y Efectos del Endeudamiento Externo en Chile." *El Trimestre Económico* (Mexico) 54 (January–March): 159–78.

Fry, Maxwell J. 1988. *Money, Interest and Banking in Economic Development*. Baltimore, Md.: Johns Hopkins University.

Giovannini, Alberto. 1985. "Saving and the Real Interest Rate in LDCs." *Journal of Development Economics* 18 (August): 197–217.

Haindl, Erik. 1986. "Trimestralización del PGB por Origen y Destino." *Estudios de Economía* (Chile) 13 (1): 117–53.

Haque, Nadeem Ul, and Peter J. Montiel. 1989. "Consumption in Developing Countries: Tests for Liquidity Constraints and Finite Horizons." *Review of Economics and Statistics* 71 (3): 408–15.

IMF (International Monetary Fund). Various years. *International Financial Statistics*. Washington, D.C.

Jadresic, Esteban. 1990. "Trimestralización sin Uso de Series Relacionadas." *Notas Técnicas CIEPLAN* (Chile).

Lagos, Luis Felipe, and Christián Aedo 1984. "Protección Efectiva en Chile 1974–1979." *Documento de Trabajo* 94, Instituto de Economía, Pontificia Universidad Católica de Chile, Santiago.

Larrañaga, Osvaldo. 1989. "El Déficit del Sector Público y la Política Fiscal en Chile." Economic Commission for Lation America and the Caribbean, Santiago.

Le Fort, Guillermo, and Cristian Ross 1985. "La Devaluación Esperada, una Aproximación Bayesiana: Chile 1964–1984." In *Serie Investigación* 72, Universidad de Chile, Departamento de Economía, Santiago.

Marshall, Jorge, and Klaus Schmidt-Hebbel. 1989a. "Economic and Policy Determinants of Public Sector Deficits." Policy Research Working Paper 321. World Bank, Country Economics Department, Washington, D.C.

———. 1989b. "Un Marco Analítico-Contable para la Evaluación de la Política Fiscal en América Latina." *Serie Política Fiscal* 1. Economic Commission for Latin America and the Caribbean, Santiago.

McNelis, Paul, and Klaus Schmidt-Hebbel. 1993. "Financial Liberalization and Adjustment: The Cases of Chile and New Zealand." *Journal of International Money and Finance* 12: 249–77.

Morandé, Felipe, and Klaus Schmidt-Hebbel. 1991. "Macroeconomics of Public Sector Deficits: The Case of Zimbabwe. Policy Research Working Paper 688. World Bank, Country Economics Department, Washington, D.C.

———, eds. 1988. *Del Auge a la Crisis de 1982: Ensayos sobre Estabilización Financiera y Endeudamiento en Chile,* ILADES-Georgetown and Instituto Interamericano de Mercados de Capital, Santiago, Chile.

Mujica, Patricio. 1990. "Financiamiento Doméstico del Déficit Fiscal y sus Efectos Macroeconómicos: Un Modelo para Chile." Economic Commission for Latin America and the Caribbean. Santiago.

Rossi, Nicola. 1988. "Government Spending, the Real Interest Rate, and the Behavior of Liquidity-Constrained Consumers in Developing Countries." *International Monetary Fund Staff Papers* 35 (1): 104–40.

Schmidt-Hebbel, Klaus. 1987. "Terms of Trade and the Current Account under Uncertainty." *Análisis Económico* 2 (1): 67–89.

———. 1988. "Consumo e Inversión en Chile (1974–1982): Una Interpretación 'Real' del Boom." In Felipe Morandé F. and Klaus Schmidt-Hebbel, eds., *Del Auge a la Crisis de 1982: Ensayos sobre Estabilización Financiera y Endeudamiento en Chile,* ILADES-Georgetown and Instituto Interamericano de Mercados de Capital, Santiago, Chile.

Schmidt-Hebbel, Klaus, and Tobias Müller. 1992. "Private Investment under Macroeconomic Adjustment in Morocco." In Ajay Chhibber, Mansoor Dailami, and Nemat Shafik, eds., *Reviving Private Investment in Developing Countries.* Amsterdam: North-Holland.

Schmidt-Hebbel, Klaus, Francisca Castro, and Iván Leng. 1990. "Una Base de Datos Trimestrales para la Economía Chilena." *Serie Investigación* 24. ILADES-Georgetown and Instituto Interamericano de Mercados de Capital, Santiago.

Schmidt-Hebbel, Klaus, Steven B. Webb, and Giancarlo Corsetti. 1992. "Household Saving in Developing Countries: First Cross-Country Evidence." *World Bank Economic Review* 6 (2): 529–47.

Servén, Luis. 1990. "A RMSM-X Model for Chile." Policy Research Working Paper 508. World Bank, Country Economics Department, Washington, D.C.

Solimano, Andrés. 1992. "How Private Investment Reacts to Changing Macroeconomic Conditions: The Case of Chile in the 1980s." In Ajay Chhibber, Mansoor Dailami, and Nemat Shafik, eds., *Reviving Private Investment in Developing Countries.* Amsterdam: North-Holland.

van Wijnbergen, Sweder. 1989. "External Debt, Inflation and the Public Sector: Toward Fiscal Policy for Sustainable Growth." *World Bank Economic Review* 3 (3): 297–320.

Vial, Joaquín, and Bárbara Marín. 1986. "Series Monetarias Chilenas 1960–1985." *Estudios de Economía* 13 (1): 191–22.

Zucker, A. 1988. "Comportamiento de la Inversión en Capital Fijo en Chile: 1974–1987." Thesis. Instituto de Economía, Pontificia Universidad Católica de Chile, Santiago.

# 5

# Colombia: Avoiding Crises through Fiscal Policy

*William Easterly*

This chapter analyzes the macroeconomic effects of public sector deficits in Colombia. The first section reviews the historical evolution of fiscal policy in Colombia, with emphasis on the adjustment program of 1985–89, and uses an econometrically estimated model of the money and credit markets to examine how fiscal deficits affect the inflation rate and the real interest rate. The subsequent section discusses the relationship between the fiscal deficit and the real exchange rate, using a reduced-form model of traded and nontraded goods.

## Historical Background

Colombia is justly celebrated in Latin America for its prudent macroeconomic management, of which careful management of fiscal deficits is the cornerstone. Even its occasional departures from conservative macroeconomic policy seem tame by Latin American standards. In this section we review the broad outlines of macroeconomic policy in Colombia and look in more detail into the adjustment that has taken place since 1985.

### Macroeconomic Management, 1960–89

From the mid-1960s through the early 1970s, macroeconomic policy in Colombia was mostly conservative, supportive of an export-oriented development strategy that was associated with high growth of both GDP and trade. (See table 5.1 for basic macroeconomic indicators; see García García 1989 for a more detailed review.) The main factor in the second half of the 1970s was the surge in coffee export revenues, which the authorities were partially successful in sterilizing. In the early 1980s the end of the coffee boom coincided with a large increase in public investment, especially in the energy sector, which led to an incipient balance of payments and debt crisis. The crisis was largely avoided thanks to a strong adjustment effort that began in 1985 and has continued to the present.

225

Table 5.1. Basic Economic Indicators, Colombia Five-Year Averages, 1960–89
(percent)

| Indicator | 1960–65 | 1965–70 | 1970–75 | 1975–80 | 1980–85 | 1985–89 |
|---|---|---|---|---|---|---|
| Real exchange rate[a] | 99.7 | 112.8 | 129.9 | 123.3 | 118.5 | 161.4 |
| Inflation rate | 11.1 | 9.3 | 16.4 | 24.2 | 23.1 | 24.0 |
| Fiscal balance as a share of GDP[b] | -1.2 | -0.5 | -1.5 | -0.1 | -3.5 | -1.8 |
| Fiscal balance as a share of GDP[c] | — | — | — | -2.1 | -5.8 | -2.6 |
| Fiscal balance as a share of GDP[d] | -2.2 | -1.8 | -2.9 | -1.5 | -5.7 | — |
| Real interest rate | | | | | | |
| On loans | — | — | 2.4 | 4.4 | 10.3 | 13.0 |
| On deposits | — | — | -2.2 | -1.0 | 4.3 | 6.1 |
| GDP growth | 4.7 | 6.0 | 5.7 | 5.4 | 2.3 | 3.6 |
| Current account balance as a share of GDP | -2.3 | -2.7 | -2.8 | 0.6 | -5.1 | -1.1 |
| Real private investment as a share of real GDP | 11.9 | 10.2 | 9.2 | 8.1 | 7.4 | 8.2 |
| Public external debt as a share of GDP[c] | — | — | 20.0 | 15.3 | 22.6 | 37.2 |
| Public external debt as a share of exports of goods and services[e] | — | — | 124.1 | 79.3 | 133.0 | 187.7 |

— Not available.
a. Depreciation is up.
b. IMF, *International Financial Statistics Yearbook 1989* (national government only): data until 1987 only.
c. World Bank data.
d. García García 1989.
e. World Bank, *World Debt Tables*; public external debt data until 1988 only.

**Figure 5.1. Central Government Balance, Colombia, 1960–87**

Percentage of GDP

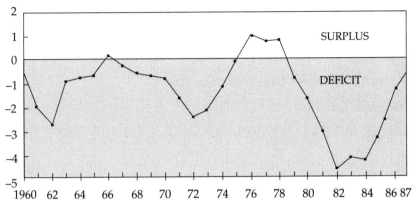

Source: IMF, *International Financial Statistics Yearbook*, 1989.

Figure 5.1 presents a historical perspective on fiscal management. The only series available for a sufficiently long period is the IMF's *International Financial Statistics* (IFS) series on the national government deficit. The IFS data are not consistent with the fiscal data used elsewhere in this chapter but do show a similar pattern. There have been three episodes of loose fiscal policy at roughly ten-year intervals—in 1961–62, in 1972–73, and in 1981–84, of which the last was by far the most severe. Each episode was followed by rapid fiscal adjustment that avoided a prolonged crisis.

The impression of relative macroeconomic stability is confirmed by the behavior of the inflation rate (figure 5.2). By Latin American standards (although not by world standards) the rate is very stable, staying within a band of about 15 to 35 percent since the early 1970s. We see roughly three distinct periods: (a) during most of the 1960s, inflation oscillated in response to periodic large adjustments of the exchange rate; (b) following the introduction of the crawling peg in 1967, the inflation rate stayed very stable for about five years; (c) after the episode of loose fiscal policy in 1972–73, inflation accelerated, remaining at more than 20 percent after the coffee boom led to some monetization of reserve inflows. After the end of the coffee boom, fiscal expansion replaced reserve inflows as a source of money creation. Subsequent fiscal contraction was just sufficient to match the reduction in external credit availability during the debt crisis, so that the need for money creation continued.

Figure 5.3 shows the evolution of money creation, roughly defined as the change in base money over GDP. We see that the reliance on seigniorage has been remarkably stable since the mid-1960s, aside from the burst of money creation associated with the coffee boom in the late 1970s.

The behavior of the real exchange rate mirrors changes in macro-economic policies, as shown in figure 5.4. The exchange rate was quite volatile in the early 1960s, but after the introduction of the

**Figure 5.2. Inflation Rate, Colombia, 1960–89**

Percent per year

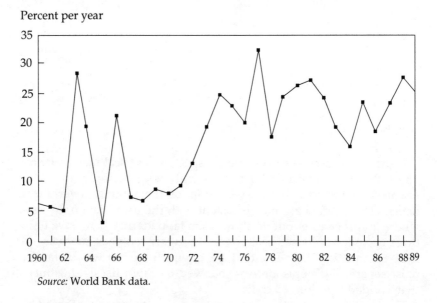

*Source:* World Bank data.

**Figure 5.3. Seigniorage, Colombia, 1960–88**

Percentage of GDP

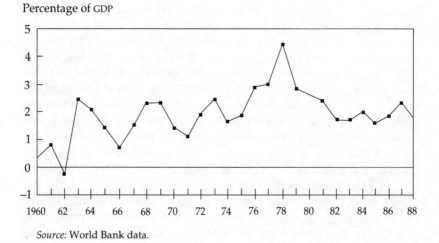

*Source:* World Bank data.

**Figure 5.4. Real Exchange Rate, Colombia, 1960–89**

Percentage deviation from 1975 value

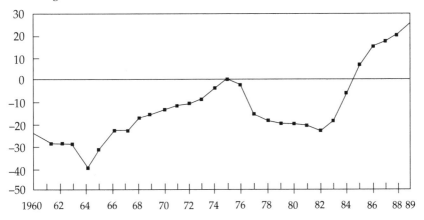

*Note:* Depreciation is up.
*Source:* World Bank data.

crawling peg, it showed a steady depreciation (depreciation is defined as up). With the onset of the coffee boom in 1975, the deterioration was reversed, and the real exchange rate appreciated steadily until the early 1980s. During the adjustment effort that began in 1985 the real exchange rate once again depreciated substantially. We see some association between the behavior of the real exchange rate and the episodes of fiscal contraction and expansion. The greatest appreciation of the currency came during the periods of expansionary fiscal policy in the early 1960s and early 1980s. The episode of loose fiscal policy during the early 1970s did not seem to have had much effect. We will examine in a subsequent section the relationship between deficits and the real exchange rate.

The real interest rate shows a more erratic path than inflation and the real exchange rate (figure 5.5).[1] This reflects the controls on interest rates prior to 1974, after which financial liberalization began by fits and starts. Interest rates were mostly market-determined during the 1980s, with occasional temporary controls such as those imposed in 1988. The fiscal expansion of the 1980s was associated with some rise in the real interest rate, although interest rates remained high even after the fiscal adjustment.

A puzzle of macroeconomic behavior in Colombia is the long-term decline in the private investment ratio, as shown in figure 5.6. (The ratio is defined in real terms to avoid any relative price effect.) The fiscal adjustment of the first half of the 1980s was associated with a decline in investment, while the adjustment program of 1985–88 was associated with a small rebound of private investment. The fall in private investment in the late 1970s and early 1980s was associated with high real interest rates and a high relative price of capital goods. But the increase in the interest rate was modest compared with the secular decline in investment. As the figure shows, public investment

**Figure 5.5. Real Deposit and Loan Rates, Colombia, 1970–88**

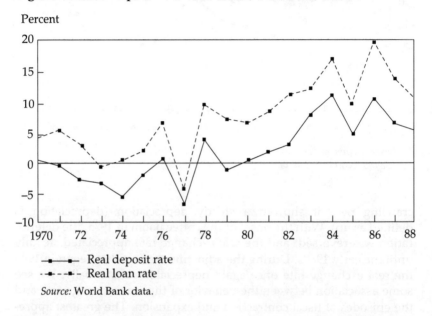

Percent

*Source:* World Bank data.

**Figure 5.6. Public and Private Investment, Colombia, 1950–88**

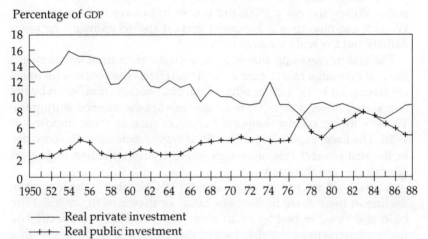

Percentage of GDP

*Source:* World Bank data.

was rising at the same time that private investment was falling, suggesting crowding-out in the long run. The effect of public capital on private investment will be examined statistically later in this chapter.

*Public Sector Accounts, 1975–85*

The public sector deficit was under control until 1981–84. Deficits were small during the heyday of the coffee boom of 1975–78. The end of the boom brought increased deficits, which worsened in 1981–84 because of a surge in public capital spending.

The genesis of the fiscal crisis of the early 1980s can be seen by separating out the effects of changes in coffee revenues and public capital spending—the two largest single influences on the deficit. As figure 5.7 shows, if public capital spending had remained constant at its 1975 value (with the coffee surplus still taking its actual values), no fiscal problem would have developed. On the contrary, the deficit would have remained relatively constant until 1982 and would then have steadily declined; by 1985 a large surplus would have been registered. Expanded public capital spending was the source of fiscal imbalances in the early 1980s. If the coffee surplus had remained the same as in 1975, the deficit would have been larger in the 1970s, implying that some fiscal problems were hidden by the good fortune of the coffee bonanza.

The increase in deficits in the early 1980s was financed for the most part from domestic sources, although external borrowing also increased. As figure 5.8 shows, the expansionary fiscal policy did lead to a rapid growth of the public debt ratio in the 1980s, which helps to explain the incipient external debt crisis of 1983–84. But the debt accumulation was from a relatively low base compared with most other Latin American countries. The debt ratio decreased during the coffee boom of 1975–79, so that the fiscal adjustment of 1985–89 was enough to avoid a full-blown debt crisis. The restrained use of external financing of the public sector in earlier periods helps to explain Colombia's avoidance of the kind of debt crisis that bedeviled its neighbors.

It helped also that Colombia's fiscal expansion during 1981–84 was comparatively modest. Unlike the situation in Mexico and Argentina, the current balance of the public sector always showed a surplus, amounting to more than 2 percent of GDP. Moreover, some of the expansion in public investment was in oil exploration and development, which paid off with a surge in oil exports after 1985.

*The Adjustment Program of 1985–89*

In late 1984 the Colombian government began a major adjustment program which succeeded in eliminating the fiscal imbalance that had

**Figure 5.7. Public Sector Deficit, with and without Coffee and Capital Spending Effects, Colombia, 1975–87**

Percentage of GDP

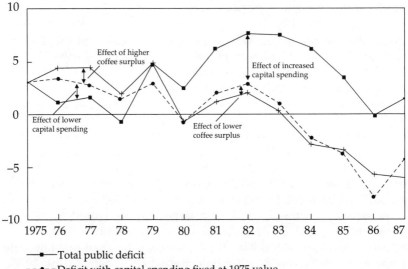

———■——— Total public deficit
– – ● – – Deficit with capital spending fixed at 1975 value
———+——— Deficit with coffee and capital spending fixed at 1975 value

*Source:* World Bank data.

**Figure 5.8. Public External Debt, Colombia, 1970–88**

Percentage of GDP

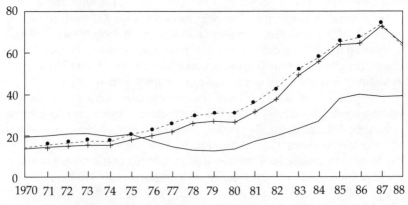

——————— Colombia
———+——— Highly indebted countries
– – – ● – – – Latin American countries

*Source:* World Bank data.

threatened to cause a macroeconomic crisis. It is useful to distinguish two main phases of adjustment. In the initial phase the overall deficit was sharply reduced, from 6.3 percent of GDP in 1984 to only 0.3 percent in 1986. In the second phase, 1987–89, the deficit increased again, to about 2 percent of GDP. The surge in coffee export revenues, a good deal of which accrued to the public sector, made 1986 an anomalous year.

Domestic financing of the deficit in the first phase was sharply reduced, from nearly 5 percent of GDP in 1984 to −2.7 percent in 1986. Domestic financing then increased again to between 1 and 2 percent of GDP. It is notable that net external financing continued to be available during the first phase of the adjustment. Colombia did not reschedule its external debt as other Latin American borrowers did. Although commercial banks were reluctant to continue lending in the wake of the global debt crisis, commercial financing was arranged with the support of two World Bank adjustment loans. A large flow of external finance in 1986, which came at the same time as the surge in coffee revenues, was successfully sterilized through public sector repayments of debt to the Central Bank and repayment of public short-term debt. In the second phase, external borrowing was reduced by the authorities in favor of domestic borrowing and money creation. This choice may help to explain the continuing high real interest rates noted earlier.

Because of the importance of the fortuitous coffee boom of 1986 for adjustment, a question arises regarding the extent to which the fiscal improvement was associated with permanent policy changes. Table 5.2 addresses this issue by decomposing the deficit into permanent policy components (structural trends) and other components. The first part of the table separates out various fiscal components whose variation is a result of nonpolicy factors or temporary policy changes. These include the overall balance of the National Coffee Fund and the transfers it made to the rest of the public sector, which are affected mainly by international coffee prices.[2] Also included is the balance and transfers of ECOPETROL, the national oil company, which became a large source of financing for the rest of the public sector by 1987, thanks to the development of a major new oil field. The fiscal improvement was helped in the first phase of the adjustment by the increased profits of the National Coffee Fund, which was the primary beneficiary of the surge in international coffee prices in 1986, and in the second phase by ECOPETROL. There were also some intrinsically temporary developments that influenced the national government's balance, the main one being a tax amnesty that increased revenues in 1987.[3]

Taking all these factors into account, we find that there was still a substantial improvement in the structural deficit. Less than 1 percent-

Table 5.2. Structural Trends in Fiscal Policy during Adjustment, Colombia, 1984–89

(percentage of GDP)

| Item | 1984 | 1985 | 1986 | 1987 | 1988 | 1989 | Change, 1984–89 |
|---|---|---|---|---|---|---|---|
| Total consolidated public sector deficit | 6.3 | 3.5 | 0.3 | 2.0 | 2.1 | 2.2 | −4.0 |
| National Coffee Fund | | | | | | | |
| Surplus | 0.4 | 1.4 | 3.2 | −0.6 | 0.0 | 0.1 | −0.3 |
| Transfers to public sector | 0.0 | 0.0 | 0.1 | 0.3 | 0.0 | 0.0 | 0.0 |
| ECOPETROL | | | | | | | |
| Surplus | 0.1 | −1.2 | −0.2 | 0.8 | 0.0 | 0.4 | 0.3 |
| Transfers to public sector | 0.1 | 0.1 | 0.2 | 0.6 | 0.6 | 0.7 | 0.6 |
| National government, temporary revenue items[a] | 0.1 | 0.1 | 0.3 | 0.5 | 0.1 | 0.1 | 0.0 |
| Structural deficit | 7.0 | 3.9 | 3.8 | 3.5 | 2.9 | 3.6 | −3.4 |
| Fixed capital formation | −8.8 | −8.3 | −6.7 | −6.1 | −6.4 | −6.7 | 2.1 |
| Structural current deficit | −1.8 | −4.4 | −2.9 | −2.6 | −3.5 | −3.1 | −1.3 |
| Wages and salaries | −6.9 | −6.3 | −5.9 | −5.7 | −5.7 | −5.8 | 1.2 |
| Structural current deficit net of wages | −8.7 | −10.7 | −8.8 | −8.3 | −9.2 | −8.9 | −0.2 |
| Import surcharge | 0.1 | 0.6 | 0.7 | 1.0 | 1.1 | 1.0 | 1.0 |
| Structural current deficit net of wages and import surcharge | −8.6 | −10.2 | −8.1 | −7.3 | −8.1 | −7.8 | 0.8 |
| Interest payments | −2.4 | −2.9 | −3.1 | −4.0 | −4.2 | −3.7 | −1.3 |
| Structural current primary deficit net of wages and import taxes | −11.0 | −13.0 | −11.3 | −11.3 | −12.3 | −11.5 | −0.5 |

Note: − indicates surplus.

a. Includes coffee tax (2.5 percent of coffee exports); backpayment of duties by ECOPETROL; Decreto 399-1986; and special revenue from the tax amnesty (in 1987).

Source: World Bank data.

age point of GDP of the improvement in the deficit during 1984–89 is explained by nonstructural factors, leaving a fiscal improvement of 3.4 percentage points of GDP to be explained. Colombia benefited from good luck, but the main part of the adjustment was achieved through its own efforts.

Another related question is to what extent the fiscal improvement was achieved at the expense of long-run growth. In the second part of table 5.2 the change in the "structural" deficit is explained by only three specific fiscal components: public investment, wages, and a surcharge on imports (although these factors were somewhat offset by the rise in interest payments). The reduction in public investment accounts for 2 percentage points of the reduction in the structural deficit. We examine the composition of this change below. The cut in spending on public wages, which accounts for 1 percent of the improvement in the fiscal deficit, largely reflects expedience, since it was achieved through a decline in real wage rates rather than a rational retrenchment of public employment. These two items alone fully explain the improvement in the structural deficit over the adjustment period.

The last two items in table 5.2 are roughly offsetting. The increase in interest payments reflects the consequences of the previous buildup of debt, the shift toward paying market interest rates on domestic debt, and the effect of the real devaluation on external interest payments. The major source of increased revenue for keeping up with the increased interest burden was the increase in the tax on imports, through a surcharge of 8 percent of import value beginning in 1985. Although increased revenue was desirable, the means chosen were again driven mainly by expedience rather than by rational long-run policy. Higher growth in the future is likely to require increased openness, which will eventually make necessary a reduction of tariff rates. (In fact, a trade liberalization was begun in 1990.)

To evaluate the reduction in public investment, we need more information about the composition of the cuts. Figure 5.9 shows the evolution of public investment since 1970. As was noted earlier, much of the fiscal expansion that triggered the near-crisis was a result of public investments in electric power and in extractive activities (coal mining and oil production). There was also some expansion in social sector investments. The sectors that had expanded earlier were precisely those that were cut during the adjustment program, with the largest adjustment in the electric power sector. In hindsight, the latter cuts were questionable, since electric power shortages developed in the early 1990s (partly as a result of mismanagement in the power sector, however).

The large reduction of investment in the mining sector mostly reflected the phasing-out of lumpy expenditures on the development

**Figure 5.9. Public Investment by Sector, Colombia, 1970–89**

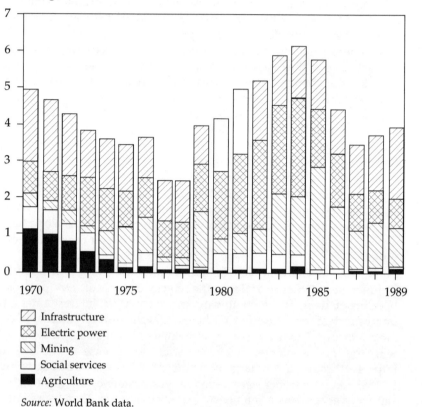

Percentage of GDP

Infrastructure
Electric power
Mining
Social services
Agriculture

*Source:* World Bank data.

of new oil and coal fields. Expenditure on infrastructure actually expanded during the adjustment program, largely because of spending in the telecommunications sector and the construction of a subway (of dubious economic value) in Medellín. More questionable for long-run prospects was the large reduction in social investments during the adjustment program and the failure to increase investment in agriculture.

The conclusions of table 5.2 are rather mixed. They suggest that some means chosen for fiscal adjustment were prejudicial to long-run growth. Improved growth performance will probably require the reversal of some of the fiscal adjustment measures of 1985–89 and their replacement by less distortionary fiscal instruments to maintain the deficit within bounds. Nevertheless, the form of adjustment was less damaging to growth than in other Latin American countries.

## Fiscal Deficits, Real Interest Rates, and Inflation

To analyze the effect of the financing of fiscal deficits on interest rates and inflation, we use the model developed for this study. For convenience, the basic structure is presented here.

### Basic Framework

The basic relationship is that of the fiscal deficit (DEF) to its means of financing:

$$(5.1) \qquad \mathrm{DEF} = \dot{L}_{cg} + \dot{L}_{dg} + \dot{F}_g^* \cdot E + \mathrm{BNOL}_g$$

where $L_{dg}$ is government borrowing from the banking system, $L_{cg}$ is credit from the Central Bank, $F_g$ is foreign debt, $E$ is the nominal exchange rate, and $\mathrm{BNOL}_g$ is net other liabilities of the government. The evolution of government borrowing from the banking system and the Central Bank determines the equilibria in the market for credit and the money market, respectively. These two markets are cleared by the domestic interest rate and the price level. Thus it is the composition of the financing of the deficit that determines the interest rate and inflation rate in this model.[4]

The equilibrium in the credit market is given by the identity that assets equal liabilities for commercial banks (see appendix 5.1):

$$(5.2) \qquad L_{dg} = (1 - r)P_c[M_1(1 - c) + QM] - L_{dp} - \mathrm{OTHD}$$

where $r$ is the reserve requirement on both narrow money and quasi money, $P_c$ is the consumption deflator, $M_1$ is real demand for narrow money, $c$ is the ratio of currency to narrow money, $QM$ is real demand for quasi money, $L_{dp}$ is private demand for credit, and OTHD is net other assets of the banking system. (As a simplification, we assume one reserve ratio for all types of deposits.) Equation 5.2 merely says that deposits in the banking system, less the amount set aside for reserves, must be equal to private and public credit demand plus net other assets of the banking system.

Private credit demand, $L_{dp}$, will evolve with private investment, $I_p$:

$$(5.3) \qquad L_{dp} = (1 + \pi)(1 - \delta)L_{dp}(-1) + P \cdot I_p \cdot \gamma$$

where $\pi$ is the inflation rate (in terms of the GDP deflator), $\delta$ is the rate of depreciation of physical capital, $P$ is the GDP deflator, $I_p$ is private investment, and $\gamma$ is the ratio of private credit to the private capital stock. This expression is derived from the assumption that the ratio of private domestic debt to the nominal value of capital (the leverage ratio) stays constant over time. This implies that the inflation component of interest payments on the debt ($\pi$ times the previous stock of debt) will always be rolled over. A constant share of gross investment will be financed by borrowing, but there will be an adjustment for depreciation on past capital, as reflected in past debt.

The equilibrium in the money market will be given by the condition that the supply of high-powered money, $H$, equals the demand:

$$(5.4) \qquad H = P \cdot M_1 \cdot [c + r(1 - c)] + rP \cdot QM$$

where the components $(cP \cdot M_1)$, $[r(1 - c)P \cdot M_1]$, and $(rP \cdot QM)$ correspond to currency demand, reserves on demand deposits, and reserves on quasi money, respectively. The supply of high-powered money, in turn, must be consistent with the outstanding credit to the government by the Central Bank:

$$(5.5) \qquad L_{cg} = H - \text{NFA}_{cb} - \text{OTHC}$$

where $\text{NFA}_{cb}$ is net foreign assets of the Central Bank and OTHC is net other assets of the Central Bank. Substituting equation 5.4 into equation 5.5 gives the condition that the assets of the Central Bank must equal its liabilities (see appendix 5.1).

### Econometrically Estimated Behavioral Equations

To complete the framework discussed above, we need to specify behavioral equations for money demand, quasi-money demand, private investment, and total output and to determine interest rate spreads. We also must estimate an equation for private consumption to determine the private saving flow that corresponds to accumulation of money and quasi money.

PORTFOLIO DEMANDS.  For private portfolio demands, we suppose a three-asset system of money, quasi money, and foreign assets. It is enough to specify behavioral equations for the first two, with foreign assets determined as a residual from the balance sheet condition for the private sector. Although the holding of foreign assets is legally restricted in Colombia, the existence of a thriving underground economy and flows of remittances from abroad make for a high degree of de facto capital mobility. The existence of a parallel-market premium implies that mobility is not perfect, but the premium is seldom more than 5 percent (and sometimes it is even negative!).

We thus model quasi-money demand as reflecting the possibility of substitution between domestic and foreign interest-bearing assets. Real quasi-money demand (deflated by the consumption deflator) is hypothesized to be a function of domestic interest rates, the inflation rate (which also reflects the possibility of substitution into real assets), the foreign interest rate plus the rate of depreciation, and real income. In the estimation process, we were not able to identify any separate effect of currency depreciation in addition to the effect of inflation.

This probably reflects the crawling peg system in Colombia, in which current inflation is the best predictor of future depreciation (the occasional large devaluations are usually unanticipated). We also constrain the income elasticity of demand for quasi money to be 1 in the estimation.[5] Thus the estimated equation shows the log of the ratio of real quasi money to GDP as a function of the real deposit rate, which is the nominal deposit rate adjusted for the ex post rise in the consumption deflator. The results are shown in table 5.3.

The real demand for narrow money (also deflated by the consumption deflator) is specified to depend on the nominal deposit interest rate and real income. Because the results for a regression in levels were unsatisfactory, we specified an error-correction format. Table 5.4 shows the first-stage levels regression of the log of real money on the log of real GDP, the nominal interest rate, and a time trend. The second regression is in differences form, with the lagged residual from the first-stage regression as one of the explanatory variables. Both real income and the nominal interest rate are significant, as is the residual from the first-stage regression. This specification has the intuitively appealing interpretation that the real growth in money demand responds to real income growth and changes in interest

**Table 5.3. Regression Results for Quasi-Money Demand, Colombia, 1973–88**

| Variable | Coefficient | Standard error | t-statistic | Significance |
|---|---|---|---|---|
| Constant | −2.47391 | 0.186360 | −13.2749 | 0.000 |
| Real interest rate | 1.43222 | 0.625282 | 2.29051 | 0.038 |
| Lagged error term | 0.849253 | 0.131997 | 6.43390 | 0.000 |
| *Equation summary* | | | | |
| No. observations | 16 | | | |
| Sum of squared residuals | 0.169050 | | | |
| $R^2$ | 0.9200 | | | |
| Adjusted $R^2$ | 0.9142 | | | |
| Standard error of regression | 0.109886 | | | |
| Durbin-Watson statistic | 1.89415 | | | |
| F-statistic (1, 14) | 160.900 | | | |
| Significance | 0.000000 | | | |
| *Autocorrelation estimation summary* | | | | |
| Standard error of rho(1) | 0.13200 | | | |
| Final rho(1) | 0.84925 | | | |
| t-value | 6.434 | | | |
| Significance | 0.000 | | | |

*Note:* The dependent variable is demand for log (real quasi money/real GDP).

rates, with a correction for the long-run relationship between levels of money and GDP. The negative time trend in the levels regression implies a secular tendency to move away from $M_1$, which could reflect technical change that economizes on the use of money in transactions.

PRIVATE INVESTMENT. As noted at the beginning of the chapter, the ratio of private investment to GDP shows a steady long-run decline in Colombia, while the ratio of public investment to GDP has been steadily rising. Table 5.5 shows the results of econometric testing to see whether there is a statistically significant relationship. The ratio of public capital stock to GDP does indeed enter with a negative sign in private investment regressions and is generally significant over both long and short sample periods and in both level regressions and error-correction specifications. This suggests that there is a high degree of substitutability between the activities in which the government invests and private sector activity. Higher public capital could either be lowering the share of private capital in the economy or actually driving down the rate of return to private capital.[6] The negative private-public capital association gives statistical confirmation to the hypothesis that the decline in private investment is linked in part to the secular growth of the state in Colombia.

The roles of other factors in the production function are not clearly resolved by the empirical results. The stock of private capital would be expected to have an ambiguous effect—increased private capital lowers the rate of return and further depresses private investment, but a higher existing stock of private capital leads to more investment to keep private capital growing at the same rate as the rest of the economy, as well as to replace depreciated capital. The coefficients on private capital are positive and significant in the error-correction results.

Similarly, the coefficient on labor, which would be expected to raise the marginal product of capital and increase investment demand, differs greatly across specifications in both sign and magnitude. The ordinary least squares (OLS) coefficient has the expected positive sign, but the error-correction specification yields a counterintuitive, significantly negative coefficient.

The other variable that might be expected to affect investment is the user cost of capital, defined as the real loan rate multiplied by the relative price of capital goods. This variable is of the "correct" negative sign in an OLS levels regression in which it is the sole variable, but it changes sign in the error-correction specification.

The clear result that emerges from the empirical analysis of investment is the negative relationship between public capital and private investment. This outcome suggests that the cuts in public investment

**Table 5.4. Regression Results for Demand for $M_1$, Colombia, Data from 1965–88**

| Variable | Coefficient | Standard error | t-statistic | Significance |
|---|---|---|---|---|
| *First stage*[a] | | | | |
| Constant | −12.9757 | 3.47126 | −3.73804 | 0.001 |
| ln (real GDP) | 1.73268 | 0.228106 | 7.59594 | 0.000 |
| Nominal interest rate | −0.218988 | 0.178090 | −1.22965 | 0.233 |
| Time trend | −0.0409874 | 0.0101437 | −4.04069 | 0.001 |
| *Equation summary* | | | | |
| No. observations | 24 | | | |
| Sum of squared residuals | 0.0431522 | | | |
| $R^2$ | 0.9727 | | | |
| Adjusted $R^2$ | 0.9686 | | | |
| Standard error of regression | 0.0464501 | | | |
| Durbin-Watson statistic | 1.18149 | | | |
| F-statistic (3, 20) | 237.672 | | | |
| Significance | 0.000000 | | | |
| *Second stage*[b] | | | | |
| Constant | −0.0342711 | 0.0272399 | −1.25812 | 0.224 |
| Difference ln(GDP) | 1.66381 | 0.561546 | 2.96290 | 0.008 |
| Difference interest rate | −0.532078 | 0.238691 | −2.22915 | 0.038 |
| Lagged residual | −0.692107 | 0.259542 | −2.66665 | 0.015 |
| *Equation summary* | | | | |
| No. observations | 23 | | | |
| Sum of squared residuals | 0.0339841 | | | |
| $R^2$ | 0.4369 | | | |
| Adjusted $R^2$ | 0.3480 | | | |
| Standard error of regression | 0.0422923 | | | |
| Durbin-Watson statistic | 1.46300 | | | |
| F-statistic (3, 19) | 4.91456 | | | |
| Significance | 0.010791 | | | |

a. The dependent variable for the regression is ln(real $M_1$), using 1965–88.
b. The dependent variable for the regression is money demand = log(real $M_1$) − log [real $M_1(-1)$], using 1966–88.

accompanying the adjustment program may not have been so damaging, as long as the allocation of the reductions at the microeconomic level was rational. However, the counterintuitive and unstable results on other variables suggest that more research on the determinants of investment in Colombia is needed.

**Table 5.5. Private Investment in Colombia**

| Regression | Constant | Private capital stock as share of GDP (percent) | Public capital stock as share of GDP (percent) | Labor as share of GDP (percent) | Real loan rate × relative price term[a] | Lagged dependent variable | Error correction term | Sample range | Durbin-Watson | Adjusted $R^2$ |
|---|---|---|---|---|---|---|---|---|---|---|
| OLS (ratio to GDP, log) | -0.68[b] (-2.58) | 0.16 (0.91) | -0.11 (-0.97) | — | — | 0.75[c] (7.15) | — | 1926–88 | 1.16 | 0.80 |
| OLS (ratio to GDP, log) | -0.87[c] (-3.3) | -0.27 (-1.1) | -0.26[b] (-2.1) | 0.21[b] (2.5) | — | 0.71[c] (6.93) | — | 1926–88 | 0.64 | 0.81 |
| Stage 1 error correction (ratio to GDP, log) | -2.09[c] (-8.44) | 1.11[c] (6.75) | -0.06 (-0.38) | — | — | — | — | 1925–88 | 0.53 | 0.60 |
| Stage 2 error correction (variables are first differences) | 0.01 (0.86) | 3.39[c] (5.07) | -0.98[c] (-2.99) | — | — | — | -0.53[c] (-4.82) | 1926–88 | 1.23 | 0.34 |
| Stage 1 error correction (ratio to GDP, log) | -2.28[c] (-9.85) | 0.30 (1.12) | -0.32[a] (-1.98) | 0.35[c] (3.68) | — | — | — | 1925–88 | 0.66 | 0.67 |
| Stage 2 error correction (variables are first differences) | -0.07[c] (-4.32) | 3.58[c] (8.45) | -0.08 (-0.32) | -3.33[c] (-6.55) | — | — | -0.71[c] (-9.13) | 1926–88 | 1.11 | 0.66 |

| | | | | | | | | | | |
|---|---|---|---|---|---|---|---|---|---|---|
| Stage 1 error correction (ratio to GDP, level) | 0.12[a] (1.94) | 0.03 (0.52) | −0.11[c] (−4.6) | — | 0.08 (1.65) | — | — | — | 1970–88 | 2.37 | 0.74 |
| Stage 2 error correction (variables are first differences) | 0.007[b] (2.43) | 0.32[c] (4.41) | −0.36[c] (−3.26) | — | 0.05[b] (2.47) | — | −1.38[c] (−6.4) | | 1971–88 | 1.94 | 0.80 |
| Stage 1 error correction (ratio to GDP, log) | −2.8[c] (−26.63) | 0.06 (0.097) | −0.61[c] (−4.58) | — | 0.698 (1.49) | — | — | — | 1970–88 | 2.14 | 0.73 |
| Stage 2 error correction (variables are first differences) | 0.09[b] (2.83) | 4.15[c] (5.4) | −2.13[c] (−3.34) | — | 0.53[b] (2.60) | — | −1.31[c] (−6.12) | | 1971–88 | 2.11 | 0.85 |
| OLS (logs) | −2.08[c] (−50.14) | — | — | — | −1.71[b] (−2.9) | — | — | — | 1970–88 | 1.56 | 0.29 |

— Not available.
a. 10 percent level of significance.
b. 5 percent level of significance.
c. 1 percent level of significance.
Note: For regressions with lagged dependent variables, the Durbin's $h$ statistic is shown in the Durbin-Watson column.

PRIVATE CONSUMPTION. Table 5.6 shows econometric results for private consumption. We regress real private consumption on real disposable income, real government savings, and the real interest rate. The inflation tax is subtracted from the conventional measure of disposable income. We experiment with different dynamic specifications, including a lagged dependent variable and error-correction equations. Degrees of freedom are scant because of limited data availability.

Disposable income is significant in all but one of the regressions.[7] In specifications using logs, the elasticity of consumption with respect to disposable income is close to unity. To test the sensitivity of the results to a larger sample size, GDP is used as a proxy for disposable income in some equations, with similar results. The real interest rate is surprisingly positive and significant in several equations, indicating that higher real interest rates tend to *lower* saving. Although this result is theoretically possible, it seems rather implausible. For the simulation, we use the second equation, in which consumption simply adjusts to disposable income. The restriction of long-run proportionality of consumption to income is not rejected by the data.

GROWTH. Table 5.6 also shows the results of productivity growth regressions, where output per worker is regressed on private and public capital stocks per worker.[8] We find that the questions on the productivity of public capital suggested by the private investment regressions are not really resolved by the growth regressions. Public capital is significant in the levels regression, but there is the usual worry about spurious correlation with nonstationary variables. The magnitudes of the coefficients would imply strong increasing returns in aggregate production.[9] The significance of public capital vanishes in an error-correction specification, and the magnitude of the coefficient on private capital is a neoclassically more plausible 0.32. Dummies for the Great Depression and World War II are also significant, while the constant term implies a rate of neutral technical progress of 1.9 percent per year. Although public capital still enters positively in the first-stage levels, the error-correction term itself is not significant, weakening confidence that the error-correction specification is appropriate. The third regression shown simply relates the growth in per worker output to the growth in private capital per worker, with a constant term gain, indicating neutral technical progress. This last equation is the one used in the simulations.

INTEREST RATE SPREAD. The other matter that must be addressed is the spread between loan and deposit interest rates, since the former enters into investment and the latter into the demand for money and quasi money. We assume that the spread between the deposit rate, $i_D$, and the loan rate, $i_L$, is explained by the reserve requirement, $r$, and an exogenous component, $i_0$, which would include profits and other costs of intermediation:

(5.6) $$i_D = i_L(1 - r) - i_0.$$

If we write the nominal loan rate as the sum of the real loan rate and the inflation rate, the nominal deposit rate can be written as a function of the real loan rate and inflation, as follows:[10]

(5.7) $$i_D = (1 - r)(r_L + \pi) - i_0$$

the real deposit rate will be given as:

(5.8) $$i_D - \pi = (1 - r)\, r_L - r\pi - i_0.$$

EQUILIBRIUM RELATIONSHIPS. We can now substitute into the equilibrium relations of equations 5.2 and 5.4 to determine the equilibrium response of real interest rates and inflation to changes in government money and domestic debt financing. Equation 5.9 shows the equilibrium relation between changes in Central Bank credit to the government, $L_{cg}$, the inflation rate, $\pi$, and the real loan rate, $r_L$:

(5.9)
$$dL_{cg} = \left\{ P \cdot M_1\,[c + r(1 - c)]\left[\frac{1}{1 + \pi} + \frac{M_1'}{M_1}(1 - r)\right] \right.$$
$$+ rP \cdot QM\left[\frac{1}{1 + \pi} - r\frac{QM'}{QM}\right] \Big\} d\pi$$
$$+ \left\{ P(1 - r)\left(\frac{M_1'}{M_1}[(c + r(1 - c)]\,M_1 + r\frac{QM'}{QM}\,QM\right) \right\} dr_L.$$

An increase in inflation will be associated with more monetary financing of the deficit as long as we have not passed the maximum point of the inflation-tax Laffer curve.[11] The first expression in equation 5.9 says that an increase in inflation is associated with less real demand for money and quasi money but a higher nominal flow of financing. The second effect is stronger than the first as long as we are on the upward-sloping part of the Laffer curve.

We can analyze the inflation-tax-maximizing inflation rate using the estimated equations of the model. A simulation of the equations for demand for money and quasi money at different inflation rates shows the relationship between the inflation rate and seigniorage revenue

**Table 5.6. Private Consumption in Colombia**

| Regression | Constant | Real interest rate | Disposable income | Government savings | GDP | Terms of trade | Lagged dependent variable | Error correction term | AR(1) | Sample range | Durbin-Watson | Adjusted $R^2$ |
|---|---|---|---|---|---|---|---|---|---|---|---|---|
| Two-stage least squares (logs)[a] | 0.47 (1.09) | — | 0.48[b] (1.96) | — | — | — | 0.49[c] (2.2) | — | — | 1971-86 | -0.74 | 0.996 |
| Two-stage least squares (logs)[a,d] | -0.056[c] (-2.73) | — | 0.72[e] (4.79) | — | — | — | 0.28[b] | — | — | 1971-86 | 1.585 | 0.640 |
| Stage-one error correction (log) | 0.32 (1.31) | 0.14[c] (2.99) | 0.80[e] (7.86) | — | — | — | 0.16 (1.69) | — | — | 1972-86 | 0.24 | 0.999 |
| Stage-two error correction: variables are first differences | -0.002 (-0.45) | 0.17[e] (3.88) | 0.79[e] (7.89) | — | — | — | 0.24[b] (2.23) | -0.98[c] (-2.8) | — | 1974-86 | -0.19 | 0.885 |
| Stage-one error correction (log) | 0.30 (1.35) | 0.18[e] (2.39) | — | — | 0.95[e] (56.10) | — | — | — | — | 1966-88 | 0.563 | 0.997 |
| Stage-two error correction: variables are first differences | -0.0007 (-0.10) | 0.09 (1.50) | — | — | 0.96[e] (6.65) | — | — | -0.20 (-1.19) | — | 1967-88 | 1.31 | 0.675 |
| Stage-one error correction (log) | 1.26[c] (2.92) | 0.29[c] (3.43) | — | -0.0005 (-0.11) | 0.88[c] (28.65) | — | — | — | — | 1974-88 | 0.937 | 0.992 |
| Stage-two error correction: variables are first differences | -0.0005 (-0.08) | 0.14[c] (2.53) | — | -0.003 (-1.15) | 0.91[e] (6.27) | — | — | -0.30 (-1.35) | — | 1975-88 | 1.159 | 0.766 |
| Two-stage least squares (log)[f] | -0.48[c] (-2.28) | — | 1.05[e] (63.01) | — | — | -0.05[b] (-2.11) | — | — | 0.09 (0.35) | 1972-86 | 1.56 | 0.997 |
| Stage-one error correction (logs) | 0.13 (0.41) | 0.13[c] (2.37) | 0.99[e] (40.3) | 0.001 (0.39) | — | -0.04 (-1.82) | — | — | — | 1974-86 | 1.689 | 0.997 |
| Stage-two error correction: variables are first differences | 0.0005 (0.13) | 0.14[c] (3.41) | 0.97[e] (9.64) | -0.002 (-0.98) | — | -0.03 (-1.38) | — | -1.01[c] (-2.59) | — | 1975-86 | 1.597 | 0.899 |
| Two-stage least squares (logs)[g] | 0.744[e] (2.99) | 0.096 (1.24) | — | — | 0.38[c] (2.45) | 0.04[b] (1.84) | 0.54[e] (3.28) | — | — | 1971-88 | 1.969 | 0.998 |

| Growth regressions (log of output per worker) | Constant | Private stock of capital per worker | Public stock of capital per worker | World War II dummy | Depression dummy | AR(1) | AR(2) | Error correction term | Sample range | Durbin-Watson | Adjusted R² |
|---|---|---|---|---|---|---|---|---|---|---|---|
| OLS | -1.27e (-3.43) | 0.7e (7.25) | 0.48e (7.0) | -0.031b (-1.92) | — | 1.4c (13.17) | -0.6e (-5.72) | — | 1927–88 | 1.97 | 0.997 |
| Stage-one error correction | -0.84e (-4.54) | 0.63e (12.33) | 0.49e (12.71) | -0.06b (-1.68) | -0.14e (-2.92) | — | — | — | 1925–88 | 0.24 | 0.980 |
| Stage-two error correction: variables are first differences | 0.02e (3.72) | 0.32e (2.47) | -0.08 (-0.60) | -0.03b (-1.73) | -0.05e (-2.7) | — | — | 0.005 (0.08) | 1926–88 | 1.47 | 0.115 |
| OLS: variables are first differences | 0.02e (4.65) | 0.31c (2.47) | — | -0.03b (1.73) | -0.05e (-2.86) | — | — | — | 1926–88 | 1.43 | 0.139 |

— Not available.

a. Instruments: government consumption, exports. Instrumented variable: disposable income.

b. 10 percent level of significance.

c. 5 percent level of significance.

d. In this equation it was imposed that the coefficients for disposable income and the lagged dependent variable sum to 1.

e. 1 percent level of significance.

f. Instruments: time trend, lagged disposable income, per capita government consumption, real interest rate. Instrumented variable: disposable income.

g. Instruments: time trend, lagged GDP, per capita government consumption. Instrumented variable: GDP.

Note: In equations with a lagged dependent variable, the Durbin's $h$ statistic is reported; in other equations, the Durbin-Watson statistic is given.

(figure 5.10). We see that maximum seigniorage (defined as the change in the money base over nominal GDP) of about 2.7 percent of GDP is achieved at inflation of a little less than 100 percent. Historically, inflation has been well below the seigniorage-maximizing rate.

The effect of a higher real interest rate on money creation is ambiguous. This is because higher interest rates have an ambiguous effect on the demand for base money: they lower demand for narrow money but raise demand for quasi money, and base money is a linear combination of the two (with the coefficients given by the currency-to-$M_1$ ratio and the reserve ratio). Base money is more likely to rise in response to an increase in the real interest rate the higher is the interest rate elasticity of quasi-money demand in relation to money demand and the higher is the ratio of existing quasi money to $M_1$. Thus, equilibrium in the money market could imply either a negative or a positive relationship between the real loan interest rate and the inflation rate for a given stock of Central Bank credit to the government.

Equation 5.10 shows the equilibrium relation in the credit market between real bank credit to the government, $L_{dg}/P$, inflation, $\pi$, and the real loan interest rate, $r_L$.

$$(5.10) \quad d\left[\frac{L_{dg}}{P}\right] = [(1 - r)(1 - c)M_1' - r(1 - r)QM']d\pi$$

$$+ [(1 - r)^2(1 - c)M_1' + (1 - r)^2QM' - \gamma l_p']dr_L.$$

An increase in the real interest rate increases credit to the government because it reduces private investment and demand for credit and increases deposits in the banking system.[12] Higher inflation reduces credit to the government because a higher inflation rate for a given real loan interest rate implies a lower real deposit rate (from equation 5.8). Quasi-money demand is therefore reduced by higher inflation. Demand deposits are also reduced, since these are a function of nominal interest rates. Thus equilibrium in the credit market for a given stock of government debt implies a positive relationship between the real loan rate and inflation.

Figures 5.11 and 5.12 show the joint determination of the real loan interest rate and inflation in the money and credit markets. Figure 5.11 illustrates the case in which the money-market equilibrium implies a negative relationship between the real interest rate and inflation. The locus of debt equilibria is always upward sloping. An increase in government borrowing from commercial banks ($L_{dg}/P$) shifts the locus of debt equilibria upward. This implies a higher real interest rate ($r_{L1}$) and a lower rate of inflation ($\pi_1$). The reason for

**Figure 5.10. Seigniorage Revenue as a Function of the Inflation Rate, Colombia**

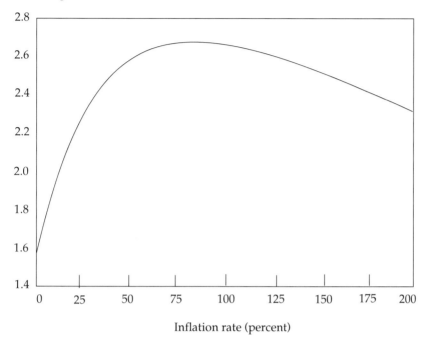

Percentage of GDP

Inflation rate (percent)

Source: World Bank data.

lower inflation is that the demand for base money is increased by higher real interest rates. High money demand implies a lower price level (and rate of inflation) for a given supply of money base. An increase in Central Bank credit to the government ($L_{cg}$) increases the money base, which shifts the money market equilibrium curve upward. Both inflation and the real interest rate ($r_{L2}$ and $\pi_2$) increase. The real interest rate increases because higher inflation represents a tax on demand and quasi-money deposits for a given real loan rate, so that deposits tend to fall unless there is an offsetting rise in interest rates.

Figure 5.12 illustrates the case in which the locus of money market equilibria is upward sloping. An increase in government borrowing causes the inflation rate, as well as the real loan rate, to increase. The higher real loan rate causes the demand for money base to fall, and inflation rises for a given supply of money base. An increase in Central Bank credit to the government still causes an increase in interest

**Figure 5.11. Effects of Changes in Deficit Financing (Case 1)**

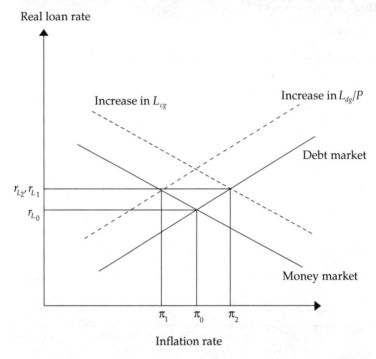

Inflation rate

rates and inflation, for the reasons given above. Although both domestic borrowing and money creation cause real interest rates and inflation to increase, it is clear from figure 5.12 that debt financing has a proportionately larger effect on real interest rates than on inflation, compared with monetary financing.

### Simulation Results

We use the model to perform counterfactual simulations within and beyond the sample period. We first calibrate the exogenous variables to reproduce the observed inflation and real interest rates over the period 1987–89.[13] This period is a mixture of within-sample and out-of-sample observations, since various regressions end in 1986, 1987, and 1988. We then consider changes in the fiscal deficit and its financing to evaluate how the deficit translates into changes in inflation and interest rates. Several of the regression equations have lagged error terms on the right-hand side. These are included in the model.

The first simulation, shown in table 5.7 as differences from the base case, is an increase in public investment financed by domestic borrowing. The fiscal expansion of 1.2 percent of GDP in 1987, 0.8 percent in 1988, and 0.7 percent in 1989 leads to an increase in the real interest rate ranging from 3 percent in 1987 to 5 percent in 1989.[14] The rise in the real interest rate causes a drop in the ratio of private investment to GDP of 0.5 percent in 1987–88 and 0.8 percent in 1989.[15] Although this is not as great as the increase in public investment, growth falls because only private investment affects growth.

As pointed out above, the effect of a debt-financed fiscal expansion on inflation is ambiguous. In this simulation there is a slight increase the first year, a fall in inflation the second year, and an increase the third year. This complicated pattern is the result of several offsetting factors. The increases in growth in 1988 and 1989 tend to lower inflation because higher growth stimulates greater demand for money, implying a lower rate of inflation for a given amount of money creation. However, the increased interest rates have two offsetting effects on demand for base money: a positive effect on reserves on quasi money, and negative effects on demand for currency and on reserves on demand deposits.

This simulation can be interpreted counterfactually as what would have happened had the fiscal adjustment described above not taken place. Thus, the difference between this simulation and the actual

**Figure 5.12. Effects of Changes in Deficit Financing (Case 2)**

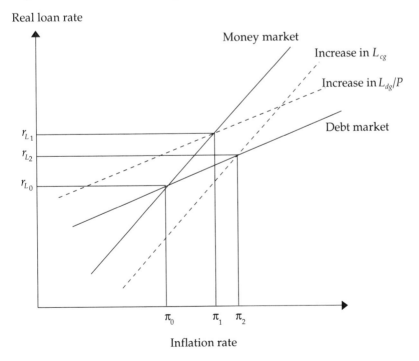

**Table 5.7.  Case of Increased Public Investment Financed by Domestic Borrowing**

(difference from base case, as percentage of GDP, unless otherwise specified)

| Change in ratio to GDP | 1987 | 1988 | 1989 |
|---|---|---|---|
| *National accounts (real)* | | | |
| Private consumption | 0.00 | 0.04 | 0.01 |
| Private investment | −0.50 | −0.68 | −0.82 |
| Public investment | 1.22 | 0.85 | 0.68 |
| Disposable income | 0.00 | 0.03 | −0.05 |
| Capital stock | | | |
| Public | 0.00 | 1.26 | 2.10 |
| Private | 0.00 | −0.33 | −0.76 |
| *Monetary accounts* | | | |
| Stocks | | | |
| Money | −0.22 | −0.13 | −0.33 |
| Quasi money | 0.38 | 0.51 | 0.63 |
| Money base | −0.01 | 0.08 | 0.02 |
| International reserves | −0.01 | 0.10 | 0.03 |
| Public sector deficit | 1.22 | 0.85 | 0.68 |
| Public deficit financing flows | | | |
| Foreign | 0.00 | 0.00 | 0.00 |
| Central Bank | 0.00 | 0.00 | 0.00 |
| Rest of financial system | 1.22 | 0.85 | 0.67 |
| Other liabilities of government | 0.00 | 0.00 | 0.00 |
| Stock of credit from | | | |
| Central Bank to government | 0.00 | 0.02 | 0.01 |
| Rest of financial system to government | 1.21 | 1.76 | 2.00 |
| Rest of financial system to private sector | −0.23 | −0.33 | −0.44 |
| *Other variables (percentage absolute change)* | | | |
| GDP growth | 0.00 | −0.12 | −0.16 |
| Inflation | 0.13 | −1.23 | 1.10 |
| Interest rates | | | |
| Loan rate | 3.84 | 3.72 | 7.65 |
| Real loan rate | 3.00 | 4.00 | 5.00 |
| Deposit rate | 3.41 | 3.29 | 6.76 |
| Real deposit rate | 2.58 | 3.59 | 4.37 |

outcome represents the consequences of adjustment compared with continuing debt-financed fiscal expansion. The implication is that the fiscal adjustment had the effect of raising growth by increasing private investment and lowering real interest rates. This effect on growth is meaningful mainly in the medium run, since the model does not incorporate any effects of demand on output.

Table 5.8 presents the results of a simulation of an increase in public investment financed by money creation. An increase in inflation of 15 percentage points per year is triggered by higher ratios of public investment to GDP and by a public deficit of 0.3 percent in 1987 and 1.1 percent in 1988–89. The reason a smaller increase in the deficit leads to the same inflation rate in the first year as that associated with higher deficits in the next two years is the portfolio shift effect. An increase in inflation causes a one-time shift out of money, which

**Table 5.8.  Case of Increased Public Investment Financed by Money Creation**
(difference from base case, as percentage of GDP, unless otherwise specified)

| Change in ratios to GDP | 1987 | 1988 | 1989 |
|---|---|---|---|
| *National accounts (real)* | | | |
| Private consumption | −0.22 | −0.18 | −0.23 |
| Private investment | −0.32 | −0.15 | −0.13 |
| Public investment | 0.26 | 1.18 | 1.09 |
| Disposable income | −0.33 | −0.21 | −0.28 |
| Capital stock | | | |
| Public | 0.00 | 0.31 | 1.44 |
| Private | 0.00 | −0.21 | −0.30 |
| *Monetary accounts* | | | |
| Stocks | | | |
| Money | −1.07 | −0.56 | −0.46 |
| Quasi money | 0.12 | 0.03 | 0.02 |
| Money base | −0.47 | −0.25 | −0.21 |
| International reserves | −1.06 | −1.98 | −2.80 |
| Public sector deficit | 0.26 | 1.18 | 1.09 |
| Public deficit financing flows | | | |
| Foreign | 0.00 | 0.00 | 0.00 |
| Central Bank | 0.26 | 1.18 | 1.09 |
| Rest of financial system | 0.00 | 0.00 | 0.00 |
| Other liabilities of government | 0.00 | 0.00 | 0.00 |
| Stock of credit from | | | |
| Central Bank to government | 0.01 | 0.97 | 1.64 |
| Rest of financial system to government | −0.08 | −0.07 | −0.05 |
| Rest of financial system to private sector | −0.15 | −0.15 | −0.15 |
| *Other variables (percentage absolute change)* | | | |
| GDP growth | 0.00 | −0.08 | −0.03 |
| Inflation | 15.00 | 15.00 | 15.00 |
| Interest rates: | | | |
| Loan rate | 19.60 | 17.98 | 18.06 |
| Real loan rate | 1.91 | 0.89 | 0.79 |
| Deposit rate | 17.42 | 15.89 | 15.96 |
| Real deposit rate | 0.84 | 0.21 | 0.11 |

sharply decreases the amount of money financing available in the first period. In the succeeding periods the quantity of money demanded grows in accordance with the new rate of inflation, without any off-setting portfolio shift.

The money-financed increase in public spending raises the real interest rate, as predicted by the comparative statics set out above. This is because the higher inflation tax on deposits for a given real loan rate requires an increase in the interest rate if deposits are to increase again and equilibrium is to be maintained. The higher real interest rates have a slight negative effect on private investment and thus on growth. The higher inflation also has a small negative effect on private consumption. Higher inflation increases the inflation tax on money balances, decreasing the after-tax disposable income of consumers. If the simulation is interpreted counterfactually, the implication is that the fiscal adjustment of the 1980s, as compared with continuing money-financed fiscal expansion, had the effect of lowering inflation and increasing private consumption and investment.

The final simulation we consider is a substitution of money for debt financing, leaving the deficit unchanged (table 5.9). The increase in money financing of 0.2 percent in 1987 and 1.1 percent in 1988–89 is again associated with increased inflation of 15 percent per year. The effect on the real interest rate is ambiguous, with offsetting effects of an increased inflation tax on deposits and a fall in government bor-rowing requirements. In the first year, when the decrease in govern-ment borrowing is small, the increased inflation tax dominates, rais-ing the real loan rate by 1.3 percent. In the second and third years the larger decrease in government borrowing dominates, and the real interest rate falls first 3 and then 5 percent.

The fall in real interest rates implies a rise in private investment of 0.8 percent of GDP by 1989. Private consumption again falls because of the increased inflation tax. Substituting money for debt finance is favorable for saving and growth. However, we must be cautious in interpreting this result: the efficiency losses associated with inflation are not captured by the model, and they could well dominate the results given here.

## The Real Exchange Rate and the Fiscal Deficit

The model in this section emphasizes that the real exchange rate is not just the outcome of the government's exchange rate policy; rather, it reflects endogenous economic forces that affect the demand for and supply of tradable and nontradable goods. Of these forces, the fiscal deficit is especially important because it represents net demand pressure that is policy induced. The Colombian adjustment

**Table 5.9.  Case of Substituting Money Creation for Debt Financing**
(difference from base case, as percentage of GDP, unless otherwise specified)

| Ratio to GDP | 1987 | 1988 | 1989 |
|---|---|---|---|
| *National accounts (real)* | | | |
| Private consumption | −0.22 | −0.19 | −0.26 |
| Private investment | −0.21 | 0.51 | 0.84 |
| Public investment | 0.00 | 0.00 | 0.00 |
| Disposable income | −0.33 | −0.21 | −0.28 |
| Capital stock | | | |
|   Public | 0.00 | 0.04 | −0.06 |
|   Private | 0.00 | −0.14 | 0.20 |
| *Monetary accounts* | | | |
| Stocks | | | |
|   Money | −1.03 | −0.28 | −0.15 |
|   Quasi money | 0.04 | −0.42 | −0.64 |
|   Money base | −0.46 | −0.25 | −0.25 |
|   International reserves | −1.06 | −1.98 | −2.82 |
| Public sector deficit | 0.00 | 0.00 | 0.00 |
| Public deficit financing flows | | | |
|   Foreign | 0.00 | 0.00 | 0.00 |
|   Central Bank | 0.26 | 1.18 | 1.06 |
|   Rest of financial system | −0.26 | −1.18 | −1.06 |
|   Other liabilities of government | 0.00 | 0.00 | 0.00 |
| Stock of credit from | | | |
|   Central Bank to government | 0.01 | 0.97 | 1.61 |
|   Rest of financial system to government | −0.34 | −1.42 | −2.03 |
|   Rest of financial system to private sector | −0.10 | 0.04 | 0.20 |
| *Other variables (percentage absolute change)* | | | |
| GDP growth | 0.00 | −0.05 | 0.13 |
| Inflation | 15.00 | 15.00 | 15.00 |
| Interest rates: | | | |
|   Loan rate | 18.72 | 12.68 | 10.31 |
|   Real loan rate | 1.27 | −2.81 | −4.67 |
|   Deposit rate | 16.63 | 11.21 | 9.11 |
|   Real deposit rate | 0.28 | −3.06 | −4.71 |

program initiated in 1985 achieved a substantial devaluation of the peso without a large acceleration of inflation as a result of the substantial fiscal adjustment that accompanied it.

### Determination of the Real Exchange Rate

This chapter follows closely the methodology given in chapter 2.[16] The model predicts that the real exchange rate will appreciate in response to an increase in the terms of trade. An increase in the terms of trade shifts supply away from nontradables and demand toward nontradables, causing a real appreciation. The model also predicts that the real exchange rate will appreciate in response to a fall in the trade surplus, which increases spending in relation to income and increases the demand for nontradables, so that their relative price increases (there is a real appreciation).

According to the Rodríguez models in chapter 2 of this volume, the real exchange rate is an ambiguous function of the level of government spending. An increase in government spending for a given resource surplus (and thus a given level of total spending) implies a redistribution of spending from the private to the public sector. If the government has a higher propensity to spend on nontradables than does the private sector, increased government spending implies a net increase in demand for nontradables, leading to a real appreciation. Conversely, if the government has a lower propensity to spend on nontradables (in other words, a higher propensity to spend on importables and exportables), increased spending results in real depreciation.

The model is estimated for Colombia over the period 1967–87; table 5.10 presents the results. We include a lagged dependent variable term to represent partial adjustment of the real exchange rate to changes in the fundamentals. The coefficient indicates that 39 percent of the long-run effect of a change in the fundamentals is realized the first year and 77 percent in the first three years. All the variables are significant (although the resource balance is not quite significant at the 5 percent level) and have the correct sign. A terms of trade increase leads to an appreciation of the real exchange rate. (As elsewhere in this chapter, an increase in the real exchange rate signifies depreciation.) An increase in the resource surplus causes a depreciation of the real exchange rate.

The sign on the government expenditure variable is positive, indicating that increased government spending causes a real depreciation. As indicated above, the sign is theoretically ambiguous. The sign found here would imply that the government devotes a lesser share of its spending to nontradables than does the private sector. In other words, government spending in Colombia is very import-intensive.

### Determination of the Resource Balance

To determine the trade surplus, we consider a variation on the Rodríguez model that incorporates some of the details on government deficit financing from the section "Fiscal Deficits, Real Interest Rates, and Inflation," above.

**Table 5.10. Regression Results for Real Exchange Rate (EXCH-RL)**

| Variable | Coefficient | Standard error | t-statistic | Significance |
|---|---|---|---|---|
| Constant | 1.45218 | 1.36576 | 1.06327 | 0.303 |
| $\ln(\text{EXCH-RL}(-1))$ | 0.612230 | 0.267986 | 2.28456 | 0.036 |
| $\ln(\text{TTRADE})$ | −0.221399 | 0.0808159 | −2.73955 | 0.015 |
| RSCBAL&GDP | 0.0299868 | 0.0146600 | 2.04548 | 0.058 |
| $\ln(\text{EXPTOT\&GDP})$ | 0.421882 | 0.131091 | 3.21824 | 0.005 |

*Equation summary*

| | |
|---|---|
| No. observations | 21 |
| Sum of squared residuals | 0.0548271 |
| $R^2$ | 0.8125 |
| Adjusted $R^2$ | 0.7656 |
| Standard error of regression | 0.0585380 |
| Durbin-Watson | 1.83021 |
| F-statistic | 17.3320 |
| Significance | 0.000011 |

*Note*: ln(EXCH-RL), log (real exchange rate); ln(TTRADE), log (terms of trade); RSCBAL&GDP, resource balance (percentage of GDP); ln(EXPTOT&GDP), log (total public expenditure/GDP); PRMFSUR&GDP-SP, primary fiscal surplus (percentage GDP). Depreciation is up. Method, two-stage least squares; dependent variable, LN(EXCH-RL); sample period, 1967–87; instrumented variables: constant, LN(ENCH-RL(−1), LN(EXCH-RL(−1)), LN(TTRADE), RSCBAL&GDP(−1), LN(EXPTOT&GDP), PRMFSUR&GDP-SP.

THEORETICAL DERIVATION. The trade surplus (*ts*) is given as the sum of the resource balance of the private sector (saving minus investment) and the primary surplus, *p*, of the public sector:

$$(5.11) \qquad ts = s_p - i_p(r) + p.$$
$$\qquad\qquad\qquad\quad (-)$$

Private investment is a negative function of the real interest rate, *r*. The real interest rate is determined by the equilibrium condition in the market for domestic government debt $d_g$:

$$(5.12) \qquad d_g = d_g(r).$$
$$\qquad\qquad\qquad (+)$$

The total derivative of equation 5.11 gives us the change in the trade surplus as a function of changes in private saving, the real interest rate, and the government primary surplus:

$$(5.13) \qquad d(ts) = d(s_p) - i'_p dr + dp.$$

We will assume in what follows that the derivative of private saving with respect to the real interest rate and the primary surplus is zero. (Ricardian equivalence does not hold.)

The real interest rate can be determined from the government financing constraint. The government borrows a fixed percentage of GDP, $f$, from abroad. This is determined exogenously, either as a government policy decision or by international capital market constraints. The residual source of financing is domestic borrowing, given by:

$$(5.14) \qquad \dot{d}_g = -p - f + (r - g)d_g + r^*d_g^*$$

where $p$ is the primary surplus, $d_g$ and $d_g^*$ are domestic and foreign debt, $r$ and $r^*$ are domestic and foreign interest rates, and $g$ is the growth rate of GDP. In discrete time, we can write the level of government debt as:

$$(5.15) \qquad d_g = -p - f + (r - g + 1)d_g(-1) + r^*d_g^*(-1).$$

The amount of government debt in equation 5.15 must equal the amount demanded by the public in equation 5.12. When $r$, $p$, and $f$ are allowed to vary, the following relationship between these variables is implied:

$$(5.16) \qquad d_g'dr = -dp - df + dr[dg(-1)]$$

from which we obtain $r$ as a function of $p$ and $f$:

$$(5.17) \qquad dr = -\frac{-dp - df}{d_g' - d_g}.$$

The real interest rate is a negative function of the primary surplus, $p$, and a negative function of foreign borrowing, $f$. Either an increase in the surplus or a shift toward foreign borrowing for a given surplus tends to relieve the pressure on domestic financial markets and decrease the interest rate. Since $p$ and $f$ enter symmetrically into the equation for $r$, a decrease in $p$ that is exactly offset by an increase in $f$ will have no effect on $r$. In other words, an externally financed fiscal expansion has no effect on domestic interest rates.

Substituting into equation 5.13, we obtain the resource balance as a function of the primary surplus and the amount of foreign borrowing:

$$(5.18) \qquad d(ts) = \left[1 + \frac{i_p'}{d_g' - d_g}\right]dp + \frac{i_p'}{d_g' - d_g}df.$$

The trade surplus is a negative function of foreign borrowing, $f$, and a positive function of the primary surplus, $p$ (if the coefficient on $f$ is less than one). Note the restriction that the coefficient on $p$ be 1 plus

**Table 5.11. Regression Results for Resource Balance
(RSCBAL&GDP)**

| Variable | Coefficient | Standard error | t-statistic | Significance |
|---|---|---|---|---|
| Constant | 2.19633 | 0.507417 | 4.32845 | 0.001 |
| PRMFSUR&GDP-SP | 0.448078 | 0.139534 | 3.21125 | 0.005 |
| FINEXT&GDP-2 | −0.412757 | 0.157042 | −2.62832 | 0.018 |

*Equation summary*
| | |
|---|---|
| No. of observations | 19 |
| Sum of squared residuals | 51.4009 |
| $R^2$ | 0.6834 |
| Adjusted $R^2$ | 0.6438 |
| Standard error of regression | 1.79236 |
| Durbin-Watson | 1.73298 |
| F-statistic | 17.2646 |
| Significance | 0.000101 |

*Note*: RSCBAL&GDP, resource balance (percentage of GDP); PRMFSUR&GDP-SP, primary fiscal surplus (percentage of GDP); FINEXT&GDP-2, external financing (percentage of GDP). Method, ordinary least squares; dependent variable, RSCBAL&GDP; sample period, 1970–88.

the coefficient on $f$. This means that a decrease in $p$ matched exactly by an increase in $f$ will reduce the resource balance one for one. Since an externally financed fiscal expansion has no effect on the real interest rate, as shown above, it can only spill into the resource balance one for one.

We use equation 5.18 as the basis for our estimated equation. As is shown in table 5.11, both variables are statistically significant and of the correct sign, and the other regression statistics are satisfactory. The coefficient on $p$ minus the coefficient on $f$ is equal to 0.86 (instead of 1, as predicted by the theory). A formal test reveals that the violation of the theoretical restriction is not statistically significant.

We can use this estimated equation to assess the behavior of the resource balance over the period 1975–88. We simulate the equation for base period (1975) values and compare the results with those of the simulation that uses actual values of the primary surplus and public external financing. As figure 5.13 shows, the resource balance went from a sizable surplus in 1977 to a large deficit in 1983 before recovering to yield a surplus in 1986–88. We see from the figure that the external financing effect is the most important factor in explaining the resource surpluses of the late 1970s, with the exception of the strong effect of the primary surplus in 1978. In the 1980s the two

**Figure 5.13. Decomposition of the Resource Balance, Colombia, 1975–88**

Percentage of GDP

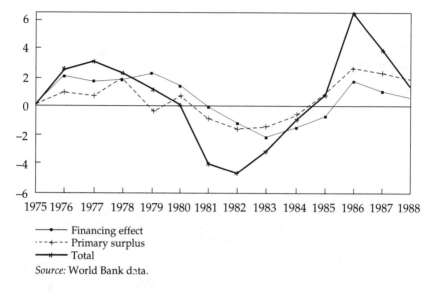

────●──── Financing effect
- - - + - - - Primary surplus
────×──── Total

*Source:* World Bank data.

effects moved together. Externally financed fiscal expansion was responsible for the deterioration of the resource balance in the early 1980s. Beginning in 1984 fiscal contraction accompanied by reduced foreign borrowing helped improve the resource balance.

## *Joint Simulation of the Real Exchange Rate and the Resource Balance*

The models in the first part of this chapter can be combined to get an idea of the role of fiscal variables in determining the real exchange rate through their effect on the resource balance. The simulated values of the resource balance, alternately varying the primary surplus and public foreign borrowing, are substituted into the real exchange rate model and simulated. The simulated real exchange rate outcome is compared with an equilibrium in which all variables are fixed, and the difference represents the change in the real exchange rate attributable to changes in the primary surplus and public foreign borrowing, respectively. The unexplained portion of the resource balance is now included under the residual.

Figure 5.14 shows the decomposition of the real exchange rate, including the indirect effects of the government primary surplus and public foreign borrowing. (As always, depreciation is up.) The effects

**Figure 5.14. Exchange Rate Decomposition, Direct and Indirect Effects, Colombia, 1975–87**

Log change since 1975

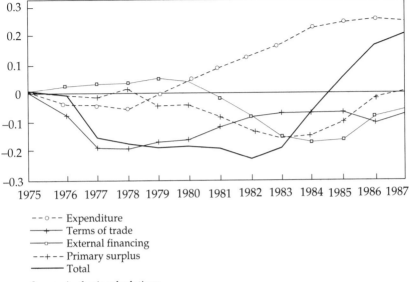

--o-- Expenditure
—+— Terms of trade
—o— External financing
--+-- Primary surplus
——— Total

*Source:* Author's calculations.

of changes in the government surplus and changes in financing tend to move together. An externally financed fiscal expansion played a major role in the appreciation of the real exchange rate in the early 1980s, just as the simultaneous reduction of the fiscal deficit and foreign borrowing supported the depreciation after 1984.

This model gives some insight into the causes for some of the most important changes in the real exchange rate. The strong appreciation of 1977 was almost entirely attributable to the sharp increase in the terms of trade. However, the appreciation was exacerbated by a reduction in the government primary surplus, contrary to what would have been a sound policy of increasing the surplus to keep the economy stable. Histories of the period often stress the attempts of monetary authorities to sterilize the reserve inflows from the coffee boom. In this model such monetary policy would work by reducing net public foreign borrowing. This, however, was not a major factor in 1977–78, at the height of the coffee boom.

The 1977 experience contrasts with the policy package implemented in 1986 during the second, and smaller, coffee boom. The appreciation implied by the change in the terms of trade was more than offset by the combination of an increased primary surplus with a large reduction in foreign borrowing, and a real depreciation was achieved.

## Conclusion

In Colombia the management of fiscal deficits and their financing has been generally sound. Episodes of loose fiscal policy have been minor in comparison with other Latin American countries. The near-crisis of the early 1980s was addressed in a timely way through a sharp fiscal adjustment. Analysis shows this adjustment to have been a combination of good luck and fundamental policy changes, with more emphasis on the latter. The means of fiscal adjustment chosen were sometimes suboptimal from the standpoint of long-run growth, and a degree of fiscal reform will eventually be needed to reverse some of the measures implemented during 1985–89. Still, the adjustment compares well with those of other Latin American countries.

Two surprising results of the study are that public capital has a negative effect on private investment and that the contribution of public investment to growth is highly uncertain. This would imply that the cuts in public investment in the late 1980s were not terribly prejudicial to growth, although some specific cuts were poorly chosen.

The analysis in this chapter shows a close relationship in Colombia between the means of financing the fiscal deficit and macroeconomic outcomes. The simulation model used here showed how money-financed and domestic-debt-financed fiscal deficits translate into inflation and changes in the real interest rate. Roughly speaking, a debt-financed deficit increase of about 1 percent of GDP translates into an increase in the real interest rate of 3 to 5 percent, and a money-financed deficit increase of about 1 percent translates into 15 percentage points of additional inflation. We then traced the relationship between externally financed fiscal deficits and the real exchange rate and found that a good part of the change in the real exchange rates over 1975–87 is attributable to fiscal policy.

Colombia's trademark is moderation: moderate budget deficits, moderate inflation, and decent, although not spectacular, growth. This moderation enabled Colombia to be the dog that did not bark— that is, the Latin country that did *not* have a debt crisis. Colombia's special character continues to be evident up to the present. During 1990–92 the Colombian authorities maintained strict fiscal discipline (budget deficits of less than 1 percent of GDP) in the face of adversity— drug-related violence, an electric power shortage, and low coffee prices. Inflation accelerated somewhat, to about 30 percent, but GDP growth continued at a decent pace, 3.2 percent over 1990–92. Colombia's long-run prospects have brightened with a major new oil

discovery. Compared with the painful adjustment of its Latin neighbors, Colombia's experience is a good advertisement for a policy of
low and stable public sector deficits.

## Appendix 5.1. Balance Sheet Identities for Money and Credit Market Equilibria

*Balance sheet of commercial banks*

| Assets | Liabilities |
|---|---|
| $L_{dg}$ = credit to government | $(1-c)M_1 \cdot P_c$ = demand deposits |
| $L_{dp}$ = credit to private sector | $QM \cdot P_c$ = quasi money |
| OTHD = net other assets | |
| $rP_c \cdot M_1(1-c)$ = reserves on demand deposit | |
| $rP_c \cdot QM$ = reserves on quasi money | |

*Balance sheet of central bank*

| Assets | Liabilities |
|---|---|
| $L_{cg}$ = credit to the government | $c \cdot P \cdot M_1$ = currency |
| NFA$_{cb}$ = net foreign assets | $r(1-c)P \cdot M_1$ = reserves on demand deposits |
| OTHC = net other assets | $r \cdot P \cdot QM$ = reserves on quasi money |

## Appendix 5.2. Definitions of Variables, Sources, and Data

*Disposable income:* disposable income less inflation tax (inflation rate times $M_1$), deflated by consumption deflator. Source: for disposable income, Departamento Administrativo Nacional Estadística (DANE), Colombia.

*Government saving:* consolidated public sector balance as a ratio to GDP, constructed from World Bank and Banco de la República data.

*Nominal interest rate:* interest rate on *certificados de depósito a término* (CDT). Source: *Revista del Banco de la República*.

*Real interest rate:* (1 plus nominal CDT rate) divided by (1 plus percentage change in consumption deflator) minus 1.

*Private capital stock:* previous year's capital stock multiplied by (1 minus depreciation rate) plus current year's private investment. (The depreciation rate is assumed to be 5 percent.) Initial year capital stock is determined as a ratio to GDP in that year, with the ratio given by the average over the period of the ratio of private investment to GDP divided by the sum of the average growth rate and the depreciation rate.

*Public capital stock:* same as for public investment. Source: for public investment in 1975 prices, DANE; for public investment, 1925–88, same as for real GDP.

*Real GDP:* for 1965–88, GDP in 1975 prices from DANE. For series 1925–88, GDP in 1950 prices spliced data from Comisión Económica para América Latina y el Caribe (CEPAL), *El Desarrollo Economico de Colombia. Anexo estadistico,* 1957, with IMF data for 1950–88.

*Real $M_1$:* $M_1$ deflated by consumption deflator. Source: for 1964–81, *Revista del Banco de la República;* for 1982–88, World Bank.

*Real private consumption:* private consumption in 1975 prices , from DANE.

*Real private investment:* private investment in 1975 prices, from DANE. For 1925–88, same source as real GDP.

*Real quasi money:* sum of CDT and *depósitos de ahorro tradicional* in the banking system, deflated by private consumption deflator. Source: for quasi money, *Revista del Banco de la República;* for private consumption deflator, DANE.

*Relative price of capital:* ratio of investment deflator to GDP deflator. Source: DANE.

### Variables for real exchange rate and resource balance regressions:

EXCH-RL: average nominal exchange rate (pesos/dollar) multiplied by (U.S. wholesale price index divided by Colombian consumer price index). Source: IMF, *International Financial Statistics* (IFS).

TTRADE: Colombian export price index divided by import price index. Source: IFS.

RSCBAL&GDP: resource balance (exports minus imports of goods and nonfactor services) as percentage of GDP. Source: DANE.

EXPTOT&GDP: total consolidated public expenditure as percentage of GDP. Source: for 1967–74, Bird (1984), table 2-1; for 1975–88, World Bank data.

PRMFSUR&GDP: primary fiscal surplus (fiscal balance excluding interest payments) as percentage of GDP. Source: for 1967–74, Garcia Garcia (1989); for 1975–88, World Bank data.

FINEXT&GDP: net external financing of public sector = net flow of medium- and long-term public debt plus net flow of short-term public debt minus change in international reserves of Central Bank. Source: for public debt, Banco de la República, "Deuda Externa de Colombia," July 1989; for reserves, IFS.

# Appendix 5.3  Data for Regressions for Consumption, Investment, and Money
(millions of pesos, except as otherwise specified)

| Year | Real private consumption | Percentage change in consumption deflator | Real GDP | Nominal interest rate | Real private investment | Real $M_1$ | Relative price of capital | Real quasi money | Real interest rate | Real disposable income |
|---|---|---|---|---|---|---|---|---|---|---|
| 1960 | — | — | — | — | — | — | — | — | — | — |
| 1961 | — | — | — | — | — | — | — | — | — | — |
| 1962 | — | — | — | — | — | — | — | — | — | — |
| 1963 | — | — | — | — | — | — | — | — | — | — |
| 1964 | — | — | — | — | — | — | — | — | — | — |
| 1965 | 168,362 | — | 235,051 | 0.052 | — | 36,147 | 1.0181 | — | — | — |
| 1966 | 180,319 | 0.16263 | 247,360 | 0.052 | — | 35,443 | 1.0531 | — | -0.09515 | — |
| 1967 | 184,146 | 0.08401 | 257,588 | 0.052 | — | 39,849 | 1.0425 | — | -0.02953 | — |
| 1968 | 196,421 | 0.07639 | 272,871 | 0.052 | — | 42,485 | 1.0538 | — | -0.02266 | — |
| 1969 | 211,216 | 0.07442 | 289,523 | 0.052 | — | 47,264 | 1.0650 | — | -0.02087 | — |
| 1970 | 224,576 | 0.08752 | 307,476 | 0.052 | 37,682 | 50,950 | 1.0422 | — | -0.03266 | 248,328 |
| 1971 | 241,733 | 0.11918 | 325,831 | 0.052 | 38,102 | 50,509 | 1.0220 | — | -0.06003 | 259,173 |
| 1972 | 255,776 | 0.12116 | 350,813 | 0.052 | 34,560 | 56,028 | 1.0306 | 10,953 | -0.06169 | 289,002 |
| 1973 | 268,183 | 0.19173 | 374,398 | 0.052 | 39,015 | 60,768 | 0.9945 | 12,577 | -0.11725 | 308,725 |
| 1974 | 284,365 | 0.26605 | 395,910 | 0.167 | 46,181 | 57,382 | 1.0045 | 15,287 | -0.07823 | 323,173 |
| 1975 | 292,779 | 0.24437 | 405,108 | 0.281 | 41,109 | 58,915 | 1.0000 | 17,507 | 0.02944 | 325,480 |
| 1976 | 313,199 | 0.20797 | 424,263 | 0.281 | 44,907 | 65,716 | 0.9907 | 21,640 | 0.06046 | 347,995 |
| 1977 | 325,847 | 0.27101 | 441,906 | 0.281 | 38,684 | 67,414 | 0.9369 | 24,078 | 0.00786 | 371,026 |
| 1978 | 353,212 | 0.17953 | 479,335 | 0.265 | 49,598 | 74,485 | 0.9840 | 29,201 | 0.07246 | 397,245 |
| 1979 | 368,439 | 0.26081 | 505,119 | 0.266 | 50,813 | 73,400 | 1.0015 | 26,108 | 0.00412 | 414,867 |
| 1980 | 384,698 | 0.26237 | 525,765 | 0.302 | 51,280 | 74,344 | 1.0019 | 38,757 | 0.03140 | 432,562 |
| 1981 | 395,910 | 0.25986 | 537,736 | 0.373 | 54,605 | 71,518 | 1.0148 | 51,500 | 0.08980 | 437,738 |
| 1982 | 401,759 | 0.24731 | 542,836 | 0.379 | 51,687 | 71,907 | 0.9842 | 47,781 | 0.10558 | 444,612 |
| 1983 | 403,572 | 0.20185 | 551,380 | 0.337 | 54,892 | 74,613 | 0.9726 | 50,462 | 0.11245 | 453,812 |
| 1984 | 415,128 | 0.20446 | 569,855 | 0.347 | 49,222 | 76,427 | 0.9802 | 52,999 | 0.11834 | 465,727 |
| 1985 | 422,917 | 0.23528 | 587,561 | 0.353 | 44,682 | 79,288 | 1.1014 | 57,716 | 0.09530 | 468,961 |
| 1986 | 435,636 | 0.24118 | 617,527 | 0.312 | 50,652 | 78,432 | 1.0794 | 63,809 | 0.05705 | 488,433 |
| 1987 | 453,196 | 0.27877 | 654,853 | 0.342 | — | 81,543 | — | 65,152 | 0.04944 | — |
| 1988 | 469,692 | 0.29768 | 679,345 | 0.318 | — | 79,040 | — | 56,128 | 0.01566 | — |

— Not available.
Source: See appendix 5.2.

**Appendix 5.4. Data for Regressions for the Real Exchange Rate and Resource Balance**
(percentage of GDP, except as indicated)

| Year | EXCH-RL (index) | FINEXT&GDP-2 | TTRADE (index) | PRMFSUR&GDP-SP | EXPTOT&GDP | RSCBAL&GDP |
|---|---|---|---|---|---|---|
| 1967 | 109 | 0.49 | 69.03 | −3.60 | 25.05 | 0.8812 |
| 1968 | 118 | −1.16 | 71.35 | −5.10 | 27.81 | −1.0454 |
| 1969 | 119 | −0.86 | 71.23 | −7.70 | 30.97 | −0.4732 |
| 1970 | 122 | 2.11 | 87.27 | −6.90 | 30.69 | −1.2842 |
| 1971 | 125 | 1.96 | 81.57 | −7.40 | 31.62 | −4.0504 |
| 1972 | 126 | 0.75 | 85.74 | −6.50 | 29.98 | 0.4567 |
| 1973 | 129 | 0.77 | 91.31 | −7.10 | 28.72 | 2.2598 |
| 1974 | 135 | 2.33 | 95.29 | −0.90 | 27.99 | −1.0903 |
| 1975 | 142 | 1.51 | 88.13 | −1.96 | 27.06 | 1.8057 |
| 1976 | 139 | −3.44 | 123.64 | 0.08 | 24.80 | 3.1512 |
| 1977 | 118 | −2.35 | 171.54 | −0.47 | 25.90 | 3.6669 |
| 1978 | 115 | −2.46 | 128.04 | 1.95 | 24.87 | 2.8274 |
| 1979 | 113 | −3.78 | 114.44 | −3.49 | 28.93 | 1.7713 |
| 1980 | 113 | −1.43 | 116.33 | −0.81 | 30.24 | 0.6210 |
| 1981 | 111 | 2.24 | 98.13 | −4.40 | 31.19 | −3.5669 |
| 1982 | 107 | 4.98 | 95.24 | −5.80 | 31.77 | −4.2781 |
| 1983 | 112 | 7.34 | 97.58 | −5.70 | 33.59 | −2.7808 |
| 1984 | 126 | 5.76 | 101.02 | −3.90 | 35.63 | −0.5792 |
| 1985 | 142 | 3.82 | 100.00 | −0.60 | 34.34 | 1.2825 |
| 1986 | 159 | −2.04 | 121.83 | 3.40 | 34.73 | 6.8532 |
| 1987 | 165 | −0.19 | 94.04 | 2.40 | 32.86 | 4.2306 |
| 1988 | 165 | 0.92 | 94.67 | 1.30 | 38.82 | 1.7062 |

Source: See appendix 5.2.

**Appendix 5.5. National Accounts, Colombia, 1950 Prices**

(millions of pesos, except as specified)

| Year | GDP | Public investment | Private investment | Public capital stock | Private capital stock | Labor force (thousands of workers) |
|---|---|---|---|---|---|---|
| 1925 | 2,721.5343 | 95.9404 | 460.4900 | 664.5795 | 2,249.5740 | 2,505 |
| 1926 | 2,981.3793 | 139.3987 | 554.9801 | 737.5203 | 2,579.5967 | 2,551 |
| 1927 | 3,249.9272 | 164.8818 | 673.2416 | 828.6500 | 2,994.8786 | 2,596 |
| 1928 | 3,488.6365 | 180.4026 | 813.0576 | 926.1877 | 3,508.4483 | 2,645 |
| 1929 | 3,614.2075 | 141.5644 | 716.2658 | 975.1333 | 3,873.8693 | 2,693 |
| 1930 | 3,583.1256 | 102.6541 | 456.0948 | 980.2741 | 3,942.5772 | 2,743 |
| 1931 | 3,525.9348 | 90.3096 | 369.9048 | 972.5563 | 3,918.2243 | 2,799 |
| 1932 | 3,759.6710 | 93.5581 | 450.1208 | 968.8588 | 3,976.5226 | 2,857 |
| 1933 | 3,971.0282 | 111.3169 | 420.7698 | 983.2897 | 3,999.6401 | 2,916 |
| 1934 | 4,220.9269 | 76.6657 | 465.8540 | 961.6265 | 4,065.5301 | 2,976 |
| 1935 | 4,324.1189 | 96.3735 | 523.8146 | 961.8374 | 4,182.7916 | 3,038 |
| 1936 | 4,552.8820 | 96.3014 | 586.4852 | 961.9550 | 4,350.9976 | 3,102 |
| 1937 | 4,623.7488 | 118.7524 | 674.1610 | 984.5119 | 4,590.0588 | 3,165 |
| 1938 | 4,924.6220 | 139.1100 | 671.1919 | 1,025.1707 | 4,802.2448 | 3,232 |
| 1939 | 5,226.7384 | 163.6546 | 761.4111 | 1,086.3082 | 5,083.4314 | 3,287 |
| 1940 | 5,339.8766 | 217.8692 | 671.2603 | 1,195.5466 | 5,246.3485 | 3,343 |
| 1941 | 5,429.3926 | 183.5068 | 668.5273 | 1,259.4987 | 5,390.2410 | 3,401 |
| 1942 | 5,440.5821 | 219.9627 | 528.8999 | 1,353.5115 | 5,380.1169 | 3,460 |
| 1943 | 5,462.9611 | 212.0940 | 581.9786 | 1,430.2543 | 5,424.0838 | 3,520 |
| 1944 | 5,832.2145 | 140.3372 | 724.4484 | 1,427.5661 | 5,606.1239 | 3,583 |
| 1945 | 6,105.7355 | 127.7040 | 950.3800 | 1,412.5135 | 5,995.8915 | 3,647 |
| 1946 | 6,692.5625 | 141.2757 | 1,132.7182 | 1,412.5378 | 6,529.0205 | 3,697 |
| 1947 | 6,952.4075 | 212.9603 | 1,324.1788 | 1,484.2443 | 7,200.2973 | 3,750 |

*(Table continues on the following page.)*

**Appendix 5.5** (*continued*)

| Year | GDP | Public investment | Private investment | Public capital stock | Private capital stock | Labor force (thousands of workers) |
|------|-----|-------------------|--------------------|----------------------|-----------------------|-----------------------------------|
| 1948 | 7,150.0886 | 213.8987 | 1,289.6227 | 1,549.7186 | 7,769.8903 | 3,805 |
| 1949 | 7,774.2139 | 149.0000 | 998.6378 | 1,543.7468 | 7,991.5391 | 3,859 |
| 1950 | 7,860.0000 | 149.0000 | 1,176.0000 | 1,538.3721 | 8,368.3852 | 3,916 |
| 1951 | 8,114.2169 | 198.8941 | 1,066.0352 | 1,583.4290 | 8,597.5819 | 3,968 |
| 1952 | 8,620.1581 | 189.2430 | 1,143.4916 | 1,614.3290 | 8,881.3153 | 4,042 |
| 1953 | 9,150.8679 | 260.9257 | 1,219.7600 | 1,713.8219 | 9,212.9438 | 4,118 |
| 1954 | 9,775.4576 | 310.7912 | 1,548.7471 | 1,853.2309 | 9,840.3966 | 4,249 |
| 1955 | 10,185.3741 | 431.5895 | 1,548.4001 | 2,099.4973 | 10,404.7570 | 4,320 |
| 1956 | 10,560.8805 | 414.3772 | 1,601.6127 | 2,303.9248 | 10,965.8940 | 4,410 |
| 1957 | 10,820.0538 | 297.8513 | 1,588.0066 | 2,371.3836 | 11,457.3113 | 4,499 |
| 1958 | 11,075.0711 | 251.6743 | 1,301.3892 | 2,385.9196 | 11,612.9694 | 4,589 |
| 1959 | 11,877.5556 | 274.6543 | 1,348.1685 | 2,421.9819 | 11,799.8409 | 4,679 |
| 1960 | 12,388.3197 | 283.0688 | 1,641.6588 | 2,462.8524 | 12,261.5156 | 4,768 |
| 1961 | 13,020.6818 | 380.1256 | 1,713.8705 | 2,596.6928 | 12,749.2346 | 4,898 |
| 1962 | 13,728.5968 | 391.9846 | 1,538.9918 | 2,729.0081 | 13,013.3030 | 5,028 |
| 1963 | 14,175.4755 | 330.2556 | 1,533.6688 | 2,786.3629 | 13,245.6415 | 5,158 |
| 1964 | 15,036.8317 | 347.4230 | 1,777.3894 | 2,855.1496 | 13,698.4668 | 5,288 |
| 1965 | 15,596.8218 | 355.9759 | 1,688.8869 | 2,925.6105 | 14,017.5070 | 5,418 |
| 1966 | 16,422.3692 | 484.3217 | 1,903.9400 | 3,117.3712 | 14,519.6963 | 5,580 |

| Year | | | | | |
|---|---|---|---|---|---|
| 1967 | 17,022.7380 | 612.2499 | 1,551.9266 | 3,417.8840 | 14,619.6533 | 5,742 |
| 1968 | 18,159.1000 | 682.9678 | 1,917.2792 | 3,759.0634 | 15,074.9672 | 5,904 |
| 1969 | 19,317.9507 | 767.4616 | 1,857.9738 | 4,150.6187 | 15,425.4442 | 6,066 |
| 1970 | 20,866.4951 | 806.3811 | 2,032.5455 | 4,541.9380 | 15,915.4453 | 6,228 |
| 1971 | 22,157.9497 | 965.5015 | 1,977.2848 | 5,053.2457 | 16,301.1856 | 6,394 |
| 1972 | 23,833.0701 | 939.4597 | 1,998.4871 | 5,487.3808 | 16,669.5541 | 6,561 |
| 1973 | 25,441.7407 | 1,029.2148 | 2,248.9155 | 5,967.8575 | 17,251.5142 | 6,727 |
| 1974 | 26,884.8266 | 981.6909 | 3,038.0139 | 6,352.7627 | 18,564.3767 | 6,893 |
| 1975 | 27,502.2109 | 994.1626 | 2,280.0540 | 6,711.6490 | 18,987.9930 | 7,060 |
| 1976 | 28,792.0105 | 1,121.3029 | 2,455.6954 | 7,161.7870 | 19,544.8892 | 7,246 |
| 1977 | 29,994.9546 | 2,051.3382 | 2,154.6881 | 8,496.9465 | 19,745.0883 | 7,433 |
| 1978 | 32,543.4991 | 1,566.7592 | 2,667.7940 | 9,214.0111 | 20,438.3735 | 7,619 |
| 1979 | 34,269.5561 | 1,395.9691 | 2,958.6151 | 9,688.5791 | 21,353.1513 | 7,806 |
| 1980 | 35,688.3380 | 1,902.9520 | 2,854.4280 | 10,622.6732 | 22,072.2641 | 7,992 |
| 1981 | 36,490.7182 | 2,163.2286 | 3,030.1966 | 11,723.6345 | 22,895.2343 | 8,233 |
| 1982 | 36,840.0653 | 2,463.1206 | 2,907.8303 | 13,014.3916 | 23,513.5411 | 8,473 |
| 1983 | 37,431.4338 | 2,652.3695 | 2,710.8030 | 14,365.3220 | 23,872.9900 | 8,714 |
| 1984 | 38,685.8233 | 2,647.7364 | 2,594.6957 | 15,576.5262 | 24,080.3867 | 8,955 |
| 1985 | 39,882.4947 | 2,313.7590 | 2,516.0213 | 16,332.6325 | 24,188.3694 | 9,195 |
| 1986 | 41,918.5275 | 2,132.5938 | 2,749.9664 | 16,831.9631 | 24,519.4988 | 9,435 |
| 1987 | 44,160.7923 | 1,789.4313 | 3,431.2046 | 16,938.1981 | 25,498.7535 | 9,675 |
| 1988 | 46,112.2827 | 1,862.4891 | 3,729.0426 | 17,106.8674 | 26,677.9208 | 9,914 |

*Source:* See appendix 5.2.

# Notes

The author is grateful to Luis Jorge Garay for advice and assistance in the construction of a historical time-series on the public sector in Colombia; to Bela Balassa, Alberto Carrasquilla, Fernando Clavijo, Vittorio Corbo, John Cuddington, Albert Fishlow, E. C. Hwa, Paolo Leme, Johannes Linn, Carmen Reinhart, and Luis Valdivieso for comments and useful discussions; and to Piyabha Kongsamut for research assistance. He also thanks participants in the World Bank Conference on the Macroeconomics of the Public Sector Deficit, June 1991. Any errors are the responsibility of the author.

1. The real interest rate is defined as [(1 + 90 day deposit rate)/(1 + CPI inflation rate) − 1] x 100.

2. Coffee prices are the main explanatory variable behind the evolution of the coffee balance, although there is some effect of real devaluation and autonomous policy changes, such as inventory investment and changes in the producer price ratio.

3. We also examined the effect on the deficit of the real exchange rate and inflation. The revaluation of the foreign currency items of government revenue net of foreign interest payments, a result of real devaluation, would have implied an improvement of about 0.4 percent of GDP in the fiscal balance. However, a regression of tax revenue on the real exchange rate finds a negative effect of devaluation, although it is of questionable significance. Apparently, the strong contraction of import volume offsets the valuation effect of real devaluation on revenue. We also tested the effect of inflation on tax revenue and found a strong Olivera-Tanzi effect, but since inflation did not change much during the adjustment program, this was not a major factor.

4. This is similar to the consistency relationship between fiscal deficits and inflation proposed by Anand and Van Wijnbergen (1989), except that we also allow for portfolio shifts between money and interest-bearing assets.

5. This restriction is emphatically rejected by the data, which would imply an income elasticity of 2.4. Because such a large elasticity would lead to implausible simulation results, we impose the income elasticity of unity. The reasons for the explosive income elasticity will be investigated in further research. For an analysis of stability of money demand, see Carrasquilla and Renteria (1990).

6. The share of private capital in output would fall with an increase in public capital if the elasticity of substitution between the two were greater than 1 in absolute value. A fall in the share of private capital would lower the share of private investment in output for a given rate of return on investment. The rate of return to private capital would fall with increases in the types of public capital that were close to being perfect substitutes for private capital. These results are discussed in chapter 1 in this volume. Note that public capital can have a negative effect on private investment as a result of crowding-out through the financial market if the loan rate is not a true measure of the cost of funds (because of financial repression and credit rationing, for example).

7. We did not find significant differences between permanent and transitory disposable income. For a careful treatment of this issue, see Cuddington and Urzua (1989) and Clavijo (1989).

8. For an alternative treatment of labor productivity trends see Clavijo (1990).

9. We experimented with a nonlinear regression of a constant elasticity of substitution (CES) function of the two types of capital (assuming that the elasticity of substitution between capital and labor was still 1), but the estimated elasticity of substitution was very close to 1, suggesting that the log-linear function is appropriate.

10. Actual inflation is used throughout as a measure for expected inflation. This implies static expectations: the current inflation is expected to continue.

11. The Laffer curve measures the relationship between tax revenue and the tax rate. It shows an inverted-U shape because the tax base shrinks as the tax rate rises. In this case, tax revenue is the printing of money (seigniorage), the tax rate is the inflation rate, and the tax base is the money stock.

12. To take account of the small negative effect of a decrease in demand deposits in the banking system, technically we must require that this effect be dominated by the quasi-money and investment effects.

13. The exogenous variables "net other assets" are adjusted to reproduce the actual equilibrium in the money and credit markets. All other exogenous variables retain their actual or estimated values over 1987–89.

14. The reason for the round number for the change in the real interest rate is that the simulation is actually run the other way around—by specifying an increase in the real interest rate and then calculating the change in inflation and the deficit consistent with an unchanged level of monetary financing. This greatly simplifies the computation, and the economic interpretation of the simulation remains the same.

15. For this simulation we use the econometric equation that relates the ratio of private investment to GDP only to the real interest rate, since this is the only specification with the "correct" sign.

16. For an interesting alternative approach to the real exchange rate in Colombia, see Clavijo (1990).

## References

Anand, Ritu, and Sweder van Wijnbergen. 1989. "Inflation and the Financing of Government Expenditure: An Introductory Analysis with an Application to Turkey." *World Bank Economic Review* 3 (1): 17–38.

Bernal, Joaquín. 1991. "La Politica Fiscal en los Años Ochenta." *Ensayos sobre Politica Economica* (Colombia) 19: 7–42.

Bird, Richard. 1984. *Intergovernmental Finance in Colombia*. Harvard Law School, International Tax Program, Cambridge, Mass.

Carrasquilla, Alberto, and Carmen Renteria. 1990. "Es Inestable la Demanda por Dinero en Colombia?" *Ensayos sobre Politica Economica* (Colombia) 17: 23–37.

Clavijo, Sergio. 1989. "Ingreso Permanente y Transitorio: Que Tanto Ahorran (o Consuman) los Colombianos?" *Coyuntura Economica* 15: 71–93.

———. 1990. "Productividad Laboral, Multifactorial y la Tasa de Cambio Real en Colombia." *Ensayos sobre Political Economica* (Colombia) 17: 73–97.

Cuddington, John T., and C. M. Urzua. 1989. "Trends and Cycles in Colombia's Real GDP and Fiscal Deficit." *Journal of Development Economics* 30: 325–43.

Easterly, William, E. C. Hwa, P. Kongsamut, and J. Zizek. 1990. "Un Modelo sobre Requisitos Macroeconomics para Adelantar Reformas de Politica." *Ensayos sobre Politica Economica* (Colombia) 18: 99–132.

Easterly, William, Carlos Alfredo Rodríguez, and Klaus Schmidt-Hebbel. 1989. "Research Proposal: The Macroeconomics of the Public Sector Deficit." World Bank, Policy Research Department, Washington, D.C.

Edwards, Sebastian. 1989. *Real Exchange Rates, Devaluation, and Adjustment.* Cambridge, Mass.: Massachusetts Institute of Technology Press.

Engle, R. F., and C. W. J. Granger. 1987. "Co-Integration and Error Correction: Representation, Estimation and Testing." *Econometrica* 55 (2): 251–76.

García García, Jorge. 1989. "Macroeconomic Policies, Crisis and Growth in the Long Run." Colombia Country Study, Revised Version, Part III. World Bank Research Project, Washington, D.C. To be published as a volume in the Comparative Macroeconomic Studies series, World Bank.

Herrera, S. 1989. "Determinantes de la Trayectoria del Tipo de Cambio Real en Colombia." *Ensayos sobre Politica Economica* (Colombia) 15 (Junio): 60–70.

IMF (International Monetary Fund). Various years. *International Financial Statistics.* Washington, D.C.

World Bank. Various years. *World Debt Tables.* Washington, D.C.

# 6

# Côte d'Ivoire: Fiscal Policy with Fixed Nominal Exchange Rates

Christophe Chamley and Hafez Ghanem

Ten years of fiscal mismanagement in Côte d'Ivoire have led to a state of virtual bankruptcy.[1] In the mid-1970s a commodity boom on the two main exports—cocoa and coffee—led the fiscal authorities to embark on a spending binge. The fiscal deficit exceeded 12 percent in 1980 and 1981, averaged 9 percent of GDP during the 1980s, and was more than 14 percent in 1990. Numerous missions of the International Monetary Fund (IMF) and the World Bank emphasized repeatedly that the level of government spending was not sustainable and suggested reforms to address the issue. The failure of fiscal policy led to a crisis of unprecedented magnitude: in April 1987 the government stopped service of its foreign debt, and public disorders forced the beginning of an emergency program in which Allassan Ouattara, former head of the Banque Centrale des Etats de l'Afrique de l'Ouest (BCEAO), had the role of a financial administrator. This chapter describes how the large deficits developed and how they eventually led to the current crisis. We focus on the impacts of these deficits on the competitiveness and growth of the economy.

The coffee and cocoa price boom of the mid-1970s allowed policymakers to throw macroeconomic policy on the wrong track. The price booms were transitory and should have been perceived as such. Instead, they stimulated an unprecedented rise of public expenditure that continued after the end of the boom in 1979. Severe imbalances became evident in 1980, when the internal and external deficits exceeded 10 percent of GDP. The numerous attempts that were made thereafter failed to put the economy back on a growth path.

With the end of the commodity boom, conditions deteriorated.

- Real GDP has been stagnant, with no difference between 1980 and 1987. Real GDP has decreased since 1983, while the population has kept growing at a rapid pace.

- Employment in the formal nonagricultural private sector decreased from 200,000 in 1982 to 146,000 (equal to 1.4 percent of the population) in 1988.
- Total investment decreased from 35 percent of GDP in 1977 to less than 15 percent since 1984. The government deficit increased from 12 to 15 percent, on an accrual basis.
- Foreign lending to the private sector has ceased for the past eight years. Since 1982 all net foreign borrowing has been carried out by the government, which has now exhausted its borrowing capability.
- The foreign debt now exceeds 100 percent of GDP. No one expects this debt to be paid back, and its value on the secondary market is between 5 and 10 percent.
- The financial system is now stalled, with many insolvent debtors.

In this chapter we investigate whether there is a relationship between the dismal performance of the economy and the large government deficits. Such linkages are difficult to establish even with a large sample of countries. Since this study is limited to one country, the analysis has to rely on a variety of data.

## External Shocks

In this section we review the impact of the celebrated commodity boom of 1976–77 and its aftermath. A quantitative evaluation of the external shocks that have affected the primary exports shows that the position of the country is now back to its initial state before the boom. The vulnerability of public revenues and of the whole economy to international fluctuations has increased.

### Commodity Prices

The economy of Côte d'Ivoire has been heavily dependent on its two main exports, cocoa and coffee. This dependency has increased over the past twenty years. These commodities show three remarkable properties:

1. Prices are highly variable (figure 6.1). The significant boom of the mid-1970s now appears as an anomaly in the time-series.
2. The comovements of the prices are remarkable, since external factors that affect the supplies of the two commodities seem to be independent. For instance, the sudden rise of coffee prices in 1976 was caused by a frost in Brazil, which had no influence on the market for cocoa, yet cocoa prices also increased.
3. There is no apparent trend in the prices of cocoa and coffee. Figure 6.1 shows no evidence of a random walk. This is con-

**Figure 6.1. External Price Shocks, Coffee and Cocoa Prices, Côte d'Ivoire, 1970–87**

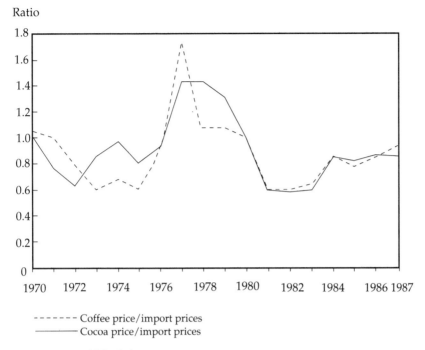

Ratio

------- Coffee price/import prices
————— Cocoa price/import prices

*Source:* World Bank data.

firmed in a more detailed analysis of the properties of the time-series of the two basic commodity prices by Melhado (1990). There is therefore no statistical support for the view that the price shocks of the mid-1970s could affect the expected prices of the commodities in the long run.[2]

It is interesting to recall some of the perceptions during the first commodity boom. A distinction was made between the two commodities: the coffee boom was understood to be the result of the frost in Brazil and was perceived as temporary, but the increase in the price of cocoa was apparently perceived as permanent. A reason for this myopia may be the dominant position of Côte d'Ivoire in the cocoa market. In 1979 authorities had experienced three years of largely above-trend prices (see figure 6.1), and they were slow to recognize that the boom had indeed been temporary.

In the mid-1980s a temporary price recovery again raised the level of expectations about the permanent level of cocoa prices. In the subsequent slump an attempt to manipulate the world price through stock retention failed. The current slump to historical lows does not so far appear to be out of line with statistical fluctuations. However, the dominant position of Côte d'Ivoire has been eroded by higher cocoa production in other countries (Malaysia, for example), and the downward price trend is likely to continue.

## Measurement of External Shocks

The continuing fiscal and economic crisis of the 1980s, which has deepened recently, has been blamed on a combination of external shocks and poorly designed and implemented fiscal policies. To disentangle these causes, we have computed some measurements of the external shocks that have affected the markets for primary commodities and the service of the foreign debt. Our purpose is to measure how variations in the prices and quantities of three primary commodities (coffee, cocoa, and wood) and the flow of the debt service affected the national income of the country. We ignore induced effects and follow an accounting approach.

For the economy of Côte d'Ivoire, the definition of national income $Y$ can be rewritten as:

$$Y_t = \sum_{i=1}^{4} P_{it}X_{it} + H_t - P_{Qt}Q_t - Z_t$$

where $P_{it}$ represents the prices and $X_{it}$ the quantities of exports of cocoa ($i = 1$), coffee ($i = 2$), wood ($i = 3$), and other commodities ($i = 4$). As a first approximation, the production of these goods, which are entirely exported, is assumed to require only domestic inputs. The value added in other sectors of the economy is $H_t$. The term $P_{Qt}Q_t$ represents the value of imports, and $Z_t$ is the interest payment on foreign borrowings. We focus on the shocks that affected the export sector and the service of the foreign debt.

Suppose that all changes in exports are attributable to external causes and that the same property applies to the service of the foreign debt. We denote by a superscript 0 a value in the base year (1975, the last year before the commodity boom). The income effect of the exogenous shocks can be decomposed into changes in $P$, $X$, and $Z$ by the accounting identity:

$$DY_t = Y_t - Y_t^0 = \sum_{i=1}^{4} (P_{it}X_{it} - P_{it}^0 X_{it}^0) - (Z_t - Z_t^0)$$

which can be rewritten as:

$$DY_t = \sum_{i=1}^{4} (P_{it} - P_{it}^0)X_{it} + \sum_{i=1}^{4} (X_{it} - X_{it}^0)P_{it}^0 - (Z_t - Z_t^0).$$

The income effects are thus the sum of the price effect, the quantity effect, and the effect of the debt service. Note that this expression is not a linear approximation and that the price effect in year $t$ is measured in combination with the actual level of output $X_t$. The output of cocoa tripled from 1975 to 1985, and a percentage point increase in prices therefore has a bigger effect in 1985 than in 1975.

To normalize the changes, we divide all terms by the level of output in the base year:

$$DY_t/Y_t^0 = \sum_{i=1}^{4} [(P_{it} - P_{it}^0)/P_{it}](P_{it}X_{it}/Y_t^0)$$

$$+ \sum_{i=1}^{4} [(X_{it} - X_{it}^0)/X_{it}^0](P_{it}^0X_{it}^0/Y_t^0) - (Z_t - Z_t^0)/Y_t^0.$$

The values of the components of the shocks for quantities, prices, and the debt service are presented in figure 6.2. The decomposition of prices is shown in figure 6.3, and quantity effects are presented in figure 6.4.

The following remarks can be made about the contribution of the primary sector to economic growth in Côte d'Ivoire.

- As we have already seen, there is no evidence of an upward trend of the prices of the three commodities. During the second boom (1984–86) prices increased less dramatically than in the first. However, the impact of the second boom was greater because output had increased since the 1970s.
- There is apparently no evidence of a contribution of the primary exporting sector to growth in the Ivorian economy over the past fifteen years. This is very striking because the principal policy-makers have always emphasized the role of primary commodities in the development of national wealth. The increase of real output in the cocoa sector was partially offset by declines in the production of coffee and wood. Between 1975 and 1987 these three primary commodities contributed to a growth of income of less than 2 percent.[3]
- The primary commodity sector became more sensitive to price fluctuations as the share of wood, which has a less volatile price, was reduced in favor of cocoa. The evolution of the activities producing primary commodities thus made the economy more vulnerable to external shocks.
- Debt service increased to 12 percent of 1975 GDP in 1989, putting a heavy burden on future income.

The structural evolution of the economy increased its vulnerability to external shocks and the fragility of the public sector's capacity to generate revenues. This fragility contributed to the severity of the current crisis, which was induced by the combination of another negative shock and the large indebtedness of the public sector.

## Public Revenues and Expenditures

Fiscal policy in Côte d'Ivoire has been characterized by wide fluctuations of revenues, expenditures, and deficits and is constrained by the rules of the CFA zone. We briefly review some of the relevant constraints and then present a brief description of the sources of public revenues and a definition of structural revenues. The level of structural revenues provides the basis for the computation of the structural surplus (or deficit). This is a useful indicator for the evaluation of the composition and the drift of expenditures that led to the present crisis.

**Figure 6.2. Contribution of External Shocks to GDP,
Côte d'Ivoire, 1970–87**

Percentage difference with respect to 1975

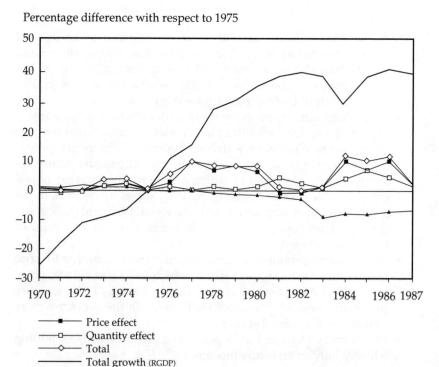

————■———— Price effect
————□———— Quantity effect
————◇———— Total
——————— Total growth (RGDP)
————▲———— Interest

*Source:* World Bank data.

**Figure 6.3. Contribution of Price Shocks to GDP,
Côte d'Ivoire, 1970–87**

Percentage difference with respect to 1975

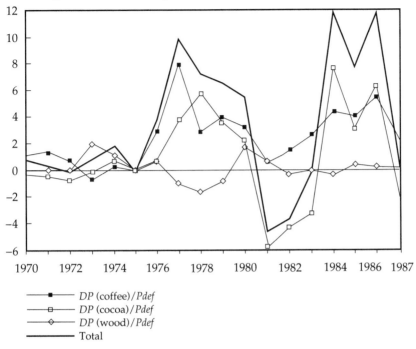

    ───■─── DP (coffee)/*Pdef*
    ───□─── DP (cocoa)/*Pdef*
    ───◇─── DP (wood)/*Pdef*
    ─────── Total

*Source:* World Bank data.

### *Constraints on Fiscal Policy*

The institutional rules of the CFA zone put severe restrictions on
domestic borrowing by the government: domestic government debt
cannot exceed 20 percent of the previous year's tax revenues. The
primary purpose of this institutional rule is to prevent the government from financing expenditures by issuing a currency that is
backed by the monetary union (and the French treasury). The use of
permanent money growth for financing is thus not feasible in Côte
d'Ivoire. The policy constraint improves the price stability of the
economy. In this environment of relative stability of prices and interest rates (compared with other Sub-Saharan African countries), the
quantity of money is also stable. Attempts to estimate a demand for
money proved fruitless.[4]

The restriction of the CFA monetary union does not eliminate the
possibility of a temporarily significant level of domestic borrowing,
and this opportunity was used immediately after the first commodity

**Figure 6.4. Contribution of Quantity Shocks to GDP, Côte d'Ivoire, 1970–87**

Percentage difference with respect to 1975

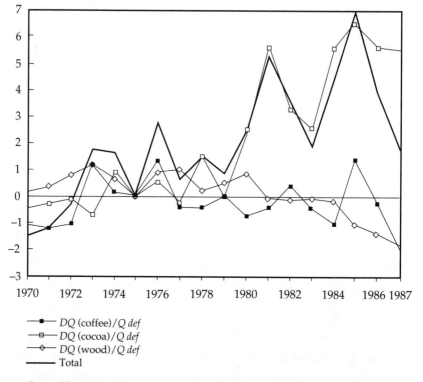

- ■— DQ (coffee)/Q def
- □— DQ (cocoa)/Q def
- ◇— DQ (wood)/Q def
- —— Total

*Source:* World Bank data.

boom in 1978–79. Once the ceiling is reached, any additional domestic borrowing is feasible only if the tax base and the formal sector are growing. There has been little or no such growth since 1980.

There is no institutional restriction on the level of foreign borrowing by the government. In the case of Côte d'Ivoire, the only limit on government borrowing is imposed by the government's creditworthiness. All public financial resources, therefore, depend on domestic taxes and foreign borrowing.

### Composition of Public Revenues

As in other Sub-Saharan African countries (see Tanzi 1987), the formal nonagricultural domestic sector represents a small share of the economy. Hence the domestic base for taxation is narrow, and most taxes are levied at the border or on the rural sector, which sells primary commodities (cocoa and coffee) to a marketing board. Domestic taxes may be implicit or explicit.

IMPLICIT TAXES. All the domestic production of cocoa and coffee is sold to a marketing board (Caisse de Stabilisation), which handles exports. The gap between the border and producer prices, net of processing costs, generates implicit taxes. Producer prices varied little from 1970 until recently (in terms of the domestic price level); the marketing board, an arm of the government, has absorbed the wide fluctuations of the border prices.

Such a policy of price smoothing seems reasonable because the public sector in Côte d'Ivoire has better access to foreign credit markets, which can provide insurance to the economy. However, the main purpose of the marketing board has been to generate revenues for the public sector. The wide fluctuations of international prices have been translated into similar fluctuations of revenues of the board. Public revenues from cocoa and coffee, plotted as a fraction of GDP in figure 6.5, match closely the variations in commodity prices illustrated in figure 6.1.

EXPLICIT TAXES. The government, pressed by a fiscal crisis since 1980, has attempted to increase the burden of explicit taxes. This burden has almost reached the maximum feasible level for the economy. Repeated changes in statutory rates on tariffs, the value added tax (VAT), and direct taxes on personal and business income have been introduced since 1980 and have failed to augment revenues noticeably. The levels of the statutory rates and the structure of the economy are such that further increases in rates lead only to higher evasion, shifts of activities toward the informal domestic sector, and smuggling.

In 1987 tariffs were increased by 30 percent across the board, and the standard VAT rate of 25 percent was extended to commerce. This reform led at first to higher revenues, both absolutely and as a share of GDP. For example, revenues from import duties rose from CFAF 233 billion in 1986 to CFAF 245 billion in 1987, despite a decline in the CIF (cost, insurance, and freight) value of imports. The ratio of tariff revenues to imports rose from 33 to 36 percent. Similarly, revenues from domestic indirect taxes rose from CFAF 195 billion (6 percent of GDP) to CFAF 224 billion (7 percent of GDP). As a result, total tax revenues increased from 18 to 20 percent of GDP. These increases proved short lived, however. The tax base shrank because of two effects. First, the general level of economic activity decreased: GDP in current prices fell 3.9 percent in 1987, 1.6 percent in 1988, and 3.3 percent in 1989. Thus, all categories of revenues fell in absolute terms.

**Figure 6.5. Sources of Public Revenue, Côte d'Ivoire, 1970–89**

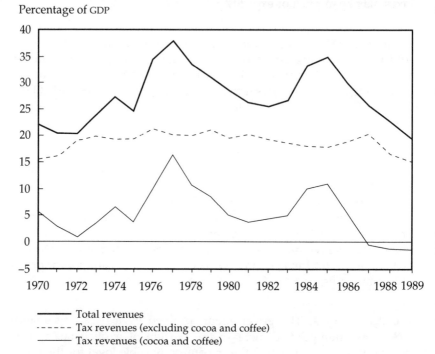

Percentage of GDP

——— Total revenues
- - - - - Tax revenues (excluding cocoa and coffee)
——— Tax revenues (cocoa and coffee)

*Source:* World Bank data.

Second, fraud increased and ad hoc tax and tariff exemptions were widened as economic agents tried to avoid the higher tax rates. In particular, total revenues from import tariffs decreased as a proportion of total revenues from 1986 to 1989, although nominal rates increased. Overall tax revenues as a proportion of GDP fell from their 1987 peak, so that by 1989 the tax-to-GDP ratio stood at the same level as in 1986 (table 6.1).

The small size of the tax base in the Ivorian economy may come as a surprise to economists who are used to industrial economies. The base has shrunk still more under the tax burden. (This issue is examined more thoroughly in appendix 6.1.) Although the tax base in the formal domestic sector is ridiculously small when compared with that of an industrial country, the level of taxes is commensurate with those observed in some industrial economies. As a consequence, the average effective rate of taxation in the formal nonagricultural sector is close to 50 percent.

The ratio between public revenues (other than for cocoa and coffee) and measured GDP has not fluctuated significantly since 1975 (see

## Table 6.1. Evolution of Tax Revenue, Côte d'Ivoire, 1986–90
(billions of CFA francs)

| Item | 1986 | 1987 | 1988 | 1989 |
|---|---|---|---|---|
| Total tax revenue | 578 | 621 | 570 | 543 |
| Import duties | 233 | 245 | 214 | 213 |
| Direct taxes | 150 | 152 | 150 | 142 |
| Indirect domestic taxes | 195 | 224 | 206 | 188 |
| GDP | 3,244 | 3,118 | 3,068 | 2,967 |
| Imports (CIF) | 709 | 672 | 620 | 658 |
| Import duties as a share of value of imports (percent) | 33 | 36 | 35 | 32 |
| Taxes as a share of GDP (percent) | 18 | 20 | 19 | 18 |

*Source:* World Bank data.

figure 6.5). This share is slowly but definitely declining because of the gradual shift toward informal activities.

STRUCTURAL LEVEL OF PUBLIC REVENUES. A key issue in the evaluation of fiscal stance is the definition of a long-run level of revenues. In an industrial country the long-run tax burden is determined by some social agreement about the fraction of the national permanent income that should be used to finance public expenditures. The value of this ratio is typically less than the maximum feasible level (at the top of the Laffer curve). It depends on the relative magnitudes of the benefits of public expenditures and the efficiency cost of taxation. In Côte d'Ivoire the situation is simplified because policy reform studies and experiments during the 1980s have shown that the level of explicit taxes on the nonagricultural sector, as a percentage of GDP, cannot be raised. We therefore assume that the ratio between explicit taxes and GDP is the maximum feasible, and this level will define the structural value of explicit tax revenues.

The main source of fluctuations in revenue is therefore to be found in implicit taxes, which are raised almost entirely by the Caisse de Stabilisation. Consider cocoa, for instance. Revenues from its implicit tax fluctuate with the international price and with the supply, which may be affected by droughts, as in the early 1980s. In a definition of the structural level of the cocoa tax, we have to take into account both effects. We define the deviation from the structural level by the formula:

$$R_t = (p_t - c_t)X_t - (p_t^0 - c_t)X_t^0$$

where $X_t$ is the actual real output of the crop, $X_t^0$ is the trend value, $p_t$ is the border price, $p_t^0$ is its trend value, and $c_t$ is the producer price, including processing costs. The value of $p_i^0$ is taken to be equal to that in the reference year 1975, adjusted for inflation:

$$p_i^0 = p_{75}P_t/P_{75}$$

where $P_t$ is the GDP deflator.

The deviation of revenues from the structural level can be expressed as fractions of GDP, and we replace the preceding formula with

$$R_t = \frac{[p_t X_t - (p_{75}P_t/P_{75})X_t^0] - c_t(X_t - X_t^0)}{\text{GDP}}.$$

The trend values of the quantities are found by a simple regression. We have made the same computation for coffee and wood.

### The Structural Surplus

We use four definitions of the actual surplus:[5]

1. The overall surplus is the difference between total revenues and total expenditures.
2. The current surplus is the difference between total revenues and current expenditures.
3. The primary surplus is the difference between the overall surplus and interest payments.
4. The primary current surplus is the difference between the current surplus and interest payments.

These definitions provide different instruments for evaluating the surplus, or the deficit, from a long-term perspective. For example, the current deficit is a reasonable indicator if it finances capital expenditures, much as when a private firm invests in a period of growth. The overall deficit is a useful indicator if public investments, unlike private investments, do not generate, at least directly, an earning capacity that is used to repay the debt.

The values of the surplus that are obtained from the preceding definitions are represented in figure 6.6. These values are subject to large fluctuations because the structure of revenue is highly sensitive to external shocks. There is no reason for the government budget to be balanced in a given year, just as there is no reason for it to be balanced on any given day or in any week. To iron out the exogenous fluctuations and evaluate the fiscal stance for the medium to long run, we use the measurements of the structural surplus.

The structural surplus is quite naturally defined as the difference between structural revenues and structural expenditures. We have already defined structural revenues. Expenditures in Côte d'Ivoire

Figure 6.6. Budgetary Surplus, Côte d'Ivoire, 1970–89

Percentage of GDP

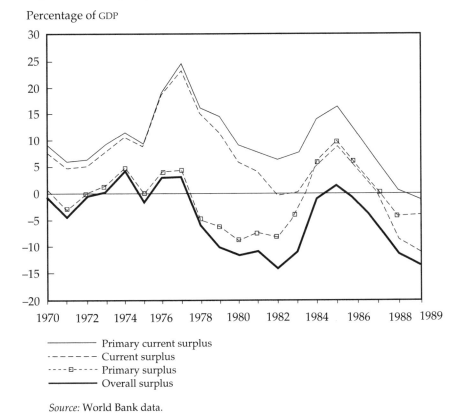

---------- Primary current surplus
- - - - - - Current surplus
- - - - ◻ - - - - Primary surplus
────── Overall surplus

*Source:* World Bank data.

are not linked to cycles as some public programs (for example, unemployment insurance) would be in industrial countries. Hence we make no correction on the expenditure side for the fluctuations of the economy. All expenditures are assumed to be discretionary and structural.

Each definition of the actual surplus corresponds to a similar definition of the structural surplus. The values of the levels of the structural surplus according to the various definitions are presented in figure 6.7, which highlights very nicely the main features of fiscal policy in Côte d'Ivoire over the past twenty years. The structural definitions of the surplus provide a clearer view than do the standard definitions, especially for the important period of structural adjustment. Three distinct regimes, separated by relatively short transition periods of about two years each, can be observed.

1. Before 1975, the structural budget was in equilibrium.
2. In the first phase (1976–82) the structural deficit jumped from

**Figure 6.7. Structural Budgetary Surplus, Côte d'Ivoire, 1972–87**

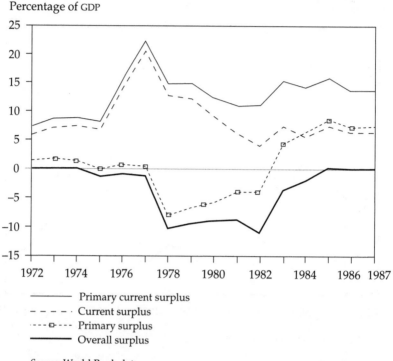

Percentage of GDP

Primary current surplus
Current surplus
Primary surplus
Overall surplus

*Source:* World Bank data.

    about 0 of GDP to 10 percent. It stayed very close to that value for six years, until 1982, and then declined until 1985.

3. Since 1985 the budget has been more or less in structural equilibrium.

    The values of the surplus are computed here on an accrual basis. However, the Ivorian government has been accumulating arrears since 1987 at an annual rate of about 3 percent of GDP. More important, the service of the foreign debt (about 8 percent of GDP) has been suspended since the spring of 1987. The budgetary deficit on a cash basis is therefore lower than its accrual value. The primary surplus is equal to the surplus when the interest service of the debt is not paid. Since the nonpayment of foreign interest is at least partially structural (nobody expects the debt to be serviced entirely), the actual value of the structural surplus is probably between those of the primary surplus and the surplus and is therefore positive. Nevertheless, the present crisis has forced the government further to reduce expenditures and to raise taxes on the agricultural sector.

To summarize, despite the most recent adjustments, the government is still running a significant deficit (see figure 6.6). This deficit is explained by adverse external conditions, which are worse than ever for the country. Our analysis, which uses the definition of the structural surplus, allows us to isolate, at least partially, the effect of these external shocks. It shows that the contractionary stance of current fiscal policy is unprecedented in Côte d'Ivoire.

### Composition of Public Expenditures

The large structural deficits during the period 1977–84 were driven by expenditures. The composition of these expenditures (figure 6.8) is one of the main issues in the discussion of the rise and fall of the deficit.

Three main features of expenditures since 1975 can be observed.

1.  Current expenditures on noninterest payments increased from an average 16 percent of GDP before the boom to about 21 percent during the 1980s (see figure 6.9). For the ten years of macroeconomic adjustment that began in 1980, one cannot detect any declining trend in the ratio of current nondebt expenditures to GDP. Attempts at structural adjustment proved to be a complete failure in this respect.
2.  The main cause of the significant deficits was the high level of

**Figure 6.8. Public Expenditure, Côte d'Ivoire, 1970–89**

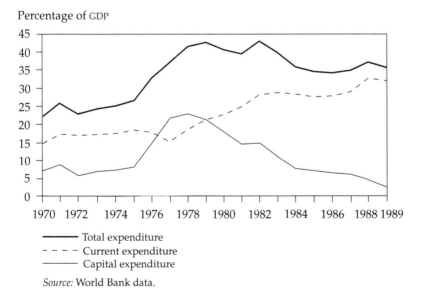

Percentage of GDP

Total expenditure
- - - - Current expenditure
Capital expenditure

*Source:* World Bank data.

**Figure 6.9. Current Public Expenditure, Côte d'Ivoire, 1970–89**

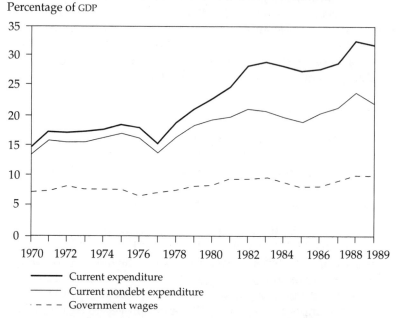

Percentage of GDP

Current expenditure
Current nondebt expenditure
Government wages

*Source:* World Bank data.

capital expenditures, which increased from 8 to 22 percent of GDP in the three years after 1975. Much of this expenditure was wasted on inefficient projects. (The composition of public investment is discussed in appendix 6.2.) The high levels were clearly unsustainable, and the ratio of capital expenditures to GDP has declined steadily since 1979. It is remarkable that all the adjustment toward a balanced budget has fallen on capital expenditures, whereas current nondebt expenditures (and the wage bill of the public sector) have continued to rise. The value of capital expenditures after the adjustment (less than 5 percent) is probably not sufficient to cover the depreciation of the public capital stock.

3. The service of the (mainly foreign) debt has increased significantly, from 3 percent of GDP in 1980 to 9 percent in 1990, because of two factors: the growth of the stock of the debt and the rise in world interest rates in the 1980s.

## Impacts of Fiscal Policy

The expansionary fiscal policy that followed the commodity boom of 1976 raises the standard issues of the "Dutch disease" and the impact of the boom on the external deficit, saving, and investment. This section begins with a brief discussion of some theoretical issues and goes on to analyze the impact on the aggregate level of prices and on the relative prices between the main sectors of production.

## The Price Level and the Real Exchange Rate

The doubling of the price level in Côte d'Ivoire between 1975 and 1980 has to be seen in the context of worldwide inflation. (For example, France's GDP deflator increased by 62 percent during the same period.) Nevertheless, a positive inflation differential appeared. This differential has been attributed by some observers to the high money growth rate and by others to the high level of government expenditure.

THE MONETARIST INTERPRETATION. The money supply in Côte d'Ivoire doubled between the end of 1975 and the end of 1977. This high growth has lent some credence to a monetarist explanation of the post-1975 inflation rate. Such arguments may have a firm arithmetic base; but what is the economic reasoning? The link between money expansion and prices is based on an ''excess supply'' of money that generates a price increase when the market for goods clears. But the policy regime is one of a fixed nominal exchange rate between Côte d'Ivoire and France, with free capital mobility—at least from Côte d'Ivoire to France (and other capital markets).[6] In this situation there can hardly be an excess supply of money; the relevant quantity of money for the country is the quantity in the whole currency zone. There may be a situation of ''excess demand'' when the country is rationed for foreign liquidities. But this situation is not relevant either for the theoretical argument or for the empirical facts of the period 1975–80. If this interpretation is correct, the quantity of money ($M_2$) in Côte d'Ivoire is driven by the supply from the banking system. Advances by the Central Bank have a strong impact on this supply. The Central Bank and the banking system act, effectively, as intermediaries for some of the savings that are deposited in foreign banks. In this case variations in the quantity of money have, in themselves, little impact on the price level.

RELATIVE PRICES IN THE BOOM. During the first commodity boom and the period that immediately followed, the level and structure of prices were significantly altered. Government expenditure expanded suddenly. Some of these expenditures were directed to imports, the prices of which are not sensitive to demand by Côte d'Ivoire. Other

**Figure 6.10. Ratio of Sectoral Prices to the GDP Deflator, Côte d'Ivoire, 1971–84**

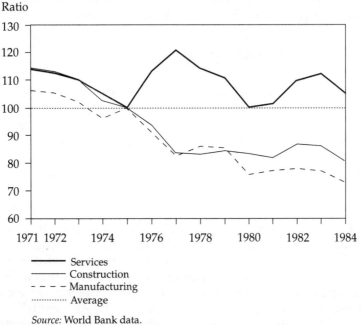

Source: World Bank data.

expenditures increased the demand for nontraded goods, the supply of which cannot be adjusted easily. Hence prices increased more for services than for manufacturing. The ratios between the prices in some sectors and the GDP deflator are presented in figure 6.10. The curves for prices of services and prices in other sectors of the economy form a pincer: during the first two years of the boom (1976–77) services led the price increases, while industrial prices lagged behind the average increase (represented on the figure by the horizontal line). In the second phase of expansionary fiscal policy (1978–81) the price increases were carried to the other sectors of the economy. The behavior of relative prices thus fits an interpretation that is based on the rise of public sector demand.

THE IMPACT OF FISCAL POLICY ON THE REAL EXCHANGE RATE. The rise of domestic prices during the first boom adversely affected the competitiveness of the economy. All the available indexes follow similar paths. The ratios between price indexes in Côte d'Ivoire and France are presented in figure 6.11. During the expansionary period, prices increased much faster in Côte d'Ivoire, but in the period of recession

Figure 6.11. Competitivity Indexes, Côte d'Ivoire, 1971–83

Ratio

Note: MPI, import price index.
a. Purchasing power parity
Source: World Bank data.

after 1980, they increased more slowly than in France.[7] Note that the inflation rate was still positive during the recession, with an average value of 7 percent. The prices in the construction and manufacturing sectors showed more variation than the average price level in the economy.

At the end of the cycle of boom and bust (about 1983–84), the relative prices in the industrial sectors were at about the same level in relation to those in France as at the beginning of the cycle. However, the general price level in Côte d'Ivoire, as measured by the consumer price index (CPI), was still 20 percent higher than in France. This difference was mainly attributable to the rise in the price of services (see figure 6.10).

The simultaneity of a large increase in public expenditure and a rise in the domestic price level in the late 1970s raises the important issue of the relation between public expenditures and the domestic price level or the real exchange rate. We have estimated for Côte d'Ivoire the same econometric relation that has been analyzed in other studies done under this project (see, in particular, Rodríguez 1990 and chapter 2 in this volume). The basic idea is that the domestic price level (with respect to some foreign prices), adjusts to clear the market for domestic goods. An exogenous increase of demand—for example, from public expenditure—increases the demand for nontradable goods and thus raises their prices with respect to those of tradable goods. Since the domestic price level is an average of the prices of nontradable and tradable goods and the price of tradable goods is determined by foreign prices, one should observe a positive relation between exogenous demand shocks (for example, public expenditure) and the ratio between domestic and foreign prices (of competing goods). This relation is embodied in the equation:

(6.1)             $\log(\text{RER}) = a_0 + aD + b\log(P_X/P_M)$

where the price indexes for exports and imports are $P_X$ and $P_M$, respectively. The real exchange rate is defined as the ratio between the domestic price level and the price of tradable goods for which the index is the price of imports. Exogenous demand shocks are represented by the variable $D$. These shocks may arise in the private or the public sector. To examine whether public sector shocks have a greater effect on the domestic price level than those of the private sector, we introduce an additional variable in the preceding equation that measures the *differential* effect of the public sector on prices:

(6.2)             $\log(\text{RER}) = a_0 + aD + b\log(P_X/P_M) + gG$

where $G$ is a variable that measures public expenditure.

Exogenous shocks to total demand, $D$, are measured by the ratio of domestic expenditures (absorption) to income. This ratio is equal to 1 minus the ratio between the trade surplus and GDP, by the identity of national income accounts. One can therefore introduce the ratio between the trade balance and GDP in equation 6.2. The variable $G$ is taken to be the ratio of total expenditures of the public sector to GDP. In this case the equation to be estimated is:

(6.3)    $\log(\text{RER}) = a_0 + a[(X - M)/Y] + b\log(P_X/P_M) + g(G/Y)$

(using variables TBY, LPXM, and GY) where $X$ is exports, $M$ is imports, $G$ is public expenditure, and $Y$ is GDP.

This equation has been estimated in levels and first differences, and results are presented in tables 6.2 and 6.3. The equation in levels has a coefficient of autocorrelation that is almost equal to 1. We will focus on the equations in first differences.

A difficult issue is how to measure the exogenous shock to overall aggregate demand. We have used the balance of trade—that is, the difference between exports and imports of goods. The level of exports

**Table 6.2. Regression Results for the Real Exchange Rate, Levels, Côte d'Ivoire, 1972–89**

| Variable | Coefficient | Standard error | t-statistic | Two-tail significance |
|---|---|---|---|---|
| DRER(−1) | 0.240 | 0.118 | 2.04 | 0.060 |
| DTBY | −1.012 | 0.233 | −4.33 | 0.001 |
| DLPXM | 0.253 | 0.061 | 4.14 | 0.001 |
| DGY | 0.295 | 0.221 | 1.33 | 0.204 |
| $R^2$ | 0.797 | | | |
| Adjusted $R^2$ | 0.753 | | | |
| S.E. of regression | 0.035 | | | |
| Durbin-Watson statistic | 1.498 | | | |
| Mean of dependent variable | −0.002 | | | |
| S.D. of dependent variable | 0.0705 | | | |
| Sum of squared residuals | 0.0171 | | | |
| F-statistic | 8.367 | | | |

*Note:* Number of observations, 18; method, ordinary least squares; dependent variable, DRER. S.E., standard error; S.D., standard deviation. DRER, real exchange rate; TBY, surplus of the trade balance; LPXM, terms of trade; GY, ratio of government expenditure to GDP.

**Table 6.3. Regression Results for the Real Exchange Rate, First Differences, Côte d'Ivoire, 1972–87**

| Variable | Coefficient | Standard error | t-statistic | Two-tail significance |
|---|---|---|---|---|
| DRER(−1) | 0.21 | 0.120 | 1.67 | 0.120 |
| DTBYR | −1.06 | 0.245 | −4.32 | 0.001 |
| DLPXM | 0.27 | 0.065 | 4.13 | 0.001 |
| DGY | 0.24 | 0.245 | 0.99 | 0.341 |
| $R^2$ | 0.80 | | | |
| Adjusted $R^2$ | 0.75 | | | |
| S.E. of regression | 0.036 | | | |
| Durbin-Watson statistic | 1.58 | | | |
| Mean of dependent variable | 0.0050 | | | |
| S.D. of dependent variable | 0.0736 | | | |
| Sum of squared residuals | 0.0156 | | | |
| F-statistic | 16.823 | | | |

*Note:* Method, ordinary least squares; dependent variable, DRER; S.E., standard error; S.D., standard deviation.

is obviously exogenous, while the level of imports is not; the level of imports probably depends on income and on the ratio of domestic prices to import prices, which is the dependent variable itself. For this reason it is recommended that this variable be omitted from the regressors and that instrumental variables be used. The results are

**Table 6.4. Regression Results for the Real Exchange Rate Using Instrumental Variables, Levels, Côte d'Ivoire, 1972–89**

| Variable | Coefficient | Standard error | t-statistic | Two-tail significance |
|---|---|---|---|---|
| DTBY | −1.29 | 0.317 | −4.09 | 0.001 |
| DLPXM | 0.26 | 0.066 | 3.92 | 0.002 |
| DGY | 0.08 | 0.234 | 0.34 | 0.733 |
| *AR*(1) | 0.33 | 0.214 | 1.54 | 0.144 |
| $R^2$ | 0.76 | | | |
| Adjusted $R^2$ | 0.71 | | | |
| S.E. of regression | 0.03 | | | |
| Durbin-Watson statistic | 1.68 | | | |
| Mean of dependent variable | −0.0027 | | | |
| S.D. of dependent variable | 0.0705 | | | |
| Sum of squared residuals | 0.0200 | | | |
| *F*-statistic | 15.07 | | | |

*Note:* Method, two-stage least squares; dependent variable, DRER; instruments, C, DLPXM, DGY, DMPI, DXPI, DM2Y. S.E., standard error; S.D., standard deviation.

**Table 6.5. Regression Results for the Real Exchange Rate Using Instrumental Variables, First Differences, 1972–87**

| Variable | Coefficient | Standard error | t-statistic | Two-tail significance |
|---|---|---|---|---|
| LPXM | 0.83 | 0.23 | 3.64 | 0.004 |
| CY | 5.13 | 1.92 | 2.67 | 0.022 |
| GY | 0.63 | 0.19 | 3.31 | 0.007 |
| C | −2.30 | 1.12 | −2.05 | 0.064 |
| *AR*(1) | −0.64 | 0.32 | −2.00 | 0.071 |
| $R^2$ | 0.536 | | | |
| Adjusted $R^2$ | 0.367 | | | |
| S.E. of regression | 0.072 | | | |
| Durbin-Watson statistic | 1.756 | | | |
| Mean of dependent variable | 0.835 | | | |
| S.D. of dependent variable | 0.091 | | | |
| Sum of squared residuals | 0.058 | | | |
| *F*-statistic | 3.17 | | | |

*Note:* Method, two-stage least squares; dependent variable, RER; instruments, C, LPXM, LPXM(−1), GY, GY(−1), DMPI, DXPI, DM2Y. S.E., standard error; S.D., standard deviation.

reported in tables 6.4 and 6.5. The estimated values of the coefficients are not very different from those of ordinary least squares, and the variations of the estimated values are in the direction that can be expected from the simultaneity bias.

Tables 6.2–6.5 show some relation between demand shocks and the real exchange rate. However, prudence is called for in interpreting the results because the surplus of the trade balance (TBY) is used as a proxy for demand shocks. Of course, it is more plausible to assume that government expenditures constitute exogenous demand shocks. Unfortunately, this variable does not appear to be significant in the preceding estimations. Although we have attempted to correct for the endogeneity through the use of instruments, it may be worthwhile to test other specifications.

Another proxy for exogenous changes in demand is obtained by two separate components of total demand—private consumption and public expenditures, as measured by their ratios to GDP (*CY* and *GY*, respectively). When these proxies are included in the equation, the estimation yields the results presented in table 6.5. The equation performs less well for curve fitting, but the coefficient of government expenditures is now significant. Overall, we conclude from these results that there is empirical evidence for a positive impact of government expenditures on the real exchange rate in the economy of Côte d'Ivoire.

## The Balance of Payments

The deficit of the balance of payments is the sum of the external deficits of the private and the public sectors. We have seen that the capacity for domestic borrowing by the public sector—albeit not insignificant—is severely limited by the regime of the CFA zone. When this capacity is exhausted, the public sector has to turn abroad for its financing.

In Côte d'Ivoire the capability of the private sector to run an external deficit depends on the willingness of foreign lenders to lend. Before and during the first commodity boom, foreign loans to the private sector were indeed significant. However, after the completion of the first cycle (1975–82), the capability of the private sector to borrow abroad seem to have been exhausted.[8]

These conditions explain the behavior of the balance of payments and its relation to the internal deficit since 1970. The raw data are presented in figure 6.12. Because the yearly reporting of the data introduces some noise, we have also given, in figure 6.13, a representation of the data after smoothing through a three-year moving average (with weights of 0.25, 0.50, and 0.25). There were four distinct regimes.

1. Before the first commodity boom the internal deficit was small (less than 3 percent of GDP). The deficit of the balance of payments was fairly large (about 8 percent of GDP) and reflected the

**Figure 6.12. Balance of Payments, Côte d'Ivoire, 1970–88**

Percentage of GDP

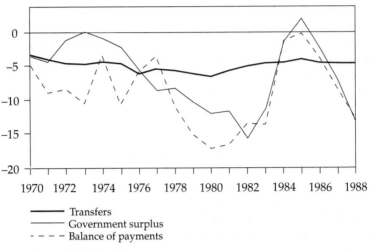

—————— Transfers
—————— Government surplus
- - - - Balance of payments

*Source:* World Bank data.

inflow of funds for investment in Côte d'Ivoire. There was apparently little relation between the internal and external deficits. The external deficit was mainly attributable to activities in the private sector.

2. During the expansionary phase of the first boom (1975–77) the external deficit stayed at the same level of about 7 percent of GDP, since the increased exports were matched by spending on foreign goods.

3. During the phase that immediately followed the boom (1978–81) the structural deficit of the government did not adjust (see figure 6.7). The external deficit increased dramatically because the private and public sectors continued to borrow as economic activity continued to increase. During this period the public sector used up all its capability for domestic borrowing.

4. Since 1982 the sources of foreign borrowing have dried up completely for the private sector. Because of the regulation of the CFA zone, the public sector could not increase its domestic borrowing. Therefore, the external and internal deficits are identical in this phase.

An important feature of the balance of payments in Côte d'Ivoire is the high level of unrequited transfers by the private sector (about 5

**Figure 6.13. Balance of Payments (Smooth Data), Côte d'Ivoire, 1970–88**

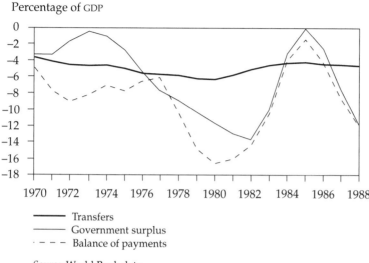

Percentage of GDP

———— Transfers
———— Government surplus
· – – – Balance of payments

*Source:* World Bank data.

percent of GDP). These transactions are probably made by expatriates and individuals with high incomes. The magnitude of the transfers indicates that a large fraction of private savings is not used for investment in the Ivorian economy.

The estimation of the impact on the trade surplus (TBY) of the terms of trade [LPXM = log(*PX/PM*)], the real exchange rate [RER = log(*P/PM*)], and the ratio of the primary government surplus to GDP (PSURY) are reported in table 6.6, for 1971–81, and table 6.7, for 1979–89. The equation has been estimated for two periods because of the structural break that probably occurred in 1982. An overlap was necessary because of the sparseness of the data. The results support the description given above:

- In the first period the primary government surplus has no significant impact on the trade balance. The terms of trade and the real exchange rate have an impact in the expected direction.
- In the second period relative prices do not have a significant effect, and the trade surplus is driven mainly by the government surplus. Some of the statistics of the estimation (residuals and the Durbin-Watson statistic) are not very satisfactory. This may be a result of the structural break that occurred during the period of estimation. There are not enough data for a post-1983 estimation.

**Table 6.6. Effect on the Trade Surplus, Côte d'Ivoire, 1971–81**

| Variable | Coefficient | Standard error | t-statistic | Two-tail significance |
|---|---|---|---|---|
| C | 0.408 | 0.087 | 4.69 | 0.002 |
| LPXM | 0.137 | 0.050 | 2.70 | 0.031 |
| RER | −0.379 | 0.109 | −3.46 | 0.010 |
| PSURY | −0.205 | 0.259 | −0.79 | 0.455 |
| $R^2$ | 0.89 | | | |
| Adjusted $R^2$ | 0.85 | | | |
| S.E. of regression | 0.012 | | | |
| Durbin-Watson statistic | 2.51 | | | |
| Mean of dependent variable | 0.075 | | | |
| S.D. of dependent variable | 0.032 | | | |
| Sum of squared residuals | 0.001 | | | |
| F-statistic | 20.30 | | | |

*Note:* Method, two-stage least squares; dependent variable, TBY; instruments, C, LPXM(−1), RER(−1), PSURY(−1), M2Y(−1). S.E., standard error; S.D., standard deviation.

**Table 6.7. Effect on the Trade Surplus, Côte d'Ivoire, 1979–89**

| Variable | Coefficient | Standard error | t-statistic | Two-tail significance |
|---|---|---|---|---|
| C | −0.055 | 0.130 | −0.421 | 0.68 |
| LPXM | −0.090 | 0.048 | −1.877 | 0.10 |
| RER | 0.17 | 0.150 | 1.165 | 0.28 |
| PSURY | 0.93 | 0.201 | 4.640 | 0.00 |
| $R^2$ | 0.89 | | | |
| Adjusted $R^2$ | 0.84 | | | |
| S.E. of regression | 0.021 | | | |
| Durbin-Watson statistic | 3.23 | | | |
| Log likelihood | 29.00 | | | |
| Mean of dependent variable | 0.10 | | | |
| S.D. of dependent variable | 0.054 | | | |
| Sum of squared residuals | 0.003 | | | |
| F-statistic | 18.90 | | | |

*Note:* Method, two-stage least squares; dependent variable, TBY; instruments, C, LPXM(−1), RER(−1), PSURY(−1), M2Y(−1). S.E., standard error; S.D., standard deviation.

The same equation was also estimated with the trade surplus replaced by the balance of payments (BOPY) and the primary surplus of the government replaced by the total surplus divided by GDP (SURY). The results, presented in tables 6.8 and 6.9, are similar to those in table 6.7.

*Saving and Investment*

The saving rate has a procyclical pattern, as one would expect from any consumption theory that is not based on hand-to-mouth behavior. We have computed and have shown in figure 6.14 two measures of the saving rate, as a share of GDP and of gross national income (GDP minus foreign interest payments). The more significant of the two

**Table 6.8. Effect on the Balance of Payments, Côte d'Ivoire, 1971–83**

| Variable | Coefficient | Standard error | t-statistic | Two-tail significance |
|---|---|---|---|---|
| C | 0.26 | 0.102 | 2.59 | 0.029 |
| LPXM | 0.20 | 0.060 | 3.30 | 0.009 |
| RER | −0.38 | 0.122 | −3.17 | 0.011 |
| SURY | 0.23 | 0.178 | 1.30 | 0.226 |
| $R^2$ | 0.85 | | | |
| Adjusted $R^2$ | 0.80 | | | |
| S.E. of regression | 0.020 | | | |
| Durbin-Watson statistic | 2.07 | | | |
| Mean of dependent variable | −0.107 | | | |
| S.D. of dependent variable | 0.0458 | | | |
| Sum of squared residuals | 0.003 | | | |
| F-statistic | 17.58 | | | |

*Note:* Method, two-stage least squares; dependent variable, BOPY; instruments, C, LPXM(−1), RER(−1), SURY(−1), M2Y(−1), GY(−1). S.E., standard error; S.D., standard deviation. BOPY, balance of payments; SURY, ratio of total surplus to GDP.

**Table 6.9. Effect on the Balance of Payments, Côte d'Ivoire, 1979–88**

| Variable | Coefficient | Standard error | t-statistic | Two-tail significance |
|---|---|---|---|---|
| C | 0.032 | 0.1180 | 0.27 | 0.795 |
| LPXM | −0.074 | 0.0617 | −1.21 | 0.271 |
| RER | −0.097 | 0.1387 | −0.70 | 0.507 |
| SURY | 0.898 | 0.1963 | 4.57 | 0.004 |
| $R^2$ | 0.922 | | | |
| Adjusted $R^2$ | 0.884 | | | |
| S.E. of regression | 0.021 | | | |
| Mean of dependent variable | −0.103 | | | |
| S.D. of dependent variable | 0.064 | | | |
| Sum of squared residuals | 0.0028 | | | |

*Note:* Method, two-stage least squares; dependent variable, BOPY; instruments, C, LPXM(−1), RER(−1), SURY(−1), GY(−1), M2Y(−1). S.E., standard error; S.D., standard deviation.

**Figure 6.14. Saving and Investment, Côte d'Ivoire, 1971–87**

Percent

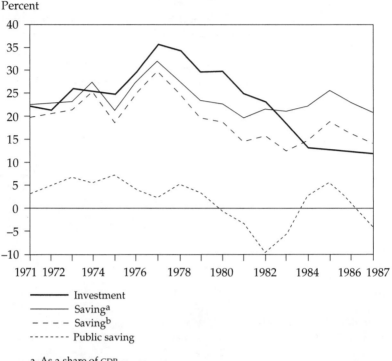

a. As a share of GDP.
b. As a share of gross national income.
*Source:* World Bank data.

indexes is the saving rate out of national income. After reaching a peak with the first commodity boom, it decreased to a historical low in the trough of the cycle (1983), recovered a little in the second boom to a level below that of the early 1970s, and reached new lows in the current crisis.

Although the data are somewhat sketchy, we have attempted to estimate a consumption function (table 6.10). The dependent variable is the ratio of consumption to GDP, which varies inversely with the saving rate. Income shocks are represented by the rate of growth of real income (GDP minus DRNY, factor payments). The variable SURNY represents the government surplus as a fraction of national income. The only variable that is significant is the terms of trade. Its positive sign can be attributed to the dominant role of the terms of trade in the business cycle of Côte d'Ivoire: during a commodity boom (with higher terms of trade), the saving rate increases.

## Table 6.10. Consumption, Côte d'Ivoire, 1972–87

| Variable | Coefficient | Standard error | t-statistic | Two-tail significance |
|---|---|---|---|---|
| C | 0.603 | 0.1321 | 4.566 | 0.001 |
| LPXM | −0.128 | 0.0444 | −2.898 | 0.014 |
| RER | 0.040 | 0.1791 | 0.226 | 0.825 |
| DRNY | −0.471 | 0.3684 | −1.280 | 0.227 |
| SURNY | 0.1962 | 0.2820 | 0.694 | 0.502 |
| $R^2$ | 0.8101 | | | |
| Adjusted $R^2$ | 0.741 | | | |
| S.E. of regression | 0.021 | | | |
| Durbin-Watson statistic | 2.242 | | | |
| Mean of dependent variable | 0.630 | | | |
| S.D. of dependent variable | 0.041 | | | |
| Sum of squared residuals | 0.004 | | | |
| F-statistic | 11.73 | | | |

*Note:* Method, two-stage least squares; dependent variable, CNY; instruments, C, LPXM, LPXM(−1), RER(−1), SURNY(−1). S.E., standard error; S.D., standard deviation. DRNY, factor payments; SURNY, ratio of government surplus to national income.

Investment was significantly higher than domestic saving only during a short period between 1979 and 1983. The gap was financed mainly by the foreign borrowing of the public sector. Since 1984 investment has been below domestic saving. This situation is explained by the external flow of transfers and the current impossibility of financing public investment through foreign borrowing.

The decomposition of investment between the private and the public sectors may be somewhat arbitrary because parastatals are substitutable with private firms. We have nevertheless attempted to provide such a decomposition in table 6.11. (There are no reliable data before 1980.)

The determinants of the investment equation are measured in table 6.12. Income shocks have a large impact on total investment. Government deficits have a negative, relatively small and marginally significant, impact on investment. However, the dependent variable is the level of total investment. If a large fraction of the deficit is used to finance public investment, the results imply that the crowding-out effect on private investment is larger. No data are available for the direct estimation of the impact of public investment and deficits on the level of private investment.

A complete analysis of the behavior of private investment would require consideration of many other factors, the most important of which is probably the profitability of capital. Such an analysis is beyond the scope of the present study. We suspect that the high level

**Table 6.11. Public and Private Investment and Consumption, Côte d'Ivoire, 1980–87**

(percentage of GDP)

| Item | 1980 | 1981 | 1982 | 1983 | 1984 | 1985 | 1986 | 1987 |
|---|---|---|---|---|---|---|---|---|
| *Investment* | | | | | | | | |
| Total | 0.260 | 0.241 | 0.213 | 0.176 | 0.123 | 0.117 | 0.118 | 0.124 |
| Public | 0.115 | 0.087 | 0.066 | 0.061 | 0.043 | 0.031 | 0.038 | 0.036 |
| *Consumption* | | | | | | | | |
| Total | 0.596 | 0.631 | 0.615 | 0.625 | 0.625 | 0.602 | 0.618 | 0.628 |
| Public | 0.177 | 0.171 | 0.169 | 0.163 | 0.153 | 0.139 | 0.151 | 0.163 |

*Source:* World Bank data.

**Table 6.12. Investment, Côte d'Ivoire, 1972–87**

| Variable | Coefficient | Standard error | t-statistic | Two-tail significance |
|---|---|---|---|---|
| C | 0.004 | 0.011 | 0.42 | 0.677 |
| IY(−1) | 0.860 | 0.059 | 14.40 | 0.000 |
| DRNY | 0.372 | 0.031 | 11.60 | 0.000 |
| SURNY | −0.102 | 0.049 | −2.05 | 0.062 |
| $R^2$ | 0.91 | | | |
| Adjusted $R^2$ | 0.89 | | | |
| S.E. of regression | 0.01 | | | |
| Durbin-Watson statistics | 2.07 | | | |
| Mean of dependent variable | 0.20 | | | |
| S.D. of dependent variable | 0.05 | | | |
| Sum of squared residuals | 0.01 | | | |
| F-statistic | 44.7 | | | |

*Note:* Method, two-stage least squares; dependent variable, *IY*, ratio of investment to GDP; instruments, *C*, *IY*(−1), DRNY(−1), SURY(−1); S.E., standard error; S.D., standard deviation.

of domestic prices (in relation to competing countries) had a strong negative impact on the marginal product of capital in the private sector. Since 1983 there has been little international lending to the private sector. We cannot determine here whether this is attributable to rationing by foreign creditors or to the low profitability of new capital in Côte d'Ivoire. The significance of the growth variable (DRNY) shows that in the Ivorian economy growth is a major factor in determining investment.

## Conclusion

The economy of Côte d'Ivoire presents a case study that highlights the effects of fiscal policy in a developing country with a fixed exchange rate. Monetary policy is constrained by the regime of the CFA zone. For the past fifteen years the growth of the Ivorian economy has been dramatically affected both by exogenous factors and by the responses of fiscal policy to these factors.

The potential impact of a commodity boom on the relative price of tradables and nontradables is well known. This relation has been observed for Côte d'Ivoire, and it has been illustrated in particular by the evolution of intersectoral prices (figure 6.10) and by the econometric results of table 6.5, in which the impacts of private and public expenditures are separated.

The adverse evolution of the real exchange rate has led to high external deficits. For the period covered in this study, the econometric results indicate that before 1982 the real exchange rate had a strong impact on the external balance. After that date, this relation was replaced with a straight identity between the government deficit and the external deficit.

The importance of the impact of government expenditure on the economy has been recognized since the beginning of the adjustment period more than ten years ago. The implementation of policy reforms has been extraordinarily slow, however. Before the crisis of 1990, which put the public sector in a state of quasi receivership, there had been no indication that the government had adjusted the level of its permanent expenditures to match the country's long-run capacity to generate revenues. Expenditures were indeed reduced significantly in 1983–84, but these reductions affected mainly capital expenditures and were not accompanied by cuts in current expenditures. For instance, from 1980 until 1989 the wage bill increased regularly.

The current situation illustrates the failure of the adjustment program of the 1980s. Since 1980 foreign indebtedness has increased from 30 percent to more than 100 percent of GDP. This foreign borrowing has not served new investment; the ratio of real investment to GDP has shrunk to a level that is barely sufficient to compensate for depreciation.

The expansionary fiscal stance has contributed to the downward rigidity of the domestic price level. Under the fixed-exchange-rate regime the competitive position of the country has worsened significantly with respect to its direct competitors in world trade, including Malaysia and neighboring countries in West Africa (for example, Ghana and Nigeria) that are not tied by fixed nominal exchange rates.

The government faced the necessity of introducing drastic changes in fiscal policy only when it had to declare insolvency. It is unfortunate that the present financial constraints severely limit the resources available for future growth.

## Appendix 6.1. The Size and Burden of the Public Sector

The share of the tax base in the Ivorian economy is much smaller than in industrial countries, and the administrative resources that are available for the collection of revenues are limited. Of the various indexes for measuring the size of the tax base, we consider two: the size of the government labor force and the average effective tax rate on the formal nonagricultural sector.

### The Labor Force in the Public Sector

According to the available data (from the 1989 Public Expenditures Survey) the number of permanent civil servants, excluding those involved in activities that provide marketed goods and services, was about 110,000 in 1989. This number was probably not very different than in 1985, when the total number of taxpayers was about 209,000. (The number of individuals who filed tax forms was about 10,000.) The ratio between the number of civil servants, defined as above, and the number of taxpayers who support them is therefore about 50 percent. To put the situation in perspective, we find it difficult to imagine an industrial country in which one-third of the formal labor force is employed in government activities that do not generate a market income.

### The Average Effective Tax Rate

The formal sector pays taxes in three stages: at the border, on firms' profits, and on personal income. A significant fraction of imports is by tax-exempt government agencies. The effective average rate of the border taxes is very high; it has been estimated at about 65 percent (Horton 1990a). Some of the goods subject to border taxation are imported for the informal sector. Correcting for these leakages, as evaluated by the World Bank, and using data on the taxes and the value added of the informal sector, the average effective tax rate on activities in the formal sector is about 48 percent. The high rate of effective taxation may explain the slow but persistent reduction of the real level of activity in the formal nongovernment sector since 1980. One branch of the formal sector has grown steadily, however—the bottling of beverages. Some would regard production by this sector as an index of domestic production (and demand) that is more accurate than GDP.

Employment in the formal private sector—already small—is not growing. In fact, it declined significantly after 1980, to 197,100 in 1982, 168,800 in 1984, 162,200 in 1986, and 146,000 in 1988. Employment decreased by 50,000 in six years, at an average rate of about 5 percent per year between 1982 and 1988. The number of African employees in managerial positions increased, however, as the number of non-Africans decreased. The entire reduction of the labor force in the formal sector fell on nonmanagement workers.

## Appendix 6.2. Public Investment

The Ivorian government reacted to the coffee and cocoa boom of the mid-1970s by increasing its capital expenditures from 8 percent of GDP in 1975 to 22 percent in 1977. Private investment remained relatively stable during that period, at about 13 to 15 percent of GDP. The increase in the share of the public sector in total investment was correlated with an apparent decline in the return on capital. In 1981 the agricultural sector received 31 percent of public investment (table 6.13), much of it in extension services, including recurrent costs (mainly wages). The rate of return on agricultural investments is difficult to measure: the data do not indicate an increase in agricultural output and value added as a result of these expenditures. Since most of the expenditures provided extension services for coffee, cocoa, and cotton, they did not encourage diversification out of traditional agricultural exports, as had been recommended by numerous World Bank missions.

The second largest share of total public investment—23 percent in 1981—was devoted to transport and roads. As a result of the public investment boom, Côte d'Ivoire today has the most developed highway and road system in Africa. However, some of the superhighways are sparsely used and have a low rate of return.

In 1981, 16 percent of the investment budget was allocated to education. This included the construction of the higher institutes (*grandes écoles*) of Yamoussoukro, in which the cost per student is higher than that in most Ivy League colleges in the United States.

**Table 6.13. Sectoral Composition of Public Investment, Côte d'Ivoire, Selected Years, 1981–90**

(percent)

| Item | 1981 | 1984 | 1987 | 1990 |
|---|---|---|---|---|
| Agriculture | 31 | 26 | 28 | 47 |
| Transport and roads | 23 | 27 | 29 | 12 |
| Housing and urban development | 13 | 18 | 17 | 20 |
| Education | 16 | 12 | 6 | 3 |
| Other | 17 | 17 | 20 | 18 |
| Public investment as a share of GDP | 17 | 8 | 5 | 4 |

*Source:* World Bank data.

When the public sector entered into productive activities—for example sugar and palm oil factories—the results were discouraging. The public companies that produce those goods do so today at a cost that is nearly double the world price.

The public investment program was drastically reduced after 1982, falling from 17 percent of GDP in 1981 to 4 percent of GDP in 1990 (see table 6.13). Its composition also changed. The share of agriculture in total investment increased from 31 to 47 percent, reflecting the difficulty of reducing employment in the extension services. Investment in transport and road construction fell from 23 to 12 percent of the reduced total, and road maintenance is now becoming an important problem. Investment in education fell dramatically, from 16 percent of total investment in 1981 to 3 percent in 1990, at the same time as the population was growing at 3.5 percent a year. Today the size of the public investment program is, in relative terms, about half of its 1975 (pre-boom) level. Moreover, about 30 percent of what is now defined as capital expenditure consists of salary payments to extension workers, an item that was not significant in 1975.

## Appendix 6.3. Costs of and Returns to Education

Government expenditures on education in Côte d'Ivoire have been estimated at 40 to 45 percent of total expenditures (Mingat and Psacharopoulos 1985). Numerous studies have found that the wages of educational staff are very high in relation to those in other countries at a similar stage of development.

High salaries in the education sector can be observed in all Francophone countries. The unit costs of public education at the primary level are at least 50 percent higher than in Anglophone Sub-Saharan African countries, where the cost is already much larger than in other developing countries. The ratio between unit costs in Côte d'Ivoire and Anglophone countries is even greater for secondary (3.4) and higher (1.5) education. Furthermore, the education system in Côte d'Ivoire places more emphasis on secondary and higher education than in other African countries.

The share of education expenditures in GDP is thus twice as high in Côte d'Ivoire as in other Sub-Saharan African countries (10 percent compared with 5 percent) or in industrial countries (about 5 percent). Since GDP per capita is higher in Côte d'Ivoire than the average in Africa, the ratio of 10 percent is quite remarkable. The high cost of staff is a consequence of institutional rigidities. When a severe budget problem occurs, the government tries to reduce expenditures on other items, such as books and supplies. Some reports indicate that the quality of education may have suffered as a result. One might expect that Ivorians are at least getting more for their money, but

**Table 6.14. Enrollment Ratios, Sub-Saharan Africa, 1983**
(percent)

| | Level of education | | |
|---|---|---|---|
| *Country or group* | *Primary* | *Secondary* | *Higher* |
| Côte d'Ivoire | 60 | 15 | 1.9 |
| Francophone countries | 46 | 14 | 2.4 |
| Anglophone countries | 77 | 17 | 1.2 |

*Source:* Mingat and Psacharopoulos 1985.

measures of output are not better than in other countries, and enrollment ratios are lower at the primary and the secondary levels than in Anglophone countries (table 6.14).

The high level of wages in the formal sector inflates the measurements of the private rate of return to education, which have been found to be high in Côte d'Ivoire (van der Gaag and Vijverberg 1987). In an environment in which institutional constraints cause the supply of jobs to be rationed because of the high wage rate, education may be an important signaling device for allocating employment. Indeed, van der Gaag and Vijverberg find that holding a diploma is a important determinant of an individual's rate. This situation may lead to an overstatement of the rate of return if the wage rate is used as an indicator of the productivity of learning.

## Notes

1. The main work on this chapter was done in 1990. Hence we do not take account of the most recent changes in economic policies, particularly the devaluation of the CFA franc.

2. Other studies on commodity prices have been undertaken by Ardeni and Wright (1990) and Cuddington and Urzua (1987).

3. The contribution of primary commodities is overstated here because the accounting procedure implies no import cost.

4. For a discussion of financial taxation in Sub-Saharan African countries, see Chamley (1990). Ivorian residents benefit from a fixed exchange rate with France and from capital mobility with European markets that is practically free for liquid assets. The fixed parity of the exchange rate (which has been left untouched since the early 1950s) is credible because of the difficulties of changing the system for all countries of the CFA zone at the same time. In this situation there are standard theoretical reasons for the nonexistence of a demand function for CFA francs in Côte d'Ivoire. This point is discussed again in the section ''The Price Level and the Real Exchange Rate.''

5. There is no inflation tax adjustment because the CFA rules effectively prevent the use of seigniorage as a significant source of revenue. The average level of seigniorage between 1979 and 1988 was less than 0.5 percent of GDP.

6. Outflows of capital are, of course, discouraged. Anecdotal evidence about the remission of CFA notes from foreign deposits (for example, in

Switzerland) indicates that controls cannot be very effective for holders of large deposits. Small depositors use money only for transactions and are probably not very sensitive to fluctuations in the opportunity cost of money in the range that has been observed under the CFA regime. As mentioned earlier, attempts to find an econometric relation between the level of money and the observed real interest rate have failed for Côte d'Ivoire.

7. The fiscal policy was more expansionary in France during the early 1980s.

8. We did not have data on the foreign indebtedness of the private sector. The constraint on foreign borrowing is mentioned in various recent reports; the data presented here can be interpreted as supporting the existence of such a constraint.

# References

Ardeni, Pier Giorgio, and Brian Wright. 1990. "The Long-Term Behavior of Commodity Prices." Policy Research Working Paper 358. World Bank, International Economics Department, Washington, D.C.

Bevan, D. L., P. Collier, and J. W. Gunning. 1987. "Consequences of a Commodity Boom in a Controlled Economy: Accumulation and Redistribution in Kenya 1975–83." *World Bank Economic Review* 1 (3): 489–513.

Clawson, Patrick. 1989. "Government Wage Policies." Report for the World Bank's Public Expenditure Review. World Bank, Occidental and Central Africa Department, Washington, D.C.

Chamley, Christophe. 1990. "Financial Taxation in Sub-Saharan Countries." World Bank, Country Economics Department, Washington, D.C.

Colclough, Christopher. 1980. *Primary Schooling and Economic Development: A Review of the Evidence.* World Bank Working Paper Series 399. Washington, D.C.

Cuddington, John T., and C. M. Urzua. 1987. "Trends and Cycles in Primary Commodity Prices." Georgetown University, Department of Economics, Washington, D.C.

Deaton, Angus, and Dwayne Benjamin. 1987. "Household Surveys and Policy Reform: Cocoa and Coffee in the Côte d'Ivoire." Discussion Paper 134. Princeton University, Research Program in Development Studies, Princeton, N.J.

Devarajan, Shantayanan, and Jaime De Melo. 1988. "Adjustment with a Fixed Exchange Rate: Cameroon, Côte d'Ivoire, and Senegal." *World Bank Economic Review* 1 (3): 447–87.

Fields, Gary S. 1988. "Labor Market Policy and Structural Adjustment in Côte d'Ivoire." Cornell University, Economics Department, New York.

Hinchliffe, Keith. 1985. "Higher Education in Sub-Saharan Africa." EDT Discussion Paper 3. World Bank, Education and Training Department, Washington, D.C.

———. 1986. "The Monetary and Non-Monetary Returns to Education in Africa." EDT Discussion Paper 46. World Bank, Education and Training Department, Washington, D.C.

Horton, Brendan. 1990a. "Côte d'Ivoire: Commerce Exterieur, Incitations et Reglementation." World Bank, Occidental and Central Africa Department, Washington, D.C.

———. 1990b. "L'emploi et les Salaires." World Bank, Occidental and Central Africa Department, Washington, D.C.

Melhado, Oscar. 1990. "A Note on the Time Series Characteristics of Cocoa Prices." Boston University, Economics Department, Boston, Mass.

Mingat, Alain, and George Psacharopoulos. 1985. "Education Costs and Financing in Africa: Some Facts and Possible Lines of Action." EDT Discussion Paper 13. World Bank, Education and Training Department, Washington, D.C.

Riveros, Luis A. 1989. "International Differences in Wage and Nonwage Labor Costs." Policy Research Working Paper 188. World Bank, Country Economics Department, Washington, D.C.

Rodríguez, Carlos Alfredo. 1990. "The Macroeconomics of the Public Sector Deficit: The Case of Argentina." Policy Research Working Paper 632. World Bank, Country Economics Department, Washington, D.C.

Tanzi, Vito. 1987. "Quantitative Characteristics of the Tax Systems of Developing Countries." In David Newbery and Nicholas Stern, eds., *The Theory of Taxation for Developing Countries*. New York: Oxford University Press.

UNDP (United Nations Development Programme) and World Bank. 1980. *African Economic and Financial Data*. New York.

UNICEF (United Nations Children's Fund). 1990. *The State of the World's Children 1990*. Oxford, U.K.: Oxford University Press.

Van der Gaag, Jacques, and Wim Vijverberg. 1987. *Wage Determinants in Côte d'Ivoire*. Living Standards Measurement Study Working Paper 33. Washington, D.C.: World Bank.

World Bank. 1983a. "Comparative Education Indicators." Education and Training Department, Washington, D.C.

———. 1983b. *Kenya: Growth and Structural Change*. 2 vols. A World Bank Country Study. Washington, D.C.

———. 1988a. *Education in Sub-Saharan Africa: Policies for Adjustment, Revitalization, and Expansion*. A World Bank Policy Study. Washington, D.C.

———. 1988b. "Côte d'Ivoire: La Mobilisation des Resources Internes en Vue d'une Croissance Stable." Occidental and Central Africa Department, Washington, D.C.

———. 1990a. "Côte d'Ivoire: Examen des Depenses Publiques." Occidental and Central Africa Department, Washington, D.C.

———. 1990b *World Tables*. Baltimore, Md.: Johns Hopkins University Press.

# 7

# Ghana: Adjustment, Reform, and Growth

*Roumeen Islam and Deborah Wetzel*

When Ghana became independent in 1957, its resource base was large, its population was well educated, and its economic infrastructure was strong. By 1983, however, the economy was a shambles, with income per capita more than 10 percent lower than in 1957. In April 1983 the government initiated the Economic Recovery Programme (ERP) in an attempt to rescue the economy. Since then, the economy of Ghana has made steady progress: the average annual growth rate of real GDP during 1984–90 was 4.5 percent, compared with −0.5 percent during 1978–83.

This chapter investigates the role that fiscal deficits have played in Ghana's economic decline and renewal. Our main conclusion is that Ghana's high money-financed fiscal deficits, in conjunction with the direct controls imposed on the economy, caused severe macroeconomic imbalances and reduced growth until 1983. Lower fiscal deficits, liberalization of the economy, and access to foreign financing led to improved economic performance after 1983.

The first section presents an overview of economic policy and the control regime that has been in place since the early 1960s. It then discusses some of the principal factors that have affected Ghana's deficit. The second section focuses on fiscal deficits and financial markets. It considers the effect of inflationary financing on the demand for money and quasi money and for long-run seigniorage, and it addresses the sustainability of the deficit. The third section assesses the impact of the fiscal deficit on private consumption and investment. The fourth section develops the relationship between fiscal deficits, the trade balance, and the real exchange rate. The fifth section presents our conclusions.

## Economic Policy and Fiscal Deficits since the 1960s

In this section we provide an overview of economic policy during the past three decades. We go on to look at the effect of economic policy on the magnitude of the fiscal deficit in Ghana.

*An Overview of Economic Policy*

From 1961 to 1966, under President Kwame Nkrumah, who had come to power with independence in 1957, Ghana's economic strategy consisted of massive investment by the state in the hope of propelling the country into higher growth.[1] The strategy focused on national planning, import substitution, and the development of public enterprises. In order to contain both the budget and current account deficits, a series of controls was put in place in 1961. Foreign exchange, price, and financial sector controls were implemented. Credit was allocated by the government, with import-substituting industries benefiting at the expense of agriculture. Quantitative restrictions on imports were used to control demand.

The first column in table 7.1 provides summary statistics on the performance of the economy during the period 1961–66. (See appendix 7.3 for annual data.) Investment during this period averaged 17.3 percent of GDP, and the average central government deficit was 6.4 percent of GDP.

During 1967–71 there was some economic liberalization, much higher growth, lower inflation, and a lower central government deficit. The National Liberation Council, or NLC (1967–69), and the Progress Party (1969–71) attempted to restore internal and external economic balance with the aid of a stabilization loan and technical assistance from the International Monetary Fund (IMF).

In 1970 the government took advantage of high world cocoa prices to permit a rapid expansion of imports and public expenditures. However, strong demand—resulting from increased income—and import liberalization led to an acute current account deficit.

After a coup by Lieutenant-Colonel I. K. Acheompong and the National Redemption Council in 1972, liberalization stopped and controls were reimposed. Another series of attempted coups led in 1975 to the establishment of the Supreme Military Council and in 1978 to the regime led by Lieutenant-General Frederick Akuffo. In 1979 a coup brought to power Lieutenant Jerry Rawlings, who implemented a currency reform to reduce money in circulation and to control inflation. After a brief period of elected government, Lieutenant Rawlings took over in 1981 and initiated the ERP in 1983.

The period 1978–83 was thus one of prolonged economic and political crisis. Annual average growth of per capita GDP during this period was −2.7 percent, while inflation averaged 73.2 percent. The real interest rate was negative, the real exchange rate became highly overvalued, and the black-market premium averaged 9.5 percent. Investment averaged only 5 percent of GDP. The results of the control regime were shortages of inputs, reduced investment and exports, rent-seeking, corruption, and the development of an extensive black-

**Table 7.1. Summary Statistics, Ghana, 1961–90**

(period averages)

| Item | 1961–66 | 1967–71 | 1972–77 | 1978–83 | 1984–90 |
|---|---|---|---|---|---|
| *Aggregate variable* | | | | | |
| Growth of real GDP | | | | | |
| (percent) | 3.0 | 4.3 | 0.3 | −0.5 | 4.5 |
| Growth of GDP per capita | | | | | |
| (percent) | 0.4 | 2.0 | −2.3 | −2.7 | 1.2 |
| Inflation (percent) | 11.8 | 3.9 | 41.4 | 73.2 | 29.7 |
| Real discount rate (percent) | −5.9 | 2.6 | −20.0 | −31.4 | −4.2 |
| Real exchange rate (cedis | | | | | |
| per U.S. dollar)[a] | 94.3 | 95.6 | 77.7 | 18.2 | 77.1 |
| Black-market premium | 1.6 | 1.7 | 2.7 | 9.5 | 1.8 |
| | | | | | |
| *Composition of output (percentage of GDP)* | | | | | |
| Private consumption | 76.5 | 75.2 | 76.3 | 86.6 | 82.9 |
| Public consumption | 12.1 | 14.4 | 12.3 | 9.4 | 9.4 |
| Gross fixed capital | | | | | |
| formation | 17.3 | 11.4 | 9.8 | 5.0 | 10.9 |
| Change in stocks | −0.5 | 0.9 | 0.5 | −0.1 | 0.1 |
| Exports | 19.2 | 18.9 | 17.7 | 7.0 | 16.0 |
| Imports | 24.6 | 20.8 | 16.6 | 7.9 | 19.3 |
| Private fixed investment | — | 8.2 | 5.7 | 3.0 | 4.5 |
| Public fixed investment | — | 4.1 | 4.7 | 1.9 | 6.5 |
| | | | | | |
| *Monetary system (percentage of GDP)* | | | | | |
| $M_1$ | 16.8 | 14.3 | 18.3 | 15.2 | 12.3 |
| Quasi money | 3.9 | 5.4 | 7.6 | 5.2 | 3.9 |
| | | | | | |
| *Balance of payments (millions of U.S. dollars)* | | | | | |
| Trade balance | −38.5 | 25.3 | 102.3 | 47.4 | −61.9 |
| Current account | −103.9 | −82.9 | −12.1 | −99.7 | −79.4 |
| Capital account | 89.6 | 54.9 | 8.7 | 85.7 | 176.0 |
| Errors and omissions | −1.6 | 0.5 | −4.4 | −76.8 | −19.1 |
| Position above the line | −15.9 | −27.6 | −7.8 | −90.7 | 77.6 |
| Total change in reserves | −15.9 | 9.8 | −15.5 | 39.0 | 46.1 |
| | | | | | |
| *Central government accounts (percentage of GDP)* | | | | | |
| Central government | | | | | |
| revenue | 17.0 | 17.6 | 13.1 | 6.6 | 13.3 |
| Central government | | | | | |
| expenditure | 23.4 | 21.0 | 23.1 | 12.8 | 13.6 |
| Central government deficit | −6.4 | −3.4 | −10.0 | −6.1 | −0.3 |
| Domestic finance | — | 1.7 | 8.9 | 5.6 | 0.0 |
| External finance | — | 1.2 | 0.1 | 0.4 | 0.3 |
| Other | — | 0.6 | 1.0 | 0.6 | 0.0 |

— Not available.

a. The real exchange rate is calculated as the cedi/dollar rate multiplied by the wholesale price index (1985 = 100) and divided by the Ghanaian CPI (1985 = 100). A fall in the real exchange rate indicates an appreciation, and a rise indicates a depreciation.

*Note:* See appendix 7.3 for details on calculations and sources. Averages are unweighted means.

market economy. The control regime also placed severe constraints on the government's ability to finance its deficits. Ghana's poor record in debt repayment, combined with a hostile attitude toward foreign capital, left the country with little access to foreign finance until the ERP was put into place.

With the onset of the ERP in 1983, a period of stabilization and adjustment under the auspices of a structural adjustment program began. The government devalued the cedi and removed controls on foreign exchange operations. Price controls and the licensing regime on imports have been eliminated, the government has called for the "rationalization" of the public sector, and trade taxes have been reformed. More recently the government has undertaken reform of the financial sector. As a result, real GDP growth averaged 4.5 percent between 1984 and 1990 (see table 7.1), and inflation has been brought down. Real interest rates are positive, the real exchange rate has depreciated significantly, and the black-market premium has been dramatically reduced. Investment has increased, but this is largely attributable to increased public investment. In summary, under the ERP Ghana's economic decline was arrested, and the country began to move toward steady economic progress.

## Measuring the Fiscal Deficit

In Ghana there are no consistent data on local government accounts and state-owned enterprises. An attempt was made in this study to construct a measure of the consolidated public sector deficit using the accounts available, including the accounts of the Social Security and National Insurance Trust, the Cocoa Marketing Board, and the Ghana Industrial Holding Corporation. (See Islam and Wetzel 1991 for details on these accounts.) The result was not significantly different from the central government accounts.

THE CENTRAL GOVERNMENT DEFICIT. Figure 7.1 shows total central government revenue and grants, total expenditure (including net lending), and the deficit, defined conventionally as revenue and grants minus expenditure of the central government.[2] Only in fiscal 1970/71 and in the years since 1986 has the central government budget been in surplus.

Note, however, that because the central government accounts presented in figures 7.1 and 7.2 exclude capital expenditure that is financed through external project loans, the expenditure category is underestimated. Data on these capital expenditures are only available from 1984; their inclusion yields the following figures (as percentages of GDP) for total expenditure and the central government fiscal deficit:

**Figure 7.1. Central Government Revenues, Expenditures, and Deficits, Ghana, 1960/61 to 1990**

Percentage of GDP

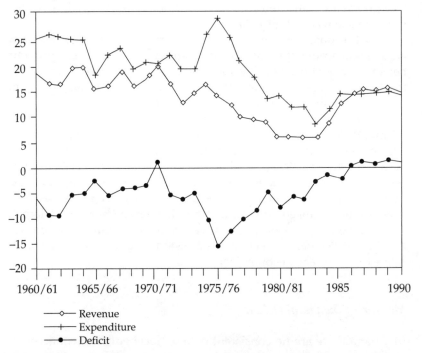

*Note:* Years through 1982 are fiscal years; calendar years are used thereafter.
*Source:* World Bank data.

|  | 1984 | 1985 | 1986 | 1987 | 1988 | 1989 | 1990 |
|---|---|---|---|---|---|---|---|
| Capital expenditure financed through project lending | 0.7 | 1.2 | 3.1 | 2.9 | 3.0 | 2.4 | 3.0 |
| Total central government expenditures | 10.9 | 15.2 | 17.4 | 17.2 | 17.3 | 16.8 | 16.6 |
| Central government surplus (+) and deficit (−) | −2.5 | −3.4 | −3.0 | −2.4 | −2.6 | −1.7 | −2.9 |

If this foreign-financed expenditure is taken into account, the fiscal surpluses that appear in the official accounts after 1986 become deficits that range from about 2 to 3 percent of GDP.

Figure 7.2 shows central government expenditure and its decomposition by economic category as a percentage of total expenditure. The breakdown of expenditure shows clearly the dominance of con-

**Figure 7.2. Decomposition of Central Government Expenditure, Ghana, 1969/70 to 1989**

Percentage of total expenditure

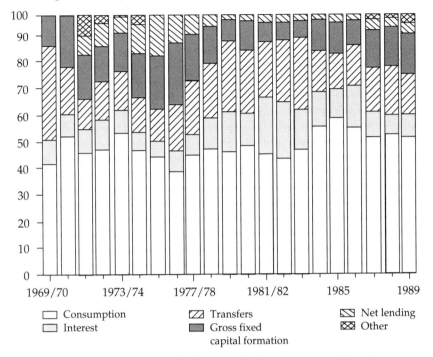

| | |
|---|---|
| ☐ Consumption | ☐ Interest |
| ▨ Transfers | ■ Gross fixed capital formation |
| ▧ Net lending | ▨ Other |

*Note:* Years through 1982 are fiscal years; calendar years are used thereafter.
*Source:* World Bank data.

sumption spending over spending on capital formation. The share of the latter continued to fall until 1983; in absolute terms this decline was the more dramatic because total spending fell during the same period. This led to a rapid decline in the country's infrastructure and helps to explain the country's negative growth over the period. Under the ERP gross fixed capital formation by the central government rose from 1.10 to 2.17 percent of GDP.[3]

As illustrated in figure 7.1, central government revenue (including grants) as a share of GDP reached a peak in 1970/71, at 20 percent of GDP and fell to a low of 5.4 percent of GDP in 1982. At the root of the instability and the decline in revenues was the government's dependence on export duties from cocoa as its principal source of revenue. Until 1981 the export duty on cocoa typically provided between 25 and 40 percent of the government's revenues.

**Figure 7.3. Direct Taxes, Taxes on Goods and Services, and International Trade Taxes, Ghana, 1966/67 to 1989**

Percentage of GDP

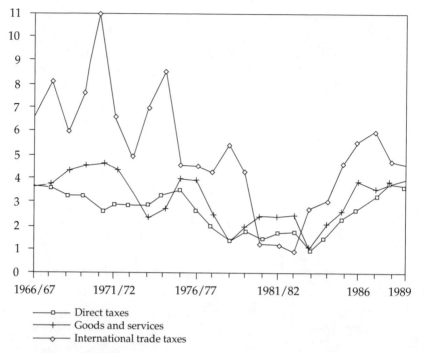

———□——— Direct taxes
———+——— Goods and services
———◇——— International trade taxes

*Note:* Years through 1982 are fiscal years; calendar years are used thereafter.
*Source:* World Bank data.

Figure 7.3 presents direct taxes, taxes on goods and services, and international trade taxes as a percentage of GDP. Each of these taxes declined as a share of GDP between the late 1960s and the early 1980s. The volatility of trade taxes is immediately obvious from the figure. The government's domestic policies were responsible for the decline in these taxes. In certain years—for example, 1978/79—free on board (FOB) prices for cocoa were higher, yet revenues from export taxes were much lower.[4] The decline in revenue is largely attributable to the decline in official cocoa production in the 1970s caused by the fall in both nominal and real producer prices: during the 1960s producers switched to new crops or smuggled cocoa out.[5] The Alien Compliance Act of 1969 forced many workers to leave the country, which in turn caused labor shortages.[6]

Appendix 7.4 presents some estimates of the revenue that has been lost as a result of smuggling. The revenue obtained by the government per metric ton of cocoa was applied to an estimate of total annual smuggling of cocoa. The average total revenue lost over the period 1970/71 to 1979/80 is 0.5 percent of GDP. Thus the primary explanation for the loss in revenue from export taxes was not smuggling but the decline in cocoa production.

Since 1984 all three tax measures—direct taxes, taxes on goods and services, and international trade taxes—have increased as a percentage of GDP. The increase in revenue from direct taxes (about 20 percent) and from taxes on goods and services (about 25 percent) is attributable to improved tax administration combined with growth.

EFFECTS OF ECONOMIC AND POLICY VARIABLES ON THE CENTRAL GOVERNMENT DEFICIT. Marshall and Schmidt-Hebbel (1989) have developed an accounting framework that allows for the decomposition of the changes in the deficit into changes attributable to feedback effects (for example, GDP growth and inflation rates), to external variables, and to fiscal policy variables. The results of the application of their methodology to the Ghanaian central government deficit are shown in appendix 7.5.

Figure 7.4 shows the change in the deficit, excluding changes attributable to domestic macroeconomic variables and changes attributable to external variables. In most years almost the entire change in the deficit is explained by fiscal policy variables, in particular the wage bill. (For example, in 1985 the wage bill caused a change in the deficit of 2.5 percent of GDP.) Transfers and spending on goods and services both had large effects on the deficit in some years; the latter had its greatest effect in 1984, when it increased the deficit by 2.0 percent of GDP. The years in which there is a divergence (for example, 1981/82) are years in which high inflation—more than 100 percent in 1977 and 1983—had feedback effects on tax revenue. In the estimation of the tax functions, inflation was found to have a significant negative impact on direct tax revenue and on "other indirect tax revenue" (indirect taxes minus import duties and export taxes, which were estimated separately). There is thus an Olivera-Tanzi effect for both direct and other indirect taxes.

MEASURING THE PUBLIC SECTOR DEFICIT FROM FINANCIAL FLOWS. Using data from the monetary survey on foreign liabilities and on the claims against the central government, nonfinancial public enterprises, and the Cocoa Marketing Board, we derived a measure of the total outstanding debt stock of the public sector. Differencing these data on stocks provides us with a measure of the financing flows to the public sector and hence with an alternative measure of the fiscal deficit.[7] As illustrated by figure 7.5, this measure shows a pattern similar to that of the central government deficit.

**Figure 7.4. Change in the Central Government Deficit, Ghana, 1972/73 to 1988**

Percentage of GDP

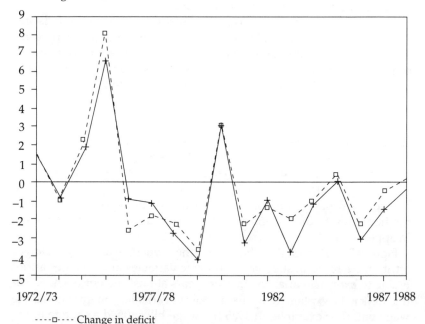

----□---- Change in deficit

——+—— Change in deficit (excluding external and fiscal policy variables)

*Note:* Years through 1982 are fiscal years; calendar years are used thereafter.
*Source:* World Bank data.

The differences between the measures are attributable to three factors. First, the total public sector measure includes the foreign financing of capital expenditure arising from project loans, which is not included in the central government accounts. Second, in the central government measure grants are treated as revenue—or as above the line—whereas in our second measure they are considered a component of financing. Third, the second measure includes financing flows to public entities and to the Cocoa Board.

A comparison of the two measures shows that prior to 1984 the financing measure of the deficit is in some years smaller than the conventional measure. Before 1983 both grants and external financing of capital expenditure are small or zero, and in some years financial flows to public entities are negative. After 1983, however, the two measures are very similar. If we move the grant component of central government revenue below the line and include capital expenditure

**Figure 7.5. Public Sector Deficit on the Basis of Financial Flows, Ghana, 1967–90**

Percentage of GDP

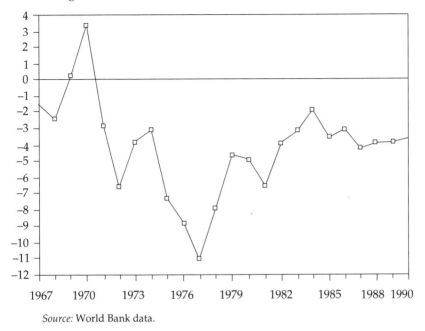

*Source:* World Bank data.

financed through external lending, our central government measure is quite close to our public sector measure.[8]

## Fiscal Deficits and Financial Markets

In this section we examine domestic financing, inflation, the demand for assets, and seigniorage. We go on to establish model equations and to look at the demand for money and quasi money. We then investigate the effect of inflation on seigniorage and the sustainability of the deficit.

### Domestic Financing of the Public Sector Deficit

Figure 7.6 illustrates the financial flows to the public sector (including nonfinancial public enterprises and the Cocoa Marketing Board), broken down into flows from the monetary authority and the domestic banks (or to the monetary authority and the domestic banks, if the flows are negative). Flows to the public sector from the private sector are negligible and have not been included in the figure. Because of the limited development of the banking system in Ghana, the govern-

**Figure 7.6. Domestic Financing of the Public Sector Deficit, Ghana, 1971–90**

Percentage of GDP

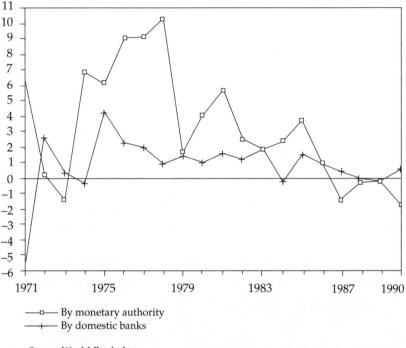

———□— By monetary authority
———+— By domestic banks

*Source:* World Bank data.

ment's ability to borrow from domestic banks has been weak, averaging about 2 percent of GDP. Until 1984, instead of borrowing from domestic banks the government relied on borrowing from the monetary authority for the greatest part of its finance.

Money financing has caused high inflation. Figure 7.7 illustrates inflation rates, measured by changes in the consumer price index, (CPI) during the 1971–90 period, with rates of more than 100 percent in 1977, 1981, and in 1983. (See appendix 7.3 for yearly rates.) These inflationary peaks also mark years in which shortages in the food supply caused food prices (which have accounted for 50 percent of the CPI) to rise. A simple regression of the inflation rate on borrowing from the central bank and dummy variables for the years of drought and fires indicates a significant relationship between monetary finance and inflation.[9]

The high levels of inflation that were brought about, at least in part, by the monetary financing of the deficit in turn affected other areas of the economy. Since the government controlled nominal interest rates,

**Figure 7.7. CPI Inflation Rate, Ghana, 1971–90**

Percent

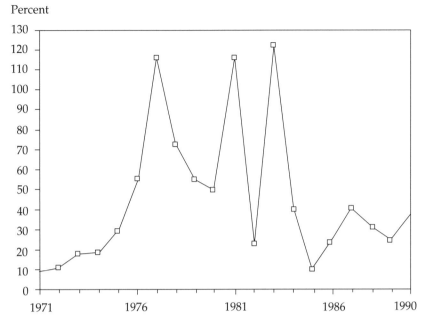

*Source*: World Bank data.

high inflation implied negative real interest rates for deposit and lending (figure 7.8) and a fall in holdings of financial assets.[10] Negative real interest rates and taxes on financial intermediation led the banking sector to institute stringent requirements for credit to the private sector.

Aryeetey and Gockel (1990) argue that the system of financial controls combined with high inflation encouraged the development of informal credit markets. They estimate that the formal sector now controls about 55 percent of the financial savings mobilized per month. Financial disintermediation and a movement into informal credit markets were also encouraged by the 1979 currency conversion. Currency notes outside banks were exchanged at a ratio of 10 old cedis to 7 new cedis for amounts of less than 5,000 cedis and at a ratio of 10 to 5 for amounts of more than 5,000 cedis. As a result of the currency conversion, the money supply was reduced by about 12 percent of broad money. The conversion, along with other measures (demonetization of the 50-cedi note, for example), undermined confidence in the banking system. Financial disintermediation and credit constraints also had important implications for economic growth: private investment was constrained by a lack of liquidity. Monetary

**Figure 7.8. Nominal and Real (Twelve–Month) Deposit Rates, Ghana, 1971–90**

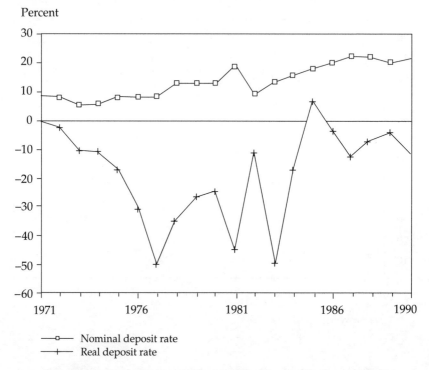

Percent

Nominal deposit rate
Real deposit rate

*Note:* Real interest rates are calculated as $\{[(1+i)/(1+p)] - 1\} \cdot 100$, where $i$ is the nominal interest rate and $p$ is the CPI inflation rate.
*Source:* World Bank data.

policy was rendered impotent—the discount rate became ineffective as a policy tool, and banks consistently held excess reserves.

*Inflation, the Demand for Assets, and Seigniorage*

As figure 7.6 shows, the government relied heavily on borrowing from the central bank through money creation and seigniorage in order to finance its deficit. This method had inflationary consequences. Figure 7.9 illustrates actual seigniorage revenue in Ghana. The traditional definition of seigniorage revenue is the change in high-powered money (currency plus reserves) divided by GDP. Chamley (1991) argues that in economies that are subject to interest controls and high inflation the government effectively has a 100 percent reserve requirement and that in this case the expansion of $M_2$ over GDP is a better measure of seigniorage revenue. As expected, we see that the second definition of seigniorage revenue is larger than

**Figure 7.9. Two Definitions of Seigniorage Revenue, Ghana, 1971–90**

Percentage of GDP

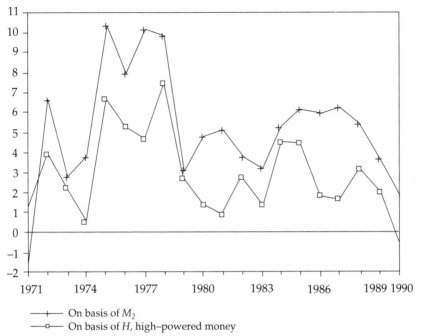

———+——— On basis of $M_2$
———□——— On basis of $H$, high–powered money

*Source:* Calculated from World Bank data.

the traditional measure, sometimes by as much as 5 percent of GDP. In general, however, the two measures follow similar patterns. (See Chamley 1991, pp. 6, 19, for a review of financial taxation in countries that are subject to financial controls.)

THE MODEL. In order to evaluate the effect of inflation on seigniorage, we use a simplified version of the Easterly, Rodríguez, and Schmidt-Hebbel (1989) framework. The equations of the model are set out in equations 7.1–7.6.[11] Equation 7.1 is the budget identity which states that the government deficit must be equal to the sum of external finance (*EF*), domestic credit to the government from the central bank (*DC_g*), and government borrowing from the banking system ($L_g$). External finance is assumed to be exogenous.[12] Equations 7.2 and 7.3 are the portfolio demands of the private sector, which determine the private sector's demand for currency and the allocation of remaining assets between quasi money and foreign currency.

(7.1) $$\text{DEF} = E\dot{F}^* + D\dot{C}_g + \dot{L}_g$$

(7.2) $$M_1 = f(\pi, i_d, Y)$$

(7.3) $$QM = g(\pi, i_d, Y)$$

where $\pi$ is the rate of inflation, $i_d$ is the real interest rate on deposits, and $Y$ is real income. Dots represent the change per unit of time. The demand for foreign currency is determined as a residual.

(7.4) $$L_g = [(1 - r)(1 - c)M_1 + (1 - r)QM] - L_p$$

(7.5) $$DC_g = H - \text{NFA} - \text{NOL}$$

(7.6) $$H = rQM + [c + r(1 - c)]M_1.$$

Equation 7.4 defines loans from the banking system to the government, which is the residual after private credit needs have been met; $L_p$, credit to the private sector, is taken as given. The reserve ratio on deposits is $r$, and $c$ is the currency-to-$M_1$ ratio. Equation 7.5 defines domestic credit to the government from the central bank as high-powered money ($H$) minus net foreign assets of the central bank (NFA) minus any additional net liabilities of the central bank (NOL). High-powered money, equation 7.6, is equal to reserves on deposits and currency.

THE DEMAND FOR MONEY AND QUASI MONEY. The basic equation used to estimate money demand is:

$$LM_t = \alpha_0 + \alpha_1\pi_t + \alpha_2 i_{dt} + \alpha_3 Y_t + e$$

where $LM$ equals the log of real $M_1$, $\pi$ is the CPI inflation rate, $i_{dt}$ is the real interest rate on twelve-month deposits, $Y$ is the log of real GDP, and $e$ is the residual.

The equation was estimated for the 1960–90 period. Dummy variables were included for the currency conversion in 1979 and for the onset of the ERP. Because these estimates indicated the presence of autocorrelation of the errors, a lagged dependent variable was incorporated into the equation. The results appear in table 7.2.

The short-run semielasticity of money demand with respect to inflation is −0.28; the long-run semielasticity is −0.77. As expected, the controlled real deposit rate has no significant effect on money demand. Real income is not significant. The currency conversion in 1979 had a negative impact on money demand, but the ERP has had no significant impact. The lagged dependent variable is highly significant.

Demand for quasi money is based on a similar equation but also incorporates the U.S. real rate on treasury bills adjusted for black-market exchange rate depreciation. Only the lagged dependent variable and the CPI were found to be significant. The short-run elasticity of quasi-money demand with respect to inflation is −0.60, and the long-run elasticity is −4.1.

**Table 7.2. Regression Results, Money and Quasi Money, Ghana**

| Variable | $M_1$ | | Quasi money |
|---|---|---|---|
| | (1) | (2) | |
| $\alpha_0$ | 1.95 | 1.61 | 1.63 |
| | (0.89) | (0.77) | (3.13) |
| $\alpha_1(\pi)$ | −0.41 | −0.28 | −0.60 |
| | (−1.07) | (−2.96) | (−8.00) |
| $\alpha_2(i_d)$ | −0.29 | | |
| | (−0.35) | | |
| $\alpha_3(Y)$ | 0.14 | 0.21 | |
| | (0.70) | (1.18) | |
| $\alpha_4(\text{DUM}79)$ | −0.24 | −0.23 | |
| | (−1.87) | (−1.87) | |
| $\alpha_5(\text{DUM}83)$ | −0.09 | | |
| | (0.70) | | |
| $\alpha_6[M1(-1)]$ | −0.67 | −0.64 | |
| | (4.18) | (−4.79) | |
| $\alpha_7[QM(-1)]$ | | | 0.85 |
| | | | (16.32) |
| $R^2$ | 0.8547 | 0.8514 | 0.9260 |
| F-statistic | 22.54 | 35.82 | 168.86 |

Note: Figures in parentheses are $t$-statistics. A blank denotes omission of the specific variable from the regression.

THE EFFECT OF INFLATION ON SEIGNIORAGE. We use the model set out above to assess the impact on seigniorage revenues of the relationship between inflation and money demand.[13] The base case and three scenarios are considered.

Figure 7.10 illustrates the base case that simulates seigniorage revenues, given the actual rate of inflation, and the case in which inflation is 20 percent *lower* than its actual rate. For most years seigniorage revenue would have been higher had inflation been lower. This is particularly the case in the late 1970s.

Figure 7.11 shows our base case with inflation maintained at 30 percent throughout the period. In some years, particularly the years of very high inflation (for example, 1981 and 1983), seigniorage revenue would have been considerably higher had the inflation rate been about 30 percent. After 1986, however, the revenues of the two converge. This is explained by the fact that during these years inflation was actually about 30 percent.

The final scenario (figure 7.12) shows that had the currency conversion measures of 1979 not taken place, seigniorage revenues would have been higher, on the order of 1 to 5 percent per year. Thus, while

**Figure 7.10. Seigniorage Revenue:  Base Case and with Inflation
20 Percent Lower than Actual Rate, Ghana, 1977–90**

Percentage of GDP

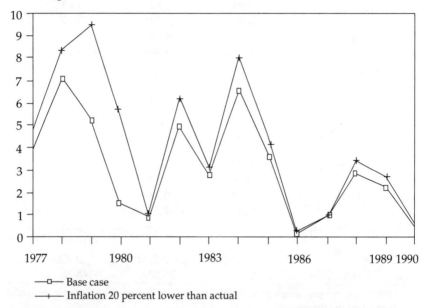

———□— Base case
———+— Inflation 20 percent lower than actual

*Source:* Calculated from World Bank data.

the currency reform had a one-time negative effect on inflation, it also
reduced seigniorage revenue. Continued deficits and the need to
finance them led to money creation after 1979 and also to higher
inflation.

Using the long-run money demand equation, we have calculated
the inflation-tax Laffer curve (figure 7.13). This figure shows the seigniorage revenue received for given inflation rates; as inflation reaches
about 125 percent in Ghana, seigniorage revenue as a function of GDP
begins to decline.[14] In most years inflation in Ghana has been below
this rate.

THE SUSTAINABILITY OF THE DEFICIT.  In recent years the public sector deficit has stabilized at about 4 percent of GDP (using our second
measure, based on financing flows to the public sector). To assess the
sustainability of the deficit given the government's macroeconomic
objectives of lower inflation and higher growth, we follow van Wijnbergen and others (1992). We start with a version of the basic equation
used in the analysis of the previous section:

$$PD + iB + Ei^*F^* = \dot{H} + \dot{D} + E\dot{F}^*$$

## Figure 7.11. Seigniorage Revenue: Base Case and with Inflation at 30 Percent, Ghana, 1977–90

Percentage of GDP

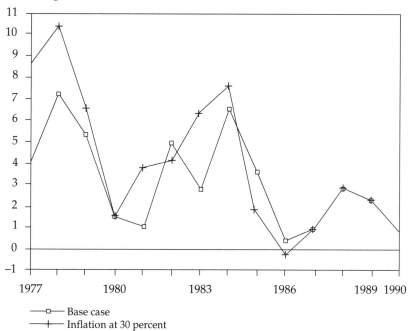

───□─── Base case
───+─── Inflation at 30 percent

*Source:* Calculated from World Bank data.

where *PD* is the nominal primary deficit of the total public sector, *i* is the domestic nominal interest rate, *B* is the domestic public debt stock, *E* is the nominal exchange rate, *i\** is the nominal foreign interest rate, *F\** is the stock of foreign debt in foreign currency units, *H* is total base money, and *D* is credit to the government from the banking system. Each component of the equation is expressed as a ratio to GDP. Dots indicate absolute changes per unit of time, and hats indicate relative changes. The equation can be rewritten as:

$$PD + (r + \hat{P})B + i^*F = \dot{H} + H(\hat{P} + \hat{Y}) + \dot{B} + B(\hat{P} + \hat{Y})$$
$$+ \dot{F}^* + F^*(-\hat{e} + \hat{P}^* + \hat{Y}).$$

The sustainable deficit is based on the condition of nonincreasing public liabilities—a condition in which $\dot{H} = 0 = \dot{B} = \dot{F}$. After imposing this condition and rearranging, the expression for the sustainable primary deficit is:

$$PD = H(\hat{P} + \hat{Y}) + B(\hat{Y} - r) + F^*(\hat{Y} - r^* - \hat{e})$$

**Figure 7.12. Seigniorage Revenue, Base Case and without the
Currency Conversion, Ghana, 1977–90**

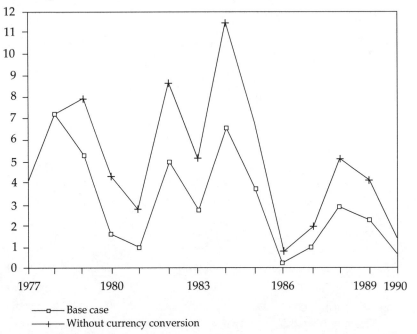

Percentage of GDP

————□———— Base case
————+———— Without currency conversion

*Source:* Calculated from World Bank data.

where $r$ and $r^*$ are real domestic and foreign interest rates and $e$ is the
real exchange rate. Using the ratios of GDP for the above liabilities for
1989, table 7.3 sets out the sustainable deficit for the base case and
three scenarios: lower inflation, high growth, and low growth combined with higher interest rates and a depreciation.

In the base case, growth is 5 percent and inflation is 25 percent.
Nominal interest rates (calculated implicitly, using interest payments
and the total stock of domestic and foreign debt) are 26 percent for
domestic debt and 8 percent for foreign debt. The real exchange rate
remains constant.

In the base case the sustainable primary deficit is 2.78 percent of
GDP, and the sustainable total public sector deficit is 5.32 percent of
GDP.[15] If we reduce inflation to 15 percent and all other variables
(with the exception of real interest rates) remain the same, the sustainable primary deficit is 1.72 percent of GDP, and the sustainable
total public sector deficit is 4.26 percent of GDP. With higher growth (8
percent), the sustainable total public sector deficit rises to 6 percent of
GDP.

**Figure 7.13. Seigniorage Revenue for Given Inflation Rates, Ghana**

Percentage of GDP

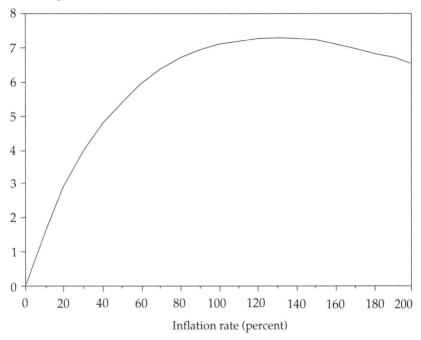

Inflation rate (percent)

*Source:* Authors' calculations.

Finally we consider a scenario in which growth is reduced to 2.5 percent of GDP, inflation remains at 25 percent, both domestic and world interest rates increase, and a 5 percent depreciation of the real exchange rate occurs. In this case the sustainable total public sector deficit is reduced to 4.52 percent of GDP. Given the results in table 7.3, the current deficit, 4 percent of GDP, appears to be sustainable. The low-inflation scenario constrains the sustainable deficit to 4.26 percent of GDP, but higher growth allows it to increase.

## Fiscal Deficits, Private Consumption, and Private Investment

Since the early 1960s private consumption has risen substantially as a share of GDP (see figure 7.14). In 1983 private consumption was 91 percent of GDP. As economic conditions worsened during the 1970s, more and more effort went into subsistence and providing for the present at the expense of saving and investment for the future. Since 1983 private consumption has fallen, but it still remains high. Private investment has been low even by developing country standards; in

**Table 7.3. Sustainability of the Public Sector Deficit, Ghana**

| Item | Base case | Lower inflation | High growth | E rate depreciation |
|---|---|---|---|---|
| GDP growth ($\hat{Y}$) | 0.05 | 0.05 | 0.08 | 0.025 |
| Domestic inflation ($\hat{P}$) | 0.25 | 0.15 | 0.25 | 0.25 |
| Domestic nominal interest rate ($i$) | 0.26 | 0.26 | 0.26 | 0.30 |
| Domestic real interest rate ($r$) | 0.01 | 0.11 | 0.01 | 0.05 |
| Foreign nominal interest rate ($i^*$) | 0.08 | 0.08 | 0.08 | 0.10 |
| Foreign inflation ($\hat{P}^*$) | 0.05 | 0.05 | 0.05 | 0.05 |
| Foreign real interest rate ($r^*$) | 0.03 | 0.03 | 0.03 | 0.05 |
| Real exchange rate depreciation ($\hat{e}$) | 0.00 | 0.00 | 0.00 | 0.05 |
| Inflation tax ($H\hat{P}$) | 0.02105 | 0.01263 | 0.02105 | 0.02105 |
| Seigniorage ($H\hat{Y}$) | 0.00421 | 0.00421 | 0.006736 | 0.002105 |
| Domestic debt effect [$B(\hat{Y} - r)$] | 0.000932 | −0.00139 | 0.001631 | −0.00058 |
| Foreign debt effect [$F(\hat{Y} - r^*)$] | 0.001804 | 0.001804 | 0.00451 | −0.00225 |
| Foreign debt capital gain ($-F\hat{e}$) | 0.00 | 0.00 | 0.00 | −0.00451 |
| Sustainable primary deficit | 0.027996 | 0.017246 | 0.033927 | 0.015807 |
| Interest on foreign debt ($i^*f$) | 0.023452 | 0.023452 | 0.023452 | 0.02706 |
| Interest on domestic debt ($ib$) | 0.001864 | 0.001864 | 0.001864 | 0.00233 |
| Sustainable public sector deficit | 0.053312 | 0.042562 | 0.059243 | 0.045197 |

*Note:* Public sector liabilities as a percentage of GDP, December 1989, are as follows: total base money, 0.0842; net foreign debt, 0.0902; net domestic debt, 0.0233.

*Source:* For public sector liabilities listed in note, see appendix 7.3.

the period 1965–88 private investment as a share of GDP peaked at about 10.5 percent (figure 7.15). The peaks and troughs have corresponded to years in which public investment was very high or there was a coup or change in government. Even after the implementation of the ERP, private investment levels remained low.

### Fiscal Deficits and Private Consumption

The equation estimated in order to assess the impact of fiscal deficits on private consumption is:

$$C_t = \alpha_0 + \alpha_1 Yp_t + \alpha_2 Cpub_t + \alpha_3 Spub_t + \alpha_4 Sfor_t + \alpha_5 r_t + \alpha_6 LC_t + e$$

**Figure 7.14. Private Consumption, Ghana, 1961–90**

Percentage of GDP

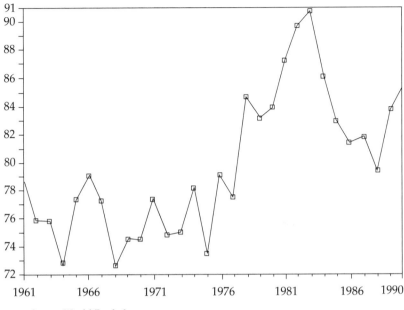

*Source:* World Bank data.

where $C_t$ is current private consumption, $Yp_t$ is current disposable income defined as gross national product (GNP) minus taxes, $Cpub_t$ is public consumption, $Spub_t$ is public savings (the central government fiscal balance is used as a proxy), $Sfor_t$ is a measure of foreign savings (exports minus imports) in the economy, $r_t$ is the real lending rate, and $LC_t$, liquidity constraints, is proxied by credit to the private sector. All variables other than $r$ are represented as percentages of GDP.

The results of our specifications are set out in table 7.4. All variables other than the real interest rate and the liquidity constraint are found to be significant. As previously noted, financial markets are weak, and Ghanaians generally do not borrow to finance consumption. It is thus not surprising that these two factors are not significant. The second equation in table 7.4 drops the insignificant variables. The relationship between current private consumption and current disposable income is almost one-to-one, which supports the Keynesian hypothesis concerning private consumption. In low-income economies, where credit markets are far from perfect, it is unlikely that most individuals will have substantial savings that act as buffers during low-income periods. Hence, current private consumption is tied closely to current disposable income.

**Figure 7.15. Private Investment, Ghana, 1965–88**

Percentage of GDP

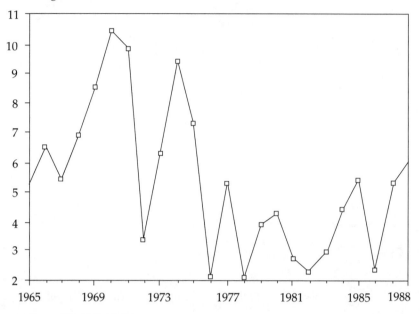

Source: World Bank data.

Public consumption is negatively related to private consumption; during the mid-1970s, when government consumption was very high, private consumption was at its lowest. This implies that some of the government's expenditure was passed on to the private sector in the form of goods or services. As this government provision stopped, private sector consumption rose to make up for the lost goods and services.

Public savings is also shown to be significantly related to private consumption. If public savings increases—for example, if the deficit goes down—private consumption increases. The results imply that deficits *do* affect private consumption decisions and that Ricardian equivalence does not hold.[16]

Foreign savings, defined as exports minus imports, has a negative effect on private consumption. This relationship captures the impact of import controls on private consumption, as well. Reducing imports through the use of controls improves the trade balance and (assuming that exports remain the same) reduces private consumption. The import regime in Ghana tended to control consumer goods more than capital goods, thus squeezing private consumption.

**Table 7.4. Regression Results, Private Consumption, Ghana**

| Variable | EQN1 | EQN2 |
|---|---|---|
| $\alpha_0$ | 11.43 | 0.966 |
|  | (0.77) | (0.09) |
| $\alpha_1(Yp)$ | 0.88 | 0.965 |
|  | (6.07) | (9.70) |
| $\alpha_2(Cpub)$ | −0.55 | −0.416 |
|  | (−2.86) | (−2.35) |
| $\alpha_3(Spub)$ | 0.117 | 0.260 |
|  | (1.97) | (3.31) |
| $\alpha_4(Sfor)$ | −0.85 | −0.837 |
|  | (−6.29) | (−6.83) |
| $\alpha_5(r)$ | 0.02 |  |
|  | (1.09) |  |
| $\alpha_6(LC)$ | −0.29 |  |
|  | (−1.68) |  |
| $R^2$ | 0.9751 | 0.9689 |
| Durbin-Watson | 2.27 | 2.02 |

*Note:* Figures in parentheses are *t*-statistics. A blank denotes omission of the specific variable from the regression.

In summary, public consumption over the years has substituted for private consumption. In the 1970s, when government expenditures were high, private expenditure remained relatively low. As the government tried to reduce its deficit by controlling public expenditure, private consumption rose sharply.

*Fiscal Deficits and Private Investment*

The investment equation that we have estimated for Ghana seems to confirm what we would expect from our analysis of Ghana's economy. In the equation, the flow of credit to the private sector was used as a proxy for liquidity constraints faced by investors.

As table 7.5 shows, public investment, corporate tax revenues, and the liquidity constraint variable are all statistically significant. The negative coefficient for public investment seems to indicate that public sector investment in Ghana has mostly substituted for private investment. This result is not surprising since, traditionally, the government's policies have not encouraged private investment.

Using corporate tax revenues collected by the government as an indicator of the tax burden faced by private firms, we find that this indicator has a positive, rather than the expected negative, sign.

**Table 7.5. Regression on Private Investment, Ghana**

| Item | Coefficient |
|---|---|
| Constant | −0.002 |
|  | (−0.10) |
| PBICDGD | −1.102 |
|  | (−2.76) |
| CDCDGDP | 0.557 |
|  | (2.42) |
| RINT3 | −0.0001 |
|  | (−0.32) |
| CPTCDGD | 3.974 |
|  | (2.27) |
| $R^2$ | 0.561 |
| Durbin-Watson | 2.04 |

*Note:* Figures in parentheses are *t*-statistics.

Since corporate revenues are correlated with profits we can think of these revenues instead as a proxy for profits rather than taxes. Unfortunately we do not have separate figures for corporate profits and cannot isolate their effects on investment.

The flow of credit to the private sector has a positive coefficient, as expected. This result tells us that one way to encourage higher levels of private investment in Ghana would be to ease the economywide credit ceilings or to reallocate credit from the government to the private sector. As noted at the beginning of this chapter, the government has controlled the allocation and overall level of credit since the early 1960s.

As expected, the real interest rate, which has been negative, does not have a substantial effect on private investment.[17]

Using the equation in table 7.5, we conducted some counterfactual simulations to determine the effect on private investment of (a) varying the supply of credit to the private sector and (b) reducing public sector investment. Figure 7.16 shows actual and simulated ratios of private sector investment to GDP. In the first simulation, public sector investment in real terms is held at 1 percent of GDP (INVSIM1); in the second, private sector credit (INVSIM2) is increased to 20 percent of GDP. In the first case investment as a percentage of GDP would have been 12.5 percent rather than 5.5 percent in 1988! From 1978 to 1983 private investment as a proportion of GDP would still have been rather low, but in most other years there would have been a definite improvement. During 1978–83 the black-market premium was very high and rising, and the parallel market was flourishing. It is possible

**Figure 7.16. Private Investment Simulations, Ghana, 1967-88**

Percentage of GDP

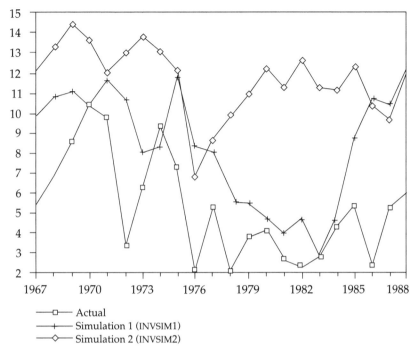

─□─ Actual
─+─ Simulation 1 (INVSIM1)
─◇─ Simulation 2 (INVSIM2)

*Source:* Estimated from World Bank data.

that during this period legal economic activities were diverted to the unofficial economy and that the low investment ratios may not really be indicative of overall private investment activities in the economy.

When the supply of credit to the private sector is raised to 20 percent of GDP and maintained there, we find that investment would have been a much higher proportion of GDP in all years. In 1983, for instance, 12.1 percent rather than 2.9 percent of GDP would have been accounted for by private investment. For the period 1977–85 availability of credit seems to be the constraining factor; during these years even lowering public sector investment does not have a great effect on private sector investment.

## Fiscal Deficits and the External Sector

This section examines the implications of Ghana's fiscal deficits for the real exchange rate, the black-market premium, and the trade balance. The fixed exchange rate system, in conjunction with high fiscal deficits and strict capital controls, led to the development of a large black market in foreign exchange in Ghana. This section incorporates models of smuggling and of the parallel market in foreign exchange. It draws on studies by May (1985), Lizondo (1987a, 1987b), and Pinto (1989) to analyze the effect of the fiscal deficit on the external sector of Ghana's economy.

## *The Parallel Market for Foreign Exchange*

Ghana's parallel market for foreign currency has been growing sub-stantially since the mid-1960s with the increase in cocoa smuggling. The decline in export earnings during the late 1970s and early 1980s, coupled with strict foreign exchange controls, exacerbated the situa-tion (see, for instance, May 1985). Figure 7.17 shows the evolution of the average and end-of-period black-market premiums from 1969 to 1987. In 1983 the black-market exchange rate was 76.58 cedis to the U.S. dollar, but the official rate was 2.75 cedis to the dollar. Table 7.6 presents estimates of the percentage of total cocoa smuggled out of Ghana during 1960–82 and the relative size of the parallel economy with respect to the formal economy. Estimates of the demand for currency in the black market show a steady increase during these years.

During the decade before the ERP was initiated, when the public sector deficit was rising, the official trade balance sank from a surplus of US$212.9 million in 1973 to a deficit of US$60.6 million in 1983. Figure 7.18 shows a positive trade balance and a negative fiscal bal-ance in years of surplus. Another important factor in the worsening of the trade balance was the deterioration in Ghana's terms of trade, which amounted to about 47 percent between 1973 and 1983. In addi-tion, an overvalued exchange rate, combined with weak demand for Ghana's main export, cocoa, led to a dramatic decline in export earn-ings, from US$1,066 million in 1979 to US$439 million in 1983.

Between 1983 and 1986 the cedi was devalued from 2.75 cedis to the U.S. dollar to 90 cedis to the U.S. dollar. A dual exchange rate system existed, with some transactions made under the fixed-rate system and others at government-managed auctions. This system was meant to, and actually did, reduce the spread between the official and paral-lel market rates.

During 1983–86 exports grew 76.1 percent in U.S. dollar terms, while the public sector deficit fell from US$476 million to US$72 mil-lion. The trade balance improved from a deficit of US$60.6 million to a surplus of US$60.9 million, and the black-market premium fell: higher fiscal deficits had also been associated with higher black-market premiums. The black-market real exchange rate depreciated

**Figure 7.17. Average and End-of-Period Black-Market Premiums, Ghana, 1969–87**

Premium, log scale $[(b - e)/e]$

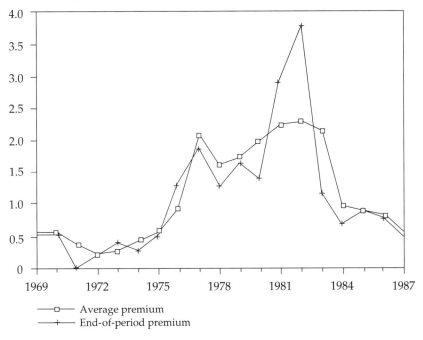

──□── Average premium
──+── End-of-period premium

*Source:* World Bank data.

during this period (figure 7.19).[18] The model on which our estimations are founded (developed in detail in appendix 7.1) is based on papers by May (1985), Lizondo (1987a, 1987b), and Pinto (1989).

THE MODEL. As in Pinto (1989), the government purchases foreign exchange from exporters at a fixed rate of $e$ cedis to the U.S. dollar and sells part of its foreign exchange to importers at rate $e$. The remainder of the foreign exchange is used to finance government consumption. With a black-market foreign exchange rate of $b$ cedis to the U.S. dollar, the marginal cost of foreign exchange is $b$; note however, that the government obtains foreign exchange at rate $e$. Thus the official exchange rate $e$ acts as a conduit for real income transfers between the government and the private sector—in particular, exporters. When the government is a net buyer of foreign exchange (as in Ghana), the black-market premium acts as a source of revenue for the government. If the implicit tax on exports is lowered (the official exchange rate is devalued so that it is brought closer to or

**Table 7.6. Cocoa Production, Smuggling, and the Parallel-Market Economy, Ghana, 1960/61 through 1982**

| Year | Cocoa production (thousands of metric tons) | Smuggled cocoa (thousands of metric tons) | Year | Illegal money (millions of cedis) | Parallel-market economy | |
|---|---|---|---|---|---|---|
| | | | | | Millions of cedis | As percentage of GDP |
| 1960/61 | 430 | 10 | | | | |
| 1961/62 | 409 | 8 | | | | |
| 1962/63 | 413 | 14 | | | | |
| 1963/64 | 428 | 11 | | | | |
| 1964/65 | 538 | 14 | | | | |
| 1965/66 | 401 | 17 | 1965 | 0.01 | 0.08 | 0 |
| 1966/67 | 368 | 17 | 1966 | 1.66 | 9.71 | 0.64 |
| 1967/68 | 415 | 21 | 1967 | 1.22 | 7.64 | 0.51 |
| 1968/69 | 323 | 17 | 1968 | 0.87 | 5.71 | 0.34 |
| 1969/70 | 403 | 25 | 1969 | 1.23 | 8.50 | 0.42 |
| 1970/71 | 413 | 31 | 1970 | 0.96 | 7.15 | 0.32 |
| 1971/72 | 454 | 37 | 1971 | 1.37 | 10.72 | 0.43 |
| 1972/73 | 407 | 42 | 1972 | 1.27 | 7.73 | 0.27 |
| 1973/74 | 340 | 34 | 1973 | 1.05 | 6.53 | 0.19 |
| 1974/75 | 376 | 30 | 1974 | 3.65 | 24.54 | 0.53 |
| 1975/76 | 396 | 38 | 1975 | 8.61 | 45.47 | 0.86 |
| 1976/77 | 320 | 40 | 1976 | 15.74 | 72.68 | 1.11 |
| 1977/78 | 271 | 45 | 1977 | 118.74 | 582.96 | 5.22 |
| 1978/79 | 265 | 50 | 1978 | 282.71 | 1,543.73 | 7.36 |
| | | | 1979 | 483.15 | 3,243.21 | 11.51 |
| | | | 1980 | 1,195.62 | 10,024.37 | 24.45 |
| | | | 1981 | 1,313.16 | 12,427.07 | 16.21 |
| | | | 1982 | 2,741.99 | 27,827.27 | 32.41 |

Source: May 1985.

**Figure 7.18. Trade Surplus and Fiscal Deficit, Ghana, 1970–87**

Millions of U.S. dollars

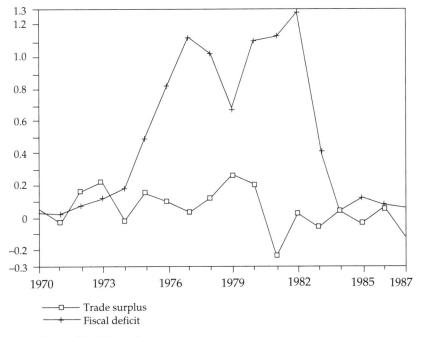

—□— Trade surplus
—+— Fiscal deficit

*Source:* World Bank data.

equals the marginal cost of foreign exchange), total exports will increase. However, the basic analysis in this section regarding the effects of higher government expenditure on the black-market premium is valid both when the government is a net purchaser of foreign exchange and when it is a net seller. This is because in this model an increase in government expenditure, other sources of finance being given, must be financed by money creation. An increased stock of money will be held only at a higher premium, regardless of whether the government is a net purchaser or seller of foreign exchange. (The formal analysis is more complex in the net-seller case.) The government also gets revenues from conventional taxes, $t$, which are fixed, and foreign aid to the government, $A$. The inflation tax finances the residual requirements of the government net of the implicit tax on exports and net of conventional taxes and foreign aid.

In this formulation, an increase in public sector expenditure increases reliance on the inflation tax. We assume that the inflation tax is given by $\pi m$, where $\pi$ is the rate of inflation and $m = M/e$. ($M$ is nominal money balances.) At the steady state the rate of inflation

**Figure 7.19. Official and Black-Market Real Exchange Rates, Ghana, 1969–87**

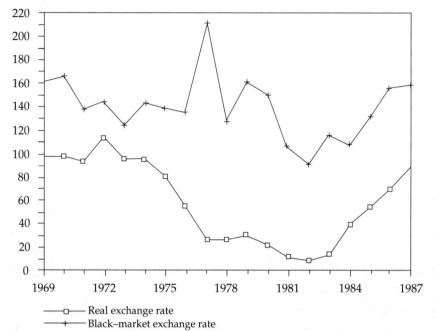

Cedis per U.S. dollars

Source: World Bank data.

equals the official rate of depreciation of the currency ($\Delta e/e$). The government's budget constraint at the steady state is:

$$(7.7) \qquad (g_T - t - A) = m\,\Delta e/e$$

where $g_T$ is total government expenditure in dollars, which is fixed and exogenously given and includes expenditure on both tradables ($g_I$) and nontradables ($g_N$). That is,

$$g_T = g_N P_N/e + g_I \qquad g_I = zg_T \qquad z < 1.$$

The real fiscal deficit is defined as $g_T - t$. Seigniorage from the inflation tax is equal to $\Delta e/em$, at the steady state. Given that the marginal cost of foreign exchange to the private sector is $b$ and that the private sector's loss is the government's gain, we can rewrite the preceding equation as:

$$(7.8) \qquad \begin{aligned} g_T &= M/b\,\Delta e/e + et/b + g(b-e)/b + eA/b \\ &= m\,\Delta e/\phi + t/\phi + g(1 - 1/\phi) + A/\phi \end{aligned}$$

(see Pinto 1989), where $\phi$ is the black-market premium, $et/b = t/\phi$ is the real tax burden on the private sector, and $eA/b = A/\phi$ is the real value of aid flows to the private sector, and similarly for the expressions involving government spending and the inflation tax. The expression $g_T(b - e)/b$ represents the implicit tax on exports (Pinto 1989). The government sets $e$ arbitrarily, and since it does not have reserves to deplete in maintaining this exchange rate, the official exchange market is rationed by capital controls and restrictions on commercial transactions. As in Lizondo (1987a, 1987b), we assume that there is no official net foreign asset accumulation by the government; thus $\dot{R} = 0$, where $\dot{R}$ is the change in official reserves. The change in the stock of money is given by:

$$(7.9) \qquad \dot{M} = \dot{R} + \dot{D}$$

where $\dot{D}$ is the change in domestic credit and $\dot{M}$ is the change in the nominal money supply.

*Production, exports, and imports.* The model is presented in detail in appendix 7.1. Agents in the economy produce exportables ($X$) and nontradables ($N$) and trade in both the official and black markets. For smuggled exports and imports, foreign exchange is valued at its marginal cost, $b$. The higher is the black-market premium, the greater are the benefits from smuggling. However, the greater is the volume of smuggling, the greater are the risks of detection and penalties. Thus exporters and importers allocate trading between the official and unofficial markets until the marginal benefits equal the marginal costs, which include the costs of bribes to evade detection and penalties. We assume that all markets are perfectly competitive.

*Balance of payments.* We assume that agents in the economy hold only non-interest-bearing assets. The balance of payments in the economy is shown in equation 7.10:

$$(7.10) \qquad \dot{F} = P_x X - I - g_I$$

where $\dot{F}$ is the total accumulation of foreign assets.

*Consumption.* We assume that agents consume only nontradable goods and that they consume a constant fraction $\gamma$ of their wealth.[19] Wealth is defined as the sum of domestic and foreign currency:

$$(7.11) \qquad M + bF = W.$$

Therefore consumption spending is given by:

$$(7.12) \qquad P_N C_N = \gamma(M + bF).$$

*Money demand.* Money demand ($M$) is given by equation 7.13:

$$(7.13) \qquad M_d = \lambda(\dot{b}/b)W \qquad \lambda < 0$$

where $\lambda$ is a decreasing function of the expected rate of depreciation in the black market (Lizondo 1987a, 1987b). Assuming that agents possess perfect foresight, the expected and actual rates of inflation are equal and, at the steady state, are equal to the rate of depreciation of the domestic currency. Assuming that private agents can adjust their portfolio composition to the desired level instantly, equation 7.13 can be rewritten as:

(7.14) $$m = \frac{\lambda(\dot{\phi}/\phi + \dot{e}/e)}{1 - \lambda(\dot{\phi}/\phi + \dot{e}/e)} \phi F$$

where $m$ is $M/e$ and $\phi$ is $b/e$.

THE STEADY STATE. The dynamic equations of our system after substitution are:

(7.15) $$\dot{F} = P[\bar{L}_x - L_1(\phi, P_x, g_I, \dot{F})] - zg_T - I(m, F, \phi, zg_T)$$

(7.16) $$\dot{m} = g_T - t - \hat{e}m - A$$

and equation 7.14. Equations 7.14–7.16 can be solved for the steady state of the economy. The steady-state solution for the black-market premium is:

(7.17) $\phi = \phi(g_T, t, \hat{e}, A)$    $\phi g_t > 0$    $\phi_t < 0$    $\phi\hat{e} < 0$    $\phi_A < 0$.

The derivative of the black-market premium with respect to government spending is of particular interest: an increase in government expenditure will raise the black-market premium. This is because an increase in government consumption reduces the supply of dollars available for the private sector and the black-market price of foreign exchange (which is the marginal cost of foreign exchange) will be pushed up. An increase in conventional tax revenues or aid flows will reduce the premium. This is because the higher the revenues from taxes or aid, the lower will be the reliance on inflationary financing and the lower will be the money stock. To restore portfolio balance, the premium must fall. An improvement in the terms of trade will increase the supply of dollars to the private sector and therefore lower the black-market premium.

An increase in the rate of official depreciation has ambiguous effects: the direction of the effect depends on the inflation elasticity of the share of domestic money in wealth, $\lambda$. There are two distinct effects of an increase in the expected rate of depreciation $\Delta e/e$. Suppose $\Delta e/e$ is raised. The differential rate of return between cedis and U.S. dollars rises, making dollars more attractive. The greater demand for foreign exchange will tend to raise the black-market premium. However, another effect works in the opposite direction: with the real deficit given, a smaller cedi base would be required for the inflation tax. The steady-state inflation tax is given by $\Delta e/eM$, where $M$ is the nominal money supply. Suppose $\Delta e/e$ rises but the inflation elasticity of $\lambda(n)$ is less than unity. Then total inflation-tax revenues will rise, the supply of dollars will increase, and the premium will fall. If, however, $\eta$ is greater than unity, inflation-tax revenues will fall and the premium will rise.

*The effects of a devaluation.* A devaluation, by raising the nominal exchange rate, will immediately lower the real money stock, $M/e = m$. At the original premium, there will be a portfolio imbalance; agents will have a greater proportion of foreign assets in their portfolio than they desire. Portfolio balance is restored because (a) the black-market premium falls immediately to reduce the share of foreign assets in wealth, and (b) as the premium rises to its new steady-state level, the rate of inflation is higher and the desired ratio of foreign assets in wealth $(1 - \lambda)$ increases.[20] Thus a devaluation, given $g_T$, will temporarily reduce the black-market premium.

*The real exchange rate and the trade balance.* The official real exchange rate $(R)$, which depends on the black-market premium, is defined as the relative price of tradable to nontradable goods: $eP_x/P_N$. Substituting for the black-market premium, we get

$$R = R(g_T, P_X, A, t, \Delta e/e).$$

An increase in government spending appreciates the real exchange rate through two channels: (a) an increase in spending that represents partly an increase in spending on nontradables will raise the relative price of nontradables, and (b) an increase in government spending will also raise the black-market premium. This again will tend to make imported inputs more expensive and will therefore raise the price of nontradables that make use of these inputs. An increase in the price of exports, $P_X$, will depreciate the real exchange rate. An increase in the rate of depreciation will lower the premium because the inflation elasticity of the demand for money in Ghana is less than unity, and the real exchange rate will appreciate.

An increase in aid flows or tax revenues lowers the premium and thus the costs of imports and nontradables. This puts downward pressure on the price of nontradables and depreciates the real exchange rate. Note that at the steady state the trade balance is zero. If agents held interest-bearing foreign assets, private agents' accumulation of foreign assets would consist of the sum of net interest payments plus the trade surplus or deficit. F would be the current account. In this case the trade balance would not be zero at the steady state.

Although it is useful to consider the steady-state determinants of the real exchange rate and the black-market premium, a look at the non-steady-state relationships between the premium, the trade balance, the real exchange rate, and public sector expenditure is particularly useful, as in equations 7.18 and 7.19. The reason is that the analysis of the steady state assumes that the rate of inflation equals the rate of official depreciation. This obviously was not true for Ghana during the past decade and a half, when the official exchange rate was kept fixed, with some discrete devaluations, but the rate of inflation was very high.

$$(7.18) \qquad \phi = \phi(\dot{F}, m, F, z, P_x, g_T).$$

An increase in the stock of money will require an increase in the black-market premium to restore portfolio balance. An increase in the stock of foreign assets implies an increase in private sector wealth and therefore an increase in private sector consumption. The higher consumption demand raises the black-market premium. An increase in government expenditure raises the black-market premium, as before.

An improvement in the terms of trade will have an ambiguous effect on the black-market premium. On the one hand, it will tend to raise the production of tradables in relation to nontradables. On the other hand, by raising wages, it will induce substitution toward imported inputs and away from labor in the nontradable sector, and imports will rise. This may put upward pressure on the black-market premium, depending on the relative short-run responses of exports and imports, given the trade balance.

An improvement in the trade balance is accompanied by a lower premium. This is because an improvement in the trade balance implies a larger supply of foreign exchange in the black market.

$$(7.19) \qquad \dot{F} = \dot{F}(P_x, m, F, z, g_T).$$

The trade balance depends positively on the terms of trade. The reason is that an increase in the terms of trade raises export revenues at every level of exports and also raises the production of tradables by inducing agents to shift away from nontradable to tradable production. An increase in the stock of money could have either a negative or a positive effect on the trade balance. In the first case, a higher stock of money implies greater wealth for the private sector and therefore greater consumption demand for nontradables. This demand puts upward pressure on the relative price of nontradables, and agents shift production toward nontradables. Since imports are used in nontradable production, imports must rise, and the trade balance tends to deteriorate. In the second case, an increase in the stock of money held by the private sector, given the trade balance, is associated with a higher black-market premium. The higher premium reduces imports—and exports, but less so—and improves the trade balance. (This is the condition for saddlepath stability.) The result makes sense if we recall that the overall trade balance reduces to the unofficial trade balance, since by construction the official sector is always in balance. All other things being equal, an increase in public sector expenditure has a negative effect on the trade balance by raising imports.

The official real exchange rate can also be written in terms of the terms of trade, the trade balance, the stock of real money balances, the stock of foreign currency, and public sector expenditure. An improvement in the trade balance leads to an appreciation of the real exchange rate. This is because a higher trade surplus implies lower imports, higher exports, or both. If the lower imports or higher exports come at the expense of production of nontradables, the relative price of nontradables will rise, causing an appreciation of the official real exchange rate.

If total public sector expenditure increases, given that the share of imports in total expenditure is fixed, part of the increase will fall on nontradables. This will lead to real exchange rate appreciation.

An increase in the stock of real balances or of foreign currency will increase wealth and therefore consumption. This will put upward pressure on the price of nontradables and will lead to real exchange rate appreciation.

*Some empirical observations.* The effect of the devaluations and other policy changes that occurred in Ghana from 1983 on can be analyzed through this model. As discussed before, a one-shot devaluation is expected to reduce the premium temporarily by lowering the real money stock instantaneously. Thus, as expected, the black-market premium fell in 1983. In subsequent years the exchange rate regime did not change significantly. There were a number of discrete devaluations in 1983–87, tax collection methods improved, and aid flows to Ghana started to rise. In this way the government gradually switched to alternative means of financing a lower level of government expenditure. All these events would, in this model, lower the black-market premium. This in fact happened; the black-market premium started to fall in 1983 and continued to fall dramatically thereafter, as the fiscal deficit decreased (see figure 7.17).

*Unification.* In 1987 the government undertook policies aimed at unifying the official and black markets for foreign exchange. In view of the fact that the fiscal deficit had been falling, as had reliance on money financing, there was not a substantial differential between the black-market and official exchange rates. In fact, by 1989 the black-market premium had fallen dramatically.

In the context of the current model, unification could lead to an increase in the rate of inflation under certain circumstances. The higher is the black-market premium before unification, the larger is the implicit tax on exports and therefore the more important is the black-market premium as a source of revenue for the government and the higher is the postunification rate of inflation. In Ghana's case $\eta <$ 1, which implies that when the exchange rate system is unified so that

$\phi = 1$, inflation may rise. If the steady-state rate of inflation is $e$ (= $b$), with $\phi_{ss} = \phi_{ss}(e, t, A, g_T, P_x)$, the rate of inflation must rise when $\phi$ falls if all other factors remain the same. If, however, other factors change (for example, if government expenditure is reduced or the terms of trade improve, either of which would reduce the steady-state premium), unification may not lead to an increase in inflation and may even be accompanied by a reduction in inflation. In fact, the slight increase in inflation after 1986 may be partly attributable to the gradual unification of the exchange rate regime, and it is also probably true that if the government had tried to unify the exchange rate without any fiscal adjustment, inflation would have been higher in 1988–89. (Inflation was about 24 percent from 1988:3 to 1989:3.)

### Estimation of the Black-Market Premium, the Real Exchange Rate, and the Trade Surplus

THE STEADY STATE. The variables used in the regressions of the first part of this chapter are:

PREM2 = average black-market premium
PREMIUM = end-of-period black-market premium
    FD2 = public sector deficit in dollars
    TOT = terms of trade
    AID = aid flows to Ghana in U.S. dollars
EXDEP = official rate of depreciation, calculated using average exchange rates
    RER = official real exchange rate, calculated using the average official exchange rate
DV83 = a dummy variable for 1983, when the ERP began.

The black-market premium is specified as:

$$\text{PREM2} = a_0 + a_1 \text{FD2} + a_2 \text{TOT} + a_3 \text{AID} + a_4 \text{EXDEP} + a_5 \text{DV83} + e.$$

In equation (a) in table 7.7 we see that both the terms of trade and the fiscal deficit are significant in determining the premium and have the expected signs. Aid flows are not econometrically significant and have been dropped from the equation.

As expected, the fiscal deficit is related positively to the black-market premium. An increase in the fiscal deficit (some of which will be spent on tradables) raises the demand for foreign exchange and depreciates the black-market exchange rate. Therefore the black-market premium increases. The official rate of depreciation is insignificant, as is the dummy variable for 1983.[21] Equation (b) in table 7.7 shows a similar regression with the end-of-period black-market premium (PREMIUM). All the variables have the expected signs. Aid flows are insignificant and have been dropped from the equation.

**Table 7.7. Regressions for the Black-Market Premium, Ghana**

| Variable | PREM2 (a) | PREMIUM (b) |
|---|---|---|
| $a_0$ | 5.1 (3.24) | 29.119 (4.17) |
| $a_1(FD2)$ | 0.006 (8.24) | 0.015 (5.002) |
| $a_2(TOT)$ | −0.028 (−2.47) | −0.219 (−4.353) |
| $a_3(AID)$ | | |
| $a_4(EXDEP)$ | −0.00007 (−0.01) | −0.016 (−2.32) |
| $a_5(DV83)$ | 3.305 (0.46) | |
| $R^2$ | 0.8646 | 0.7425 |
| Durbin-Watson | 2.28 | 2.02 |

*Note:* Figures in parentheses are *t*-statistics. A blank denotes omission of the specific variable from the regression.

Equation (c) in table 7.8 shows the regression of the official real exchange rate against the relevant explanatory variables. The public sector deficit is significant. The expected rate of depreciation and the terms of trade are not econometrically significant determinants of the real exchange rate and have been dropped from the equation. First-differencing the relationship gives the same results, with only the fiscal deficit significant.

An increase in aid flows to Ghana is expected to depreciate the official real exchange rate, but the coefficient on aid flows is negative. We expect a positive coefficient because higher aid means lower reliance on money financing and therefore a lower steady-state money stock and a lower premium to achieve portfolio balance. The negative coefficient can be explained if we assume that part of the foreign aid flows represents transfers to the private sector. In such a case private consumption of nontradables would increase, as would the price of nontradables, and the real exchange rate would appreciate.

THE NONSTEADY STATE. Equations (d),(e) and (f) in table 7.9 show the results of the estimation of the non-steady-state relationships, with the variables defined as follows:

**Table 7.8. Regression of Official Real Exchange Rate, Ghana**

| RER | (c) |
| --- | --- |
| $b_0$ | 114.24 |
|  | (15.37) |
| $b_1(FD2)$ | −0.06 |
|  | (−8.08) |
| $b_2(\text{TOT})$ |  |
| $b_3(\text{AID})$ | −0.13 |
|  | (−3.81) |
| $b_4(\text{EXPDEP})$ |  |
| $b_5(DV83)$ | −57.26 |
|  | (−3.49) |
| $R^2$ | 0.82 |
| Durbin-Watson | 1.81 |

*Note:* Figures in parentheses are *t*-statistics. A blank denotes omission of the specific variable from the regression.

$TS2$ = trade surplus in dollars
$\text{PSD2}$ = public sector expenditure in dollars
$\text{TILMOND}$ = stock of privately held foreign assets in dollars
$\text{TOT}$ = terms of trade
$\text{MONEYD}$ = stock of money in dollars ($M_1$)
$\text{WLTHD2}$ = wealth in dollars; ($L$ denotes lagged values).[22]

Data on the stocks of foreign currency held by the private sector were not available. The data up to 1982 were taken from May's (1985) estimates of the parallel-market economy in Ghana (or the volume of illegal money, as he refers to it). The data for the stock of foreign assets after 1982 were derived using May's 1982 stock figure and the estimated change in the size of the black market in relation to the legal market for currency, based on the change in the black-market premium. It was assumed that the proportion of illegal-to-legal activities fell and rose in the same proportion as the black-market premium.

Since the data for privately held foreign assets must be subject to a great deal of measurement error, a proxy variable was initially estimated and was used in the regressions. The use of a proxy for private sector wealth did not improve the results, and the final equations presented are therefore based on the initial estimated figures.

Equation (d) in table 7.9 shows the results for the trade balance equation. Both the terms of trade and public sector expenditure are

**Table 7.9. Non-Steady-State Regression Results for the Trade Surplus, the Real Exchange Rate, and the Black-Market Premium, Ghana**

| Variable | TS2 (d) | RER (e) | BMP (f) |
|---|---|---|---|
| $c_0$ | −266.293 (−2.05) | 15.165 (14.07) | −0.218 (−0.56) |
| $c_1$(PSD2) | −0.080 (−2.09) | −0.034 (−7.59) | 0.002 (6.84) |
| $c_2$(TILMONDL) | −0.882 (−0.85) | −0.238 (−3.19) | |
| $c_3$(TOT) | 2.608 (3.05) | | |
| $c_4$(MONEYDL) | 0.076 (1.82) | | |
| $c_5$(WEALTH2) | | | 0.001 (4.19) |
| TS2 | | | −0.002 (−0.88) |
| AR(1) | −0.485 (−1.74) | 0.836 (3.31) | −0.237 (−0.89) |
| AR(2) | | −0.643 (−2.49) | |
| $R^2$ | 0.5391 | 0.9295 | 0.926 |
| Durbin-Watson | 2.36 | 2.19 | 2.04 |

*Note:* Figures in parentheses are t-statistics. A blank denotes omission of the specific variable from the regression.

significant, while private sector wealth is not. Considering that Ghana's export base is very narrow, the significance of the TOT coefficient is cause for concern over the longer term. A long-term downward trend in cocoa prices could lead to a prolonged period with a deteriorating trade balance. We see that the foreign asset component of private sector wealth is negatively correlated with the trade balance surplus, while the money stock enters with a positive sign. An increase in private sector wealth has two effects on the trade balance: it raises imports, which has a negative effect, and it raises the black-market premium, which has an overall positive effect.

Estimation of the equation for the official real exchange rate (e) shows that the public sector deficit and private sector wealth were the two most important factors affecting the official real exchange rate.[23] Public sector expenditure has the expected negative effect on the

**Figure 7.20. Trade Balance Simulations, Ghana, 1972–87**

Millions of U.S. dollars

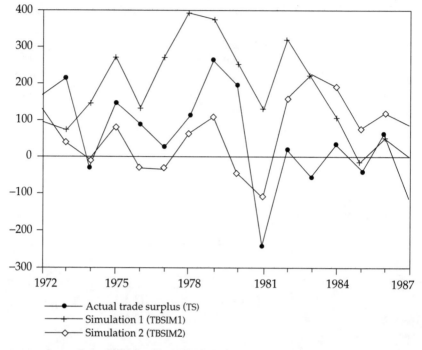

————•———— Actual trade surplus (TS)
————+———— Simulation 1 (TBSIM1)
————◇———— Simulation 2 (TBSIM2)

*Source:* Estimated from World Bank data.

official real exchange rate. Thus we can attribute some of the real exchange rate appreciation that occurred from the early part of the 1970s to 1982 to rising public sector expenditures. Private sector wealth, represented by the stock of foreign assets in this equation, also appreciates the real official exchange rate by raising domestic consumption demand.[24]

Equation (f) in table 7.9 shows that the black-market premium is significantly affected by both public sector expenditure and the stock of real wealth. As expected, both variables have a positive effect on the premium.

SOME COUNTERFACTUAL SIMULATIONS. In this section we use our regression estimates to evaluate the effects of counterfactual fiscal policies on the trade balance and the real exchange rate. We used equation (d) in table 7.9 to estimate what the trade balance would have been if public sector expenditure had been kept at its 1970 level. Figure 7.20 shows the actual trade surplus (TS2), the estimated trade surplus with public sector expenditure constant at the 1970 level

**Figure 7.21. Real Exchange Rate Simulations, Ghana, 1972–87**

Cedis per U.S. dollars

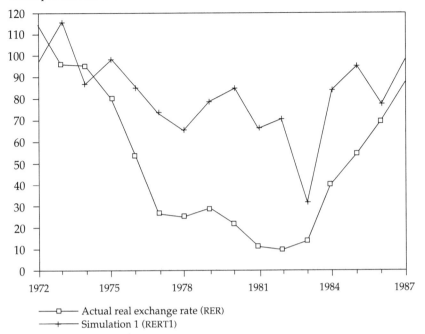

———□——— Actual real exchange rate (RER)
———+——— Simulation 1 (RERT1)

*Source:* Estimated from World Bank data.

(TBSIM1), and the estimated trade surplus with terms of trade held at the 1970 level (TBSIM2). In almost every year the trade surplus would have been substantially larger if public sector expenditure had remained constant. We see, for instance, that in 1981 instead of a $244 million deficit there would have been a $313 million surplus.

The significant negative effect of public sector expenditure on the official real exchange rate seems to indicate that if the government had allowed the nominal exchange rate to depreciate with the increase in domestic public sector expenditure, Ghanaian exports would have fared better. Figure 7.21 shows how the estimated official real exchange rate (RERT1) would have evolved if public sector expenditure had been held at its 1970 level. (These estimates are based on equation [e] of table 7.9.) In all cases in which public sector expenditure was higher, the actual (official) real exchange rate was more appreciated than the simulated value, holding public expenditure constant. We see, in fact, that the appreciation of the official exchange rate in 1972–87 would not have occurred if public sector expenditure had remained constant.

**Figure 7.22. Black-Market Premium Simulations, Ghana, 1970–87**

Premium, log scale [$(b - e)/e$]

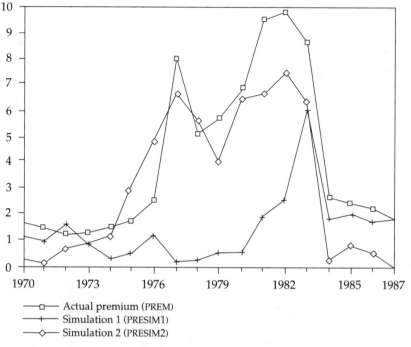

———□——— Actual premium (PREM)
———+——— Simulation 1 (PRESIM1)
———◇——— Simulation 2 (PRESIM2)

*Source:* Estimated from World Bank data.

The empirical results also show that rising public sector expenditure, along with stringent restrictions on foreign exchange transactions, did seem to lead to very high black-market premiums, which were partly responsible for the real exchange rate appreciation. Figure 7.22, which presents simulation results for two cases—one in which the terms of trade are kept fixed at their 1974 (high) level (PRESIM2) and the other in which the fiscal deficit is kept at its 1970 level (PRESIM1)—shows how in both cases the premium would have been lower than it actually was. In 1980, for example, the black-market premium (plus 1) would have been 0.54 instead of 6.9 if public sector expenditure had remained at its low 1970 level. (Estimates are based on equation [c] in table 7.8.) From 1977 to 1983 the public sector deficit seems to have been the most important factor affecting the black-market premium. Even with very favorable terms of trade (the 1974 level), the premium would have been high during these years if the fiscal deficit had remained at its actual level.

To summarize, Ghana's policy of maintaining a fixed exchange rate and the monetization of the government's budget deficit were two of the most important features of public sector policy. It is these policy measures, as well as high government expenditures, that led to an overvalued real exchange rate and a deteriorating (official) trade balance during the early and mid-1980s. The official trade balance was probably worse than the overall trade balance, since much activity was unreported.

## Conclusions

Fiscal deficits in Ghana have clearly had important effects throughout the economy, not only because of the size of the deficits but also because of the constraints that the government faced in financing the deficit. The government had little access to foreign finance, and domestic financial markets were too weak to provide the levels of financing needed. Therefore the government turned to money financing. It was the resulting high inflation, in the context of a controlled economy, that made fiscal deficits so harmful to the economy even though the inflation rate was below its long-run revenue-maximizing level.

The government's measures for countering high inflation, such as the currency conversion in 1979, had a negative impact on the demand for money in the long run and hence reduced seigniorage revenue for a given inflation rate in the long run. High inflation combined with financial sector controls meant that real interest rates were negative throughout the period. These factors precipitated a general movement toward financial disintermediation, which further reduced the government's capacity to turn to the banking system for finance. Since 1984 the public sector deficit has stabilized at a sustainable level, about 4 percent of GDP.

Fiscal deficits affected private sector consumption negatively. As the government reduced its provision of goods and services after 1976, private consumption rose sharply. Private consumption was found to have an almost one-to-one correspondence with disposable income. Import controls also reduced private consumption.

Public sector investment was found to crowd out private investment. The government's extensive credit needs and its regulation of credit allocation greatly reduced the private sector's access to credit. Over time, the investment rate fell so low as to be insufficient for replacing capital stock. This, in turn, implied reduced growth.

Perhaps the most important effect that the government's policies had on the economy was on the exchange rate. High inflation combined with the fixed exchange rate regime led to a highly overvalued exchange rate. Fiscal deficits thus contributed to deterioration in the trade balance and to a high black-market premium. As the black market grew, incentives to export through official channels fell, and so did government revenues, particularly from cocoa taxes. The differential between the official exchange rate and the black-market exchange rate created substantial incentives for rent-seeking activity, further reducing the productivity of the economy.

Fiscal deficits in the context of the control regime thus contributed extensively to Ghana's economic decline up to 1983. The ERP provided the government with much-needed external finance, reduced its reliance on money creation, and emphasized improved revenue mobilization. Most important, however, the ERP began to dismantle the extensive control regime that has been in place for so long. In reducing the deficit and removing economic controls, the government is laying the foundation for sustained growth and development in the future.

## Appendix 7.1. The Model

### Production, Exports, and Imports

Exportables and importables are traded on both official and unofficial (illegal) markets. Agents maximize the following:

$$(7.20) \qquad P_N N + e P_x X_o + b P_x X_u - b C_1(X_u) - e I_o - b I_u$$
$$- e R(I_o) - b C_2(I_u) - W(L_1 + L_o + L_u)$$

subject to:

$$(7.21) \qquad N_1^a = L^{1-a} I$$

$$(7.22) \qquad \bar{L} = L_1 + L_o + L_u$$

$$(7.23) \qquad I, L_i \geq 0$$

$$(7.24) \qquad X_i = L_i$$

where

$P_x$ = world price of exports
$P_I$ = world price of imports, normalized to unity
$P_N$ = price of nontraded goods
$X$ = total exports
$I$ = total imports
$I_o$ = imports through official channels
$X_o$ = exports through official channels
$C_1(X_u)$ = cost of smuggling exports
$C_2(I_u)$ = cost of smuggling imports
$R(I_o)$ = cost of importing through official channels, including rent-seeking

$L_1$ = labor employed in the production of nontradable goods.
$L_o$ = labor employed in the production of exports going through official channels
$L_u$ = labor employed in the production of exports going through illegal channels.

The cost functions have the following properties:

(7.25)   $C_1'(X_u) > 0$      $C_1''(X_u) > 0$      $R'(I_o) > 0$      $C_2'(I_u) > 0$

$$C_2''(I_u) > 0 \qquad R_o''(I) > 0.$$

The first-order conditions for the maximization problem are:

(7.26)      $\delta/\delta L_1 = 0 \Rightarrow aP_N \ N = WL_1$

(7.27)      $\delta/\delta I = 0 \Rightarrow P_N \ N = [b/(1-a)][1 + C_2'(I - I_o)]I$

(7.28)      $\delta/\delta I_o = 0 \Rightarrow (\phi - 1) = R' - \phi C_2'$

(7.29)      $\delta/\delta Lu = 0 \Rightarrow -eP_x + bP_x - bC_1'(X_u) \geq 0$

or

$$X_u = X_u(\phi, P_x)$$

(7.30)      $\delta/\delta L_o = 0 \Rightarrow eP_x = w.$

## Balance of Payments

The balance of payments equation in the text can be rewritten as:

(7.31)            $\dot{F} = P_x X - I - g = P_x(\bar{L} - L_1) - I - g_I.$

Equations 7.31 and 7.23, and first-order conditions 7.26 and 7.27 can be used to solve for $L_1$ and $I$ in terms of $F$, $g$, $I$, $P_x$, and $\phi$:

(7.32)                        $L_1 = L_1(Px, g_I, \phi, \dot{F})$

(7.33)                        $I = I(Px, g_I, \phi, \dot{F}).$

## The Real Exchange Rate and the Trade Balance

The official real exchange rate, $eP_x/P_N$, can be rewritten as $a(L_1^{q-1}/I)$, since $P_N = eP_x L_1/aN = eP_x/a(L_1/I)^{1-a}$.
  The black-market real exchange rate is defined as:

(7.34)        $bP_x/P_N = (bP_x a/eP_x)(L_1^{(a-1)}/I) = \phi a[-L_1^{(a-1)}/I].$

Using equation 7.15 and first-order conditions 7.26 and 7.27, we can express the trade balance as:

(7.35)      $\dot{F} = P_x[\bar{L} - L_1 (\phi, P_x, \dot{F}, g_T)] - gz - I(m, F, \phi, z, g_T).$

## Appendix 7.2. Dynamics

The system of equations $(\dot{F}, \dot{m}, \dot{f})$ is linearized around the steady state:

$$
\begin{bmatrix} \dot{F} \\[2mm] \dot{m} \\[2mm] \dot{\phi} \end{bmatrix} = \begin{bmatrix} -\dfrac{\partial I}{\partial F} & \dfrac{-\partial I}{\partial m} & \dot{F}_\phi \\[3mm] 0 & -\hat{e} & 0 \\[3mm] \dot{\phi}_F & \dot{\phi}_m & \dot{\phi}_\phi \end{bmatrix} \begin{bmatrix} dF \\[2mm] dm \\[2mm] d\phi \end{bmatrix}
$$

$$
\dot{\phi} = P_x \frac{\partial L_1}{\partial \phi} \frac{\partial I}{\partial \phi} \gtreqqless 0
$$

$$
\dot{\phi}_F = -\frac{\lambda(1-\lambda)}{\lambda'} \frac{\phi}{F} > 0
$$

$$
\dot{\phi}_m = \frac{(1-\lambda)^2}{\lambda'} < 0
$$

$$
\dot{\phi}_\phi = \frac{-\lambda(1-\lambda)}{\lambda'} > 0.
$$

For saddlepath stability, the determinant of the preceding matrix must be greater than 0. A sufficient condition for saddlepath stability is that $\dot{F}_\phi > 0$.

**Appendix 7.3. Time-Series Data for Principal Variables, Ghana, 1961–90**

| Item | 1961 | 1962 | 1963 | 1964 | 1965 | 1966 | 1967 | 1968 | 1969 |
|---|---|---|---|---|---|---|---|---|---|
| *Aggregate variables* | | | | | | | | | |
| GDP (millions of current cedis) | 932 | 997 | 1,101 | 1,237 | 1,466 | 1,518 | 1,504 | 1,700 | 2,001 |
| Growth of real GDP | 6.17 | 4.83 | 3.47 | 2.15 | 1.36 | 0.09 | -3.00 | 6.43 | 5.88 |
| Growth of GDP per capita | 5.09 | 3.62 | 2.29 | -3.23 | -3.09 | -2.06 | -5.04 | 4.11 | 3.62 |
| Inflation[a] | 6.25 | 5.88 | 5.56 | 15.79 | 22.73 | 14.81 | -9.68 | 10.71 | 6.45 |
| Nominal discount rate[b] | 4.5 | 4.5 | 4.5 | 4.5 | 4.5 | 7.0 | 6.0 | 5.5 | 5.5 |
| Nominal lending rate[b] | 8.25 | 8.25 | 8.25 | 8.25 | 8.25 | 8.25 | 8.5 | 8.5 | 8.5 |
| Nominal deposit rate[b] | 3.25 | 3.25 | 3.25 | 3.25 | 3.25 | 3.375 | 3.375 | 3.375 | 3.375 |
| Real discount rate[b] | -1.65 | -1.31 | -1.00 | -9.75 | -14.85 | -6.81 | 16.80 | -4.71 | -0.89 |
| Real lending rate[b] | 1.88 | 2.24 | 2.55 | -6.51 | -11.80 | -5.72 | 20.13 | -2.00 | 1.92 |
| Real deposit rate[b] | -2.82 | -2.49 | -2.18 | -10.83 | -15.87 | -9.96 | 14.45 | -6.63 | -2.89 |
| Official exchange rate[c] | 0.71 | 0.71 | 0.71 | 0.71 | 0.71 | 0.71 | 0.86 | 1.02 | 1.02 |
| Real exchange rate[c] | 117.02 | 110.83 | 104.70 | 90.43 | 75.15 | 67.64 | 90.54 | 99.23 | 96.88 |
| Real black-market exchange rate[c] | 121.23 | 155.17 | 168.57 | 149.38 | 148.33 | 130.67 | 186.06 | 180.87 | 160.46 |
| Black-market premium[c] | 1.04 | 1.40 | 1.61 | 1.65 | 1.97 | 1.93 | 2.06 | 1.82 | 1.66 |

*(Table continues on the following page.)*

**Appendix 7.3** *(continued)*

| Item | 1961 | 1962 | 1963 | 1964 | 1965 | 1966 | 1967 | 1968 | 1969 |
|---|---|---|---|---|---|---|---|---|---|
| *Composition of output (percentage of GDP)* | | | | | | | | | |
| Private consumption | 78.6 | 75.8 | 75.8 | 72.7 | 77.3 | 79.1 | 77.2 | 72.6 | 74.6 |
| Public consumption | 10.8 | 11.1 | 11.4 | 11.8 | 14.5 | 13.0 | 15.0 | 16.8 | 14.2 |
| Gross fixed capital formation | 20.5 | 16.8 | 18.0 | 17.1 | 18.1 | 12.9 | 11.6 | 11.0 | 9.8 |
| Change in stocks | -1.9 | -1.0 | -0.6 | 1.0 | -0.2 | -0.1 | -1.3 | 0.1 | 2.0 |
| Exports | 23.9 | 21.9 | 19.4 | 18.2 | 17.1 | 14.6 | 17.4 | 20.3 | 19.7 |
| Imports | 31.9 | 24.7 | 24.0 | 20.9 | 26.7 | 19.6 | 19.9 | 20.8 | 20.3 |
| *Memoranda:* | | | | | | | | | |
| Private fixed investment | 0.0 | 0.0 | 0.0 | 0.0 | 5.2 | 6.5 | 5.3 | 6.9 | 8.5 |
| Public fixed investment[d] | 0.0 | 0.0 | 0.0 | 0.0 | 12.7 | 6.3 | 5.0 | 4.2 | 3.2 |
| *Monetary summary (millions of cedi, end of period)[e]* | | | | | | | | | |
| Net foreign assets | — | — | — | — | — | — | -60 | -67 | -77 |
| Net domestic assets | — | — | — | — | — | — | 468 | 488 | 513 |
| Claims on government | — | — | — | — | — | — | 251 | 271 | 255 |
| Claims on public entities | — | — | — | — | — | — | 25 | 25 | 47 |
| Cocoa financing | — | — | — | — | — | — | 128 | 126 | 120 |
| Claims on private sector | — | — | — | — | — | — | 64 | 67 | 92 |
| Net other assets | — | — | — | — | — | — | 0 | 0 | 0 |

| | | | | | | | | | |
|---|---|---|---|---|---|---|---|---|---|
| Revaluation account | — | — | — | — | — | — | — | — | — |
| $M_1$ (percentage of GDP) | 14.66 | 15.53 | 14.06 | 17.69 | 16.37 | 16.23 | 15.96 | 15.17 | 14.42 |
| Quasi money (percentage of GDP) | 2.93 | 3.65 | 3.31 | 3.68 | 4.09 | 4.43 | 5.20 | 5.48 | 4.95 |
| *Balance of payments (millions of U.S. dollars)* | | | | | | | | | |
| Trade balance | −52.2 | 9.8 | −29.6 | −0.1 | −118.2 | −40.4 | 18.8 | 39.8 | 49.7 |
| Current account | −103.5 | −33.7 | −80.1 | −55.4 | −222.9 | −127.8 | −84.9 | −56.1 | −60.1 |
| Capital account | 86.1 | 45.9 | 51.3 | 48.3 | 228.3 | 77.7 | 8.0 | 48.3 | 67.2 |
| Errors and omissions | −4.4 | −8.1 | −12.0 | 2.9 | 9.8 | 2.0 | 4.6 | 1.3 | −34.0 |
| Position above the line | −21.8 | 4.1 | −40.8 | −4.2 | 15.2 | −48.1 | −72.3 | −6.5 | −26.9 |
| Total change in reserves | 21.8 | −4.1 | 40.8 | 4.2 | −15.2 | 48.1 | 49.0 | −1.5 | 18.6 |
| *Central government accounts (percentage of GDP)*ᶠ | | | | | | | | | |
| Central government revenue | 16.41 | 15.69 | 19.35 | 19.27 | 15.16 | 15.98 | 18.75 | 15.36 | 16.92 |
| Central government expenditure | 25.91 | 25.3 | 24.96 | 24.66 | 17.68 | 21.63 | 23.1 | 19.35 | 20.6 |
| Central government deficit | −9.50 | −9.61 | −5.61 | −5.39 | −2.52 | −5.65 | −4.35 | −3.99 | −3.68 |
| Domestic financing | — | — | — | — | — | 1.82 | 3.41 | 2.74 | 2.58 |
| External financing | — | — | — | — | — | 4.35 | 1.38 | 0.89 | 1.49 |
| Other | — | — | — | — | — | −0.52 | −0.44 | 0.36 | −0.39 |

(Table continues on the following page with years 1970–79.)

**Appendix 7.3** (continued)

| Item | 1970 | 1971 | 1972 | 1973 | 1974 | 1975 | 1976 | 1977 | 1978 | 1979 |
|---|---|---|---|---|---|---|---|---|---|---|
| *Aggregate variables* | | | | | | | | | | |
| GDP (millions of current cedis) | 2,259 | 2,501 | 2,815 | 3,501 | 4,660 | 5,283 | 6,526 | 11,163 | 20,986 | 28,222 |
| Growth of real GDP | 6.76 | 5.56 | -2.49 | 15.25 | 3.39 | -12.86 | -3.52 | 2.26 | 8.48 | -3.17 |
| Growth of GDP per capita | 4.65 | 2.58 | -4.96 | 11.57 | 1.03 | -15.16 | -7.64 | 1.27 | 5.05 | -0.67 |
| Inflation[a] | 3.03 | 8.82 | 10.81 | 17.07 | 18.75 | 29.82 | 55.41 | 116.52 | 73.09 | 54.52 |
| Nominal discount rate[b] | 5.5 | 8.0 | 8.0 | 6.0 | 6.0 | 8.0 | 8.0 | 8.0 | 13.5 | 13.5 |
| Nominal lending rate[b] | 8.5 | 13.0 | 13.0 | 10.0 | 10.0 | 12.5 | 12.5 | 12.5 | 18.5 | 18.5 |
| Nominal deposit rate[b] | 3.38 | 8.0 | 8.0 | 5.5 | 5.5 | 8.0 | 8.0 | 8.0 | 13.0 | 13.0 |
| Real discount rate[b] | 2.40 | -0.76 | -2.54 | -9.46 | -10.74 | -16.81 | -30.50 | -50.12 | -34.43 | -26.55 |
| Real lending rate[b] | 5.31 | 3.84 | 1.98 | -6.04 | -7.37 | -13.34 | -27.61 | -48.04 | -31.54 | -23.31 |
| Real deposit rate[b] | 0.33 | -0.76 | -2.54 | -9.89 | -11.16 | -16.81 | -30.50 | -50.12 | -34.72 | -26.87 |
| Official exchange rate[c] | 1.02 | 1.03 | 1.33 | 1.16 | 1.15 | 1.15 | 1.15 | 1.15 | 1.76 | 2.75 |
| Real exchange rate[c] | 97.80 | 93.55 | 114.35 | 96.05 | 95.36 | 80.24 | 53.80 | 26.38 | 25.19 | 28.62 |
| Real black-market exchange rate[c] | 166.20 | 137.77 | 143.73 | 123.84 | 143.12 | 138.71 | 135.95 | 211.06 | 127.90 | 161.96 |
| Black-market premium[c] | 1.70 | 1.47 | 1.26 | 1.29 | 1.50 | 1.73 | 2.53 | 8.00 | 5.08 | 5.66 |

*Composition of output (percentage of GDP)*

| | | | | | | | | | | |
|---|---|---|---|---|---|---|---|---|---|---|
| Private consumption | 74.4 | 77.4 | 74.8 | 75.0 | 78.2 | 73.3 | 79.2 | 77.4 | 84.7 | 83.1 |
| Public consumption | 12.8 | 13.0 | 12.6 | 10.9 | 12.2 | 13.0 | 12.2 | 12.6 | 11.3 | 10.3 |
| Gross fixed capital formation | 12.0 | 12.4 | 8.7 | 7.7 | 11.9 | 11.6 | 9.8 | 9.4 | 5.1 | 6.7 |
| Change in stocks | 2.1 | 1.7 | -1.6 | 1.4 | 1.1 | 1.1 | -1.0 | 1.7 | 0.3 | -0.2 |
| Exports | 21.3 | 15.8 | 20.7 | 21.4 | 18.3 | 19.4 | 15.7 | 10.5 | 8.4 | 11.2 |
| Imports | 22.7 | 20.2 | 15.2 | 16.4 | 21.8 | 18.4 | 16.0 | 11.6 | 9.7 | 11.2 |
| *Memoranda:* | | | | | | | | | | |
| Private fixed investment | 10.4 | 9.8 | 3.4 | 6.3 | 9.4 | 7.3 | 2.1 | 5.4 | 2.1 | 3.8 |
| Public fixed investment$^d$ | 3.7 | 4.3 | 3.7 | 2.7 | 3.6 | 5.4 | 6.8 | 5.7 | 3.3 | 2.7 |
| *Monetary summary (millions of cedi, end of period)*$^e$ | | | | | | | | | | |
| Net foreign assets | -57 | -149 | -46 | 98 | -91 | 49 | -157 | -149 | -898 | -529 |
| Net domestic assets | 514 | 701 | 799 | 839 | 1,215 | 1,560 | 2,106 | 3,229 | 6,056 | 6,540 |
| Claims on government | 295 | 283 | 349 | 372 | 570 | 903 | 1,584 | 2,780 | 4,500 | 4,906 |
| Claims on public entities | -12 | 108 | 107 | 122 | 196 | 101 | 124 | 141 | 234 | 290 |
| Cocoa financing | 97 | 152 | 170 | 119 | 175 | 188 | 259 | 271 | 867 | 1,340 |
| Claims on private sector | 127 | 175 | 174 | 207 | 294 | 365 | 395 | 572 | 751 | 799 |
| Net other assets | 7 | -17 | -1 | 19 | -554 | 3 | -256 | -535 | -296 | -795 |
| Revaluation account | — | — | — | — | — | — | — | — | — | — |
| $M_1$ (percentage of GDP) | 13.35 | 12.71 | 16.31 | 15.30 | 14.08 | 18.76 | 21.96 | 21.37 | 19.59 | 16.59 |

*(Table continues on the following page.)*

**Appendix 7.3** (continued)

| Item | 1970 | 1971 | 1972 | 1973 | 1974 | 1975 | 1976 | 1977 | 1978 | 1979 |
|---|---|---|---|---|---|---|---|---|---|---|
| Quasi money (percentage of GDP) | 5.37 | 6.14 | 7.29 | 6.56 | 6.60 | 7.15 | 7.26 | 5.83 | 4.79 | 4.48 |
| *Balance of payments (millions of U.S. dollars)* | | | | | | | | | | |
| Trade balance | 51.9 | -33.6 | 161.4 | 212.9 | -29.2 | 150.4 | 88.8 | 29.4 | 112.5 | 262.6 |
| Current account | -67.7 | -145.8 | 108.2 | 126.7 | -171.5 | 17.6 | -74.0 | -79.7 | -45.9 | 122.0 |
| Capital account | 53.6 | 97.2 | -80.6 | -12.6 | 40.6 | 82.4 | -36.5 | 59.0 | 103.0 | 19.8 |
| Errors and omissions | 16.6 | 14.0 | 5.4 | -9.5 | -14.1 | 6.3 | -26.7 | 12.3 | -119.7 | -106.0 |
| Position above the line | 2.5 | -34.7 | 33.1 | 104.6 | -144.9 | 106.3 | -137.3 | -8.4 | -62.7 | 35.8 |
| Total change in reserves | 8.0 | -25.0 | -77.7 | -86.8 | 83.3 | -10.6 | 59.7 | -61.1 | -117.5 | 51.9 |
| *Central government accounts (percentage of GDP)*$^f$ | | | | | | | | | | |
| Central government revenue | 20.43 | 16.36 | 12.39 | 14.31 | 16.18 | 13.79 | 12.15 | 9.57 | 8.89 | 8.52 |
| Central government expenditure | 19.68 | 22.10 | 19.02 | 19.38 | 25.89 | 28.98 | 24.82 | 20.49 | 17.65 | 13.47 |
| Central government deficit | 0.75 | -5.74 | -6.63 | -5.07 | -9.71 | -15.19 | -12.67 | -10.92 | -8.76 | -4.95 |
| Domestic financing | -2.39 | 1.98 | 2.79 | 5.75 | 4.62 | 17.95 | 10.64 | 11.84 | 4.25 | 4.13 |
| External financing | 1.46 | 0.63 | 1.04 | -0.34 | 0.05 | -0.01 | 0.06 | -0.04 | -0.03 | 0.82 |
| Other | 0.18 | 3.13 | 2.80 | -0.34 | 5.04 | -2.74 | 1.97 | -0.89 | 4.55 | 0.01 |

| Item | 1980 | 1981 | 1982 | 1983 | 1984 | 1985 | 1986 | 1987 | 1988g | 1989g | 1990g |
|---|---|---|---|---|---|---|---|---|---|---|---|
| *Aggregate variables* | | | | | | | | | | | |
| GDP (millions of current cedis) | 42,852 | 72,626 | 86,451 | 184,038 | 270,561 | 343,048 | 511,373 | 746,000 | 1,051,196 | 1,417,214 | 1,956,090 |
| Growth of real GDP | 0.00 | -1.79 | -7.20 | 0.70 | 2.64 | 5.09 | 5.20 | 4.80 | 5.60 | 5.11 | 3.00 |
| Growth of GDP per capita | -2.33 | -4.81 | -10.43 | -3.10 | -1.26 | 2.37 | 2.54 | 2.14 | 0.66 | 1.34 | 0.38 |
| Inflation[a] | 50.15 | 116.50 | 22.31 | 122.85 | 39.65 | 10.31 | 24.56 | 39.81 | 31.40 | 25.20 | 37.20 |
| Nominal discount rate[b] | 13.5 | 19.5 | 10.5 | 14.5 | 18.0 | 18.5 | 20.5 | 23.5 | 26.0 | 26.0 | 33.0 |
| Nominal lending rate[b] | 18.0 | 18.0 | 18.0 | 19.0 | 22.5 | 23.0 | 23.0 | 26.0 | 30.3 | 30.3 | 30.3 |
| Nominal deposit rate[b] | 13.0 | 19.0 | 9.0 | 12.5 | 16.0 | 18.0 | 20.0 | 22.3 | 22.0 | 20.0 | 21.5 |
| Real discount rate[b] | -24.41 | -44.80 | -9.66 | -48.62 | -15.51 | 7.42 | -3.26 | -11.67 | -4.11 | 0.64 | -3.06 |
| Real lending rate[b] | -21.41 | -45.50 | -3.52 | -46.60 | -12.28 | 11.50 | -1.25 | -9.88 | -0.84 | 4.07 | -5.03 |
| Real deposit rate[b] | -24.74 | -45.03 | -10.88 | -49.52 | -16.94 | 6.97 | -3.66 | -12.56 | -7.15 | -4.15 | -11.44 |
| Official exchange rate[c] | 2.75 | 2.75 | 2.75 | 8.83 | 35.99 | 54.36 | 89.20 | 153.73 | 202.35 | 270.00 | 326.33 |
| Real exchange rate[c] | 21.77 | 10.97 | 9.15 | 13.35 | 39.89 | 54.37 | 69.53 | 87.96 | 91.61 | 102.53 | 93.56 |
| Real black-market exchange rate[c] | 150.40 | 104.73 | 205.09 | 115.79 | 107.14 | 131.25 | 144.12 | 124.74 | 114.08 | 119.24 | — |
| Black-market premium[c] | 6.90 | 9.55 | 22.43 | 8.67 | 2.69 | 2.41 | 2.07 | 1.42 | 1.24 | 1.16 | — |
| *Composition of output (percentage of GDP)* | | | | | | | | | | | |
| Private consumption | 83.9 | 87.2 | 89.8 | 90.8 | 86.1 | 83.0 | 81.2 | 81.8 | 79.4 | 83.7 | 85.2 |
| Public consumption | 11.2 | 8.8 | 6.5 | 8.6 | 7.3 | 9.4 | 11.1 | 10.0 | 10.0 | 10.3 | 8.1 |
| Gross fixed capital formation | 6.1 | 4.7 | 3.5 | 3.8 | 6.9 | 9.5 | 9.3 | 10.4 | 10.9 | 13.5 | 15.8 |
| Change in stocks | -0.5 | -0.2 | -0.2 | 0.0 | 0.0 | 0.0 | 0.1 | 0.1 | 0.1 | 0.1 | 0.1 |
| Exports | 8.5 | 4.8 | 3.3 | 6.1 | 7.5 | 9.7 | 16.0 | 21.2 | 20.7 | 20.6 | 16.7 |
| Imports | 9.2 | 5.3 | 3.0 | 9.3 | 7.7 | 11.6 | 17.7 | 23.4 | 21.0 | 28.1 | 25.8 |

*(Table continues on the following page.)*

363

**Appendix 7.3** (continued)

| Item | 1980 | 1981 | 1982 | 1983 | 1984 | 1985 | 1986 | 1987 | 1988g | 1989g | 1990g |
|---|---|---|---|---|---|---|---|---|---|---|---|
| *Memoranda:* | | | | | | | | | | | |
| Private fixed investment | 4.2 | 2.7 | 2.3 | 2.9 | 4.4 | 5.4 | 2.0 | 2.5 | 2.9 | 5.6 | 8.4 |
| Public fixed investment[d] | 1.4 | 1.9 | 1.1 | 0.9 | 2.5 | 4.2 | 7.3 | 7.9 | 8.0 | 7.9 | 7.4 |
| *Monetary summary (millions of cedi, end of period)[e]* | | | | | | | | | | | |
| Net foreign assets | −358 | −1,160 | −1,024 | −16,676 | −31,440 | −48,702 | −130,656 | −141,707 | −141,303 | −127,903 | −86,376 |
| Net domestic assets | 8,404 | 12,943 | 16,004 | 22,601 | 39,237 | 66,464 | 94,590 | 105,335 | 116,631 | 102,918 | 76,149 |
| Claims on government | 6,526 | 10,655 | 11,064 | 21,059 | 24,170 | 27,176 | 29,647 | 22,220 | 11,008 | −10,454 | −34,012 |
| Claims on public entities | 400 | 441 | 527 | 813 | 676 | 4,795 | 5,274 | 8,610 | 10,422 | 19,674 | 24,592 |
| Cocoa financing | 1,559 | 2,950 | 5,553 | 521 | 3,580 | 13,545 | 16,889 | 16,471 | 21,000 | 23,747 | 17,058 |
| Claims on private sector | 943 | 1,345 | 1,562 | 2,841 | 12,153 | 21,178 | 37,455 | 46,946 | 58,183 | 61,140 | 80,218 |
| Net other assets | −1,024 | −2,448 | −2,702 | −2,633 | −1,342 | −230 | 5,325 | 11,088 | 16,018 | 8,811 | −11,707 |
| Revaluation account | — | — | — | 16,158 | 29,930 | 41,884 | 133,135 | 179,768 | 224,990 | 276,189 | 332,332 |
| $M_1$ (percentage of GDP) | 14.20 | 12.96 | 12.96 | 8.94 | 10.12 | 12.81 | 12.87 | 12.74 | 13.23 | 13.06 | 11.09 |
| Quasi money (percentage of GDP) | 4.32 | 3.06 | 4.20 | 2.22 | 2.68 | 3.40 | 3.93 | 4.99 | 4.77 | 3.88 | 3.39 |
| *Balance of payments (millions of U.S. dollars)* | | | | | | | | | | | |
| Trade balance | 195.3 | −243.6 | 18.3 | −60.6 | 32.9 | −36.3 | 60.9 | −124.7 | −112.4 | −191.8 | — |
| Current account | 29.2 | −420.8 | −108.6 | −174.1 | −38.8 | −134.2 | −43.0 | −96.9 | −65.8 | −97.5 | — |
| Capital account | 40.9 | 108.1 | 123.6 | 119.1 | 206.9 | 84.9 | 63.4 | 255.7 | 209.0 | 236 | — |
| Errors and omissions | −100.4 | 24.0 | −32.8 | −125.8 | −132.5 | 63.4 | −81.2 | −18.7 | 37.9 | 17.1 | — |
| Position above the line | −30.3 | −288.7 | −17.8 | −180.8 | 35.6 | 14.1 | −60.8 | 140.1 | 181.1 | 155.6 | — |
| Total change in reserves | 93.4 | −16.1 | −23.8 | 246.4 | 7.3 | 54.8 | 24.1 | 394.0 | −76.2 | −127.4 | — |

*Central government accounts (percentage of GDP)*[f]

| | | | | | | | | | | | |
|---|---|---|---|---|---|---|---|---|---|---|---|
| Central government revenue | 5.68 | 5.72 | 5.37 | 5.58 | 8.37 | 11.75 | 14.40 | 14.89 | 14.63 | 15.14 | 13.68 |
| Central government expenditure | 13.76 | 11.57 | 11.72 | 8.27 | 10.16 | 13.96 | 14.34 | 14.34 | 14.26 | 14.41 | 13.51 |
| Central government deficit | -8.08 | -5.85 | -6.35 | -2.69 | -1.79 | -2.21 | 0.06 | 0.55 | 0.37 | 0.73 | 0.17 |
| Domestic financing | 7.32 | 7.06 | 5.65 | 2.16 | 1.12 | 1.18 | 1.04 | -0.39 | -0.59 | -1.08 | -1.39 |
| External financing | 0.83 | 0.00 | 0.25 | 0.53 | 0.67 | 1.03 | -1.10 | -0.16 | 0.21 | 0.35 | 1.22 |
| Other | 0.00 | -1.20 | 0.45 | — | — | — | — | — | — | — | — |

— Not available.

a. Annual change in consumer price index.

b. The discount rate is the Central Bank rate, the lending rate used is that for unsecured loans, and the deposit rate used is that on twelve-month deposits. Real rates are calculated as $[(1 + i)/(1 + p) - 1]100$, where $i$ is the nominal interest rate and $p$ is the CPI inflation rate. Interest rate data are from the *Quarterly Digest of Statistics* (QDS) and World Bank data.

c. The official exchange rate given is cedis per U.S. dollar (period average). The real exchange rate is calculated as the cedi/dollar rate multiplied by the U.S. wholesale price index (1985 = 100) and divided by the Ghanaian CPI (1985 = 100). A fall in the real exchange rate indicates an appreciation, and a rise indicates a depreciation. The real black-market exchange rate is the black-market cedis per U.S. dollar rate (from Pick's currency yearbook) multiplied by the U.S. wholesale price index and divided by the Ghanaian CPI. The black-market premium is the ratio of the black-market rate to the official rate.

d. Data prior to 1965 are from the 1958 System of National Accounts. Prior to 1981, estimates of private and public fixed investment are from Huq (1989); thereafter they are based on World Bank estimates.

e. Monetary data are from World Bank data and include secondary banks after 1983.

f. Data for 1961–81 are for fiscal years (that is, 1962 represents the 1962/63 fiscal year). Data for 1961 and 1963 represent fifteen-month periods. Data for 1961 through 1965 are from Killick (1978), p. 150. Data for all other years are from World Bank data and the *Quarterly Digest of Statistics*.

g. Data for 1988 through 1990 are subject to revision.

**Appendix 7.4. Central Government Revenue Lost as a Result of Smuggling, Ghana, 1960/61 through 1981/82**

| Fiscal year | Estimated total smuggled cocoa (thousands of metric tons) | FOB price (new cedis per metric ton) | CMB costs (new cedis per metric ton) | Official producer price (new cedis per metric ton) | Revenue to government (new cedis per metric ton)[a] | Revenue lost through smuggling (thousands of metric tons)[b] | Revenue lost through smuggling (percentage of GDP) |
|---|---|---|---|---|---|---|---|
| 1960/61 | 10 | 342 | 35 | 224 | 83 | 855 | 0.1 |
| 1961/62 | 9 | 318 | 35 | 224 | 59 | 555 | 0.1 |
| 1962/63 | -3 | 337 | 39 | 220 | 78 | -226 | 0.0 |
| 1963/64 | -12 | 357 | 43 | 202 | 112 | -1,344 | -0.1 |
| 1964/65 | -9 | 278 | 42 | 187 | 49 | -451 | 0.0 |
| 1965/66 | 1 | 262 | 43 | 187 | 32 | 42 | 0.0 |
| 1966/67 | 57 | 396 | 47 | 224 | 125 | 7,138 | 0.5 |
| 1967/68 | 47 | 562 | 46 | 254 | 262 | 12,262 | 0.8 |
| 1968/69 | 27 | 761 | 49 | 284 | 428 | 11,342 | 0.6 |
| 1969/70 | 34 | 818 | 45 | 293 | 480 | 16,272 | 0.8 |
| 1970/71 | 34 | 643 | 45 | 293 | 305 | 10,462 | 0.4 |
| 1971/72 | 32 | 688 | 88 | 293 | 307 | 9,947 | 0.4 |
| 1972/73 | 22 | 824 | 146 | 366 | 312 | 6,833 | 0.2 |
| 1973/74 | 14 | 1,294 | 211 | 439 | 644 | 9,080 | 0.2 |
| 1974/75 | 27 | 1,688 | 314 | 489 | 885 | 23,541 | 0.5 |
| 1975/76 | 34 | 1,526 | 397 | 585 | 544 | 18,714 | 0.3 |
| 1976/77 | 35 | 2,596 | 336 | 732 | 1,528 | 53,480 | 0.6 |
| 1977/78 | 64 | 3,942 | 1,071 | 1,333 | 1,538 | 98,893 | 0.6 |
| 1978/79 | 61 | 10,396 | 1,492 | 2,667 | 6,237 | 379,210 | 1.5 |
| 1979/80 | 56 | 9,120 | 2,120 | 4,000 | 3,000 | 168,000 | 0.5 |
| 1980/81 | 62 | 6,300 | 3,573 | 4,000 | -1,273 | 0 | 0.0 |
| 1981/82 | 46 | 5,000 | 5,200 | 12,000 | -12,200 | 0 | 0.0 |

Note: FOB, free on board; CMB, Cocoa Marketing Board.
a. FOB price minus CMB costs minus official producer price.
b. Revenue to government per metric ton (column 5) multiplied by total smuggled cocoa (column 1).
Source: Stryker 1990, pp. 266, 308, 313.

**Appendix 7.5. Decomposition of the Changes in the Central Government Deficit According to Changes in Economic and Policy Variables, Ghana, 1972/73 through 1988**

(ratio to GDP)

| Item | 1972/73 | 1973/74 | 1974/75 | 1975/76 | 1976/77 | 1977/78 | 1978/79 |
|---|---|---|---|---|---|---|---|
| Changes attributable to domestic variables | 0.000 | −0.003 | 0.008 | 0.015 | −0.016 | −0.007 | 0.006 |
| Real GDP ($\hat{Y}$) | −0.001 | −0.011 | 0.003 | 0.006 | −0.026 | −0.011 | 0.013 |
| Exports ($\widehat{EX}$) | 0.003 | −0.004 | −0.005 | 0.002 | −0.003 | −0.008 | −0.003 |
| Imports ($\widehat{IM}$) | −0.001 | 0.008 | 0.002 | −0.002 | 0.001 | 0.000 | −0.003 |
| Domestic nominal interest rate ($di$) | −0.001 | −0.002 | 0.004 | 0.002 | 0.000 | 0.005 | 0.003 |
| Domestic inflation ($d\hat{P}$) | 0.000 | 0.001 | 0.001 | 0.000 | 0.003 | 0.003 | −0.002 |
| Real exchange rate ($R\hat{E}R$) | 0.001 | 0.006 | 0.004 | 0.007 | 0.008 | 0.004 | 0.000 |
| Changes attributable to foreign variables | 0.002 | 0.001 | 0.000 | −0.001 | 0.000 | 0.000 | 0.001 |
| Foreign nominal interest rate ($di^*$) | 0.002 | 0.001 | 0.000 | −0.001 | 0.000 | 0.000 | 0.001 |
| Changes attributable to fiscal policy variables | −0.007 | 0.020 | 0.006 | 0.055 | −0.021 | −0.030 | −0.019 |
| Foreign debt ($D^*/\hat{P}^*$) | 0.001 | 0.000 | 0.000 | 0.000 | 0.000 | 0.000 | 0.000 |
| Domestic debt ($D/\hat{P}$) | 0.000 | 0.000 | 0.000 | 0.002 | 0.001 | −0.001 | −0.002 |
| Wage bill ($WB/\hat{P}$) | 0.003 | −0.001 | 0.013 | −0.007 | −0.021 | 0.003 | −0.001 |
| Goods and services ($GS/\hat{P}$) | −0.004 | 0.019 | −0.002 | 0.004 | −0.012 | −0.001 | −0.010 |
| Transfers and subsidies ($TS/\hat{P}$) | 0.005 | 0.001 | 0.006 | 0.026 | −0.013 | −0.006 | −0.015 |
| Investment ($I/\hat{P}$) | 0.005 | 0.001 | −0.007 | 0.003 | 0.017 | −0.013 | 0.013 |
| Policy dummy: direct taxes ($dD7478$) | 0.000 | 0.000 | 0.003 | 0.000 | 0.000 | 0.000 | −0.003 |
| Policy dummy: post-1984 direct taxes ($dD84$) | 0.000 | 0.000 | 0.000 | 0.000 | 0.000 | 0.000 | 0.000 |
| Policy dummy: import duties ($dD7275$) | −0.010 | 0.000 | 0.000 | 0.009 | 0.000 | 0.000 | 0.000 |
| Policy dummy: import duties ($dD77$) | 0.000 | 0.000 | 0.000 | 0.000 | 0.000 | −0.015 | 0.014 |
| Policy dummy: 1980–82: export tax ($dD8082$) | 0.000 | 0.000 | 0.000 | 0.000 | 0.000 | 0.000 | 0.000 |
| Policy dummy: export tax ($dD7374$) | 0.000 | −0.018 | 0.000 | 0.017 | 0.000 | 0.000 | 0.000 |
| Policy dummy: export tax ($dD7879$) | 0.000 | 0.000 | 0.000 | 0.000 | 0.000 | 0.000 | −0.016 |
| Policy dummy: other indirect taxes ($dD73$) | 0.000 | 0.012 | −0.011 | 0.000 | 0.000 | 0.000 | 0.000 |
| Nontax revenues ($NT/\hat{P}$) | −0.007 | 0.007 | 0.003 | 0.000 | 0.007 | 0.002 | 0.001 |
| Grants (grants/$\hat{P}$) | 0.000 | −0.001 | 0.001 | 0.000 | 0.000 | 0.000 | 0.000 |
| Sum of explained changes | −0.005 | 0.018 | 0.015 | 0.069 | −0.037 | −0.037 | −0.012 |
| Unexplained changes | 0.024 | −0.028 | 0.006 | 0.014 | 0.013 | 0.020 | −0.009 |
| Change in deficit | 0.018 | −0.010 | 0.020 | 0.083 | −0.025 | −0.018 | −0.022 |

*(Table continues on the following page with years 1980/81–88.)*

## Appendix 7.5 (continued)

| Item | 1980/81 | 1981/82 | 1982 | 1983 | 1984 | 1985 | 1986 | 1987 | 1988 |
|---|---|---|---|---|---|---|---|---|---|
| **Changes attributable to domestic variables** | | | | | | | | | |
| Real GDP ($\hat{Y}$) | 0.001 | 0.010 | −0.003 | 0.019 | 0.002 | 0.005 | 0.011 | 0.010 | 0.005 |
| Exports ($\hat{EX}$) | −0.003 | 0.015 | 0.009 | 0.000 | −0.001 | −0.001 | −0.001 | 0.000 | 0.001 |
| Imports ($\hat{IM}$) | −0.001 | 0.000 | 0.000 | 0.000 | 0.000 | 0.004 | 0.003 | 0.002 | 0.004 |
| Domestic nominal interest rate ($di$) | 0.000 | −0.001 | −0.002 | 0.001 | 0.003 | 0.001 | 0.003 | 0.004 | 0.001 |
| Domestic inflation ($d\hat{P}$) | 0.005 | −0.002 | −0.007 | 0.009 | 0.003 | 0.000 | 0.001 | 0.002 | −0.001 |
| Real exchange rate ($R\hat{ER}$) | 0.002 | −0.009 | 0.002 | 0.011 | −0.009 | −0.003 | −0.005 | 0.000 | 0.000 |
| **Changes attributable to foreign variables** | | | | | | | | | |
| Foreign nominal interest rate ($di^*$) | 0.001 | 0.000 | 0.000 | 0.000 | 0.000 | −0.002 | −0.002 | 0.000 | 0.001 |
| **Changes attributable to fiscal policy variables** | | | | | | | | | |
| Foreign debt ($D^*/\hat{P}^*$) | 0.028 | −0.030 | −0.006 | −0.043 | 0.008 | 0.038 | 0.006 | 0.007 | 0.004 |
| Domestic debt ($D/\hat{P}$) | 0.000 | 0.000 | 0.000 | −0.001 | 0.004 | 0.003 | 0.004 | 0.001 | 0.000 |
| Wage bill ($WB/\hat{P}$) | −0.003 | −0.001 | 0.001 | −0.005 | −0.002 | 0.000 | 0.000 | −0.003 | −0.003 |
| Goods and services ($GS/\hat{P}$) | 0.004 | −0.012 | −0.003 | −0.006 | 0.000 | 0.025 | 0.012 | −0.001 | 0.002 |
| Transfers and subsidies ($TS/\hat{P}$) | −0.003 | −0.002 | 0.002 | −0.004 | 0.020 | 0.001 | −0.005 | 0.003 | 0.003 |
| Investment ($I/\hat{P}$) | −0.003 | −0.009 | −0.002 | −0.006 | −0.005 | 0.007 | −0.002 | 0.002 | 0.006 |
| Policy dummy: direct taxes ($dD7478$) | 0.001 | 0.000 | 0.000 | −0.002 | 0.006 | 0.009 | −0.002 | 0.006 | 0.006 |
| Policy dummy: post-1984 direct taxes ($dD84$) | 0.000 | 0.000 | 0.000 | 0.000 | 0.000 | 0.000 | 0.000 | 0.000 | 0.000 |
| Policy dummy: import duties ($dD7275$) | 0.000 | 0.000 | 0.000 | 0.000 | −0.007 | 0.000 | 0.000 | 0.000 | 0.000 |
| Policy dummy: import duties ($dD77$) | 0.000 | 0.000 | 0.000 | 0.000 | 0.000 | 0.000 | 0.000 | 0.000 | 0.000 |
| Policy dummy: 1980–82: export tax ($dD8082$) | 0.015 | 0.000 | 0.000 | −0.017 | 0.000 | 0.000 | 0.000 | 0.000 | 0.000 |
| Policy dummy: export tax ($dD7374$) | 0.000 | 0.000 | 0.000 | 0.000 | 0.000 | 0.000 | 0.000 | 0.000 | 0.000 |
| Policy dummy: export tax ($dD7879$) | 0.016 | 0.000 | 0.000 | 0.000 | 0.000 | 0.000 | 0.000 | 0.000 | 0.000 |
| Policy dummy: other indirect taxes ($dD73$) | 0.000 | 0.000 | 0.000 | 0.000 | 0.000 | 0.000 | 0.000 | 0.000 | 0.000 |
| Nontax revenues ($NT/\hat{P}$) | 0.001 | 0.001 | −0.005 | −0.001 | −0.005 | −0.005 | 0.002 | 0.001 | 0.000 |
| Grants (grants/$\hat{P}$) | −0.001 | 0.000 | 0.000 | 0.000 | −0.003 | −0.002 | −0.003 | −0.001 | −0.004 |
| Sum of explained changes | 0.030 | −0.020 | −0.009 | −0.024 | 0.010 | 0.041 | 0.015 | 0.017 | 0.010 |
| Unexplained changes | 0.002 | −0.003 | −0.003 | 0.005 | −0.019 | −0.037 | −0.038 | −0.022 | −0.008 |
| Change in deficit | 0.032 | −0.023 | −0.013 | −0.019 | −0.009 | 0.004 | −0.023 | −0.005 | 0.002 |

*Note:* As an example, a figure such as 0.021 should be interpreted as increasing the deficit by 2.1 percent of GDP.

# Notes

At the time of writing, Deborah Wetzel was at St. Anthony's College, Oxford, U.K. She participated in this project as a consultant to the World Bank.

1. For a detailed account of economic policy through 1972, see Leith (1974) and Killick (1978). Rimmer (1992) provides an in-depth account of economic policy from preindependence days to the present. Stryker (1990) provides a detailed account of the political economy, with particular emphasis on agricultural policies.

2. Net lending has primarily been to public sector entities, and very little of it has been repaid. It therefore is considered expenditure rather than financing.

3. Adding foreign-financed capital expenditure increases central government gross fixed capital formation from 1.8 percent of GDP in 1984 to 5.27 percent of GDP in 1990.

4. The export tariff that goes to the central government is determined as the FOB price minus Cocoa Marketing Board costs minus the price paid to producers.

5. A newly planted cocoa tree takes five years to begin producing and about eight years to reach maximum yield. Depending on maintenance, a tree will produce for thirty years. Hence, policies in the 1960s affected the volume of production in the 1970s.

6. Revenue from the export tax on cocoa was zero in 1980/81 through 1982 because the FOB price during these years, when converted into cedis, was not sufficient to meet Cocoa Marketing Board costs and payments to producers. Hence there was nothing left to go to the government.

7. See the appendix tables for the data on financial stocks. Note also that since 1983 a revaluation account has been included in the monetary survey. This account essentially serves as credit to the monetary authorities to offset losses brought about by the devaluation of the cedi. In 1990 the government issued long-term bonds to cover this liability.

8. Making these changes gives the following figures (as a percentage of GDP):

|  | 1984 | 1985 | 1986 | 1987 | 1988 | 1989 | 1990 |
|---|---|---|---|---|---|---|---|
| Central government deficit | −2.83 | −3.88 | −3.79 | −3.16 | −3.73 | −3.52 | −4.36 |
| On basis of financing flows | −1.86 | −3.43 | −3.05 | −4.17 | −3.88 | −3.89 | −3.70 |

(Central government deficits exclude grants but include capital expenditure financed through project lending.)

9. The estimation results are:

$$\pi = 27.92 + 1.81\text{BORCB} + 81.3\text{DUM.}$$
$$\qquad\quad (2.05) \qquad\quad (7.8)$$

Adjusted $R^2 = 0.828$

Durbin-Watson statistic $= 1.69$

The data used are the CPI inflation rate and World Bank data on public sector borrowing from the central bank for 1971–90.

10. For a detailed review of the Ghanaian financial sector and the controls to which it is subject, see Aryeetey and others (1990).

11. Note that although the economy is credit-constrained, the identities in the equations still hold. Lending from the banking system to the private sector, $L_p$, is taken as given (for example, credit to the private sector is allocated).

12. For simplicity we have assumed that domestic and foreign interest payments are zero.

13. Note that we define seigniorage in these simulations as the change in high-powered money over GDP, rather than as the change in $M_2$ over GDP.

14. The seigniorage-maximizing inflation rate is based on the long-run (steady-state) equation and does not incorporate any short-term influences. As a result, it is higher than the seigniorage-maximizing rate implicit in the scenarios presented earlier, which incorporate short-run influences.

15. These figures are based on the assumption that there was no increase in the end-1989 stocks of total base money, net foreign debt, and net domestic debt.

16. Ricardian equivalence assesses whether an increase in the deficit as a result of a decrease in (neutral) taxes affects private consumption. See Haque and Montiel (1987), Haque (1988), Leiderman and Blejer (1988), and Rossi (1989) for tests of Ricardian equivalence in developing countries. To date, the results have been mixed.

17. The real rate of interest was estimated by the relationship $r = \{[(1 + i)/(1 + \pi)] - 1\}100$, where $r$ is the real interest rate, $i$ is the nominal interest rate, and $\pi$ is the rate of inflation calculated using the CPI.

18. For studies on black markets and their effects on the economy (and vice versa) see May (1985), Pinto (1989), and Chhibber and Shafik (1990). The real exchange rate has been defined as the relative prices of tradable and nontradable goods.

19. The assumption that agents consume only nontradables is a simplifying one; the conclusions do not depend on this assumption, but the algebra is less complex.

20. $\lambda = \lambda(\phi + e)$: with $e$ given, any increase in $\phi$ will lower $\lambda$.

21. We have used rational expectations in this model. Thus the official rate of depreciation equals the expected and actual rate of inflation at the steady state.

22. For variables related to the stock of wealth, lagged values are used, since it seems reasonable that the stock of wealth at the end of the preceding period determines consumption in the present period.

23. Terms of trade was found to be insignificant and was dropped from the equation. If the sum of the real money stock and the private sector's stock of foreign assets (wealth) is used, that sum is significant. When the money stock and the stock of foreign assets are entered separately, only the stock of foreign assets is significant.

24. The error process in this equation has the form $e_t = \theta_1 e_{t-1} + \theta_2 e_{t-2} + v_t$. This process is stationary if $\theta_1 + \theta_2 < 1$, $\theta_2 - \theta_1 < 1$, and $-1 < \theta_2 < 1$. These conditions are satisfied for our equation.

# References

Ahmad, Naseem. 1970. *Deficit Financing, Inflation and Capital Formation: The Ghanaian Experience 1960–65.* Munich, Germany: Weltform Verlag.

Amoako-Adu, Ben. 1991. "Demand for Money, Inflation and Income Velocity: A Case Study of Ghana (1956–1986)." *Savings and Development* 1 (15): 53–65.

Aryeetey, Ernest, and Fritz Gockel. 1990. "Mobilizing Domestic Resources for Capital Formation in Ghana: The Role of Informal Markets." African Economic Research Consortium, Nairobi.

Aryeetey, Ernest, Yaw Asante, Fritz Gockel, and Alex Kyei. 1990. "Mobilizing Domestic Savings for African Development and Diversification: A Ghanaian Case-Study." International Development Centre, Oxford University, Oxford, U.K.

Buiter, Willem H. 1983. "Measurement of the Public Sector Deficit and Its Implications for Policy Evaluation and Design." *International Monetary Fund Staff Papers* 30 (June): 307–49.

———. 1985. "A Guide to Public Sector Debt and Deficits." *Economic Policy* 1 (November): 13–79.

———. 1988. "Some Thoughts on the Role of Fiscal Policy in Stabilization and Structural Adjustment in Developing Countries." NBER Working Paper 2603. National Bureau of International Research, Cambridge, Mass.

Chamley, Christophe. 1991. "Taxation of Financial Assets in Developing Countries." World Bank, Country Economics Department, Washington, D.C.

Chhibber, Ajay, and Nemat Shafik. 1990. "Exchange Reform, Parallel Markets and Inflation in Africa: The Case of Ghana." Policy Research Working Paper Series 427. World Bank, Country Economics Department, Washington, D.C.

Easterly, William. 1989a. "A Consistency Framework for Macroeconomic Analysis." Policy Research Working Paper Series 234. World Bank, Country Economics Department, Washington, D.C.

———. 1989b. "Fiscal Adjustment and Deficit Financing during the Debt Crisis." Policy Research Working Paper Series 138. World Bank, Country Economics Department, Washington, D.C.

———. 1989c. "How Much Fiscal Adjustment Is Enough? The Case of Colombia." Policy Research Working Paper Series 201. World Bank, Country Economics Department, Washington, D.C.

Easterly, William, Carlos A. Rodríguez, and Klaus Schmidt-Hebbel. 1989. "Macroeconomics of the Public Sector Deficit: Research Proposal." World Bank, Country Economics Department, Macroeconomic Adjustment and Growth Division, World Bank, Washington, D.C.

Ghana, Central Bureau of Statistics. Various years. *Quarterly Digest of Statistics.* Accra.

Green, Reginald. 1987. *Stabilization and Adjustment Policies and Programmes: Ghana Case Study.* Helsinki: World Institute for Development Economics Research.

Haque, Nadeem Ul. 1988. "Fiscal Policy and Private Sector Saving Behavior in Developing Economies." *International Monetary Fund Staff Papers* 35: 316–35.

Haque, Nadeem Ul, and Peter J. Montiel. 1987. "Ricardian Equivalence, Liquidity Constraints, and the Yaari-Blanchard Effect: Tests for Developing

Countries." IMF Working Paper. International Monetary Fund, Washington, D.C.

Huq, M. M. 1989. *The Economy of Ghana.* New York: St. Martin's Press.

Islam, Roumeen, and Deborah Wetzel. 1991. "The Macroeconomics of Public Sector Deficits: The Case of Ghana." Policy Research Working Paper 672. World Bank, Country Economics Department, Washington, D.C.

Ize, Alain, and Guillermo Ortiz. 1987. "Fiscal Rigidities, Public Debt, and Capital Flight." *International Monetary Fund Staff Papers* 34 (2): 311–32.

Killick, Tony. 1978. *Development Economics in Action: A Study of Economic Policies in Ghana.* London: Heinemann.

Leiderman, Leonardo, and Mario I. Blejer. 1988. "Modeling and Testing Ricardian Equivalence: A Survey." *International Monetary Fund Staff Papers* 35 (March): 1–35.

Leith, J. Clark. 1974. *Foreign Trade Regimes and Economic Development: Ghana.* Vol. 2. New York: Columbia University Press.

Lizondo, J. Saul. 1987a. "Exchange Rate Differential and Balance of Payments under Dual Exchange Markets." *Journal of Development Economics* 26: 37–53.

———. 1987b. "Unification of Dual Exchange Markets." *Journal of International Economics* 22: 57–77.

Marshall, Jorge, and Klaus Schmidt-Hebbel. 1989. Economic and Policy Determinants of Public Sector Deficits. Policy Research Working Paper 321. World Bank, Country Economics Department, Washington, D.C.

May, Ernesto. 1985. *Exchange Controls and Parallel Market Economies in Sub-Saharan Africa: Focus on Ghana.* World Bank Staff Working Paper 711. Washington, D.C.

Mansfield, Charles. 1980. "Tax-Base Erosion and Inflation: The Case of Ghana." *Finance and Development* 17 (3): 31–34.

Pinto, Brian. 1989. "Black Market Premia, Exchange Rate Unification and Inflation in Sub-Saharan Africa." *World Bank Economic Review* 3 (September): 321–38.

Rimmer, Douglas. 1992. *Staying Poor: Ghana's Political Economy, 1950–90.* Oxford, U.K.: Pergamon Press.

Rossi, Nicola. 1989. "Government Spending, the Real Interest Rate, and Liquidity Constrained Consumers' Behavior in Developing Countries." In Mario I. Blejer and Ke-young Chu, eds., *Fiscal Policy, Stabilization, and Growth in Developing Countries.* Washington, D.C.: International Monetary Fund.

Stryker, J. Dirck. 1990. "Trade, Exchange Rate and Agricultural Pricing Policies in Ghana." In Anne O. Krueger, Maurice Schiff, and Alberto Valdés, eds., *The Political Economy of Agricultural Pricing Policy.* Vol. 3. Baltimore, Md.: Johns Hopkins University Press.

Tanzi, Vito. 1982. *The Underground Economy in the U.S. and Abroad.* Lexington, Mass.: Lexington Books.

Tanzi, Vito, and Mario I. Blejer. 1983. "Fiscal Deficits and Balance of Payments Disequilibrium in IMF Adjustment Programs." In Joaquín Muns, ed.,

*Adjustment, Conditionality and International Financing.* Washington, D.C.: International Monetary Fund.

van Wijnbergen, Sweder, Ritu Anand, Ajay Chhibber, and Roberto Rocha. 1992. *External Debt, Fiscal Policy, and Sustainable Growth in Turkey.* Baltimore, Md.: Johns Hopkins University Press.

# 8

# Morocco: Reconciling Stabilization and Growth

Riccardo Faini

The recent economic history of Morocco resembles the histories of many African countries. The origins of Morocco's economic difficulties can be traced to the commodity (phosphate) boom of the mid-1970s, which coincided with rising government expenditure and an unprecedented expansion of the public investment program. This in turn signaled the end of the conservative fiscal policies of the past. The sudden reversal in the terms of trade in the late 1970s—a result of the plunge in phosphate prices and the second oil shock—prompted Morocco to resort increasingly to external capital markets in order to maintain an unabated level of public expenditure. However, the continued deterioration of the terms of trade, the unanticipated rise in international interest rates, and the severe drought of 1980–84 eroded debt service capacity and precipitated a major foreign exchange crisis in 1983. In response to this crisis, Morocco launched a medium-term program of economic reform and, in consultation with the International Monetary Fund (IMF) and the World Bank, introduced a comprehensive set of stabilization and structural adjustment measures. Since 1983 Morocco has made great progress in alleviating both internal and external disequilibria. The overall budget deficit has been reduced from 9.2 percent of GDP in 1982 to 4.9 percent in 1988; during the same period, the current account shifted from a deficit of 12 percent of GDP to a surplus of 2 percent.[1]

There are several interesting features in Morocco's adjustment. Despite the severity of the crisis and the size of the adjustment undertaken, growth has remained fairly high, at least in relation to other highly indebted countries, and inflation has been subdued. As measured by the consumer price index (CPI), inflation was equal to 6.2 percent in 1983. It increased to 12.5 percent in 1984 but then steadily declined, to 2.3 percent in 1988 and 3.1 percent in 1989. During the same period (1983–89) real GDP growth averaged 4.3 percent.[2] This performance seems to contradict the conventional wisdom that large budget deficits foster inflation. The inflation record is particularly surprising if we also consider that Morocco achieved a 20 percent real depreciation during the 1980s.

In this chapter we try to uncover the reasons underlying the performance of the Moroccan economy. We argue that wage moderation and judicious monetary policies were instrumental in restraining inflation. With one brief exception, in 1983, the monetary authorities eschewed any inflationary financing of the budget deficit. The strategy of avoiding inflationary financing could succeed only because the wide-ranging system of credit and monetary regulations worked to channel domestic funds toward the treasury at relatively low cost. The prospects for continuing such a strategy are not favorable, however. It appears that the growth performance can be attributed to an exceptional export response to the new trade regime and to a set of favorable supply shocks, including a string of record agricultural harvests and the collapse of real oil prices.

In the next section we study the evolution of the budget and its different components and examine how the deficit was financed. In the subsequent section we argue that the reluctance of Morocco's policymakers to monetize existing budget deficits is well explained by the sharply unfavorable tradeoffs between higher monetization and inflation in Morocco, as is suggested by the estimation of a system of asset demand. Next, we analyze the implications of continuing budgetary disequilibria for investment and saving decisions. We find that such implications may be substantial, even though they may not work exclusively through traditional interest rate channels. We then assemble the various pieces of econometric evidence collected in the chapter to study the impact of fiscal policy in the context of a macro-econometric model. The final section offers conclusions.

## The Budget Deficit: Evolution and Financing

The budget in Morocco continued to deteriorate during the second half of the 1970s. Some timid attempts at macroeconomic stabilization only temporarily restrained the burgeoning financial needs of the treasury. However, at the beginning of the 1980s the fiscal repercussions of the fall in Morocco's terms of trade, the increasing burden of foreign debt attendant on the steep rise of international interest rates, and the growing reluctance of foreign commercial banks to provide continuing financing to the treasury brought the issue of fiscal responsibility to the forefront. The budget deficit, measured on a cash basis, showed a slow but steady decline starting in 1981. After increasing (as a percentage of GDP) from an average annual value of 3.2 percent during 1971–74 to 13 percent during 1976–81, it fell to 8 percent in 1984–85, stabilized at about 5 percent at the end of the decade, and declined again in more recent years (table 8.1).

Cash-based measures of the budget deficit can be quite misleading when the treasury resorts to the use of financial arrears to cope with

**Table 8.1. Budget Deficit (Cash Basis, before Debt Relief),
Morocco, 1971–91**
(percentage of GDP)

| Year | Deficit | Year | Deficit |
|------|---------|------|---------|
| 1971 | 2.99 | 1981 | 13.64 |
| 1972 | 4.01 | 1982 | 9.22 |
| 1973 | 2.01 | 1983 | 11.52 |
| 1974 | 3.91 | 1984 | 8.07 |
| 1975 | 9.51 | 1985 | 8.57 |
| 1976 | 18.09 | 1986 | 5.71 |
| 1977 | 15.84 | 1987 | 6.28 |
| 1978 | 11.28 | 1988 | 5.81 |
| 1979 | 10.14 | 1989 | 4.84 |
| 1980 | 9.01 | 1990 | 4.44 |
|  |  | 1991 | 3.81 |

*Source:* World Bank data.

mounting financial difficulties. This is what happened in Morocco up
to 1985. Because cash-based deficits do not allow for the accumulation
of financial arrears, they fail to reflect the full pressure that fiscal
policy exerts on available resources. Conversely, they overestimate
the size of the fiscal problem when financial arrears are being
decumulated. After 1985 adjustment programs in Morocco involved a
sizable reduction in the stock of arrears. It is more appropriate to treat
the accumulation of arrears as a source of (involuntary) finance and to
measure the deficit on a payment-order basis instead. If this is done
for Morocco, the deficit picture changes substantially. Both the initial
increase and the subsequent decline in the budget deficit are more
pronounced when a payment-order basis rather than a cash basis is
used (table 8.2). The reason is simply that arrears were accumulated
up to 1985 and decumulated afterward (except in 1989).

A further correction relates to interest payments. As column (3) of
table 8.2 shows, interest payments have increased steadily as a share
of GDP, thereby compounding the budget difficulties. Yet allowance
should be made for the fact that under inflationary conditions a
potentially conspicuous share of interest payments represents early
amortization of outstanding debt and, as such, should be considered
a (negative) financing component. For Morocco historically low
inflation rates mean that this correction is not going to produce dramatic effects. Its impact is nonetheless substantial, albeit declining.
From column (4) we can see that capital gains on domestic debt attributable to inflation were equal to 3.1 percent of GDP in 1984. Because of
the drop in inflation, they declined to 1.3 percent in 1988. As a result,
the operational budget, which only includes real interest payments,
registered a relatively modest deficit in 1986 but showed some im-

**Table 8.2. Budget Deficit, Morocco, under Various Definitions, 1983–91**

(percentage of GDP)

| Year | Cash basis (1) | Payment-order basis (2) | Interest payments (3) | Capital gains on domestic debt (4) | Capital gains on foreign debt (5) | Operational deficit (6) | Primary deficit (7) |
|---|---|---|---|---|---|---|---|
| 1983 | 11.5 | 12.1 | 4.9 | 0.0 | 0.0 | 12.1 | 7.2 |
| 1984 | 8.1 | 11.2 | 6.1 | 3.1 | 0.5 | 8.1 | 5.1 |
| 1985 | 8.6 | 9.6 | 6.2 | 3.1 | 1.9 | 6.6 | 3.4 |
| 1986 | 5.7 | 5.4 | 5.9 | 3.6 | 1.8 | 1.8 | −0.5 |
| 1987 | 6.3 | 5.9 | 5.9 | 2.0 | 3.0 | 3.9 | −0.1 |
| 1988 | 5.8 | 4.7 | 6.3 | 1.3 | 4.4 | 3.4 | −1.6 |
| 1989 | 4.8 | 6.0 | 6.2 | 1.7 | 8.5 | 4.2 | −0.2 |
| 1990 | 4.4 | 3.4 | 6.2 | 2.2 | 5.1 | 1.2 | −2.8 |
| 1991 | 3.2 | 3.0 | 5.6 | 2.0 | 4.3 | 1.0 | −2.6 |

*Source:* Author's computations, based on World Bank data.

provement over the following years. We have also corrected for valuation effects in the stock of foreign debt by multiplying the outstanding stock at the beginning of the period by the excess of domestic inflation (measured by the change in the GDP deflator) over the rate of devaluation (taken with respect to the nominal effective exchange rate). We feel much less confident about this correction insofar as it overlooks valuation changes resulting from, for instance, cross-currency fluctuations, which appear to have played a substantial role in determining the evolution of foreign debt indicators for Morocco. We nonetheless report the results of such correction (column 5). Its impact was basically insignificant in 1984 but grew over time until 1989, as domestic inflation was no longer matched by a corresponding devaluation.

The last measure we consider is the primary deficit, which excludes all interest payments. By excluding a component that is to a large extent beyond the control of fiscal authorities, the primary deficit provides a more accurate indicator of the effort to redress existing fiscal imbalances. The evolution of the primary deficit highlights the adjustment effort on the fiscal front. The primary deficit was equal to 7.2 percent of GDP in 1983 and has steadily declined since then, reaching a surplus of 1.6 percent of GDP in 1988 and 2.6 percent in 1991. This amounts to a turnaround of almost 10 percentage points of GDP. The improvement in the overall and the operational deficits is less significant because of the increased burden of nominal and real interest payments, respectively.

In summary, Morocco has undeniably taken a decisive step on the road to redressing its fiscal imbalances. The substantial achievements in the overall budget reflect an even more significant improvement in the primary deficit that more than compensates for the rising burden of nominal interest payments. The favorable evolution of the budget was also facilitated by low and even negative real interest rates on domestic debt. This situation, however, is coming to an end, and the treasury is now forced to offer positive and high real rates of return to convince domestic investors to finance its deficit. Despite the drop in inflation, the average nominal interest rate on domestic debt increased from 4.3 percent in 1983 to 5.7 percent in 1986 and reached 7.6 percent in 1989. The possibility that the average cost of servicing domestic debt will keep rising rapidly in the near future underscores the need for continuing fiscal restraint.

### The Primary Deficit

In this section we examine how the sizable reduction in the primary deficit was achieved. It is apparent from the data in tables 8.3 and 8.4 that the burden of adjustment fell mostly on public investment, which registered a major decrease, from 11.5 percent of GDP at the onset of the adjustment program in 1982 to only 3.6 percent in 1986. It then recovered, but only marginally, to about 6 percent in 1987 and 1988 and 7.5 percent in 1989.[3] The drop in public investment was therefore the major contributing factor in the process of fiscal retrenchment. This is an unfortunate feature of the adjustment process that Morocco shares with many other developing countries. We shall see later how the decline in public investment may have negatively affected the propensity of the private sector to invest.

Table 8.3 shows how other items contributed to fiscal restraint. Expenditure on goods and services fell somewhat, reflecting mostly the drop in public employees' real wages and the more sober trend in public employment. Simple econometric analysis shows that expenditure on goods and services is not related to inflation (see note 5), suggesting that inflation was not paramount in reducing public sector real wages. Finally, we find that the share of subsidies in GDP fell substantially, mainly because the decline in imported food prices reduced the need for government intervention. This is, incidentally, an interesting example of the *direct* impact of terms of trade fluctuations on the budget.

A comparison between the evolution of the primary deficit, as reported in table 8.2, and the behavior of public expenditure in table 8.3 suggests that taxation did not contribute noticeably to the improvement in the budget. As a matter of fact, taxation as a percentage of GDP fluctuated around 21 percent until 1987 (table 8.4). Had it not been for the windfall revenue attendant on the petroleum levy introduced in 1986, fiscal revenues would actually have registered a

## Table 8.3.  Central Government Expenditures (before Debt Relief), Morocco, 1971–89

(percentage of GDP)

| Year | Current | Capital | Goods and services | Interest | Subsidies |
|------|---------|---------|--------------------|----------|-----------|
| 1971 | 14.12 | 3.72 | 12.51 | 1.01 | 0.06 |
| 1972 | 14.35 | 4.06 | 12.61 | 1.14 | 0.61 |
| 1973 | 14.14 | 4.03 | 12.25 | 1.08 | 0.80 |
| 1974 | 19.25 | 5.77 | 12.20 | 0.85 | 6.21 |
| 1975 | 20.18 | 12.66 | 13.62 | 0.95 | 5.62 |
| 1976 | 19.48 | 18.89 | 14.93 | 1.20 | 3.35 |
| 1977 | 18.58 | 18.93 | 14.44 | 1.50 | 2.63 |
| 1978 | 18.89 | 13.36 | 15.00 | 1.90 | 1.99 |
| 1979 | 19.46 | 12.89 | 15.05 | 2.19 | 2.21 |
| 1980 | 21.24 | 9.31 | 14.19 | 2.54 | 4.51 |
| 1981 | 24.60 | 12.33 | 15.42 | 3.88 | 5.30 |
| 1982 | 22.91 | 11.54 | 14.75 | 3.61 | 4.55 |
| 1983 | 23.72 | 9.64 | 15.05 | 4.86 | 3.81 |
| 1984 | 24.00 | 8.09 | 13.63 | 6.08 | 4.29 |
| 1985 | 23.09 | 7.18 | 12.63 | 6.23 | 4.23 |
| 1986 | 20.56 | 3.63 | 12.17 | 5.90 | 2.49 |
| 1987 | 20.38 | 6.10 | 12.64 | 5.89 | 1.84 |
| 1988 | 20.83 | 6.47 | 12.63 | 6.17 | 2.03 |
| 1989 | 21.51 | 7.54 | 13.08 | 6.22 | 2.21 |

Source: World Bank data.

major decline over the period 1983–88. There are several reasons for this unsatisfactory evolution. This was a time when, with the support of the World Bank and the IMF, the Moroccan government implemented far-reaching reforms in the system of both direct and indirect taxation. For instance, the value added tax (VAT), introduced in 1986, was expected to provide in the medium run a more efficient and more reliable source of revenue. Because of implementation problems, it was accompanied on its introduction by a revenue shortfall. Similarly, the overhaul of the system of direct taxation with the phasing-out of a set of income taxes differentiated according to income sources and their consolidation into a unique tax was not likely to contribute to the budget in the very short run. Overall, the impact of reforms on the tax system was most likely to be felt only in the medium run. Some promising signs can already be detected from the evolution of fiscal revenues in 1988 and 1989.[4]

Other factors may also have helped to determine the evolution of tax revenues. For instance, it is often argued that high inflation reduces the real value of tax revenues because of delays in revenue collection. This effect is unlikely to be significant in Morocco because of relatively low inflation levels, and it may also have been offset by the bracket creep mechanism whereby taxpayers in an unindexed

**Table 8.4. Tax Revenue, Morocco, 1971–89**
(percentage of GDP)

| Year | Total revenues | Income taxes | Taxes on goods and services | Taxes on international trade | Petroleum levy | OCP contribution |
|------|------|------|------|------|------|------|
| 1971 | 14.85 | 2.83 | 7.60 | 2.58 | 0.00 | 0.00 |
| 1972 | 14.40 | 2.81 | 7.40 | 2.41 | 0.00 | 0.00 |
| 1973 | 16.16 | 3.21 | 7.45 | 2.88 | 0.00 | 0.35 |
| 1974 | 21.11 | 3.29 | 6.80 | 3.62 | 0.00 | 5.77 |
| 1975 | 23.33 | 7.02 | 7.32 | 4.33 | 0.00 | 2.72 |
| 1976 | 20.29 | 4.01 | 7.69 | 4.18 | 0.00 | 1.85 |
| 1977 | 21.67 | 5.14 | 8.33 | 5.19 | 0.00 | 0.94 |
| 1978 | 21.20 | 5.34 | 8.01 | 4.67 | 0.00 | 0.61 |
| 1979 | 22.25 | 5.36 | 7.99 | 5.48 | 0.00 | 0.61 |
| 1980 | 20.51 | 4.51 | 7.84 | 5.56 | 0.00 | 0.90 |
| 1981 | 22.57 | 4.77 | 7.64 | 6.19 | 0.00 | 1.52 |
| 1982 | 22.05 | 4.12 | 8.41 | 6.23 | 0.00 | 0.76 |
| 1983 | 21.27 | 4.38 | 8.77 | 5.31 | 0.00 | 0.91 |
| 1984 | 20.89 | 4.44 | 8.51 | 5.01 | 0.00 | 1.16 |
| 1985 | 20.65 | 4.58 | 8.38 | 4.46 | 0.00 | 1.27 |
| 1986 | 18.84 | 4.03 | 7.07 | 3.62 | 2.44 | 0.13 |
| 1987 | 20.68 | 4.48 | 7.66 | 3.56 | 2.78 | 0.39 |
| 1988 | 22.74 | 4.81 | 7.78 | 4.07 | 3.26 | 0.25 |
| 1989 | 23.06 | 5.13 | 8.07 | 4.48 | 2.71 | 0.08 |

Note: OCP, Office Cherifien des Phosphates.
Source: World Bank data.

system are taxed at increasing rates because of the impact of inflation on nominal incomes. Again, econometric evidence does not suggest a significant relation between the share of tax revenues in GDP and the level of inflation.[5]

Another noticeable factor that affected the evolution of government revenues was trade reform. In 1984, with the support of the World Bank, the Moroccan government, in an attempt to rationalize the trade regime and reduce protection for import-substituting productions, chose to reduce substantially import duties and gradually phase out the "special import tax," an across-the-board tariff. The revenue shortfall caused by this measure was estimated to reach 4 percent of GDP. Continuing budgetary difficulties prompted a reassessment of the situation. In 1987 the special import tariff was increased from 5.0 to 12.5 percent, and its name was changed to "fiscal import duty." This policy shift is reflected in the increase in trade taxes between 1987 and 1989 (see table 8.4).

The last column of table 8.4 highlights the impact of terms of trade fluctuations on fiscal revenues. In Morocco production of and trade in phosphate products (the main foreign exchange earners for the country) are controlled by the Office Cherifien des Phosphates (OCP), a public enterprise. The latter contributes to the treasury budget through tax and dividend payments. As shown in table 8.4, these contributions reached a peak in 1974–75, concomitantly with the boom in phosphate prices. They have been declining since then (except for a small rebound in 1981), adding substantially to budgetary difficulties.

## Financing the Budget Deficit

The impact of budget deficits on macroeconomic conditions is to a significant extent a function of the mode of financing of the deficit itself. It is therefore essential to take a closer look at the way budget imbalances have been financed in the past. Prior to 1983 foreign borrowing financed nearly 60 percent of the treasury deficit. The availability of foreign finance came to an abrupt halt in 1983, precipitating a major payment crisis. Between 1985 and 1989, despite a substantial amount of debt relief, foreign finance accounted, on average, for only 33.3 percent of the treasury's financial needs. The government was forced to rely to an unprecedented extent on Central Bank borrowing during 1983 and, to a lesser extent, 1984. This was also the period when the government increasingly resorted to the accumulation of arrears as a source of involuntary finance from the private sector. Yet the increased monetization of the deficit would unavoidably have provided fuel for inflation. It was the firm commitment by monetary authorities to eschew inflationary financing that prompted a major revision in the financing strategy. Starting in 1984 the government increasingly began tapping noninflationary domestic sources of financing. Nonbank sources, which accounted for a negligible share of the treasury's financial requirements in 1983–84, increased their share to 38.1 percent, on average, between 1986 and 1989. The treasury's reliance on voluntary domestic lending was exacerbated by the need to reduce the sizable stock of financial arrears. The partial liberalization of domestic interest rates, which was associated with a major shift in the private sector portfolio composition from currency to time deposits, assisted this strategy. At the same time the ceiling on credit to the economy forced commercial banks to channel a substantial part of these financial resources to the treasury, thereby creating a steady source of finance for government deficits. Another important component of the financing strategy was the direct sales of treasury bills and bonds to the nonbank sector.

The disadvantage of this strategy was its increased cost to the treasury. Time deposits were excluded from the base on which obligatory investments in treasury bonds are calculated, to make these deposits more palatable to commercial banks. Together with the reduction of arrears and the need to offer attractive (posttax) returns on direct issues of government bonds, this measure induced a significant increase in the cost of domestic financing for the treasury. This is clearly reflected in the rapid increase in the average real interest rate on government domestic debt. Further institutional developments are likely to reinforce this trend if existing proposals to reduce liquidity and portfolio requirements for the banking sector are implemented. In the final section of this chapter, we assess the macroeconomic impact of furthering the process of financial liberalization.

Overall, Morocco has been quite successful in coping with formidable financial difficulties. The foreclosure of foreign borrowing in 1983 represented a major shock to the economy and to the treasury, more significant perhaps than the rise in international interest rates. The two-pronged response of Moroccan policymakers relied on a sharp reduction in the treasury's financing needs and a shift in the composition of finance in hopes of avoiding inflationary pressures. A key to the success of this strategy was the availability of relatively cheap sources of domestic finance, mostly determined by a complex array of financial regulations. However, the difficulties of relying at the margin on financial repression as a cheap source of funds have already prompted a further reassessment of existing strategies and, perhaps more crucially, highlighted the sharp tradeoffs that continuing budget disequilibria entail. Today the major question facing Morocco's policymakers is whether the country's fiscal policy stance is consistent with the maintenance of low inflation, the resumption of investment and growth, and the external payments constraint.

Consider the plausible case in which foreign finance will not increase substantially in the medium run. Under these circumstances, the failure to persevere on the road to fiscal discipline may entail severe macroeconomic consequences. Even the perpetuation of present financing strategies with a relatively unchanged budget deficit would soon run into severe problems. First, the budget deficit may not be compatible with the maintenance of low inflation. Simple calculations show that in 1989 the sustainable primary budget should register at least a surplus of 2.2 percent of GDP. Similar calculations for the overall budget indicate a maximum deficit equal to 3.3 percent.[6] Second, the budget is still extremely vulnerable to international interest rates and terms of trade shocks. For instance, the revenue from the petroleum levy, which in 1988 still represented 3.3 percent of GDP, is likely to vanish under increasing oil prices. Third, the cost of domestic finance is likely to rise steeply in the future. Again, the treasury budget would be extremely vulnerable to such an evolution. If the cost of servicing domestic debt were to increase to competitive market rates, this would add to the budget an extra burden equal to about 2 percent of GDP.[7]

Could Morocco resort at the margin to larger monetization of the deficit? By international standards Morocco's usual inflation is low, and an increase in the inflation tax might be a palatable alternative. However, besides the danger of tampering with monetary policy and damaging the credibility of the Central Bank's commitment against inflation, there is actually little room for a substantial contribution to budget financing from increased monetization of the deficit. We will show in the next section that, for Morocco, estimated elasticities from a system of asset demand indicate an extremely unfavorable tradeoff between inflation and monetary financing. Continuing reliance on noninflationary sources of domestic finance is therefore essential for keeping inflation in check. However, this strategy could lead to large increases in domestic real rates of interest. Besides putting into jeopardy budget equilibria, this would most likely crowd out investment. Furthermore, as discussed in a later section, even if, under the most optimistic scenario, the increase in interest rates did not materialize and the perpetuation of a system of financial regulations guaranteed a source of inexpensive finance for the treasury, the impact of unabated budget deficits on investment and growth would still be significant because of their effects on credit markets.

## Asset Demand, Seigniorage, and the Inflation Tax

According to the standard wisdom, when conventional lending dried up after 1982, developing countries had to make a twin transfer of resources: abroad, to foreign creditors (insofar as new lending had fallen much below the required service on outstanding external debt); and at home, to the government, which in several developing countries had assumed the burden of servicing foreign debt. Together with the increase in interest rates and the fall in terms of trade, this has often meant the disruption of budgetary equilibria, already jeopardized in many cases by unsustainable fiscal policies. The inflexibility of the tax system, the downward rigidity of fiscal expenditure (with the notable exception of public investment), and the thinness of domestic financial markets left local policymakers with little choice but to monetize the fiscal deficits, with sometimes calamitous consequences for the inflation rate.

Under this interpretation, Morocco's exceptional inflation record is undoubtedly puzzling. In the preceding section we saw how financial repression, by channeling low-cost funds to the treasury, was instrumental in a strategy of reliance on noninflationary sources of budget financing. Yet the question remains: why do Moroccan policymakers remain staunchly opposed to even a limited monetization of the deficit and a greater reliance on the inflation tax? Inflation in Morocco is extremely low—3.1 percent in 1989—even by the most stringent international standards. A recent report (United Nations–World Bank 1990), while not explicitly advocating an increase in inflation, argues nonetheless that there is some scope for boosting revenues from the inflation tax. The issue, however, cannot be solved on purely theoretical grounds because the inflationary implications of deficit monetization depend on the response of money base demand to changes in inflation and interest rates. In what follows, we argue that the outlook for greater monetization of the budget deficit in Morocco is altogether unfavorable. If we also consider the destabilizing effects on Morocco's social fabric and the loss of Central Bank credibility that higher inflation would imply, we can perhaps understand the firm commitment of the country's policymakers to price stability.

The scope for increased government revenue from seigniorage and the inflation tax is determined by the private sector's choice of assets. We rely on a standard Tobin's portfolio approach to analyze the demand for currency, demand deposits, and time deposits. Other assets are not included in our analysis on the grounds that, with the exception of real assets, they play a minor role in the private sector's choices. In what follows we assume that the demand for each asset is related to its own return, the other assets' returns, and the level of income. As a proxy for the return on real assets, we use the (expected) inflation rate, which, however, is not observable. We assume that expectations are formed rationally and depend on the set of information available at time $t - 1$, which is defined to include the lagged level of prices, the money supply, the wage rate, and the exchange rate. According to our estimates, the last variable does not contribute significantly to the prediction of future prices. This preliminary evidence suggests that for Morocco the passthrough from the nominal exchange rate to domestic prices is fairly weak, which accounts for the fact that a sizable real depreciation did not translate into higher inflation. Similarly, the significant impact of wages on prices supports the claim that wage moderation (the real minimum wage declined by 7.3 percent from 1983 to 1988, after increasing by 17 percent in the three previous years) was instrumental in containing inflation.

The fitted value from the price equation is used as an estimate for expected inflation. In the process of estimation, we rely on Pagan's (1984) procedure and use our proxy for expected inflation as an instrument for actual inflation. This procedure is designed to yield consistent standard errors for the coefficients. We still have to determine whether the system of asset demand should be expressed in nominal terms (with prices included among the explanatory variables and the relevant homogeneity assumptions tested) or in real terms.

We take the first course for the sake of generality and assume that asset demand does not respond fully and instantaneously to changes in prices. We also allow for the possibility of lagged adjustment to variations in income and interest rates. Therefore the estimating equation for a generic asset $M^j$ reads:

$$(8.1) \quad \ln M_t^j = a_0 + a_1 \ln p_t + a_2 \ln p_{t-1} + a_3 \ln Y_t + a_4 \ln Y_{t-1}$$
$$+ a_5 i_t + a_6 i_{t-1} + a_7 \pi_t^e + a_8 \ln M_{t-1}^j$$

where $p$ and $Y$ denote the price level and the income level, respectively, $i$ represents the vector of asset returns, and $\pi^e$ is expected inflation. We would expect that in the long run the price elasticity of asset demand is equal to 1; asset demands are homogeneous of degree 1 in prices. This implies that $a_1 + a_2 = 1 - a_8$. We can reparametrize the previous equation to test this restriction:

$$(8.2) \quad \ln(M^j/p)_t = a_0 + (a_1 - 1) \ln p_t + a_3 \ln Y_t + a_4 \ln Y_{t-1}$$
$$+ a_5 i_t + a_6 i_{t-1} + a_7 \pi_t^e + (a_8 + a_2) \ln p_{t-1}$$
$$+ a_8 \ln(M^j/p)_{t-1}.$$

If long-run price homogeneity holds, we have:

$$(8.3) \quad \ln(M^j/p)_t = a_0 - (a_2 + a_8) \Delta \ln p_t + a_3 \ln Y_t + a_4 \ln Y_{t-1} a_5 i_t$$
$$+ a_6 i_{t-1} + a_7 \pi_t^e + a_8 \ln(M^j/p)_{t-1}.$$

Equation 8.3 shows that imposing long-run price homogeneity in a context in which assets are expressed in real terms adds a backward-looking inflation term, $\Delta \ln p$, to the equation itself. This term however does not reflect any substitution effects between real and financial assets but only lagged adjustment to price changes. Traditional money demand equations, in which inflationary expectations are simply modeled in a backward-looking fashion, may therefore mistakenly interpret the significance of $\Delta \ln p$ as evidence of substitution toward real assets, whereas only a process of dynamic adjustment is involved. Our approach allows us to distinguish between these two effects by separately evaluating the statistical significance of $\Delta \ln p$ and $\pi^e$. Admittedly, the coexistence of these two effects may be difficult to justify on purely theoretical grounds.

In the estimation process we include in the vector of asset returns ($i$) only the interest rate on time deposits, $i^{td}$. Some demand deposits are also remunerated, but their return moves in an almost perfectly collinear way with that on time deposits. We have therefore included the remunerated component of demand deposits in the aggregate of time deposits. We can now describe the results of the estimation of the system of asset demand. We start from the more general dynamic specification in equation 8.3 and restrict the model to obtain a parsimonious representation of the data-generating process. We report only the final equation.

All equations are estimated over the 1974–88 period. The last observation is saved to test for out-of-sample stability. Given the limited number of degrees of freedom, all our results should be interpreted with considerable caution. We begin with the demand for currency (*t*-statistics are in parentheses):

$$\Delta \ln(\text{CUR}_t/p_t) = 0.46 - 1.10i_t^{td} + 0.30[\ln Y_t - \ln(\text{CUR}_{t-1}/p_{t-1})]$$
$$\qquad (2.69) \ (-6.39) \qquad (3.95)$$

$$\qquad - 0.60 \ln(p_t/p_{t-1}).$$
$$\qquad (-4.30)$$

Durbin-Watson statistic = 2.01
Standard error of regression (SER) = 0.013
Lagrange multiplier = 0.25
Hendry = 1.47

The demand for currency (CUR) is significantly related to the interest rate on time deposits ($i_{td}$) and to income.[8] Actual inflation appears in the equation with a short-run elasticity of 0.60. As mentioned earlier, this is not necessarily an indication of significant substitution possibilities between currency and real assets but may simply reflect the fact that nominal demand for currency fully adjusts to changes in the price level only in the long run. The coefficient on expected inflation is not significantly different from zero and bears the wrong (positive) sign. Jointly estimating the price and the currency equations (and testing for the expectational restrictions) did not improve the results. The variable $\pi^e$ therefore has been eliminated from the final equation. Finally, the hypothesis that nominal demand for currency is unit elastic in the long run with respect to the price level and to income is not rejected by the data ($t_{10} = 0.94$). As diagnostic tools we rely on the Hendry test for out-of-sample stability and the Lagrange multiplier test for serial correlation. Both are distributed as $X(1)$. They do not provide any indications of misspecification.

We now turn to the equation for demand deposits (*DD*).

$$\ln(DD_t/p_t) = -11.65 - 8.46i_t^{td} - 1.03 \ln(p_t/p_{t-1}) + 1.20 \ln Y_t$$
$$\qquad (2.73) \ (-2.85) \quad (-2.49) \qquad\qquad (3.22)$$

$$\qquad + 0.76 \ln Y_{t-1}.$$
$$\qquad (2.97)$$

Durbin-Watson statistic = 1.43
SER = 0.033
Lagrange multiplier = 1.56
Hendry = 0.81

Demand deposits have been defined to exclude savings deposits. Following this modification, we find that both inflation and the return on time deposits significantly affect demand deposits. The long-run income elasticity is equal to 1.96 and is significantly different from 1. Once again, expected inflation does not contribute in a statistically significant way to the equation. The usual battery of tests shows no sign of misspecification.

The last component of our menu of assets is time deposits (*TD*):

$$\Delta \ln(TD_t/p_t) = -1.54 + 6.94 i_t^{td} + 0.43[\ln Y_t - \ln(TD_{t-1}/p_{t-1})].$$
$$(1.93) \quad (1.64) \qquad (2.43)$$

Durbin-Watson statistic = 1.97

SER = 0.09

Lagrange multiplier = 0.02

Hendry = 0.18

As expected, time deposits respond positively, albeit not very significantly, to an increase in their own rate of return. The dynamic specification of the equation is very simple. There is no evidence of lagged adjustment to prices, as indicated by the insignificant coefficient on actual inflation. On the same ground, expected inflation has been excluded from the final equation. It was not possible to reject the hypothesis that the demand for time deposits is unit elastic with respect to income ($F_{1,9} = 0.54$). The Lagrange multiplier and Hendry tests fail to point to any misspecification problems.

These results indicate that the demand for monetary assets in Morocco is strongly influenced by the pattern of returns. The estimated semielasticities on the interest rate suggest potentially conspicuous shifts in portfolio composition in response to variations in the structure of interest rates. There is, however, less indication of strong substitution possibilities with respect to real assets. Our proxy for expected inflation did not prove significant in any of the three equations. It is not apparently a problem of multicollinearity, insofar as the coefficient on actual inflation was quite well determined even in the more general specifications, nor is it a problem of statistical methodology. Our approach should provide, at a minimum, consistent estimates of the coefficients and of their standard errors. As mentioned earlier, more efficient simultaneous estimation methods, which also allow for the expectational restrictions, fail to change the basic findings. Perhaps our proxy for expected inflation is not well specified. Alternatively, expectations may to a large extent have an adaptive form.

To compute the amount of monetary financing that the government can count on, we allow for the distinction that must be made in Morocco between reserve requirements, which apply only to demand deposits, and liquidity requirements, which force commercial banks to invest a fixed share of their deposits in low-yield treasury bills. Recent financial sector reforms have substantially increased the interest rate paid on liquidity requirements, which now approaches market rate levels. We therefore include liquidity requirements in the domestic public debt. The narrowly defined monetary base (*MB*) is equal to:

$$(8.4) \qquad MB_t = \text{CUR}_t + rr^{dd}DD_t$$

where $rr^{dd}$ is the reserve requirement coefficient for demand deposits. At any point of time the amount of monetary financing is equal to the change in the monetary base. As a share of nominal GDP, the amount of monetary financing ($\Delta MB_t/p_tY_t$) is equal to the rate of change of monetary base ($\Delta MB_t/MB_t$), multiplied by the GDP share of *MB*, ($MB_t/p_tY_t$).

We can now use equation 8.4, together with our estimates of the demand for currency and demand deposits, to evaluate the relationship between inflation and the amount of monetary financing. We focus on a steady-state situation and arbitrarily impose the condition that demand deposits be unit elastic with respect to income. As a result, the growth rate of *MB* will be equal to the sum of output growth ($n$) and inflation ($\pi$). Similarly the GDP share of *MB* will depend on the level of nominal interest rates and, possibly, of inflation:

$$(8.5) \qquad \Delta MB/pY = (\pi + n)[(\text{CUR} + rr^{dd}DD)/pY]$$

$$= (\pi + n)[Ae^{-\alpha\pi} rr^{dd}Be^{-\beta\pi}]$$

where the coefficients $A$ and $\alpha$ ($B$ and $\beta$) are computed from our estimates of the currency (demand deposits) equation. In assessing the values of $\alpha$ and $\beta$, we face a basic ambiguity. Changes in the inflation rate can have a direct impact on the demand for monetary base to the extent that they lead to corresponding variations in the nominal interest rate. We assume this to be the case. But changes in $\pi$ also have a direct effect on the demand for currency and deposits. If we interpret the significant coefficient of actual inflation in our estimates simply as evidence of lagged adjustment, this effect should not reasonably play a role in a steady-state analysis. If, however, we believe that expectations are adaptive and our results reflect the existence of significant substitution possibilities with real assets, inflation should have an independent effect. Note that in the latter case changes in inflation will exert a larger impact on the demand for monetary base. In what follows we allow for both possibilities.

We can use equation 8.5 to compute the effect of a 1 percent change in the inflation rate on the quantity of monetary financing as a share of GDP. We take GDP growth to be equal to 4 percent. From equation

**Table 8.5. Impact on Inflation of a 1 Percent Increase in the Share of Monetary Financing in GDP, Morocco**
(percent)

| Initial inflation | Low elasticities | High elasticities |
|---|---|---|
| 4 | 7.5 | 7.6 |
| 6 | 9.8 | 11.1 |
| 8 | 12.9 | 16.9 |
| 10 | 16.7 | 26.4 |

8.5 it is clear that this derivative is a function of the initial share in GDP of currency and demand deposits, which in turn depends on the inflation rate itself. We can also compute the inverse derivative, which measures the increase in the inflation rate attendant on a 1 percent increase in the GDP share of monetary financing (table 8.5). Our estimates suggest that the inflationary impact of higher monetization increases very rapidly and that the tradeoff between inflation and inflation tax worsens substantially as inflation increases. This provides considerable support to the choice of Moroccan policymakers not to rely on inflationary forms of deficit finance.

## Investment and Saving Decisions

### The Investment Choice

A high rate of investment represents a basic condition for sustained increases in economic growth over the long term. From 1982 to 1987, however, the share of investment in (current prices) GDP in Morocco has been steadily falling, causing increasing concern about the long-run perspectives of the economy. Looking at the constant-price ratio between investment and GDP would only accentuate the fall in investment because of the increase in the real price of investment goods attendant on the real depreciation. Perhaps of greater concern is that the drop in capital accumulation is generalized, involving both public and private investment. Contrary to initial expectations, the fiscal retrenchment that took a toll on public investment was not compensated by a matching increase in private investment. Is the fall in investment a significant cause for concern? It could be argued that most of the fall in investment can be predicated on the higher cost of capital. By encouraging less capital-intensive projects (and removing the previous bias against labor-intensive production), an increase in the cost of capital would allow the same level of growth to be achieved with a lower volume of investment. Under this interpretation the drop in investment would only represent the outcome of adjusting to a new constellation of factor prices. Empirical evidence for a relatively large sample of developing countries (Faini and de Melo 1990) suggests that cost of capital considerations can account for only a small fraction of the fall in investment and that other factors must therefore be at work. In what follows, we shall assess the relevance of this approach for the case of Morocco.

We rely on a simple model of a firm. This firm is assumed to maximize its net worth, subject to a standard neoclassical production function. In contrast to the traditional setup however, financial choices are assumed to have a significant bearing on real decisions.[9] We model the impact of financial variables by assuming that external equity financing is unavailable and that the firm must rely on two alternative sources of investment finance: retained earnings and bank debt. Given that the entrepreneurs' discount rate is assumed to be larger than the risk-free interest rate—otherwise the firm would accumulate financial assets—debt is the favored source of investment finance. (Another possible reason for this choice is tax considerations.) An internal solution for the optimal debt choice of the firm still exists, though, provided we assume that higher debt, in relation to the firm's capital stock, is associated with increasing agency costs. Finally, because of constraints in the credit market, at each point of time the firm's ratio of debt to capital stock is bounded by an upper constraint. Expressing this formally:

(8.6)    $\max \Sigma[(1)^t/(1 + i)][p_t Y_t - w_t N_t - q_t I_t + B_t$

$- (1 + r)B_{t-1} - c(B_t, q_t K_t)]$

subject to

(8.7)                      $B_t \leq B_t^* q_t K_t$

(8.8)                      $Y_t = F(K_t, N_t)$

(8.9)                      $K_t = (1 - \delta)K_{t-1} + I_t$

where, in standard notation, $Y$, $N$, $K$, and $I$ represent the levels of production, employment, capital stock, and investment, respectively, and $B$ denotes the outstanding stock of debt. The function $c(\ )$, with $c_1 > 0$, $c_{11} > 0$, and $c_2 < 0$, is the agency cost function, which is assumed to be quadratic. The output price is indicated as $p$, the wage as $w$, and the price of investment goods as $q$; $r$ and $i$ denote the interest rate and the discount rate, respectively. We assume that $1 + i = (1 + r)(1 + k)$, where $k$ is a multiplicative risk premium. Finally, $\delta$ is the depreciation rate.

Equation 8.7 is the leverage constraint that defines the maximum amount of debt as a time-varying proportion ($B^*$) of the capital stock, while equations 8.8 and 8.9 describe the production relationships and the capital accumulation identity, respectively. Suppose first that the debt constraint is not binding. It can be easily shown that, at an optimum, debt will be a fixed proportion $\gamma$ of nominal capital stock. In turn, $\gamma$ is a function of the risk premium $k$ and of the parameters affecting the position of the agency cost function.[10] Note that $\gamma$ does not depend on the interest rate. As a matter of fact, variations in $r$ (or in the discount rate $i$) affect the level of investment, not the composition of its financing. The choice of $\gamma$ in turn affects the demand for capital, which is otherwise determined in a standard way. If we consider the case in which the debt constraint is binding ($B^* < \gamma$), financial conditions have a direct impact on the demand for capital. Under linear homogeneity conditions in production, it can be shown that the optimal capital stock will be a function of output, the real cost of capital, and the availability of debt. Represented formally:

$$(8.10) \qquad K_t = K\,(Y_t,\, c_t/p_t,\, B_t^*)$$

where $c_t = q_t\,(1 - \mu)\,[i(1 - r) + \delta]$. The parameters $r$ and $\mu$ denote the corporate tax rate and the percentage reduction in $q$ induced by the system of fiscal and financial incentives available to investors in Morocco. Full details for the calculations of $r$ and $\mu$ are provided in World Bank (1990).

A problem with the formulation of equation 8.10 is that it contains one obviously endogenous variable, the level of output ($Y_t$). It is assumed that because of delivery lags, firms must determine their desired level of output, and therefore their investment, one period in advance. They will therefore need to predict, on the basis of available information, the optimal level of capacity output for the following period. This will determine in turn their demand for investment goods. We also assume that expectations about the determinants of the capacity decision can be simply modeled by a first-order autoregressive process. The expected optimal level of output ($_{t-1}Y_t$) will therefore be equal to:

$$(8.11) \qquad {}_{t-1}Y_t = Y\,(p_{t-1},\, w_{t-1},\, c_{t-1},\, Ipub_{t-1},\, MS_{t-1})$$

where the information set has been augmented to include both *Ipub* (the level of public investment), on the ground that this may affect the production relationship, and *MS* (the stock of money), as a further predictor of prices. In estimating equation 8.11 we must allow for the fact that $_{t-1}Y_t$ is not observable. However, under rational expectations, it will differ from actual output $Y$ only for a random term uncorrelated with any available information at $t - 1$. We can therefore use the actual production level $Y_t$ as the dependent variable and take the fitted value of equation 8.11 as the estimate of expected output.

By estimating equation 8.11, it is now possible to take into account the endogeneity of $Y$ in equation 8.10. We again follow Pagan's (1984) suggestion and take the estimated value of $Y(Yf)$ in equation 8.11 as an instrument for the actual level of output in equation 8.10. Both equations 8.10 and 8.11 are estimated over the 1972–88 period. We first present the estimates of equation 8.11:

$$\ln Y_t = 7.01 - 0.30 \ln(w/p)_{t-1} - 0.26 \ln(c/p)_{t-1} + 0.43(MS/Y)_{t-1}$$
$$(29.9) \ (-0.48) \qquad\qquad (-1.81) \qquad\qquad (1.63)$$

$$+ \ 0.52(Ipub/Y)_{t-1} + 0.59time - 0.04D1$$
$$(1.72) \qquad\qquad (8.08) \qquad (-2.84)$$

$$R^2 = 0.99$$
Durbin-Watson statistic = 2.02
SER = 0.02
Lagrange multiplier = 0.21
Hendry = 0.01

where $D1$ is a dummy variable that takes a value of 1 in correspondence to the agricultural negative supply shocks in 1981, 1983, and 1987. The wage rate does not contribute significantly to the equation. This may be attributed to measurement errors (we use an indicator of the minimum wage) or to the fact that labor is not a significant constraint. Instead, the cost of capital appears to play a more significant role, together with public investment and the money supply. As a diagnostic tool, we rely on the Lagrange multiplier test for serial correlation; this test does not indicate any significant problems.

Moving now to the investment equation, we find that:

$$\Delta \ln I_t = -3.55 + 1.94 \, \Delta \ln Y_t - 0.69 \ln(I/Y)_{t-1}$$
$$(-1.96) \ (2.21) \qquad\qquad (-2.24)$$

$$- \ 0.76 \ln(c/p)_t + 5.67(B_p/Y)_t$$
$$(-2.0) \qquad\qquad (1.84)$$

$$R^2 = 0.50$$
Durbin-Watson statistic = 1.11
SER = 0.12
Godfrey = 0.02
Sargan = 0.18

where $I$ represents private investment and $B_p/Y$, the ratio of firms' credit to GDP, is used as a proxy of the stringency at an aggregate level of the debt constraint. The equation has been estimated by an instrumental variable procedure, with the fitted value from equation 8.11 as an instrument for $Y$. The restriction that investment be unit elastic with respect to output has been tested ($t_{12} = 0.5$) and imposed on the data.[11]

The estimation of equation 8.10 yields two interesting results. First, investment is significantly affected by the real cost of capital. Fiscal policy can therefore affect the investment decision through its impact on interest rates or, more directly, by changing the set of fiscal and financial incentives available to investors. Second, the level of investment in the economy also depends on the availability of credit. The joint presence of both the cost of capital and credit availability may appear redundant (or even contradictory), until we recall that investment in this model is not fully determined by the stock of credit that financial intermediaries are willing to extend to firms. Even if firms are credit constrained, a change in the interest rate $r$ will still affect investment through its impact on the discount rate $i$ and thus on the choice of retained earnings. By varying their retention behavior, firms can relax the credit constraint somewhat. Fiscal policy therefore will affect investment by influencing either the interest rate or credit availability. This latter channel, as we shall see later, can play a crucial role in determining the macroeconomic outcome of different fiscal policies. It is also plausible, of course, to interpret the joint significance of credit and interest rates in equation 8.10 as implying that only a subset of firms is credit constrained. Finally, by putting together the estimation results for equations 8.10 and 8.11, we can argue that public investment bears a complementary relationship to private investment. Indeed, an increase in public investment will lead to higher-capacity output (equation 8.11) and, through this channel, to higher private investment. By severely cutting public investment, fiscal policy may have contributed in the past to the stagnation of private investment.

### The Saving Decision

A steady supply of domestic savings will ensure that a sustained rate of investment would not be incompatible with existing constraints on external payments. It is therefore essential to gather an adequate understanding of the determinants of saving behavior. Unfortunately, the measurement of saving is beset with difficulties, to the extent that consumption is computed residually in Morocco's national accounts. We can rely on two alternative measures of private saving, one derived as the difference between private disposable income and private consumption, the other based on the saving-investment identity for the economy. Because of statistical inconsistencies, the two procedures do not yield the same result. In what follows we use the first measure. We find that after peaking during the phosphate price boom and declining afterward, the average propensity to save increased steadily during the 1980s.

To model the behavior of saving, we first take a simple permanent income approach. Under well-known conditions, the maximization of intertemporal utility by the representative consumer will imply that aggregate consumption (denoted as *C*) is simply equal to a proportion of permanent income ($Y^p$). That is,

(8.12)                                      $C = kY^p$.

Note that *C* should be defined to include only the consumption of nondurable commodities. Lack of data precludes this important refinement. For estimation purposes, at least two issues need to be addressed. First, we must specify an indicator of permanent income. In what follows, we simply regress the actual value of real disposable income on a time trend and take the fitted value from such an equation as an estimate of $Y^p$. Second, we need to recognize that the parameter *k* will not in general be fixed but will depend on the real interest rate and, possibly, the real exchange rate. The impact of both variables on the propensity to consume, however, is theoretically ambiguous. Our estimated equation reads:

$$\Delta \ln C_t = 0.13 + 0.48(\ln Y_t - C_{t-1}) - 0.17r_t - 0.044 \ln \lambda_t$$
$$(1.08) \quad (6.27) \qquad\qquad (-1.65) \quad (-2.41)$$

$$R^2 = 0.73$$
$$\text{Durbin-Watson statistic} = 2.25$$
$$\text{SER} = 0.015$$
$$\text{Godfrey} = 0.75$$
$$\text{Sargan} = 8.59$$

where *r* and $\lambda$ denote the real interest rate and the real exchange rate, respectively.[12] The latter is defined so that an increase in $\lambda$ implies a real depreciation. The equation is estimated over the 1972–88 period. In the estimation, the restriction that the long-run elasticity of consumption with respect to permanent income be equal to 1 has been tested and imposed in the equation ($t_{14} = 0.81$).[13] Our results indicate that both an increase in the real interest rate and a real depreciation will lead to a decline in the propensity to consume.

This approach highlights two channels through which fiscal policy affects private saving and consumption behavior. First, fiscal policy may influence, through taxes and transfers, the volume of disposable income that consumers can spend. In this framework a temporary tax increase will have a more limited effect than a permanent increase. Second, fiscal policy may affect private saving by influencing the level of interest rates. Our estimates suggest that an expansionary fiscal policy will crowd out both consumption and investment demand through the interest rate channel. A third channel (not allowed for in the preceding estimates) may be at work if rational private agents take fully into account the future tax liabilities associated with bond-financed deficits. The implication is that a shift from tax to debt finance of a given volume of public expenditure should be neutral insofar as it would be matched by offsetting behavior by private agents. This is because private agents would be perfectly aware of the future tax liabilities and the consequent reduction in their permanent disposable income that the increased deficit entails and would reduce their consumption correspondingly (according to the Ricardian equivalence hypothesis). There are several reasons why this proposition may not hold. For example, private agents may discount the future at a different rate than the government, and the presence of capital market imperfections may hinder the intertemporal smoothing of consumption by private agents. Evidence on these issues for developing countries is limited (see, however, Haque 1988; Rossi 1988; Haque and Montiel 1989; Nam 1989; Deaton 1990).

In what follows we amend the previous specification to allow for the possibility that by reducing perceived disposable income, government deficits may have a negative impact on consumption. We do not expect this effect to be particularly strong in Morocco because of the absence of a consumer credit market and the likely pervasiveness of liquidity constraints on households. To capture this effect, we introduce liquid assets (currency plus bank deposits) into our equation. We also allow for inflationary effects on the grounds that a high level of inflation will lead to capital losses on liquid assets and impart an upward bias to the traditional measure of disposable income. Furthermore, an increase in inflation should be associated with greater uncertainty and should lead, through this channel, to an increase in precautionary saving. If we estimate this more complete model, we find that:

$$\Delta \ln C_t = \underset{(8.19)}{0.46}(\ln Y_{t-1} - \ln C_{t-1}) + \underset{(3.92)}{0.06}(\ln M_t - \ln C_{t-1})$$

$$\underset{(-4.66)}{- 0.09} \ln \lambda_t \underset{(-1.64)}{-0.12} \Delta \ln p_t \underset{(-2.40)}{- 0.13}(def/\text{GDP})_t$$

$$R^2 = 0.83$$
$$\text{Durbin-Watson statistic} = 2.16$$
$$\text{SER} = 0.012$$
$$\text{Godfrey} = 0.08$$
$$\text{Sargan} = 6.26$$

where, in standard notation, $p$ indicates the price level (so that $\Delta \ln p$ approximates the inflation rate), $M$ denotes the stock of liquid assets, and $def/$GDP indicates the ratio of the budget deficit to GDP. In this specification, which closely follows Hendry and von Ungern Sternberg (1981), consumers adjust their expenditures to ensure constant steady-state equilibrium ratios of consumption to disposable income and to liquid assets. These long-run equilibrium ratios are affected by the real exchange rate, by inflation, and by budget deficits. The significant impact of budget deficits on consumption in Morocco may seem surprising, but it was already found by Haque and Montiel (1989) and Schmidt-Hebbel and Müller (1990). We have tried to assess whether this effect could be attributed to the fact that government expenditure on goods and services substitutes for private consumption, but the high collinearity between the deficit variable and government expenditure prevented such a test.

Finally, both higher inflation and a depreciating real exchange rate have a negative impact on consumption. The real interest rate is no longer statistically significant and has been excluded from the equation. We can only speculate why the real interest rate no longer seems to be a significant determinant of consumption behavior. One explanation might be that in the previous specification the real interest rate variable was actually picking up the effect of budget deficits. Indeed, as our last estimates suggest, a larger budget deficit lowers consumption and is also likely to be associated with rising interest rates. Overall, this new specification appears to indicate an even stronger role for fiscal policy in influencing consumption behavior.

## Modeling the Impact of Fiscal Policy

A comprehensive analysis of fiscal policy requires an economywide model. In this section, building on the estimates already presented and on previous work by Faini, Porter, and van Wijnbergen (1989), we rely on a simple macroeconometric model to evaluate the impact of fiscal policy in Morocco. A full presentation of the model goes beyond the scope of this chapter but can be found in Faini, Porter, and van Wijnbergen (1989). Table 8.9 in the appendix provides some details about the model. The model is based on a simple variant of the aggregate supply–aggregate demand open-economy framework. Its salient features include the emphasis on supply behavior, the modeling of import demand under rationing, and the analysis of both external and domestic debt dynamics. In what follows, we discuss only some of the main comparative static properties of the model. We then present a number of simulations on the impact of fiscal policy.

Throughout this section, it is assumed that the economy is rationed on international capital markets. As a result, the financeable current account deficit is given at each point of time. As in the model by van Wijnbergen (1989), two equilibrium conditions play a crucial role in determining the equilibrium in the economy. The first is the goods market equilibrium (equation 33 in table 8.9), which can be read as requiring that demand and supply for domestically produced goods be equal. The second condition is the current account constraint, whereby the excess of private and public saving over total investment must be equal to the exogenously given level of the current account. This condition can be derived by combining equations 12–13, 25–26, and 33–34 in table 8.9. The real exchange rate and the real interest rate will move to equilibrate the goods markets and ensure that the current account constraint is not violated. Following van Wijnbergen (1989), the model admits a simple graphical representation. A decline in the interest rate will lead to an excess demand for domestic goods and a current account deficit. Under plausible conditions (fulfilled in our model), a real depreciation (an increase in λ) will be required to meet the current account constraint, while for the goods market equilibrium, λ will need to decrease. As a result, the current account constraint has a negative slope and the goods market equilibrium has a positive slope (figure 8.1).

The impact of fiscal policy will depend to a substantial extent on whether the credit constraint on investment is binding. Consider the first case, in which firms are rationed on the credit market (that is, $B^* < \gamma$). Desired private investment cannot be fully financed from existing sources. Equilibrium can be achieved through (a) an increase in private saving, (b) a drop in notional private investment, or (c) a fall in the demand for financial assets, which frees private saving for investment finance. As shown in the preceding section, changes in private saving and investment will be brought about by variations in

**Figure 8.1. Macroeconomic Equilibrium with a Current Account Constraint**

*r* (real interest rate)

Current account constraint

Goods market equilibrium

λ (real exchange rate)

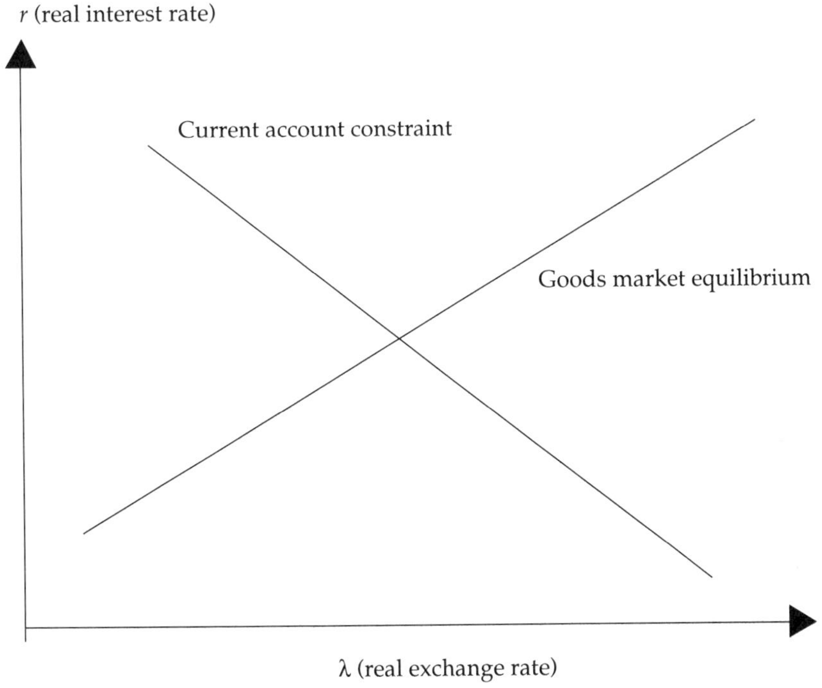

the discount rate, $i$. Indeed, the excess demand for funds will put pressure on the rate at which households and firms make their saving and investment decisions. The discount rate will rise, presumably through the operations of an informal credit market, so as to bridge the gap between retained earnings, household saving, and investment demand. Under this interpretation it is the rate on the secondary credit market that plays a crucial role in the attainment of macroeconomic equilibrium. Changes in the official interest rate are much less effective. Actually, by increasing the demand for financial assets these changes may aggravate the situation of excess demand for funds (van Wijnbergen 1983). In what follows, we also assume that the differential between the secondary market and the official interest rate is kept constant, presumably by monetary authorities, in an attempt to keep the official rate from falling too much off line with respect to actual credit market conditions. Even under these assumptions, the official credit market will not reach equilibrium. As shown in the preceding section, the composition of investment finance (as desired by the firm) does not depend on the interest rate but only on the risk premium. The firm, therefore, is still off its notional demand for credit, since it is unable to achieve the desired composition of its investment finance. Finally, both changes in secondary market interest rates and availability of bank credit will affect investment demand.

The situation is, of course, much simpler if we abstract from the possibility of credit rationing. Under this condition the interest rate on bank loans will adjust to clear the credit market. What is more crucial is that investment demand will depend on a price signal—the level of interest rates—and not on quantity signals (the availability of credit). We begin by considering such a case through three simple simulation exercises.[14]

Column (1) in table 8.6 presents the base solution. In the first comparative static exercise (column 2) we study the impact of an increase in government spending by 3,000 million dirhams (DH), approximately 1.7 percent of GDP. Foreign saving and monetary financing are assumed to be unchanged. From equation 26 in table 8.9 we see that most of the financing will rely on domestic debt. The increase in government spending will cause an excess demand for domestic goods, leading to a decline in saving and thus a current account imbalance. The goods market schedule will shift upward, as will the current account locus. Indeed, for a given real exchange rate, the interest rate needs to increase to stimulate net saving and eliminate excess demand for domestic goods. The current account schedule, however, will shift less. As a matter of fact, higher interest rates lead to a fall in absorption, which is fully reflected in a current account improvement, whereas the reduction in the excess demand for domestic goods depends also on the marginal propensity to spend on

**Table 8.6. Simulation of the Impact of Fiscal Policy, Morocco**

| Variable | Base run (1) | Domestically financed increase in G (2) | Externally financed increase in G (3) | Financial liberalization (4) | Domestically financed increase in Ipub (5) |
|---|---|---|---|---|---|
| r | 3.0 | 4.3 | 4.2 | 4.7 | 4.0 |
| λ | 100.0 | 99.2 | 94.9 | 94.6 | 100.5 |
| I/Y | 18.8 | 17.7 | 18.4 | 18.0 | 17.9 |
| BD/Y | −5.3 | −7.1 | −7.1 | −8.1 | −6.9 |
| Y' | 6.3 | 6.4 | 6.9 | 7.0 | 6.3 |

*Note:* G, government expenditure; r, real interest rate; λ, real exchange rate; I/Y, private investment as share of GDP; BD/Y, budget deficit as share of GDP; Y', growth rate; Ipub, public investment; DH, Moroccan dirhams.

Simulations: 1, base case; 2, increase in government current consumption of DH 3,000 million; 3, as in 2, plus a current account deterioration of DH 3,000 million; 4, as in 3, plus abolition of the liquidity requirement on demand deposits; 5, increase in public investment of DH 3,000 million.

such goods, which is less than 1. As expected, in the new equilibrium (column 2), the real exchange rate will appreciate (because of the excess demand for domestic goods) and the real interest rate will increase (because of the fall in saving). Growth increases somewhat, but the impact of the expansionary fiscal policy is reflected mostly in a crowding-out of investment. Capital accumulation falls because of higher interest rates, although the effect of these rates on the cost of capital is mitigated by the real appreciation. Table 8.6 only presents the effect of a fiscal expansion. Over the medium run the supply impact of the drop in investment offsets the expansionary demand effects of higher public spending: it takes only three years for the initial positive output effect to be reversed. Table 8.7 provides some details about the simulation results over a four-year period.

Suppose now that the increase in government expenditure is financed from abroad. We model this by assuming that the current account is allowed to deteriorate by an amount equal to the increase in public consumption. The outcome of this simulation is presented in column (3) of table 8.6. The higher current account deficit is equivalent to an increase in foreign saving, which fully offsets the fall in domestic saving. Diagrammatically, the current account schedule will not shift. We would then expect the interest rate to increase less than in the case in which no extra financing from abroad was available. This is indeed what we find in column (3). The real interest rate is now equal to 4.2 percent, as against 4.3 percent in the previous simulation. As expected, the real exchange rate appreciation is more pronounced (5.1 compared with 1 percent in the previous simula-

**Table 8.7. Simulation of the Impact of Fiscal Policy, Morocco, 1988–92**

| Variable | 1988 | 1989 | 1990 | 1991 | 1992 |
|---|---|---|---|---|---|
| *Base case* | | | | | |
| Y | 167,235.922 | 177,819.613 | 187,583.305 | 197,081.872 | 207,245.196 |
| ER | 1.000 | 0.999 | 1.030 | 1.061 | 1.046 |
| R | 5.0 | 3.0 | 3.6 | 3.8 | 4.3 |
| IPRIV/Y | 16.3 | 18.8 | 18.2 | 17.4 | 16.9 |
| BD/Y | −3.9 | −5.3 | −5.9 | −6.3 | −7.3 |
| Y' | 10.4 | 6.3 | 5.5 | 5.1 | 5.2 |
| *Domestically financed increase in government current expenditure* | | | | | |
| Y | 167,235.922 | 177,978.036 | 187,673.768 | 197,033.446 | 207,042.953 |
| ER | 1.000 | 0.991 | 1.018 | 1.047 | 1.031 |
| R | 5.0 | 4.3 | 4.8 | 5.1 | 5.8 |
| IPRIV/Y | 16.3 | 17.7 | 16.9 | 16.1 | 15.6 |
| BD/Y | −3.9 | −7.1 | −7.8 | −8.3 | −9.4 |
| Y' | 10.4 | 6.4 | 5.4 | 5.0 | 5.1 |
| *Externally financed increase in government current expenditure* | | | | | |
| Y | 167,235.922 | 178,844.371 | 188,661.661 | 198,082.640 | 208,168.847 |
| ER | 1.00 | 0.948 | 0.976 | 1.010 | 1.001 |
| R | 5.0 | 4.2 | 4.5 | 4.6 | 5.1 |
| IPRIV/Y | 16.3 | 18.4 | 17.8 | 17.1 | 16.6 |
| BD/Y | −3.9 | −7.1 | −7.5 | −7.8 | −8.7 |
| Y' | 10.4 | 6.9 | 5.5 | 5.0 | 5.1 |
| *Externally financed increase in government current expenditure plus financial liberalization* | | | | | |
| Y | 167,235.922 | 178,903.272 | 188,682.372 | 198,059.977 | 208,099.632 |
| ER | 1.000 | 0.945 | 0.972 | 1.006 | 0.995 |
| R | 5.0 | 4.7 | 5.0 | 5.1 | 5.7 |
| IPRIV/Y | 16.3 | 18.0 | 17.4 | 16.7 | 16.1 |
| BD/Y | −3.9 | −8.1 | −8.3 | −8.8 | −9.9 |
| Y' | 10.4 | 7.0 | 5.5 | 5.0 | 5.1 |
| *Domestically financed increase in public investment* | | | | | |
| Y | 167,235.922 | 177,722.467 | 187,580.965 | 197,162.728 | 207,392.948 |
| ER | 1.000 | 1.005 | 1.035 | 1.066 | 1.050 |
| R | 5.0 | 4.0 | 4.4 | 4.7 | 5.4 |
| IPRIV/Y | 16.3 | 17.9 | 17.1 | 16.3 | 15.7 |
| BD/Y | −3.9 | −6.9 | −7.6 | −8.2 | −9.3 |
| Y' | 10.4 | 6.3 | 5.5 | 5.1 | 5.2 |

*Note: Y*, real GDP at factor cost (millions of real dirhams); *ER*, real exchange rate (1988–91); *R*, real interest rate (percent); IPRIV, private investment as a percentage of GDP; *BD*, budget deficit as a percentage of GDP; *Y'*, growth rate of real GDP.

tion). Both the greater real appreciation and the more limited increase in the real interest rate stimulate investment, in comparison with the case in which foreign financing was unavailable. GDP growth is therefore more sustained. We find that the level of GDP is always higher than its base-case value over the full simulation period (see table 8.7).

We next consider the case in which a foreign-financed increase in government expenditure is accompanied by a process of financial liberalization. We model this process as a reduction in the liquidity requirement imposed on demand deposits. There are certainly beneficial influences from a more competitive and less heavily regulated financial sector, but these are not modeled in our setup. There are also some macroeconomic costs, as shown in column (4) of table 8.6. The direct impact of the reduction in the liquidity requirement is to increase interest expenses on public domestic debt and force the government to rely to a larger extent on domestic credit markets. Interest rates increase, with a further expansionary effect on public spending. The simulation indicates that the budget deficit would increase as a share of GDP by a full percentage point, with a crowding-out effect on private investment. Against the long-run benefits from financial liberalization, therefore, should be placed the macroeconomic costs stemming from a further deterioration of the government budget.

The results would change somewhat if the expansionary fiscal policy stance were characterized by higher capital rather than current government expenditure. Column (5) of table 8.6 shows the impact of an increase of DH 3,000 million in public investment. The resulting deficit is assumed to be financed domestically. Comparison of column (5) with column (2) shows that the impact on output of larger capital expenditures is smaller in column (5). Indeed, because of the relatively larger import content of investment, demand for domestic goods increases by less, and the current account deterioration is more pronounced. The real exchange rate therefore must depreciate rather than appreciating as it did in column (2). In turn, the real depreciation has a negative impact on output. In the medium run, though, the larger volume of public investment boosts both investment and supply. The contrast with the effect of an increase in current expenditure is worth stressing. Following an increase in public investment, output in the medium run is systematically larger than in the base case (see table 8.7).

Consider now the case in which the interest rate on the primary credit market is not allowed to clear the loan market. It is assumed that $B^* < \gamma$; that is, firms are rationed on the official credit market. As shown in the preceding section, both credit availability and the (secondary credit market) interest rate will affect investment demand. The base-case simulation is presented in column (1) of table 8.8. Suppose again that government current consumption is increased by DH 3,000 million. The new equilibrium is reported in column (2). As expected, an expansionary fiscal policy brings both a real appreciation and higher interest rates. Compared with table 8.6 (where credit mar-

**Table 8.8. Simulation of the Impact of Fiscal Policy, Morocco: The Credit Rationing Case**

| Variable | Base run (1) | Domestically financed increase in G (2) | Externally financed increase in G (3) | Financial liberalization (4) |
|---|---|---|---|---|
| $r$ | 0.9 | 1.2 | 1.9 | 1.9 |
| $\lambda$ | 100.0 | 98.6 | 94.6 | 94.0 |
| $I/Y$ | 17.1 | 15.2 | 16.5 | 15.6 |
| $BD/Y$ | −5.6 | −7.4 | −7.4 | −8.4 |
| $Y'$ | 6.5 | 6.6 | 7.1 | 7.2 |
| $B_p/Y$ | 18.7 | 17.0 | 18.5 | 17.5 |

Note: G, government expenditure; $r$, real interest rate; $\lambda$, real exchange rate; $I/Y$, private investment as share of GDP; $BD/Y$, budget deficit as share of GDP; $Y'$, growth rate; $B_p/Y$, credit to the private sector as share of GDP.

Simulations: 1, base case; 2, increase in government current consumption of DH 3,000 million; 3, as in 2, plus a current account deterioration of DH 3,000 million; 4, as in 3, plus abolition of the liquidity requirement on demand deposits.

kets were allowed to clear), however, the impact on the level of interest rates is more limited, whereas the effect on investment is substantial (almost 2 percentage points of GDP compared with a 1 percentage point drop in table 8.6). The reason behind this result is relatively simple. Under credit rationing, the crowding-out effect of a larger volume of government spending takes place mostly through lower availability of credit to the private sector. As a matter of fact, we find that the share of private sector credit in GDP declines by almost 2 percentage points following the fiscal expansion. There is, therefore, less of a need for interest rates to increase to restore macroeconomic equilibrium.

Consider, finally, the case in which the increase in government expenditure is financed from abroad. We would again expect the interest rate to increase less than when foreign financing was unavailable. This is not, however, what we find in column (3). The real interest rate is now equal to 1.9 percent, compared with 1.2 percent in the preceding simulation. The reason for this somewhat surprising result again hinges on the crucial role the credit market plays in this model. What is happening is that the possibility for the government to finance a larger portion of its deficit abroad reduces the pressure that government borrowing puts on credit markets, leaving a larger share of total domestic credit for the private sector. Investment, therefore, is not crowded out through credit rationing and declines less than in the previous simulation. A larger increase in the interest rate and a more sustained real appreciation are required to achieve equilibrium in the goods market. When, as in the first simulation, the higher government expenditure was financed domestically, most of the crowding-out of investment took place through credit rationing, with more limited effects on the interest rate. The differences in outcome between the two simulations highlight the crucial role that conditions in the credit market play in determining the impact of fiscal policy.

## Conclusions

Morocco has made great progress toward macroeconomic and fiscal stability. Yet the need to consolidate and broaden the achievements to date remains paramount. We have argued that even the financing of a relatively unchanged budget deficit may pose major problems. It is unlikely that foreign finance will increase substantially in the medium run. Monetary financing does not seem to be a palatable alternative, given its highly inflationary implications. Finally, increasing reliance on domestic financial markets is likely to lead to a steep increase in interest costs for the treasury, with a destabilizing effect on the evolution of the main public debt indicators. Our simulations also suggest that an increase in government current expenditure crowds out investment, so that the short-run benefits on output of an expansionary fiscal policy will be outweighed by its long-run negative impact on growth. It is essential, therefore, that the commitment toward fiscal discipline remain unshaken. At the same time a determined effort is required to implement effective reforms in the tax and public expenditure system, to avoid having the brunt of fiscal adjustment again fall mostly on public investment. Our results indicate that because of the complementarity between public and private investment, reductions in public capital expenditures would lead to lower growth, creating a detrimental effect on tax revenues and the budget.

At a more general level, we have argued that the effects of budget deficits cannot be measured by looking only at their impact on aggregate demand through changes in the interest rate. As shown by Blinder and Stiglitz (1983), Blinder (1987), and Bernanke and Blinder (1988), a large part of the impact of fiscal policy may be felt through the credit market. An expansionary fiscal policy may exacerbate the pervasiveness of credit-rationing effects on investment demand, with a limited impact on the level of the interest rate. Our estimates support the claim that, even after controlling for the cost of capital, the availability of credit plays a significant role in influencing the demand for investment. As suggested by our simulations, the credit channel in turn will be a crucial factor in determining the macroeconomic outcome of different fiscal policies. Finally, we argue that the impact on macroeconomic equilibria, in particular on the government budget, should be a relevant factor in assessing the speed of financial liberalization.

## Appendix. The Model

In this appendix we present a brief description of the macro-econometric model used to simulate the impact of fiscal policy. A full presentation of the model goes beyond the scope of this chapter and can be found in Faini, Porter, and van Wijnbergen (1989). The presentation is organized by economic agent (firms, households, government, and so forth) rather than by the more usual approach based on a distinction between markets (goods, labor, and money markets). This should permit a more critical evaluation of the microeconomic foundations of the model.

Table 8.9 can be used as a guide to the main relationships included in the model. The first block of equations focuses on the firm. The approach used to specify the behavior of the representative firm was described in "Investment and Saving Decisions," above, when analyzing the investment decision. In equation 1 in the table, investment ($I$) is a function of expected output ($Y^e$), the cost of capital ($c$) and, possibly, depending on the way the model is specified, the availability of credit ($B_p$). The cost of capital (equation 2) is a function of the tax ($\mu$ and $r$) and depreciation ($\delta$) parameters, as well as of the interest rate, $r$, and the price of investment goods, $p_I$. In turn $p_I$ (equation 3) is equal to a weighted average of the price of domestic ($p_{di}$) and imported capital goods, with the latter depending on the real exchange rate ($\lambda$, defined as the ratio of foreign to domestic prices) and on the tariff rate of domestic capital goods ($r_{mi}$). The specification of expected output is taken from equation 8.11 in the text. After choosing total investment, the firm decides how to allocate it between domestic and foreign capital goods as a function of their relative prices and of the extent of quantitative restrictions on imports of capital goods, $q_I$ (equation 5).

Short-run choices can be described as follows. Output supply ($Y$) depends on total capital stock and the real exchange rate, where the latter acts as a proxy for variable costs. A real depreciation, that is, an increase in $\lambda$, will boost wage and intermediate input costs and lower supply. Given that capital stock data are not available in Morocco, we take a first quasi difference of the original supply function and estimate equation 6 of table 8.9. In the estimation we take the primary and the government sector output levels to be exogenous. After output supply is determined, the demand for intermediate inputs ($M_n$) and the supply of exports ($X$) can be simply described in equations 7 and 8, respectively, as a function of $Y$, the wage rate ($w$), and the relevant price variables ($p_{mn}$ and $p_x$). We do not, however, assume the country to be small in export markets. The price of exports, $px$, depends therefore on export volume, world demand ($WD$), and foreign competitors' prices ($p^*$).[15]

The choices of households are described in equations 10–13 of table 8.9. We make a separability assumption in which consumers first determine the level of saving and aggregate consumption and then allocate total consumption between domestic and foreign commodities. Aggregate private consumption (equation 10) is modeled as in the previous section. In the present version of the model we take the specification in which private consumption depends on disposable income ($Y^d$), the real interest rate ($r$), and the real exchange rate ($\lambda$). The allocation of aggregate consumption between domestically and foreign produced commodities is described by a simple (constrained) linear expenditure system. We allow for the fact that a subset of foreign consumption goods cannot be freely imported, and we study the impact of these rationing measures on total imports of consumption goods. Our approach parallels that used for investment imports and draws on Bertola and Faini (1990). As with investment goods, we do not allow for any rationing effect on aggregate consumption. The resulting equation is highly nonlinear but can be simply described (equation 11) as stating that imports of consumption goods depend on total consumption, relative prices, and the extent of quotas on such import categories ($q_c$). Equation 12 defines disposable income as the sum of GDP, net factor income from abroad (NFI), interest on public domestic debt (GDINT) and foreign debt (GXINT); transfers ($TR$) minus total taxes ($T$); and monetary financing ($MF$). Note that interest payments on foreign debt (GXINT) are already included, with a negative sign, in NFI. Then NFI + GXINT is a measure of net factor payments from abroad accruing to the private sector (mainly workers' remittances). Finally, equation 13 defines private saving.

The behavior of the government sector is described very simply by a set of accounting and technical identities. As far as taxation is concerned, we distinguish between trade (equation 14) and income (equation 16) taxes. Import duties are estimated endogenously in the model by applying the relevant duty rate to imports of consumption, investment, intermediate, and other commodities. We also allow for the temporary levy on petroleum (equation 15), which was introduced in 1986 and should be gradually phased out over the next few years. Information on the revenue likely to be generated by this levy was provided by IMF sources. Finally, the income tax rate is determined residually from total revenue data. Equation 17 defines total tax revenue. Government expenditures are also fairly disaggregated. We distinguish between government investment (equation 19), government current spending on goods and services (equation 20), current transfers (equation 18), and interest payments on both domestic

## Table 8.9. The Complete Model, Morocco

| Equation | Item |
| --- | --- |
| *Firms* | |
| (1) $I = I(Y^e, c, B_p)$ | Private investment demand |
| (2) $c = p_I(1 - \mu)\,[r(1 - \tau) + \delta]$ | Cost of capital |
| (3) $p_I = \beta_d p_{di} + (1 - \beta_d)(1 + \tau_{mi})\lambda$ | Price of investment goods |
| (4) $Y^e = Y^e[w(-1), c(-1), Ipub(-1), MS(-1)]$ | Expected output[a] |
| (5) $M_I = M_I(I, p_{di}/p_{mi}, q_I)$ | Imports of investment goods |
| (6) $Y = Y[Y(-1), I, \lambda, \lambda(-1)]$ | Supply of goods |
| (7) $M_n = M_n(Y, w/p_{mn})$ | Imports of intermediate goods |
| (8) $X = X^s(Y, p_x/w)$ | Export supply of manufactured goods |
| (9) $p_x = p_x(X, WD, p^*)$ | Export demand of manufactured goods |
| *Households* | |
| (10) $C = C(Y^d, r, \lambda)$ | Private consumption |
| (11) $M_c = M_c(C, p_{dc}/p_{mc}, q_c)$ | Imports of consumption goods |
| (12) $Y^d = Y + \text{NFI} + \text{GDINT} + \text{GXINT} + TR$ $\qquad - T - MF$ | Disposable income |
| (13) $S_p = Y^d - C$ | Private saving |
| *Government* | |
| (14) $TM_j = \tau_j \lambda M_j$ | Import taxes ($i = n, c, o, i$) |
| (15) $TP = t_p Y$ | Petroleum tax |
| (16) $TY = t_y Y$ | Other taxes |
| (17) $T = \Sigma\, TM_i + TP + TY$ | Total taxes |
| (18) $TR = t_r Y$ | Transfers |
| (19) $Ipub = i_{pub}\, Y$ | Public investment |
| (20) $G = gY$ | Government expenditure on goods and services |
| (21) $\text{GDINT} = i_{dd} B_g(-1)$ | Interest payments on domestic debt |
| (22) $\text{GXINT} = i_{xd} XD(-1)$ | Interest payments on foreign debt |
| (23) $\Delta i_{dd} = \Delta r + \Delta \pi$ | Interest rate on domestic debt |
| (24) $S_g = T - \text{GDINT} - \text{GXINT} - TR - G$ $\qquad + MF$ | Government saving |
| (25) $XD + XD(-1)\lambda/\lambda(-1) - CA$ | External debt |
| (26) $\Delta B_g - CA = I_{pub} - S_g$ | Government budget constraint |
| *Credit and the money markets* | |
| (27) $CUR = CUR(Y, i^{td}, \pi)$ | Demand for currency |
| (28) $DD = DD(Y, i^{td}, \pi)$ | Demand for demand deposits |

**Table 8.9** *(continued)*

| Equation | Item |
|---|---|
| *Credit and the money markets (continued)* | |
| (29) $TD = TD(Y, i^{td}, \pi)$ | Demand for time deposits |
| (30) $MF = \Delta CUR + rr^{dd} \Delta DD$ | Monetary financing |
| (31) $B_p = DD + TD - R + B_g$ | Credit to the private sector |
| (32) $\Delta i^{td} = \Delta r + \Delta \pi$ | Interest rate on time deposits |
| *National income identities* | |
| (33) $Y = C + G + X - M + I + I_{pub} - \Sigma\, TM_i$ | Goods market equilibrium |
| (34) $CA = X - M + $ NFI | Current account |

*Notes:* $\lambda$, real exchange rate; $r$, real interest rate; $MS$, money supply; $q_i$, extent of quantitative restrictions (QRs) on imports of type $i$; $B_p$, credits to firms; $p^*_{mj}$, foreign currency price of imports of type $j$; $p_{mj}$, domestic currency price of imports of type $j$; $p_{dj}$, domestic price of good $j$; $p_x$, export price; $r_j$, tariff rate on good $j$; $r$, corporate tax rate; $w$, wage rate; $WD$, world demand; NFI, net factor income from abroad; $\pi$, inflation rate; $R$, bank reserves; $rr^{dd}$, reserve requirement for demand deposits.
   a. Not implemented in the present version of the model.

debt ($B_g$ in equation 21) and foreign debt ($XD$ in equation 22). We can therefore account in our simulations for debt dynamics. The accumulation of foreign debt is determined by the current account deficit and by real exchange rate variations that induce a capital gain or loss in the value of outstanding external debt (equation 25). The change in domestic debt ($\Delta B_g$) is equal to that part of the budget deficit which cannot be financed abroad or by seigniorage and the inflation tax (equation 26). Simple manipulations show that $\Delta B_g$ can also be expressed as the difference between private saving and private investment. The last item in the consolidated government Central Bank budget identity is revenue from seigniorage and the inflation tax (equation 24).

Government revenue from seigniorage and the inflation tax is determined by the private sector's choice of assets. Following the analysis in "The Budget Deficit: Evolution and Financing," above, demands for currency (equation 27), for demand deposits (equation 28), and for time deposits (equation 29) are simply modeled as functions of income ($Y$), inflation ($\pi$), and the interest rate on time deposits ($i^{td}$). In equation 30, monetary financing ($MF$) is defined as the change of narrowly defined monetary base—that is, of currency and of demand deposits, the latter multiplied by the required reserve ratio ($rr^{dd}$). Finally, conditions in the money market will help determine, through the banks' balance sheet identity, the equilibrium in the credit market. The sum of demand and time deposits defines total commercial bank liabilities, which are allocated on the assets side among compulsory reserves ($R$), credit to the private sector ($B_p$), and credit to the government ($B_g$) (equation 31). It is assumed that banks do not hold free reserves. Credit to the government has already been determined by the government budget constraint for a given current account deficit and given monetary financing. From equation 31 we find that credit to the private sector is determined residually. Depending on the way the model is specified, the availability of credit may in turn influence investment demand. In such a case fiscal policy will affect investment not only through its impact on interest rates, but also, more directly, by influencing the availability of credit.

To close the model, we need to specify the equilibrium condition in the goods markets (equation 33) where total demand for domestic goods (the sum of consumption, investment, government spending on goods and services, and the resource balance) must equal aggregate supply ($Y$). A second condition relates to the current account constraint (equation 34), which is identically defined as the sum of the resource balance and net factor income from abroad (NFI). It is easy to verify that

$$S_p + S_g - CA = I + Ipub.$$

That is, the sum of private, government, and foreign saving (the negative of $CA$) is, in equilibrium, equal to total investment.

Table 8.9 presents the complete model in summary form. There are, altogether, 34 equations and 35 endogenous variables—32 endogenous left-hand side variables (allowing for the fact that $Y$ appears twice and the monetary base, $MB$, is exogenous), plus the real exchange rate, the real interest rate, and the inflation rate. In the text, we present a number of simulations on the impact of fiscal policy. The common assumption underlying all the simulations is that the economy is rationed on international capital markets. As a result the financeable current account deficit is given at each point of time; that is, $CA$ is exogenous.

## Notes

This chapter is an offspring of joint work with John Porter and Sweder van Wijnbergen. I am very grateful to both of them for stimulating discussions and suggestions. The chapter has also greatly benefited from the extensive comments of Klaus Schmidt-Hebbel. I would also like to thank Tobias Müller and Klaus Schmidt-Hebbel for providing me with crucial data and a preview of their paper, which has been a constant source of ideas and insights. Finally, I am very grateful to Roberto Fumagalli for skillful research assistance. The responsibility for any remaining errors is mine alone.

1. The current account worsened in 1989, mostly because of delivery problems with phosphoric acid exports. Similarly, the budget deficit registered a small slippage in the same year because of an increase in public investment. Both the current account and the budget deficit improved substantially in the

following years. In 1991 the current account deficit was 1.0 percent of GDP and the budget deficit was 3.2 percent of GDP.

2. Between 1989 and 1991 GDP increased at an annual average rate of 4.4 percent. There was, however, a significant increase in inflation, with the consumer price index increasing by 8.0 percent in 1991. Much of the worsening inflation performance can be attributed to the depreciation of the nominal exchange rate.

3. Public investment fell again to 7.1 percent of GDP in 1990 and 6.1 percent of GDP in 1991.

4. Similarly, in 1990 total revenues as a percentage of GDP increased to 23.8 percent of GDP, accounting for a large share of the improvement in the budget. However, most of the increase in tax revenue must be attributed to higher taxes on international trade flows, which climbed from 4.1 percent of GDP in 1988 to 4.9 percent in 1991.

5. We have run two simple regressions of total government revenues ($R$) and government expenditure on goods and services ($GS$) on the following variables: real GDP ($Y$), the real exchange rate ($\lambda$), and the inflation rate ($\pi$). For both regressions, the hypothesis of a unit-elasticity with respect to $Y$ was not rejected at very comfortable significance levels ($F_{1,10} = 0.04$ and $F_{1,10} = 0.33$, respectively):

$$\ln T = 2.59 + \ln Y - 0.40\,\pi - 0.24\lambda.$$
$$(3.81) \qquad (-1.0) \quad (-1.57)$$

$$R^2 = 0.22$$
$$\text{Durbin-Watson statistic} = 2.56$$
$$\text{SER} = 0.03$$

$$\ln GS = -5.3 + \ln Y + 0.06\pi - 0.74\lambda.$$
$$(-7.99) \qquad (0.15)\ (-4.99)$$

$$R^2 = 0.70$$
$$\text{Durbin-Watson statistic} = 1.75$$
$$\text{SER} = 0.03$$

(Figures in parentheses are $t$-statistics.) These results indicate that while inflation plays no significant role in both equations, a real appreciation (that is, a drop in $\lambda$) will lead to higher revenues and expenditures. Presumably, if the real appreciation is brought about by higher tariffs and better terms of trade, both factors should also lead to higher revenues. In turn, more buoyant revenues may prompt the government to increase the expenditure on goods and services ($GS$).

6. The calculations are based on standard formulations (Buiter 1985; Anand and van Wijnbergen 1989). In computing the sustainable aggregate deficit ($ds$), we follow Haque and Montiel (1991):

$$d^s = (\pi + n)(b + b^* + m) - e'\,b^*$$

where, with standard notation, $\pi$ and $n$ denote the inflation rate and the growth rate of GDP, $b$, $b^*$, and $m$ indicate, respectively, the ratios of domestic debt, foreign debt, and monetary base to GDP, $e$ is the nominal exchange rate, and a prime denotes a proportional rate of change. The sustainable primary deficit ($pd^s$) is equal to:

$$pd^s = (n - r)b + (n - r^* - \lambda')b^* + (\pi + n)m$$

where $r$ and $r^*$ are the domestic and foreign real interest rates, respectively, while $\lambda$ indicates the real exchange rate (defined so that an increase in $\lambda$ implies a real depreciation).

7. Recall that at the end of 1988, the stock of domestic debt was equal to approximately 40 percent of GDP. The computation assumes, quite conservatively, that the average cost of servicing domestic debt at market rates could rise by 5 percentage points from its present level of 6 percent.

8. Equation 8.3 has been reparametrized to test for a unitary elasticity of currency with respect to income. Take the simple case in which $a_4 = 0$. The long-run elasticity of $M/p$ with respect to $Y$ is then equal to $a_3/(1 - a_8)$. By subtracting the lagged value of $\ln(M/p)$ from both sides of equation 8.3, we can test whether $a_3/(1 - a_8) = 1$, that is, whether $a_3 = 1 - a_8$, by simply checking whether the coefficients of $\ln Y_t$ and $\ln(M/p)_{t-1}$ sum to zero. The same specification is used for the time-deposit and the consumption equations.

9. For a seminal contribution to the empirical evaluation of the relationship between capital market imperfections and investment decisions, see Fazzari, Hubbard, and Petersen (1988).

10. The formal solution to the firm's optimization problem is available from the author.

11. As diagnostic tools we rely on the Godfrey and Sargan tests for residual autocorrelation and overidentifying restrictions, respectively. These are the appropriate procedures in the estimation context of instrumental variables. Both tests are distributed as $\chi^2$ with 1 degree of freedom. They do not provide any indications of misspecification.

12. The equation was estimated by an instrumental variable procedure to allow for the possible endogeneity of income, $r$, and $\lambda$. The Sargan test is distributed as $\chi^2$ with 5 degrees of freedom.

13. See note 8. The lagged value of disposable income was never statistically different from zero and was therefore dropped from the final equation.

14. We do not examine in the simulations the effect of financing the larger volume of government spending through money creation. The impact of monetary financing has already been investigated in "Asset Demand, Seigniorage, and the Inflation Tax," which pointed to the fairly unfavorable tradeoff between inflation deficit monetization in Morocco. The results were derived from a simple partial equilibrium rather than an economywide model. It should be recalled, however, that in our macroeconometric model the level of real interest rates is virtually determined without reference to the money market, the only exception being the impact of the inflation tax on disposable income and, thereby, on consumption. Empirically, this effect appears to be negligible. As a result, changes in the growth of money supply will be reflected mostly in inflation.

15. In the empirical implementation of the external sector block of the model, only manufacturing exports, net tourist receipts and consumption, investment, and intermediate goods imports are endogenous. All the other components of the resource balance (mainly exports of phosphate rocks and agricultural goods and imports of food and petroleum) are assumed, in the

present version of the model, not to depend on price incentives and are, as a result, projected exogenously. In a future version of the model it would be important to endogenize the flow of workers' remittances (a component of net factor income from abroad), by relating it to the level of the interest rate and the (expected) movements of the exchange rate.

## References

Anand Ritu, and Sweder van Wijnbergen. 1989. ''Inflation and the Financing of Government Expenditure: An Introductory Analysis with an Application to Turkey.'' *World Bank Economic Review* 3 (1): 17–38.

Bernanke, Ben S., and Alan S. Blinder. 1988. ''Credit, Money, and Aggregate Demand.'' NBER Working Paper 2534. National Bureau of Economic Research, Cambridge, Mass.

Bertola, Giuseppe and Riccardo Faini. 1990. ''Import Demand and Non-Tariff Barriers: The Impact of Trade Liberalization.'' *Journal of Development Economics* 34: 269–86.

Blinder, Alan S. 1987. ''Credit Rationing and Effective Supply Failures.'' *Economic Journal* 91: 327–52.

Blinder, Alan S., and J. Stiglitz. 1983. ''Money, Credit Constraints and Economic Activity.'' *American Economic Review, Papers and Proceedings* 73 (May): 297–302.

Buiter, Willem H. 1985. ''A Guide to Public Sector Debt and Deficits.'' *Economic Policy* 1 (November): 13–79.

Deaton, Angus. 1990. ''Saving in Developing Countries: Theory and Review.'' *Proceedings of the World Bank Annual Conference on Development Economics 1989*, pp. 61–96. Washington, D.C.: World Bank.

Faini, Riccardo, and Jaime de Melo. 1990. ''Adjustment, Investment and the Real Exchange Rate.'' *Economic Policy* 5: 492–519.

Faini, Riccardo, John Porter, and Sweder van Wijnbergen. 1989. ''Trade Liberalization, Budget Deficits and Growth.'' World Bank, Europe, Middle East, and North Africa Department, Washington, D.C.

Fazzari, F. M., G. Hubbard, and C. Petersen. 1988. ''Financing Constraints and Corporate Investment.'' *Brookings Papers on Economic Activity* 1: 141–206.

Haque, Nadeem Ul. 1988. ''Fiscal Policy and Private Saving Behavior in Developing Economies.'' *International Monetary Fund Staff Papers* 35: 316–35.

Haque, Nadeem Ul, and Peter Montiel. 1989. ''Consumption in Developing Countries: Tests for Liquidity Constraints and Finite Horizons.'' *Review of Economics and Statistics* 71 (August): 408–15.

————. 1991. ''Macroeconomics of the Public Sector Deficit: The Case of Pakistan.'' Policy Research Working Paper 633. World Bank, Washington, D.C.

Hendry, David, and T. von Ungern Sternberg. 1981. ''Liquidity and Inflation Effects in Consumers Expenditures.'' In Angus Deaton, ed., *Essays in Theory and Measurement of Consumer Behaviour*. Cambridge, U.K.: Cambridge University Press.

Nam, Sang-Woo. 1989. "What Determines National Saving? A Case Study of Korea and the Philippines." Policy Research Working Paper 205. World Bank, Washington, D.C.

Pagan, Adrian. 1984. "Econometric Issues in the Analysis of Regressions with Generated Regressors." *International Economic Review* 25 (February): 221–47.

Rossi, Nicola. 1988. "Government Spending, the Real Interest Rate and the Behavior of Liquidity-Constrained Consumers in Developing Countries." *International Monetary Fund Staff Papers* 35:104–40.

Schmidt-Hebbel, Klaus, and Tobias Müller. 1990. "Private Investment under Macroeconomic Adjustment in Morocco." In Ajay Chhibber, Mansoor Dailami, and Nemat Shafik, eds., *Reviving Private Investment in Developing Countries*. Amsterdam: North-Holland.

United Nations–World Bank. 1990. "Morocco 2000: An Open and Competitive Economy." Trade Expansion Program.

van Wijnbergen, Sweder. 1983. "Interest Rate Management in LDCs," *Journal of Monetary Economics* 12 (3): 433–52.

———. 1989. "Growth, External Debt and the Real Exchange Rate in Mexico." Policy Research Working Paper 257. World Bank, Europe, Middle East, and North Africa Department, Washington, D.C.

World Bank. 1990. "Morocco: Sustained Investment and Growth in the 1990s." Europe, Middle East, and North Africa Department, Washington, D.C.

# 9

# Pakistan: Fiscal Sustainability and Macroeconomic Policy

*Nadeem Ul Haque and Peter J. Montiel*

Over the past two decades Pakistan has experienced fiscal deficits that have been very large in relation to the size of its economy, by international standards. The country's authorities have made repeated attempts, including several adjustment programs, to deal with fiscal imbalances over this period, but they have achieved only temporary successes. The deficit of the federal and provincial governments combined averaged about 6.75 percent of gross national product (GNP) during the 1980s and amounted to 7.75 percent in fiscal 1991/92, the most recent year for which data are available.[1] As in many developing economies, the deficit remains high because of the government's political and administrative inability to mobilize additional resources and cut current expenditures. Weaknesses in the tax system have led to an inelastic tax structure and a heavy reliance on trade taxes for revenues. Moreover, with defense expenditures constituting about 25 percent of expenditures, interest payments 15 percent, and administration (including social services) another 15 percent, a large fraction of expenditures is not amenable to large cuts. The burden of expenditure cuts, therefore, has often fallen on development expenditure, at the cost of much-needed investments in infrastructure.

In spite of the similarities to conditions elsewhere, the macroeconomic consequences of fiscal deficits in Pakistan have apparently been quite dissimilar to those in other developing countries with fiscal deficits of comparable magnitude. Specifically, Pakistan has experienced neither hyperinflation nor debt rescheduling. As measured by official figures, growth has remained quite strong through the past two decades, inflation has not been high, and the current account deficit, at an average of about 2.5 percent of GNP, has remained largely financeable and has not posed debt-servicing problems for the country. For these reasons, Pakistan presents an interesting contrast to some of the other countries in this book. This chapter focuses on why Pakistan's extremely high observed fiscal deficits proved relatively benign during the decade of the 1980s—at least in comparison with the experience of several Latin American countries.

413

The analysis concentrates on the 1980s because the political setting was relatively stable during this period under the martial law government that assumed power in 1977 and held it until the election of a new democratic government in 1988. In addition, Pakistan undertook a comprehensive revision of its macroeconomic data in 1989, and available data compiled under the new methodology go back only to fiscal 1980/81. Thus, for the purpose of empirical work, time series of useful length are available only under the old methodology, which was applied up to fiscal 1987/88. Moreover, although macroeconomic data are available for the period after 1988, as of the time of writing, information for the period 1989/90 to 1991/92 remains provisional or incomplete.

The first section of this chapter begins with a brief overview of macroeconomic developments in Pakistan over the past two decades. This is followed by a more detailed look at fiscal developments that is intended to address the question of why fiscal deficits have remained consistently high. The macroeconomic consequences of these deficits are examined in the subsequent three sections, which describe calculations of equilibrium deficits, estimates of the effects of fiscal policy variables on other macroeconomic relationships, and counterfactual simulations of alternative historical fiscal policies. The main conclusions are presented in the last section.

## Overview

The distinguishing feature of Pakistan's recent macroeconomic history has been the country's relatively good macroeconomic performance, in spite of fiscal deficits that were high by international standards. This is not to say that fiscal deficits of the magnitudes observed have not had harmful effects or that performance could not have been improved with lower deficits. But there is no evidence in Pakistan of the recurring acute macroeconomic crises—as manifested in extended periods of negative growth of income per capita, hyperinflation, and inability to service external debt—that have characterized many other developing countries with comparable fiscal performance.

As figure 9.1 shows, the deficit (as a percentage of GNP) remained very high over a period of nearly two decades, amounting, on average, to about 7 percent of GNP during 1972–88.[2] Although some fiscal adjustment was achieved during fiscal 1977/78 to 1981/82, when the extremely high deficits of the 1974–78 period were nearly cut in half, the fiscal deficit was on a rising trend during most of the 1980s. Over these years Pakistan's fiscal deficit averaged nearly twice that of Asian countries as a group, according to the International Monetary Fund's *World Economic Outlook* (IMF 1990).

**Figure 9.1. Consolidated Deficit of the Federal and Provincial Governments, Pakistan, Fiscal 1972/73 to Fiscal 1987/88**

Percentage of GNP

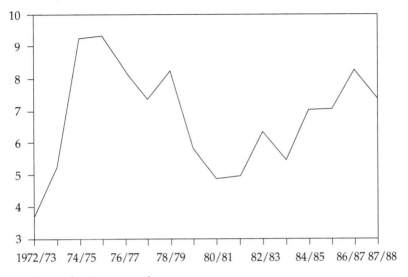

*Source:* Country case study.

Figure 9.2 shows that inflation performance in Pakistan appears to have been remarkably good, whether measured by the consumer price index (CPI), the wholesale price index (WPI), or the GDP deflator. The peak inflation rate during this period, in 1973/74, amounted to 33 percent, as measured by the GDP deflator. After 1976 the inflation rate averaged less than 10 percent per year, falling in recent years to about 5 percent. At the same time, economic growth has been robust (figure 9.3), averaging more than 7 percent during the 1980s and never falling below 3 percent per year over the entire 1972–88 period. As a result, real GNP per capita exhibited a continuously rising trend during this time, with a cumulative increase of about 60 percent.[3]

Turning to the external sector, figure 9.4 depicts the ratio of the current account to GNP. After a peak deficit of more than 8 percent of GNP in 1974/75, Pakistan achieved a substantial current account adjustment, registering a small surplus by 1982/83. Although the country benefited during this period from a substantial increase in external receipts in the form of worker's remittances (primarily from Pakistani workers employed in Middle Eastern oil-exporting countries), it is noteworthy that this boon contributed to current account adjustment rather than to an import binge. Expressed as a proportion of exports of goods and nonfactor services, Pakistan's external debt service ratio remained relatively low, even by Asian standards—about

**Figure 9.2 Inflation, Pakistan, Fiscal 1972/73 to Fiscal 1987/88**

Percent

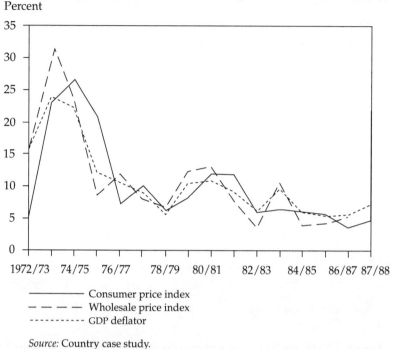

——————— Consumer price index
— — — — Wholesale price index
----------- GDP deflator

*Source:* Country case study.

17.5 percent until around 1981/82 (see figure 9.5). Expressed in proportion to a more relevant measure—exports of goods and services,
which includes remittances—the average debt service ratio for the
same period was substantially lower, about 12 percent. This reflects
not only the current account adjustment but also Pakistan's access to
external funds at concessional rates.

After 1981/82, however, Pakistan's debt service ratio increased
sharply, with the higher measure stabilizing at about 23 percent of
exports of goods and nonfactor services after 1983/84. As is shown
below, this reflects a shift in the composition of external financing
toward increased borrowing at market rates.

A final notable feature of Pakistan's macroeconomic performance
over the period examined concerns the rather unusual behavior of
national saving and investment. These variables, expressed in proportion to GNP, are plotted in figure 9.6. In spite of a substantial
increase in saving and investment rates over the 1974-76 period,
levels of saving and investment in Pakistan have been rather low, not
only by developing country standards but also in view of the coun-

**Figure 9.3. Rate of Growth of GDP, Pakistan, Fiscal
1972/73 to Fiscal 1987/88**

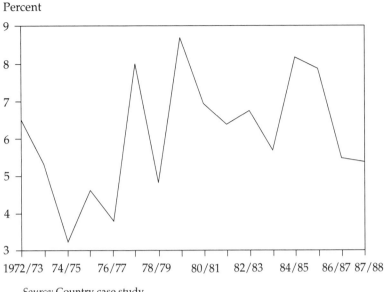

*Source:* Country case study.

**Figure 9.4. Ratio of Current Account to GNP, Pakistan,
Fiscal 1972/73 to Fiscal 1987/88**

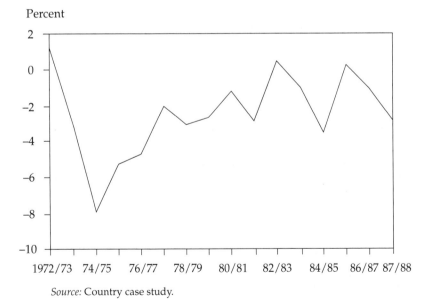

*Source:* Country case study.

**Figure 9.5. Ratio of Debt Service, Pakistan,
Fiscal 1972/73 to Fiscal 1987/88**

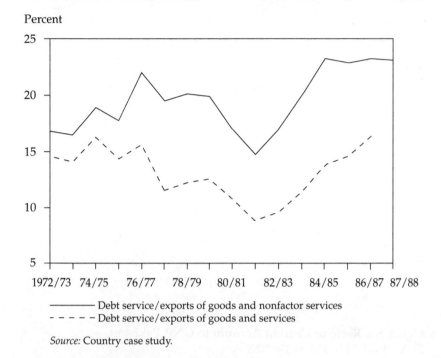

Percent

1972/73  74/75     76/77     78/79     80/81       82/83     84/85     86/87  87/88

——————— Debt service/exports of goods and nonfactor services
− − − − − Debt service/exports of goods and services

*Source:* Country case study.

try's high growth rate. The implication is that measured incremental
capital-output ratios (ICORs) have been remarkably low (figure 9.7).

## Fiscal Policy in Pakistan

The origin of fiscal deficits in Pakistan is similar in many ways to that
in other developing countries. In brief, an upsurge of externally
financed development spending during the early to mid-1970s, primarily in the form of investment by public enterprises, proved relatively permanent, and the public sector was unable to generate the
revenues—either through taxation or from the direct return to the
investments undertaken—to close the fiscal gap thereby created.

For the purposes of this section, it is useful to split the description
of fiscal policy in Pakistan into three periods: the period of democratic
rule under the Bhutto government (1972/73 to 1976/77), the early
years of the subsequent martial law government (1977/78 to 1981/82),
and 1982/83 to 1987/88. The section closes with a brief description of
events since 1988.

## Figure 9.6. National Saving and Investment, Pakistan, Fiscal 1972/73 to Fiscal 1987/88

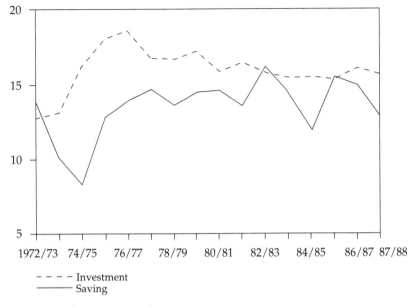

Percentage of GNP

- - - - Investment
——— Saving

*Source:* Country case study.

### The Bhutto Period

The Bhutto government, which took power after the 1971 war with India and the separation of Bangladesh, saw as its mandate the rapid economic and social development of the former West Pakistan. It was given a strong impetus in this direction by the combination of military defeat and plentiful external financing at concessional terms—primarily from Middle Eastern oil producers, which at that time were reaping the windfalls of the first oil price shock.

To an extent that is difficult to quantify, spending during these years was influenced by factors such as flood relief and the attempt to provide a countercyclical offset to the negative output effects of the oil price shock. (Pakistan is a net oil importer.) However, in view of the circumstances described above, the bulk of the spending increase (from 18 percent of GNP in 1972/73 to 24 percent in 1975/76; see table 9.1) must be understood as the result of a conscious policy choice—that is, as an explicit intention of fiscal policy. During these years the share in GNP of development expenditures—consisting of investment by the federal and provincial governments, capital transfers to local

**Figure 9.7. Incremental Capital-Output Ratio, Pakistan, Fiscal 1972/73 to Fiscal 1987/88**

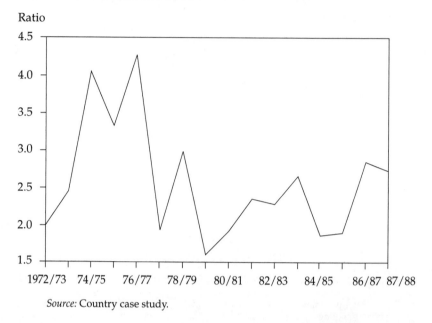

Ratio

*Source:* Country case study.

governments and state enterprises, and production subsidies—more than doubled. Public sector investment was devoted to the development of the chemical industry, as well as to cement, fertilizers, engineering, petroleum, steel, and production of a vegetable substitute for *ghee*, or clarified butter. This period also saw the nationalization of the seven principal manufacturing and industrial groups, as well as of banking, insurance, shipping, and educational institutions. Within the category of current expenditures, subsidies (classified as "other" current expenditure in table 9.1) increased by an average annual rate of 39 percent in the first half of the 1970s. The principal subsidy over the period was on *atta* (whole wheat flour), which was distributed at ration depots, primarily in urban areas, at prices below those required to cover the cost of wheat imports. By the mid-1970s, therefore, the public sector had a significantly larger role in Pakistan than at the beginning of the decade.

Unfortunately, this expansion of the role of the public sector was not matched by an corresponding rise in revenues. In fact, although expenditures of the federal and provincial governments increased by more than 6 percent of GNP from 1972/73 to 1975/76, total revenues increased by only 1 percent of GNP during the same period. (As table 9.1 indicates, the gain came primarily from an increase in domestic indirect taxes.)[4] The result, of course, was an increase in the fiscal deficit, to about 5 percent of GNP. Since foreign loans were plentiful and were available at favorable terms, almost three-quarters of the deficit financing during these years was external.

### Early Years of the Martial Law Government

The martial law government that assumed office in July 1977 initiated a short-term stabilization program supported by an IMF standby facility. This program attempted to address some of the structural problems of the economy and to correct perceived financial imbalances, the most significant of which was the fiscal deficit. An important goal of the new government was to deemphasize the role of the public sector in the economy. This change in regimes coincided with an external shock—the sharp curtailment of financial assistance from the United States because of Pakistan's nuclear program—and these developments ushered in a period of relative fiscal retrenchment. Between 1976/77 and 1980/81 the fiscal deficit fell from more than 8 percent of GNP to less than 5 percent.

The fiscal improvement during this period took the form of both cuts in expenditure and increases in revenue. The spending reductions can safely be treated as exogenous policy measures, since they represented an avowed policy goal of the new government. Moreover, the bulk of the spending cuts (more than 2 percent of GNP over the period) occurred in development spending, consistent with the government's explicit goal of curtailing public involvement in productive activities and leaving these to the private sector. An effort to reduce current expenditures met with only limited success; current spending remained roughly constant as a share of GNP over the period.[5]

The total contribution of revenue increases to the fiscal adjustment (amounting to 1 percent of GNP) was about half that of spending cuts. However, despite the serious effort to improve tax administration, only about half of the revenue increase appears to have been the result of exogenous fiscal policy measures. To assess the extent to which discretionary revenue measures may have contributed to the fiscal adjustment, we report in table 9.2 the results of very simple regressions that relate the three components of total tax revenue (direct taxes, taxes on international trade, and other indirect taxes) to their primary determinants. In the case of direct taxes and indirect taxes other than trade taxes, we took the primary determinant to be the tax base (proxied by nominal GNP). For trade taxes the determinants consist of the tax base (in the form of exports and imports) and the share of workers' remittances in GNP.[6] The latter variable greatly improves the fit of the regression. We take it as a proxy for the compo-

**Table 9.1. Consolidated Accounts of the Federal and Provincial Governments, Pakistan, Fiscal 1972/73 to 1987/88**
(percentage of GNP)

| Item | 1972/73 | 1973/74 | 1974/75 | 1975/76 | 1976/77 | 1977/78 | 1978/79 | 1979/80 | 1980/81 | 1981/82 | 1982/83 | 1983/84 | 1984/85 | 1985/86 | 1986/87 | 1987/88 |
|---|---|---|---|---|---|---|---|---|---|---|---|---|---|---|---|---|
| Total expenditures | 17.9 | 20.9 | 24.1 | 24.2 | 22.7 | 21.7 | 23.4 | 21.6 | 21.2 | 20.5 | 21.6 | 21.7 | 22.4 | 22.8 | 24.2 | 24.1 |
| Current | 13.4 | 14.7 | 15.5 | 15.0 | 13.0 | 13.6 | 14.6 | 13.0 | 12.6 | 12.8 | 14.3 | 15.6 | 16.1 | 16.1 | 17.3 | 17.4 |
| Consumption | 11.4 | 9.6 | 10.6 | 11.4 | 10.8 | 10.1 | 9.7 | 9.3 | 9.7 | 9.9 | 10.5 | 11.2 | 11.2 | 11.3 | 12.1 | 12.7 |
| Total interest | 1.8 | 1.8 | 1.6 | 1.8 | 1.8 | 1.8 | 1.9 | 2.0 | 2.0 | 2.2 | 2.8 | 3.1 | 3.2 | 3.4 | 3.7 | 3.8 |
| Foreign | 1.1 | 0.9 | 0.9 | 0.8 | 0.9 | 0.9 | 1.0 | 0.9 | 0.8 | 0.8 | 1.1 | 1.1 | 1.1 | 1.1 | 1.2 | 1.1 |
| Domestic | 0.7 | 0.9 | 0.7 | 1.0 | 0.9 | 0.9 | 0.9 | 1.1 | 1.2 | 1.4 | 1.7 | 1.9 | 2.0 | 2.3 | 2.5 | 2.7 |
| Other | 0.2 | 3.3 | 3.3 | 1.8 | 0.4 | 1.6 | 3.0 | 1.7 | 1.0 | 0.7 | 1.0 | 1.4 | 1.8 | 1.4 | 1.5 | 0.9 |
| Development | 4.5 | 6.2 | 8.6 | 9.3 | 9.7 | 8.1 | 8.8 | 8.6 | 8.6 | 7.6 | 7.3 | 6.1 | 6.4 | 6.8 | 6.9 | 6.7 |
| Investment | 2.0 | 2.7 | 3.0 | 3.3 | 3.5 | 2.8 | 2.9 | 2.4 | 2.6 | 3.1 | 2.7 | 2.7 | 2.7 | 2.6 | 2.8 | 3.0 |
| Other | 2.4 | 3.5 | 5.6 | 5.9 | 6.2 | 5.4 | 5.9 | 6.3 | 6.0 | 4.6 | 4.6 | 3.4 | 3.7 | 4.1 | 4.1 | 3.7 |
| Total revenues | 14.2 | 15.7 | 14.8 | 15.4 | 14.4 | 14.3 | 15.1 | 15.8 | 16.4 | 15.5 | 15.2 | 16.2 | 15.4 | 15.8 | 15.9 | 16.7 |
| Tax revenue | 11.0 | 11.8 | 11.4 | 11.7 | 11.4 | 11.5 | 12.0 | 12.9 | 12.9 | 12.4 | 12.2 | 11.6 | 10.8 | 10.7 | 10.2 | 10.9 |
| Direct taxes | 2.2 | 1.9 | 1.7 | 2.1 | 2.0 | 1.8 | 1.9 | 2.2 | 2.5 | 2.6 | 2.3 | 2.0 | 1.9 | 1.7 | 1.7 | 1.9 |
| Indirect taxes | 8.8 | 9.9 | 9.7 | 9.6 | 9.4 | 9.7 | 10.1 | 10.7 | 10.4 | 9.8 | 9.9 | 9.6 | 8.9 | 9.0 | 8.5 | 9.0 |
| Export duties | 1.6 | 2.1 | 0.9 | 0.6 | 0.1 | 0.2 | 0.1 | 0.2 | 0.2 | 0.1 | 0.1 | 0.1 | 0.1 | 0.2 | 0.0 | 0.2 |
| Import duties | 2.3 | 2.6 | 3.4 | 3.3 | 3.9 | 4.2 | 4.7 | 4.8 | 4.5 | 4.2 | 4.5 | 4.6 | 4.4 | 4.1 | 4.1 | 4.2 |
| Other | 5.0 | 5.3 | 5.4 | 5.8 | 5.5 | 5.3 | 5.2 | 5.7 | 5.7 | 5.5 | 5.3 | 5.0 | 4.4 | 4.7 | 4.4 | 4.6 |

| | | | | | | | | | | | | | | | | |
|---|---|---|---|---|---|---|---|---|---|---|---|---|---|---|---|---|
| Nontax revenue | 3.2 | 3.7 | 3.3 | 3.3 | 2.7 | 2.6 | 2.7 | 2.4 | 2.7 | 2.6 | 2.5 | 4.0 | 4.1 | 4.6 | 5.2 | 5.1 |
| Interest receipts | 0.8 | 0.8 | 0.7 | 0.9 | 1.3 | 1.0 | 0.9 | 1.0 | 1.0 | 0.9 | 1.0 | 0.9 | 1.0 | 0.9 | 1.4 | 1.2 |
| Other | 2.4 | 2.9 | 2.6 | 2.4 | 1.4 | 1.6 | 1.7 | 1.4 | 1.7 | 1.7 | 1.6 | 3.1 | 3.2 | 3.7 | 3.9 | 3.9 |
| Surplus of autonomous bodies | 0.0 | 0.1 | 0.2 | 0.4 | 0.3 | 0.3 | 0.5 | 0.6 | 0.7 | 0.5 | 0.6 | 0.6 | 0.5 | 0.5 | 0.4 | 0.7 |
| Overall deficit | 3.7 | 5.2 | 9.3 | 8.8 | 8.3 | 7.4 | 8.3 | 5.8 | 4.8 | 4.9 | 6.4 | 5.5 | 7.1 | 7.1 | 8.3 | 7.4 |
| Bank financing | 0.0 | 0.5 | 1.8 | 2.9 | 3.9 | 2.6 | 4.1 | 2.5 | 2.0 | 0.4 | 1.5 | 1.4 | 1.8 | 3.3 | 0.9 | 1.6 |
| State Bank of Pakistan | -1.1 | -0.6 | 3.1 | 0.9 | 3.7 | 3.0 | 3.5 | 1.2 | -0.6 | 3.0 | -1.3 | 4.5 | 0.8 | 0.6 | 0.9 | 1.3 |
| Scheduled banks | 0.0 | 1.1 | -1.3 | 2.1 | 0.2 | -0.4 | 0.6 | 1.3 | 2.5 | -2.7 | 2.8 | -3.2 | 1.0 | 2.7 | 0.0 | 0.3 |
| External financing | 3.2 | 3.7 | 7.0 | 5.1 | 3.8 | 3.3 | 3.2 | 2.8 | 2.6 | 1.5 | 1.3 | 1.1 | 1.0 | 1.5 | 1.8 | 1.6 |
| Domestic nonbank | 0.9 | 1.1 | 0.5 | 1.4 | 0.6 | 1.5 | 1.0 | 0.6 | 0.3 | 3.0 | 3.6 | 3.0 | 4.3 | 2.3 | 5.7 | 4.2 |
| *Memorandum items:* | | | | | | | | | | | | | | | | |
| Deficit/GDP (percent) | 4.60 | 6.96 | 12.00 | 10.59 | 9.84 | 9.28 | 9.82 | 7.52 | 6.20 | 6.19 | 7.97 | 6.91 | 8.73 | 8.64 | 9.81 | 8.70 |
| Composition of deficit financing | 1.00 | 1.00 | 1.00 | 1.00 | 1.00 | 1.00 | 1.00 | 1.00 | 1.00 | 1.00 | 1.00 | 1.00 | 1.00 | 1.00 | 1.00 | 1.00 |
| Bank financing | 0.00 | 0.09 | 0.20 | 0.33 | 0.47 | 0.36 | 0.49 | 0.43 | 0.40 | 0.08 | 0.23 | 0.25 | 0.25 | 0.47 | 0.11 | 0.21 |
| External financing | 0.87 | 0.70 | 0.75 | 0.57 | 0.46 | 0.44 | 0.39 | 0.47 | 0.53 | 0.31 | 0.20 | 0.20 | 0.14 | 0.21 | 0.21 | 0.22 |
| Domestic nonbank | 0.23 | 0.20 | 0.05 | 0.15 | 0.08 | 0.20 | 0.12 | 0.10 | 0.07 | 0.61 | 0.57 | 0.55 | 0.61 | 0.33 | 0.68 | 0.57 |

Note: Items may not sum to totals because of rounding.
Source: *Pakistan Economic Survey*, various issues.

**Table 9.2. Determinants of Tax Revenues, Pakistan**

| Item | Direct taxes (1) | Trade taxes (2) | Other indirect taxes |
|---|---|---|---|
| Constant | −3.72 | −0.47 | −1.93 |
| | (6.35) | (−0.76) | (−3.13) |
| GNP | 0.98 | 0.93 | 0.92 |
| | (21.00) | (29.80) | (20.33) |
| Inflation rate | | | −0.22 |
| | | | (−0.32) |
| Ratio of imports to GNP | | 0.50 | |
| | | (3.58) | |
| Ratio of exports to GNP | | 0.21 | |
| | | (2.12) | |
| Ratio of remittances to GNP | | 0.14 | |
| | | (3.22) | |
| $R^2$ | 0.97 | 0.99 | 0.99 |

*Note:* All variables except the inflation rate are in log form. The data are annual, and the regressions are estimated for the period 1972/73 to 1987/88. Figures in parentheses are $t$-statistics. A blank denotes the omission of the specific variable from the regression.

sition of imports, on the hypothesis that an increase in remittances increases the share in total imports of dutiable imports such as consumer durables.

These simple regressions account for almost all of the variation in tax receipts. When their residuals are taken as measures of discretionary tax changes, total discretionary tax measures account for about 5.5 percent of total tax receipts in 1979/80 and about 7.5 percent in 1980/81 and in 1981/82. At their peak, these increases in discretionary revenue amounted to less than one-sixth of the total increase in tax revenues. The remainder of the increase in tax revenues is accounted for by an increase in trade taxes; both exports and imports rose rapidly during these years. A real exchange rate depreciation fueled by the depreciation of the U.S. dollar against the currencies of Pakistan's trading partners in the late 1970s gave a boost to exports, while a substantial increase in remittances as a consequence of the second oil shock gave rise to an import boom at the end of the 1970s.

### The Period 1982/83 to 1987/88

The improvements in Pakistan's fiscal stance did not turn out to be permanent. After 1981/82 the fiscal deficit began to increase once again and by 1986/87 had reached 8.5 percent of GNP, a level comparable to those of the mid-1970s. Although the composition of revenues changed somewhat during the period—slippage in the collection of direct taxes and of domestic indirect taxes tended to be offset by increased nontax revenues, primarily in the form of profits on the distribution of oil products[7]—the share of public sector revenue in GNP showed no trend. The increase in the deficit arose from the expenditure side.

This was so in spite of a continued contraction in the share of development expenditures in GNP; this type of spending peaked at 9.75 percent of GNP in 1976/77 and then declined to 8.5 percent in 1980/81 and to 6.75 percent in 1987/88. Part of the increase in spending was from higher public consumption in the form of public sector wages and salaries and defense spending. The share of defense spending in total public expenditure rose from 21 percent in 1978/79 to 27 percent in 1984/85 (see Kemal 1987). But by far the most rapidly increasing category of spending during the 1980s was total public sector interest payments.

The increase in the share of interest payments reflects a conscious change in the composition of deficit financing after 1980/81. In an effort to keep inflation in check and to tap directly what was perceived to be a plentiful supply of private saving originating with remittance inflows, the government limited its borrowing from the domestic banking system after 1980/81, and domestic nonbank borrowing became the residual source of finance. When the international debt crisis curtailed the availability of external financing in 1981/82, external funds were also replaced by domestic nonbank borrowing. Thus the combination of an aversion to inflationary finance, reduced availability of external funds, and increased public consumption brought about a substantial increase in domestic nonbank borrowing. As will be shown later, the rising stock of internal debt could be absorbed domestically only by offering higher interest rates, and the combination of higher debt stock and increasing interest rates caused interest payments to mount over time. Total public sector interest payments, which were 2 percent of GNP in 1980/81, had almost doubled as a percentage of GNP by 1987/88, accounting for about two-thirds of the increase in the deficit-to-GNP ratio over that period. For any given year, therefore, this component of the deficit increase reflected past financing decisions rather than current policy.

In summary, the upsurge in fiscal deficits in Pakistan during the 1980s was the result of two policy choices: an increase in public consumption in the face of a political inability to raise commensurate revenues, and a change in the financing mix from domestic bank borrowing and external financing to domestic nonbank borrowing.

## Recent Developments

As indicated earlier, recent revisions in Pakistan's macroeconomic data complicate any attempt to present a cohesive picture of the entire period 1972–92. Data after 1987/88 are compiled under a different methodology than was used earlier and therefore are not directly comparable. Nevertheless, recent developments can be described by using 1987/88 as a benchmark and focusing on changes since that time.

The most significant feature of the recent period is a notable improvement in the fiscal picture between 1988/89 and 1990/91. As a fraction of GNP, the fiscal deficit declined from about 8.5 percent in 1987/88 (as against 8.75 percent under the earlier methodology) to about 5.75 percent in 1990/91. Most of the adjustment came from the expenditure side, as total spending of the federal and provincial governments contracted by about 2 percent of GNP. Three-quarters of this adjustment was achieved by restraining current expenditures, offering some hope for a permanent fiscal adjustment. However, in 1991/92 the government appears to have missed its fiscal deficit target of 5 percent of GDP by a wide margin, and recent figures suggest a return to the range of 7 to 8 percent that characterized the period up to 1987/88.

The period of fiscal retrenchment was accompanied by a growth slowdown. The growth rate of real GDP, which amounted to 6.5 percent in 1987/88 under the new methodology, slowed to 4.75 percent in 1988/89 and 1989/90 before rising to more than 5.5 percent in 1990/91. Preliminary figures for 1991/92 suggest that growth has again risen above 6 percent. It would be simplistic, however, to attribute the growth slowdown exclusively to the fiscal retrenchment, since Pakistan was buffeted by several shocks during these years. Not the least of these shocks were a continuous and severe contraction in worker remittances from abroad after 1986/87 (associated with low oil prices and war-related dislocations in the Middle East) and political uncertainty at home, culminating in 1990 in the replacement of the government of Benazir Bhutto by that of Nawaz Sharif.

Other broad macroeconomic indicators behaved in a more stable fashion after this period. The rate of inflation has remained low (in the 10 percent range, as measured by the CPI), and the ratios of the current account of the balance of payments to GNP and of debt service to exports of goods and nonfactor services have remained broadly unchanged since 1990.

Overall, recent developments appear to be consistent with the experience of the 1980s. The latest data suggest that no permanent change was achieved in Pakistan's fiscal policy under the two democratic governments that followed the end of martial law, and the country's macroeconomic performance reflects a continuation of previous trends.

## Deficits and Inflation

As indicated previously, Pakistan has operated for the better part of two decades with fiscal deficits that, by international standards, are quite large in relation to GNP. In other developing countries fiscal deficits of smaller magnitude have been blamed for a number of adverse macroeconomic developments, chief among them being a high rate of inflation. By contrast, Pakistan has performed relatively well in a macroeconomic sense, with a high average rate of economic growth, low inflation, and a relative absence of major external imbalances. The key questions that arise in connection with Pakistan's fiscal policy are thus the following: What macroeconomic effects have Pakistan's sustained high fiscal deficits had on its economy? Why in Pakistan have high deficits not been associated with inflation of Latin American proportions?

In popular discussions, the link between deficits and inflation in developing countries is quite direct. In the absence of secondary securities markets, open-market operations are not an important monetary policy tool in such countries. Since government borrowing from the central bank expands the supply of base money, the rate of growth of the money supply is taken to depend primarily on the size of the fiscal deficit. With the rate of inflation in turn being determined by the rate of growth of the money supply, the link between deficits and inflation follows.

Although variations in velocity, the availability of other modes of financing, and several other factors tend to complicate matters, there is nevertheless a valid long-run relationship, emerging from the solvency constraint of the public sector, between fiscal deficits and inflation. This is easiest to show formally. (The discussion that follows draws heavily on Buiter 1985.) Let $b$ denote the real stock of the debt of the public sector (including the central bank) to the domestic private sector. Let $F_G$ denote public external debt; $d$, the real primary fiscal deficit; and $m$, the real stock of base money, all measured as ratios to GNP. Also, let $r_B$ denote the real interest rate on domestic borrowing; $r_F$, the real interest rate paid on foreign debt; $r^*$, the real interest rate prevailing in international capital markets (external interest rate plus rate of depreciation minus domestic inflation); $n$, the rate of growth of real GNP; and $s$, the nominal exchange rate. The public sector's budget constraint can then be written:

(9.1)  $\dot{b} + s\dot{f}_G + \dot{m} = d + (r_B - n)b + (r_F + \hat{s} - n)sf_G - (\pi + n)m$

where $\pi$ is the rate of inflation. (A dot [˙] over a variable denotes a time derivative; a hat [^] denotes a proportional rate of change.) This can be transformed into:

(9.2)     $\dot{b} + s\dot{f}_G = \tilde{d} + (r^* - n)(b + sf_G) - [\dot{m} + (\pi + n)m]$

where $\tilde{d} = d + (r_B - r^*)b + (r_F + \hat{s} - r^*)sf_G$ is the adjusted primary deficit—that is, the primary deficit, plus the excess interest paid on domestic debt and foreign debt over that prevailing in international capital markets.

The initial net worth of the public sector is given by $-(b + sf_G)$, and the public sector will be solvent if the present value (calculated using the growth-corrected interest rate, $r^* - n$) of its anticipated future debt service is at least equal to its net debt; that is:

(9.3)          $PV[\dot{m} + (\pi + n)m - \tilde{d}] \geq b + sf_G.$

The resources available to the public sector for servicing debt consist of future seigniorage revenue, given by $\dot{m} + (\pi + n)m$, and future adjusted primary surpluses, given by $-\tilde{d}$. Note that, other things being equal, an increase in the present value of the stream of future deficits requires an increase in the present value of the inflation tax, $\pi m$. It is in this present-value sense that higher fiscal deficits are related to higher inflation.

Note also that a number of factors influence the present value of the inflation tax associated with a given path of the primary deficit. The following observations are germane:

- The relevant value of the primary deficit is the adjusted deficit— that is, the deficit adjusted to take account of differences between the interest rate on domestic borrowing and the actual interest paid on external borrowing, on the one hand, and the marginal cost of external funds, on the other. Access to domestic or external funds at favorable rates reduces the present value of the inflation tax associated with a given unadjusted deficit.
- The amount of seigniorage required to finance a given path of the adjusted deficit is smaller, the smaller is the initial net stock of debt.
- Given the amount of seigniorage required, the larger are the rates of growth of output and the greater is the secular growth in the money-to-income ratio, the smaller is the requisite inflation tax.

### Application to Pakistan

During 1980–88 the consolidated deficit of Pakistan's federal and provincial governments averaged about 6.5 percent of GNP (see table 9.1).

**Figure 9.8. Inflation and Growth of Base Money, Pakistan, Fiscal 1980/81 to Fiscal 1987/88**

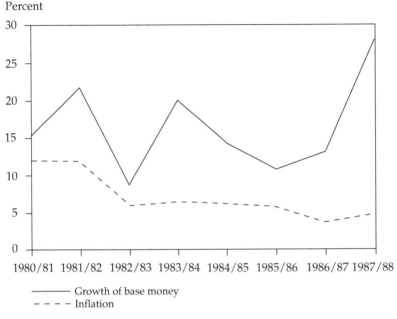

Percent

— Growth of base money
- - - - Inflation

*Source:* Country case study.

Since the 1980 stock of base money amounted to about 12 percent of GNP, financing of this deficit solely through the issuance of base money would have implied an inflation rate of more than 54 percent per year during the 1980s, even before allowing for erosion of the monetary base through a rise in velocity. After allowing for such erosion, reliance on the inflation tax to finance a deficit of this magnitude would imply inflation rates of Latin American proportions. The actual growth of base money and domestic prices (measured by the GDP deflator) in Pakistan is depicted in figure 9.8. During 1980–88 both the growth of base money and the domestic inflation rate fell substantially short of what the simple analysis would predict. The observations of the preceding section can be used to explain why.

Rearranging equation 9.2 permits us to write:

$$(9.4) \qquad \dot{b} + s\dot{f}_G + \dot{m} = \bar{d} - [(\pi + n)(b + sf_G + m) - \hat{s}sf_G]$$
$$= \bar{d} - d^*$$
$$= d^A$$

where $\bar{d} = d + (r_B + \pi)b + (r_F + \pi)sf_G$ is the conventional deficit–GNP ratio. The variable $d^A$ represents the inflation- and growth-adjusted

**Figure 9.9. Actual and Equilibrium Deficits as a Share of GNP, Pakistan, Fiscal 1980/81 to Fiscal 1987/88**

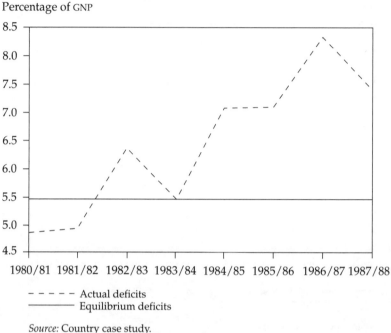

Percentage of GNP

| | |
|---|---|
| 1980/81 1981/82 1982/83 1983/84 1984/85 1985/86 1986/87 1987/88 | |

– – – – – Actual deficits
————— Equilibrium deficits

*Source:* Country case study.

deficit ratio—that is, the actual deficit minus the portion that can be financed without altering the ratio of total debt to GNP. When the value of this expression is 0, the conventional deficit can be accommodated without requiring macroeconomic adjustments—including adjustment of the rate of inflation—because the requisite amount of financing will presumably be forthcoming. In other words, $d^* = [(\pi + n)(b + sf_G + m) - \hat{s}sf_G]$ is the deficit ratio that is consistent with macroeconomic equilibrium, with inflation rate $\pi$ and real growth rate $n$. Figure 9.9 depicts the ratios of the actual deficit, $d$, and estimated "equilibrium" deficit, $d^*$, for Pakistan in the years 1980–87, using smoothed values of the growth rate, $n$, and the rate of inflation, $\pi$, and setting the rate of exchange rate depreciation equal to $\pi$ to calculate the equilibrium deficit. The equilibrium deficit estimate should be seen as the midpoint of a range of estimates. It is based on our estimate of the end-1979/80 stock of net (interest-bearing and non-interest-bearing) public sector debt, as well as on the assumption that all such debt was willingly held.[8] As figure 9.9 shows, the equilibrium deficit was indeed at a relatively high value for Pakistan during this period (about 5.5 percent of GNP, on average), in spite of an

**Figure 9.10. Public Sector Liabilities as a Share of GNP, Pakistan, Fiscal 1980/81 to Fiscal 1987/88**

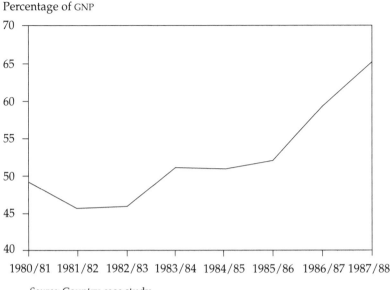

Percentage of GNP

*Source:* Country case study.

inflation rate (for the GDP deflator) averaging a little more than 7 percent. This is primarily attributable to a very high rate of growth of real output (about 6 percent per year), which permitted a fairly rapid expansion of both interest-bearing and non-interest-bearing debt without recourse to inflationary finance.

The ratio of the actual deficit to GNP fell below this equilibrium value for the first two years of the period 1980/81 to 1987/88 (see figure 9.9). For the remainder of the period the deficit averaged about 7 percent of GNP. Thus the adjusted deficit was significantly smaller than the actual deficit during the period; it amounted to about 1.5 percent of GNP, on average. It is not surprising, therefore, that, as indicated in figure 9.10, the ratio to GNP of the liabilities of Pakistan's consolidated public sector (consisting of base money, domestic debt, and external debt) was fairly stable for the first half of the 1980s, in spite of a fiscal deficit amounting to almost 6 percent of GNP. Only in the last two years covered by the analysis (1986/87 to 1987/88) did this ratio increase rapidly, consistent with a substantial increase in the adjusted deficit during the second half of the 1980s.

In short, this analysis suggests that low inflation can be reconciled with large deficits in Pakistan because of the economy's very high growth rate. Rapid economic expansion made it possible to finance large fiscal deficits by issuing debt, without extensive reliance on the inflation tax. Because of the economy's high growth rate, the relationship of the stock of public sector liabilities to GNP remained fairly stable until recent years in spite of the large fiscal deficits recorded during the 1980s.

### Some Qualifications

Although the foregoing analysis follows fairly conventional lines, its application to Pakistan—and possibly to many other low-income developing countries—is problematic. The reason is that the calculation of an equilibrium deficit is done from the financing side of the budget on the assumption that if private agents willingly hold an initial stock of claims on the public sector, they would be willing to expand the real value of claims at the rate of growth of real output. Complications arise when the initial stock of debt is not in fact willingly held by private agents on market terms. It would then be misleading to assume that the market would be willing to accept a steady growth of claims on the public sector under prevailing macroeconomic conditions—that is, without disrupting macroeconomic equilibrium.

In Pakistan's case, as with many other developing countries, there are at least two reasons why the stock of public sector debt cannot be treated as willingly held by market agents. First, as pointed out above, much of Pakistan's external debt was acquired at concessional terms on a bilateral basis and thus contains a substantial grant element. Second, some of Pakistan's domestic debt is held by financial institutions, partly to satisfy reserve requirements. During much of the period with which we are concerned, for example, commercial banks faced a required "liquidity" ratio of 30 percent, which had to be satisfied with government securities (see Morshed 1987). To the extent that such institutions would have required a higher rate of return to hold these securities willingly, this requirement subjects them to an implicit tax. As figure 9.11 illustrates, interest rates paid by Pakistan on both external and domestic debt were much below international interest rates (measured by the London interbank offered rate, or LIBOR) during most of this period. It cannot be safely assumed that either set of creditors—foreign governments or domestic financial institutions—would be willing to see their claims on the government of Pakistan expand in real terms at the rate of growth of domestic output. To the extent that such an assumption is unwarranted, the equilibrium deficit reported above would be overestimated, and thus the inflation- and growth-adjusted deficit would be underestimated.

To assess the potential importance of these factors, we have performed some rough calculations to correct our estimate of $d^*$ for nonmarket lending. To do so, we require estimates of the total amount of

**Figure 9.11. Interest Rates on Domestic and External Debt, Pakistan, Fiscal 1980/81 to Fiscal 1987/88**

Percent

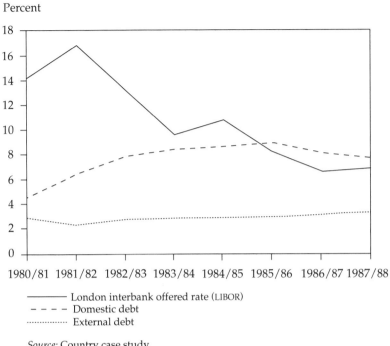

London interbank offered rate (LIBOR)
- - - - Domestic debt
················ External debt

*Source:* Country case study.

claims on the Pakistan government that domestic and foreign market agents would have willingly held initially (that is, at the beginning of fiscal 1980/81). We assume that to hold this debt willingly external agents would have required a return equal to LIBOR and that domestic agents would have required an interest rate corresponding to uncovered interest parity (UIP), using LIBOR as the foreign counterpart rate. The proportional differences between the average values of these rates and the average interest rates paid by the government on its external and on its domestic debt during the 1980/81 to 1987/88 period were taken as estimates of the grant element associated with foreign loans and the implicit tax rate on domestic securities, respectively. Applying these proportions to the debt stocks outstanding at the end of 1980/81 yields our estimates of the claims that would have been willingly held by market agents. Using these as the basis for our equilibrium deficit calculations yields a corrected equilibrium deficit, denoted $d_c^*$, of 2.1 percent of GNP for the period 1980/81 to 1987/88, reflecting a scaling down in the estimate of voluntary lending from market agents if there is no change in prevailing macroeconomic conditions.

**Figure 9.12. Actual and Corrected Deficits as Shares of GNP, Pakistan, Fiscal 1981/82 to Fiscal 1987/88**

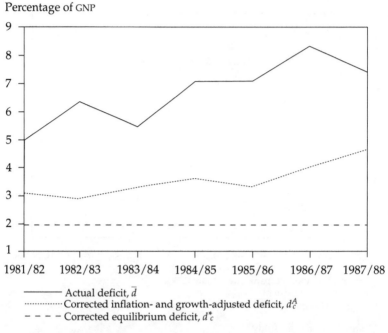

Percentage of GNP

——— Actual deficit, $\bar{d}$
·············· Corrected inflation- and growth-adjusted deficit, $d_c^A$
– – – – Corrected equilibrium deficit, $d_c^*$

*Source:* Country case study.

It would be incorrect, however, to recalculate the adjusted deficit as $\bar{d} - d_c^*$ because Pakistan *was* in fact able to borrow at favorable rates during the 1980s. In other words, the grant component of foreign lending and the implicit tax associated with domestic borrowing would in any event have reduced the government's need to raise funds at market rates. In essence, these funds simply represent unconventional sources of revenue. We therefore corrected the actual deficit by adding as revenues the grant component of foreign borrowing and the implicit tax component on domestic borrowing.[9] Because this step essentially involves reclassifying financing as revenue entries, the corrected actual deficit, $\bar{d}_c$, is substantially reduced in relation to $\bar{d}$.

The corrected inflation and growth-adjusted deficit, calculated as $d_c^A = \bar{d}_c - d_c^*$, as well as $\bar{d}_c$ and $d_c^*$, are plotted in figure 9.12. As is clear from the figure, these corrections decrease the size of both the actual and equilibrium deficits but have little effect on the difference between them. In other words, the corrected deficit is not substantially affected. Whereas our previous estimates may exaggerate the

**Figure 9.13. Composition of Deficit Financing, Pakistan, Fiscal 1980/81 to Fiscal 1987/88**

Percentage of GNP

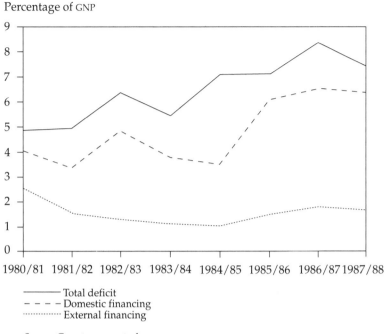

————— Total deficit
– – – – Domestic financing
·············· External financing

*Source:* Country case study.

amount of financing that private agents would have been willing to supply to the government during the 1980s, they also exaggerate—by an approximately equal amount—the size of the deficits that required financing. Our previous conclusions and their implications for the macroeconomic effects of Pakistan's fiscal deficits therefore continue to hold.

*Conclusions*

As shown in figures 9.9 through 9.12, fiscal deficits in the second half of the 1980s have begun to substantially exceed the equilibrium values calculated for 1980–87. Deficits of such magnitude can indeed be expected to exert significant effects on financial markets. However, even these higher deficits have not been associated with an inflationary upsurge. An explanation for this is suggested in figure 9.13, which shows how recent deficits have been financed. The height of each curve measures the amount of financing from specific sources. The lowest curve measures the flow of external financing, the middle curve adds to this domestic financing, and the top curve, which represents the total deficit, adds money financing. As is evident from the figure, recent years have witnessed a rapid increase in domestic borrowing (the gap between the lowest curve and the middle one). As a

**Figure 9.14. Base Money, Domestic Debt, and External Debt as Shares of GNP, Pakistan, Fiscal 1980/81 to Fiscal 1987/88**

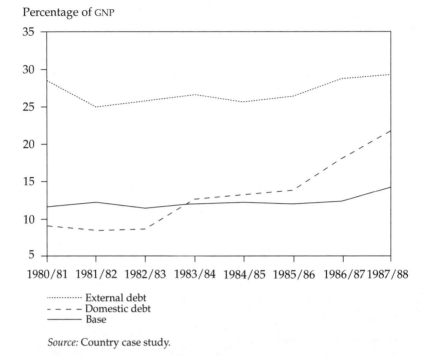

Percentage of GNP

*Source:* Country case study.

result, while the ratios of base money and external debt to GNP have remained roughly stable, the ratio of domestic debt to GDP has risen noticeably (figure 9.14). The macroeconomic effects of this financing policy are investigated in the next section.

To summarize, a simple explanation linking fiscal deficits to inflation fails to hold in Pakistan. Three reasons emerge. First, the measured deficit overstates the "true," economically meaningful deficit because foreign financing contained a substantial grant element and domestic financing contained an implicit tax element that should be treated as above the line in the general government budget. This aspect of deficit financing in Pakistan reduces the pressure exerted by fiscal deficits on financial markets. Second, the economy grew very rapidly during the 1980s, so that equilibrium deficits proved extremely large by international standards. Third, when the corrected inflation and growth-adjusted deficits in fact became large in the second half of the 1980s, primary reliance was placed on domestic debt financing rather than on the inflation tax. Nevertheless, deficits of the magnitude experienced during the second half of the 1980s can be expected to have other macroeconomic effects, and these are addressed in the remainder of the chapter.

**Figure 9.15. Private and Public Investment, Pakistan, Fiscal 1964/65 to Fiscal 1987/88**

Percentage of GNP

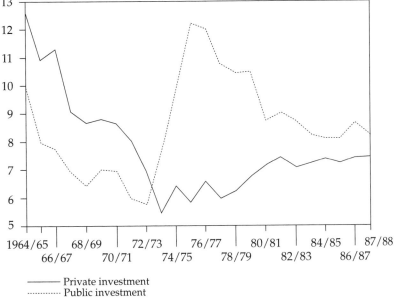

——— Private investment
············ Public investment

*Source:* Country case study.

## The Effects of Fiscal Policy on Economic Behavior

This section examines the effects of fiscal policy in general (rather than fiscal deficits specifically) on the behavior of economic agents in Pakistan. Recent research has shown that both the investment and the consumption decisions of economic agents may be directly affected by policy variables such as government consumption and investment. Such decisions also are known to respond to financial variables that are themselves affected by the mode of government financing.

*Consumption and Investment*

As was shown in the Overview to this chapter, total investment as a ratio of GNP in Pakistan has averaged about 15 percent. Figure 9.15 shows the separate behavior of the ratio of public and private investment to GNP over the period 1964/65 to 1986/87. The strong investment drive initiated by the Bhutto government in 1973 is clearly discernible, and it is evident that private investment, although it

**Figure 9.16. Private and Public Consumption, Pakistan,
Fiscal 1964/65 to Fiscal 1987/88**

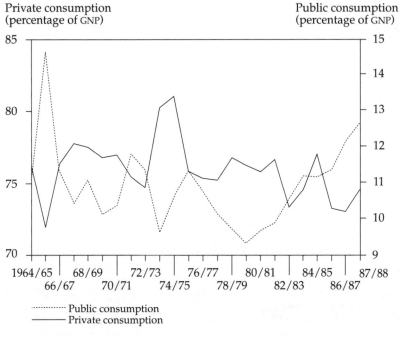

Private consumption
(percentage of GNP)

Public consumption
(percentage of GNP)

··········· Public consumption
——— Private consumption

*Source:* Country case study.

increased steadily after 1973/74, did not return to the high levels that
it had attained prior to the period under study. Thus, since 1972/73
public investment has represented the bulk of total investment.
Whether the high level of public investment has served to catalyze
private investment since 1973/74 by means of infrastructural development or whether private investment was crowded out during most of
the period by the elevated level of public investment is an issue that
will be examined econometrically.

Figure 9.16 shows the movements of private and public consumption as a ratio to GNP between 1963/64 and 1987/88. The first significant point to note is that Pakistan has exhibited a fairly high average
propensity to consume: the ratio of total consumption to GNP averaged about 85 percent per year over the period. It may be suspected
that the high fiscal deficit played a role in producing this result.
However, the data also suggest an underlying relationship between
public and private consumption. As public consumption increases,
private consumption appears to decline. Such compensating behavior would appear to lend credence in the case of Pakistan to the
Ricardian view that there is a tendency for aggregate consumption to

**Table 9.3. Tests for Unit Roots, Pakistan: Consumption**

| Indicator | Durbin-Watson | Dickey-Fuller | Augmented Dickey-Fuller |
|---|---|---|---|
| Consumption | 0.088 | −0.629 | −0.427 |
| Disposable income | 0.094 | −0.086 | 1.049 |
| Government consumption | 0.196 | −0.843 | 0.996 |
| Government deficit | 0.268 | −0.196 | −3.052 |
| Permanent income | 0.082 | 0.316 | 1.621 |
| Permanent public sector saving | 0.736 | −1.529 | −1.19 |

*Note:* All variables are logs of real per capita values. Critical values are, for the cointegrating residuals Durbin-Watson (CRDW) test, 1.1, and for the Dickey-Fuller test, −2.61.

maintain its level regardless of the level of public consumption (see David and Scadding 1974; Haque and Montiel 1989).

CONSUMPTION. The estimation of the consumption function was conducted with annual data for the period 1963–87 derived from various issues of the *Pakistan Economic Survey*. We assume that consumption can be explained by permanent income, disposable income, and possibly several fiscal variables (public consumption, the fiscal deficit, and permanent public sector saving). Disposable income (denoted $y^D$ in per capita terms) is defined as GNP minus total tax revenue. Public sector saving (PBS) is defined as government revenues minus current expenditures, which include interest payments on government debt. Preliminary investigations revealed that the disposable income process could be represented by an autoregressive moving average (ARMA) (1, 1) specification. Consequently, permanent income per capita, $y^p$, was generated as the predicted value of the following equation:

$$(9.5) \qquad y_t^p = -2730.86 + 1.07 y_{t-1}^d - 0.26 MA(1).$$

Before estimating the consumption function, unit root tests were conducted on the aggregate variables that were to be included in the consumption function. Table 9.3 presents the results of the unit root tests for consumption, disposable income, general government consumption, the public sector deficit, permanent income, and the permanent public sector deficit. Test results for the Durbin-Watson, Dickey-Fuller, and augmented Dickey-Fuller tests, as well as the critical values for the tests, are presented in the table. The null hypothesis that a unit root is present was rejected only in the case of the public sector deficit (by the augmented Dickey-Fuller test). Since the presence of unit roots in almost all the variables suggests that conventional estimation approaches may yield misleading results, an alternative approach, which resulted in an error-correction specification, was used.

The cointegrating regression, which captures long-run equilibrium relationships between consumption and some of its determinants, was based on the permanent income approach, incorporating both the income variables mentioned above and intertemporal relative price variables such as the rate of interest and the rate of inflation. Additional fiscal policy variables—as indicated above—were also included. Instrumental variables were used for estimation to contend with endogeneity issues. The instruments used included the lagged values of permanent income, the government deficit, government consumption, and the rate of inflation. The estimated cointegrating equation is as follows (with $t$-ratios in parentheses):

$$(9.6) \qquad c_t^p = -0.45 + 1.35 y_t^p - 0.56 c_t^g - 1.11 \pi_t$$
$$\qquad\qquad (-0.41) \quad (6.33) \quad (-2.86) \quad (-2.83)$$

$$R^2 = 0.93; \text{Durbin-Watson statistic} = 1.8$$

where $c_t^p$ is private consumption, $y_t^p$ is permanent income obtained using equation 9.5, and $c_t^g$ is government expenditure, all measured in log real per capita terms, while $\pi$ is the annual inflation rate measured by the CPI. The equation appears to fit well—it explains 93 percent of the variation in the log of private consumption per capita— and, as the Durbin-Watson statistic shows, the residuals are non-autocorrelated. The signs and magnitudes of the coefficients are in keeping with economic theory. The coefficient of permanent income, for example, is not significantly different from unity. Increases in government consumption do appear to lead to a reduction in private consumption, as do increases in the rate of inflation. We could not find significant roles for the other fiscal variables listed above or for the real interest rate.

Using the residuals from the cointegrating regression in equation 9.6 as the error-correction term (denoted $ec$), an error-correction specification of the consumption function produced the following result (see Davidson and others 1978):

$$(9.7) \quad \Delta c_t^p = 0.23 + 0.19 \Delta y_t^p + 0.58 \Delta y_t^d - 0.41 \Delta c_t^g - 0.22 ec_{t-1}.$$
$$\qquad (1.40) \quad (1.10) \qquad (2.45) \qquad (-3.27) \quad (-1.67)$$

$$R^2 = 0.93; \text{Durbin-Watson statistic} = 2.01$$

Instrumental variables included the error-correction term, government consumption, lagged inflation, lagged disposable income, and lagged permanent income. Judging by the usual criteria, the equation seems to fit well. The error-correction term, $ec_{t-1}$, is significant at the 10 percent level, suggesting that private consumption does not adjust fully to its long-run desired level in the first period. Only a fifth of the total adjustment is completed in the first year. The negative and significant coefficient of $\Delta c_t^g$ shows that even in the short run, increases in government consumption are negatively related to private consumption. The coefficient of the change in disposable income is positive and statistically significant, suggesting that consumption behavior may be influenced by liquidity constraints.

The evidence suggests that fiscal policy may have affected private consumption in Pakistan primarily through direct substitutability between private and public consumption, through tax policy, and through indirect effects operating through macroeconomic variables such as the level of real income and the rate of inflation.

INVESTMENT. The private investment equation was estimated using annual data for the period 1972/73 to 1987/88 from the *Pakistan Economic Survey*. Capital stock series for both the public and private sectors, denoted $K_t^g$ and $K_t^p$, respectively, were constructed using an initial (1971/72) economywide capital-output ratio of 2, an initial share of 30 percent for public capital stock, and depreciation rates of 10 percent overall and 5 percent for public capital stock. A rental cost of capital ($r_t^k$) series was constructed by dividing the product of the real rate of interest and the investment deflator by the GDP deflator. We take private investment in Pakistan to be determined by the sizes of the private and public capital stocks, the level of real output, and the rental rate on capital. We could find no evidence of a credit availability effect, in spite of the repressed financial conditions prevailing in the economy for most of the period. Unit root tests were again conducted for the level variables and are presented in table 9.4. The tests suggest that unit roots cannot be ruled out for most of the relevant variables. Consequently, the approach adopted in this case was similar to that for private consumption. The cointegrating regression was estimated as follows:

(9.8)     $K^P/Y = -0.07 - 1.26r^K + 2.09K^G/Y - 0.09\text{DUM}$
             $(-0.25) \; (-5.33) \qquad (3.78) \qquad (-2.79)$

$$R^2 = 0.90; \text{Durbin-Watson statistic} = 2.26$$

where DUM is a dummy variable for the immediate post-Bhutto (1977–81) period.[10]

The estimates support the hypothesis that the government capital stock is positively correlated with private sector capital accumulation. The infrastructural buildup brought about by government investment appears to facilitate private investment. At the same time, an increased real rental cost of capital depresses private investment. The remaining variables did not prove statistically significant.

Once again, the residuals from the cointegrating regression were used in computing the tests for cointegration. For our cointegrating regression, the cointegrating residuals Durbin-Watson (CRDW), the Dickey-Fuller, and the augmented Dickey-Fuller statistics were esti-

**Table 9.4. Tests for Unit Roots, Pakistan: Investment**

| Test | Durbin-Watson | Dickey-Fuller | Augmented Dickey-Fuller |
|---|---|---|---|
| Private capital stock | 0.026 | 5.31 | −1.41 |
| Government capital stock | 0.099 | −1.74 | — |
| Real rate of interest | 0.51 | −1.51 | −1.33 |
| Real output | 0.15 | −0.29 | −0.65 |

— Not available.

Note: See note to table 9.3.

mated to be 1.8, −4.85, and −3.93, respectively. These values imply that the null hypothesis of unit roots in the residuals can be rejected and that equation 9.8 is a cointegrating form. Consequently, an error-correction specification is warranted in this case as well.

The results for the error-correction estimation are:[11]

$$(9.9) \quad \Delta(K^P/Y) = -0.02 - 0.27ec_{-1} + 0.90\Delta(K^G/Y)$$
$$(-4.10) \ (-1.53) \qquad (1.94)$$

$$- 0.11\Delta r^K - 0.03\Delta\text{DUM}.$$
$$(-0.71) \quad (-2.20)$$

$$R^2 = 0.45; \text{ Durbin-Watson statistic} = 1.64$$

This equation obviously leaves much of the variation in the ratio of the capital stock to output unexplained, and some of the individual coefficients are not estimated very precisely. The coefficient of the error-correction term $ec_{-1}$ suggests that the private capital stock adjusts slowly to its long-run desired level. The positive and significant coefficient of $\Delta(K^G_t/Y_t)$ indicates that even in the short run, an increase in the government capital stock or a positive level of government investment induces an increase in private investment. Consistent with the theoretical hypothesis, the rental return on capital is negatively related to private investment.

Our examination of investment behavior, therefore, shows that fiscal policy has both direct and indirect effects on private investment in Pakistan. The direct effect appears to operate by expanding domestic infrastructure through public investment. More indirect effects are traceable through the effects of fiscal policy on interest rates.

*Output*

To complete the simulation model used in the next section, a specification of the determinants of real output growth is required. Production is assumed to follow a Cobb-Douglas technology with three inputs: the two capital stocks—government and private—and labor. Using population as a proxy for labor and assuming constant returns to scale, the production function was estimated in per capita form as:

$$(9.10) \qquad y = 5.838 + 0.076k^g + 0.268k^p + 0.82MA(1)$$
$$\qquad\qquad (17.732) \quad (2.304) \quad (13.048) \quad (3.842)$$

$$R^2 = 0.90; \text{ Durbin-Watson statistic} = 1.87$$

where $k^g$ and $k^p$ denote per capita values of the public and private capital stocks. As expected, the coefficients of both the public and the private capital stocks are significant and positive. The magnitudes of the coefficients are also reasonable, suggesting that the total share of capital in output is about 30 percent, which leaves 70 percent of total output as the share of labor.

### Financial Sector

In order to model the domestic financial sector, it is assumed that the household sector's total financial assets (denoted as $A$) consist of domestic currency ($C$), domestic deposits ($D$), government bonds ($B^P$), and foreign currency assets ($F^P$). Household financial wealth ($W$) therefore consists of the holdings of these four assets minus household debt to the banking system ($L_P^{CB}$):

$$(9.11) \qquad\qquad W = A - L_P^{CB}$$

where

$$(9.12) \qquad\qquad A = C + D + B^P + sF^P.$$

Since these assets are substitutes in individual portfolios, individual demand for each asset is a function of the asset's own rate of return as well as of the returns available on other assets.[12] Domestic currency, which pays no interest, is demanded for transaction purposes. All other assets are assumed to be held for portfolio reasons. The total size of the portfolio to be allocated among these remaining assets thus consists of financial wealth net of currency plus credit from the banking system. As is well known, in this framework only three asset demand equations need be estimated, since the adding-up constraint (equation 9.12) yields the properties of the demand for the remaining asset. The demand for currency was specified as a function of the nominal interest rate on deposits—the closest substitute for currency—and income (real GNP). Demand for domestic government bonds as a share of the allocable portfolio was taken to be a function of the rate of return on those bonds, as well as on competing assets— the nominal rate of interest on domestic deposits and the return on foreign assets (that is, the foreign market interest rate corrected for changes in the exchange rate).[13] The household asset demand system

## Table 9.5. Financial Asset Demand Functions, Pakistan

| Item | Log of the ratio of currency in circulation to wealth | Log of the ratio of domestic public debt to wealth | Log of the ratio of domestic deposits to foreign currency holdings |
|---|---|---|---|
| Interest rate on deposits | −0.079 (−5.21) | | 0.271 (5.098) |
| Interest rate on public sector debt | | 0.058 (2.186) | |
| Interest rate on foreign currency assets[a] | | −0.002 (−0.354) | |
| Log of the ratio of income (GNP) to wealth | 0.996 (3.875) | | |
| Exchange rate depreciation | | | −0.008 (−2.95) |
| Dummy for 1970–72 | | | 1.505 (4.67) |
| Lagged dependent variable | | 0.932 (5.787) | |
| MA(1) | 0.97 (3.962) | | |
| Constant | −1.517 (−1.373) | −0.464 (−1.130) | 1.635 (3.269) |
| $R^2$ | 0.81 | 0.91 | 0.64 |
| Durbin-Watson | 1.85 | 2.34 | 2.27 |

Note: Figures in parentheses are $t$-statistics. A blank denotes the omission of the specific variable from the regression.

a. London interbank offered rate (LIBOR) plus expected exchange rate depreciation.

is completed with a currency-substitution equation (in which the ratio of domestic deposits to foreign currency-denominated assets is taken to be a function of the three rates of return). This is equivalent to the alternative approach of estimating either the demand for domestic deposits or foreign currency assets and determining the remaining demand as a residual. The approach adopted was preferred, however, because of our interest in obtaining direct estimates on currency substitution comparable to those existing in a wide body of empirical literature.

The results of the estimations for the financial sector have been collected in table 9.5. The nominal interest rates on both deposits and income are significant and carry the expected signs in the currency demand equation. Moreover, as expected, the income elasticity is close to unity. The demand for government bonds is positively and significantly related to the interest rate on those bonds and is negatively related to the deposit interest rate. Interestingly enough, deposit interest rates do not significantly affect the demand for these bonds, and they were therefore dropped from the equation. Strong evidence of partial adjustment behavior was found for this equation. In the currency-substitution equation both the deposit rate and the expected change in the exchange rate are significant and of the expected sign. The effect of the exchange rate change is surprisingly small. In the deposit equation a dummy variable for the period of the Bangladesh war was found to be significant and is therefore included.

To investigate the indirect effects of fiscal deficits on the real sector operating through the financial variables described above, the next section embeds the equations just estimated in a general equilibrium model.

## Policy Simulations

The preceding section identified several direct channels through which fiscal policy may have affected macroeconomic outcomes in Pakistan. These include the direct crowding-out of private by public consumption, the effects of taxation on household resources, and the effects of increases in the public capital stock on economic growth, both directly and indirectly, through a complementary relationship with private investment. The nature of these relationships suggests one set of reasons why fiscal deficits may not have greatly inhibited Pakistan's growth performance: public dissaving in the form of con-sumption may have been in part offset by private saving, thereby limiting the effects of changes in public consumption on the resources available for investment. Simultaneously, public investment has itself been directly productive and may have tended to stimulate private investment.

These relationships, however, are only part of the picture. Not only are these direct channels of influence dynamic in nature, so that their long-run implications may differ from contemporaneous effects cap-tured in regression coefficients, but indirect channels of influence must also be taken into account. These channels include, in particu-lar, the possibility of financial crowding-out because domestic bor-rowing to finance fiscal deficits may tend to raise domestic interest rates, thereby raising the rental cost of capital. To capture dynamic effects as well as these indirect interactions, in this section we analyze counterfactual scenarios under two alternative fiscal policies with a view to assessing how the fiscal adjustment would have affected the performance of Pakistan's economy during the 1980s.

The model used for the simulations embodies the behavioral equa-tions estimated in the preceding section. These include the perma-nent income equation (9.5), the consumption function (9.7), a private investment function derived from equation 9.9, the growth equation (9.10), and a set of financial sector equations consisting of identities 9.11 and 9.12, as well as the three asset demand functions reported in table 9.5. The model is completed with several additional identities. For brevity, these are reported in table 9.6.

The workings of the model can be described as follows: public sector consumption, investment, and tax revenues are taken to be policy-determined fiscal variables.[14] Monetary policy variables consist of the supply of base money, borrowing by the public sector from the commercial banks, lending by the central bank to the commercial banks, and the required reserve ratio. We treat public external borrowing as an exogenous variable. As can be seen from the public sector budget constraint in table 9.6, the implication is that domestic nonbank borrowing is the residual mode of financing for the public sector. This would seem to be the appropriate assumption for Pakistan during the 1980s.

Two other policy variables deserve mention. Pakistan maintained a system of administered interest rates in the commercial banking system until July 1, 1985, when all deposit and new financing operations of the banks were placed on an Islamic noninterest basis. The new profit-loss system implied a move toward fairly flexible market-related rates of remuneration for deposits and charges for bank loans. However, although for most of the period over which our asset demand functions were estimated bank interest rates were essentially a policy instrument, we treat them as endogenous in the simulation exercises. Implicitly, we are assuming that the monetary authorities managed these interest rates so as to maintain financial market equilibrium. In fact, these rates do exhibit substantial year-to-year variation over the sample period.

Finally, Pakistan maintained a fixed exchange regime for most of the period under review. In 1982 this regime was modified in favor of a managed exchange rate, under which the authorities undertake frequent small devaluations of the rupee. While the exchange rate has been managed with an eye on the effective real exchange rate of the rupee (and a substantial real depreciation has been achieved since 1982), price stability has also been an objective of the authorities. For the purpose of the simulations, therefore, we treat the nominal exchange rate as a policy instrument.

The model is solved as follows: at the beginning of each period real output is a predetermined variable, given as a function of the inherited private and public capital stocks. Beginning-of-period asset stocks are also predetermined because they are given by last period's government financing decisions and allocations of the private sector portfolio. The domestic-currency value of the private sector's stock of foreign assets, however, is also affected by the official exchange rate

## Table 9.6. Identities for the Simulation Model, Pakistan

1. *Public sector budget constraint*

$$\Delta H + \Delta B + S\,\Delta F_G + (\Delta L_G^{CB} - \Delta L_{CB}^G = \text{DEF})$$

2. *Public sector deficit*

$$\text{DEF} = (C^G + I^G - T)P + i^B B_{-1} + i^F S F_{-1}^G + i^{CB}(L_G^{CB} - L_{CB}^G)_{-1}$$

3. *Public sector capital accumulation*

$$K^G = I^G + (1 - \delta^G)K_{-1}^G$$

4. *High-powered money*

$$H = C + rrD$$

5. *Commercial banks' balance sheet*

$$L_P^{CB} = (1 - rr)D - L_G^{CB} - B^{CB} + L_{CB}^G$$

6. *Deposit interest rate*

$$i^D = [1/(1 - rr)]i^{CB}$$

7. *Household disposable income*

$$Y^D = Y + Z - T + \frac{i^B B_{-1}^P + i^* S F_{-1}^P + i^D D_{-1} - i^{CB} L_{P-1}^{CB}}{P}$$

8. *Household budget constraint*

$$\Delta W = (Y^D - C^P - I^P)P + (S - S_{-1})F_{-1}^P$$

9. *Private investment*

$$I^P = (K^P/Y)Y - (1 - \delta^P)K_{-1}^P$$

10. *Rental cost of capital*

$$r^K = \frac{(i^B - \pi + \delta^P)P^K}{P}$$

11. *Relative price of capital*

$$P^K/P = \alpha_0$$

12. *Equilibrium condition for public-sector securities*

$$B = B^P + B^{CB}$$

13. *Trade balance*

$$TB = Y - C^P - C^G - I^P - I^G$$

*Note:* Variables not previously identified are defined as follows: $F_G$ = foreign debt of the public sector; $L_G^{CB}$ = commercial bank lending to the public sector; $L_{CB}^G$ = Central Bank lending to commercial banks; $H$ = high-powered money; $B$ = total public sector securities outstanding; DEF = public sector deficit; $I^G$ = public investment; $T$ = taxes; $i^B$ = interest rate on public sector securities; $i^F$ = interest rate on foreign debt; $i^{CB}$ = commercial bank lending rate; $rr$ = reserve ratio; $Y^D$ = household disposable income; $Z$ = foreign remittances; $\delta^P$ = rate of depreciation on private capital stock; $\delta^G$ = rate of depreciation on public capital stock; $P^K$ = price of capital goods; $TB$ = trade balance; $\alpha_0$ = fixed positive parameter.

during the period. For these assets to be willingly held, the price level (which affects the demand for currency), the interest rate on public sector securities, and the deposit interest rate all adjust endogenously to achieve equilibrium levels. The interest rate on public sector securities, in turn, determines the rental rate on capital. (The expected rate of inflation is treated as an exogenous variable in these simulations.) The rental rate on capital, together with public consumption and investment decisions and other contemporaneous exogenous determinants of private disposable income, determine private consumption, investment, and saving, as well as the fiscal deficit and the trade balance. Public sector financing decisions will then determine the increments during the period to the domestic components of the private sector's asset portfolio that are to be carried over to the next period. The total size of the portfolio depends on private saving and the amount of lending that banks are able to make available to the private sector after satisfying the public sector's financing needs. Any discrepancy between the increase in the private sector's portfolio and the total new liabilities issued by the public sector and the banks is accumulated by the private sector in the form of foreign assets. With private and public capital stocks determined from this period's net investment by the respective sectors, next period's output is determined, and the model is ready to be solved again.

We have used this model to undertake two different simulation exercises. The first represents an alternative way of financing historical fiscal deficits, while the other represents a deficit reduction scenario.

In the first simulation we examine the macroeconomic consequences of limiting the buildup of domestic debt by the public sector after 1982/83 by increasing the use of money financing. Specifically, we reduce the flow of new domestic debt in each year from 1983/84 to 1987/88 by 10 percent and assign the role of residual financing to the issuance of base money. The results are presented in figure 9.17.

It is obvious that reduction of the flow of debt by 10 percent in each period keeps the stock of debt below its baseline value, but the percentage deviations from that value vary over time. Additional money financing would have implied larger price increases than historically observed (panel E of figure 9.17). But since domestic interest rates would have been lower (panel C), private investment would have increased (panel F), and as a result real GDP would have attained higher levels in the short run (panel D). Although factor income would have been higher, lower domestic interest rates and higher prices would have squeezed private disposable income, leading to lower private consumption (panel D). With higher output and lower private consumption, the trade deficit would have fallen, in spite of the increase in private investment (panel B). In fact, lower debt and

# Figure 9.17. Macroeconomic Effects of a 10 Percent Reduction in Domestic Debt: Simulations for Pakistan

### A. Domestic debt of the government

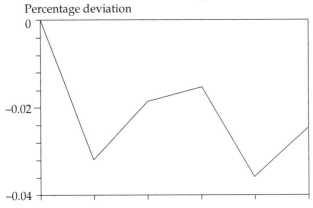

### B. Ratio of the deficit and trade deficit to GDP

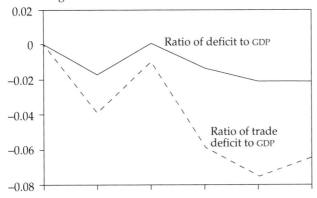

### C. Interest rate on government debt

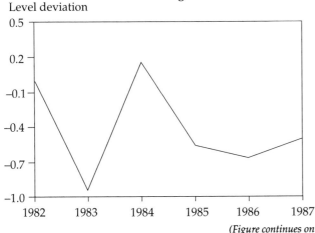

*(Figure continues on the following page.)*

**Figure 9.17** *(continued)*

### D. Real private consumption and real GDP

Percentage deviation

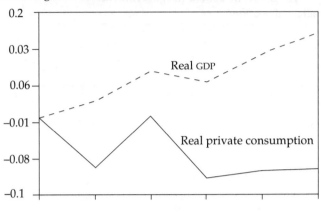

### E. Price level

Percentage deviation

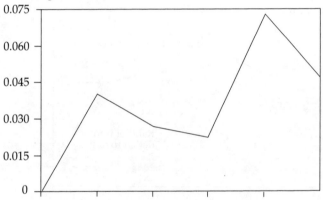

### F. Real private investment

Percentage deviation

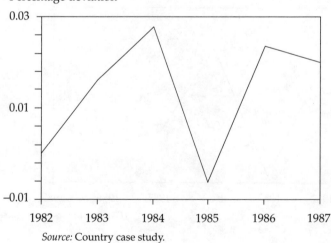

*Source:* Country case study.

lower interest rates would have implied a lower value of the ratio of the fiscal deficit to GNP. In short, the mode of financing actually chosen seems to have operated as intended to contain the consequences of the deficit for the price level, but at the expense of somewhat lower investment—and therefore slower economic growth—than would otherwise have been observed during this period.

Our remaining simulation covers the same period but assumes a 10 percent reduction in the fiscal deficit. As in the baseline simulation, debt issuance is the residual mode of financing. Deficit reduction is brought about by means of a reduction in public sector investment. Figure 9.18 shows that reducing the deficit by 10 percent in each year would have required larger and larger reductions in public investment in relation to the baseline. As is evident from panel D, this would have implied progressively larger reductions in real output, both because of the lower public capital stock and because of the induced decrease in the private capital stock, since the smaller public capital stock would have depressed private investment (panel F). Crowding-in through lower interest rates does not materialize in this case because the lower public capital stock represents a substantial negative supply shock, which raises prices (panel E) and thus actually *increases* the domestic interest rate (panel C). Both reduced output and higher prices depress private consumption (panel D). However, the reductions in public and private investment, together with the decrease in private consumption, do succeed in reducing the trade deficit (panel B), in spite of the lower level of output.

Overall, the macroeconomic effects of Pakistan's deficits from 1983/84 to 1987/88 depend on the nature of the counterfactual fiscal policy. It appears that reducing the deficit by cutting public investment, which has tended to be a favorite vehicle for deficit control in Pakistan, could have had favorable trade balance effects, but at a cost to economic growth and with little payoff in terms of price-level objectives. The way in which the historical deficits were financed also seems to have had an important effect on the economy's macroeconomic performance during the 1980s. According to our simulation results, altering the composition of deficit financing from domestic borrowing to the issuance of money would have had fairly predictable effects: shifting to more money financing would have meant lower interest rates and higher growth in the short run.[15]

## Conclusions

The underlying causes of Pakistan's high fiscal deficits during the period 1972/73 to 1987/88 were not dissimilar to those in other developing countries. The deficits reflected explicit policy choices, as well as political and administrative problems, and they were facili-

# Figure 9.18. Reduction in Government Deficit through a Reduction in Government Investment Expenditure, Pakistan

## A. Real government investment

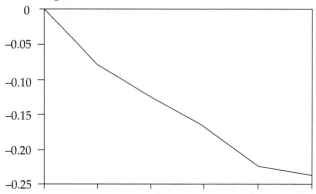

## B. Ratio of the deficit and trade deficit to GDP

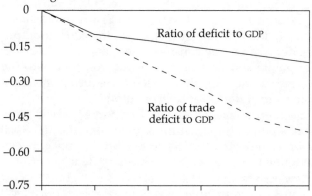

## C. Interest rate on government debt

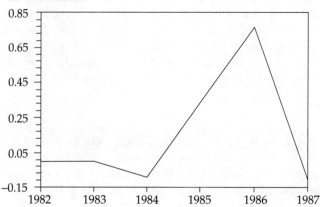

## D. Real private consumption and real GDP

Percentage deviation

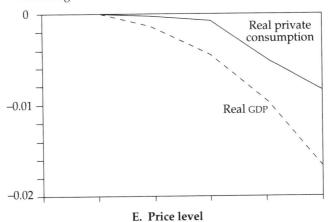

## E. Price level

Percentage deviation

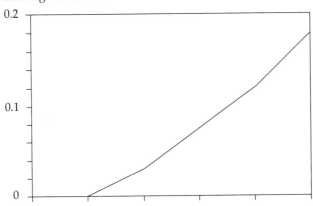

## F. Real private investment

Percentage deviation

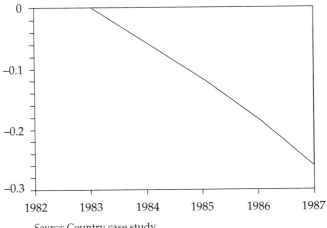

*Source:* Country case study.

tated by the availability of external financing. Among the key policy decisions contributing to Pakistan's fiscal performance during the period were the Bhutto government's decision in the early 1970s to substantially enlarge the role of the public sector and the decisions of subsequent governments to maintain high levels of defense spending and consumer subsidies and, in the 1980s, to rely heavily on nonbank domestic borrowing as a source of finance. Throughout the period, a key political problem has been the inability to levy significant taxes on various politically powerful economic sectors. Coupled with the administrative difficulties that would be posed by greater reliance on income taxation, this inability has prevented the emergence of a revenue base for financing the chosen expenditure levels. Nevertheless, the ready availability of external finance at concessionary rates has permitted the resulting deficits to be financed without a fiscal explosion.

As has been emphasized throughout the chapter, a distinctive feature of the Pakistani experience during the period under review has been the coexistence of very large fiscal deficits for long periods of time with an economic performance that was relatively satisfactory and crisis-free with respect to growth, inflation, and the external accounts. To some extent, this reflects a statistical illusion. The grant element associated with external borrowing and the tax element attached to domestic borrowing are, in effect, sources of revenue for the general government but are treated in official statistics as financing items. Although our crude estimates for the magnitude of the effects of these elements on the measured deficits should be considered as an upper bound, this factor may in fact have accounted for sizable proportions of the deficit at various times during the past two decades. Growth itself also explains the absence of high inflation, since the associated expansion of the base for both conventional taxes and seigniorage made it possible to finance in a noninflationary way equilibrium deficits that were significantly larger than could have been financed in a slow-growth economy.

Nevertheless, since the early 1980s fiscal deficits in Pakistan have clearly exceeded such equilibrium values. These deficits have been financed by domestic nonbank borrowing, resulting in increasing ratios of domestic public debt to GNP and, until quite recently, in rising interest rates on such debt. Our simulations indicate that although relying on this source of finance may have mitigated the inflationary consequences of the deficits, it was done at the expense of some crowding-out of private investment, implying slower growth than would otherwise have been observed. Controlling the deficit over the period would perhaps have made it possible to generate more favorable macroeconomic outcomes, at least with respect to growth and the external accounts. This outcome, however, would not have been possible if the deficit reduction were brought about in a manner commonly relied on both in Pakistan and elsewhere—that is, through reducing public investment.

The question of alternative modes of deficit reduction has become an issue of increasing relevance to Pakistan in recent years because fiscal deficits of the magnitudes observed up to 1987/88 have become more difficult to finance. Not only are the sources of external financing at concessional rates dependent on the vagaries of the world oil market and political developments in the Middle East, but the accumulation of domestic debt and the increased costs of borrowing at home both require the generation of lower primary deficits. The alternative, money financing, runs the risk of moving Pakistan's macroeconomic performance closer to that of the countries with high fiscal deficits that are discussed elsewhere in this book.

## Notes

At the time of writing, both authors were on the staff of the International Monetary Fund (IMF). The views expressed are the sole responsibility of the authors and do not represent the views of the IMF. The authors are grateful to William Easterly, E. Ahmed, Mohsin S. Khan, and Malcolm D. Knight for their comments on an earlier draft and to Ravina Malkani for providing excellent research assistance.

1. The dates in this chapter refer to fiscal years, which in Pakistan run from July 1 to June 30.

2. Deficit as a percentage of GNP is the broadest deficit measure available for Pakistan and will be used throughout the chapter. Because of the importance of workers' remittances in Pakistan's economy, GNP rather than gross domestic product (GDP) is used as the scale variable.

3. To some extent the official price and output figures could be misleading. The quality of the output data is open to question, particularly in the presence of what is said to be a substantial underground economy. As for prices, the consumer price index encompasses a number of goods that have been subject to price controls for some time. Data problems, however, are unlikely to account for more than a very minor part of the discrepancy between the macroeconomic performance of Pakistan and that of, say, several large countries in Latin America.

4. Indirect taxes account for about 80 percent of total tax revenues in Pakistan, and foreign trade taxes represent about half of total indirect tax revenue. Administrative problems have hampered the collection of direct taxes, and the taxation of agricultural incomes has not been politically feasible.

5. The previous government, however, had already achieved a substantial reduction in public consumption during its last year in office.

6. Since the coefficient of log GNP in equation 2 in table 9.2 is essentially unity, this equation in effect regresses the share of trade taxes in GNP on the share of exports and imports in GNP, as well as on the share of remittances in GNP.

7. These show up as "other nontax revenues" in table 9.1.

8. Uncertainty surrounds not just the netting-out procedure used to calculate net debt for 1979/80 but also the degree to which the debt can be considered to be "willingly held" by creditors. Complications arise in the latter regard because some of the domestic debt was held by domestic banks subject to asset supervision, and much of the external debt consisted of bilateral lending.

9. More precisely, the procedure involved subtracting the product of the grant element and the flow of foreign loans, as well as the product of the implicit tax rate and the flow of domestic borrowing, from the actual deficit.

10. This variable is intended to capture confidence effects associated with the change in economic policies as a result of the change in political regimes.

11. The instruments used in the estimation included the error-correction term, the lagged growth rate, the ratio of the government capital stock to GDP, and the time trend.

12. The framework used is a variation of the Tobin (1969) general-equilibrium approach.

13. The economy is assumed to be fairly open, in accordance with the findings of Haque and Montiel (1990, 1991).

14. This implicitly assumes that tax rates are adjusted to offset deviations in the tax base from baseline values.

15. For additional information on the influence of foreign financing on fiscal policy in Pakistan, see Haque, Husain, and Montiel (1991).

## References

Buiter, Willem H. 1985. "A Guide to Public Sector Debt and Deficits." *Economic Policy* 1 (November): 13–79.

David, Paul, and John Scadding. 1974. "Private Saving, Ultrarationality, and Denison's Law." *Journal of Political Economy* 82 (March–April): 225–49.

Davidson, James F. H., David F. Hendry, Frank Srba, and Stephen Yeo. 1978. "Econometric Modelling of the Aggregate Time-Series Relationship Between Consumers' Expenditure and Income in the UK." *Economic Journal* 88 (December): 661–92.

Haque, Nadeem Ul, and Peter J. Montiel. 1989. "Consumption in Developing Countries: Tests for Liquidity Constraints and Finite Horizons." *Review of Economics and Statistics* 71 (August): 408–15.

———. 1990. "How Mobile Is Capital in Developing Countries?" *Economic Letters* 33: 359–62.

———. 1991. "Capital Mobility in Developing Countries—Some Empirical Tests." *World Development* 19 (10): 1391–98.

Haque, Nadeem Ul, Aasim Husain, and Peter J. Montiel. 1991. "An Empirical 'Dependent Economy' Model for Pakistan." IMF Working Paper 91/102. International Monetary Fund, Washington, D.C.

IMF (International Monetary Fund). 1990. *World Economic Outlook*. Washington, D.C.

Kemal, A. R. 1987. "Fiscal System of Pakistan." World Bank, Development Research Department, Washington, D.C.

Khan, Mohsin S. 1990. "Macroeconomic Policies and the Balance of Payments in Pakistan: 1972–1986." IMF Working Paper 78. International Monetary Fund, Washington, D.C.

Morshed, Nader. 1987. "Pakistan Banking System and Capital Markets." World Bank, Development Research Department, Washington, D.C.

Pakistan. Various issues. *Pakistan Economic Survey*. Islamabad.

State Bank of Pakistan. Various issues. *Annual Report*. Islamabad.

Tobin, James. 1969. "A General Equilibrium Approach to Monetary Theory." *Journal of Money, Credit, and Banking* 1 (February): 15–29.

# 10

# *Zimbabwe: Fiscal Disequilibria and Low Growth*

*Felipe Morandé and Klaus Schmidt-Hebbel*

Since independence in 1980, Zimbabwe has shown significant improvements in areas such as education, health, and smallholder agriculture. However, these social achievements have not been matched by improvements in the overall standard of living. Large fiscal imbalances, a vast and loss-making state-enterprise sector, and lack of an adequate market-based incentive structure have severely affected macroeconomic stability and growth during the past twelve years. This inadequate policy framework has also hampered Zimbabwe's ability to cope with recurrent droughts, of which the recent 1992 drought was the most severe. As a result of policy mistakes and bad luck, GDP growth per capita was zero during 1980–92.

Public sector imbalances have been at the heart of the country's difficulties. The deficit of either the central government or the consolidated nonfinancial public sector (CNFPS, comprising general government and public enterprises) shows double-digit levels in almost every year since 1981 (table 10.1). Some sustained fiscal adjustment was achieved in 1987, when the deficit was brought down from about 14 percent of GDP to 10 percent. Additional fiscal adjustment has been pursued since 1991 in the framework of a structural adjustment program but has not been reflected in declining fiscal deficits because of the massively adverse budgetary impact of the 1992 drought.

Public deficit financing has gone through various phases: foreign financing in the early 1980s, domestic debt financing in the mid- to late 1980s, and a resurgence of foreign financing since 1990. In the absence of significant private capital flows, the public sector deficit has been the driving force behind Zimbabwe's current account deficits and the derived steep increase in total foreign debt, from 15 percent of GDP in 1980 to 69 percent in 1992.

Strongly declining current account deficits were associated with a cumulative 56 percent depreciation of the real exchange rate between 1981 and 1985. Between 1985 and 1990 the real exchange rate did not change much, while the current account deficit was small. As a result of the partial trade liberalization since 1991 and, principally, the recent drought, the current account deteriorated in 1991 and reached a record deficit of 20.7 percent of GDP in 1992. The real exchange rate depreciated by a cumulative 23.1 percent between 1990 and 1992.

Monetary policy followed a relatively prudent course until 1990, with low to moderate inflation rates. An expansionary monetary policy in 1991 (reversed in late 1991) started an inflationary cycle fueled by subsequent drought-induced price increases, leading to an unprecedented 46 percent inflation rate in 1992.

Macroeconomic policy is only partly to blame for Zimbabwe's unsatisfactory economic performance. In fact, a large array of trade and factor market distortions has hampered investment and growth. Domestic price controls have been prevalent in agriculture, utilities, and transport. Interest rate controls have kept interest rates on bank deposits and public debt at levels that have rarely exceeded domestic inflation rates. Foreign trade has been subject to large barriers in the form of quantitative and nonquantitative trade restrictions. Rationing of foreign exchange by the government is institutionalized through a system of foreign exchange allocation to importing sectors on the basis of noneconomic criteria. These barriers have maintained significant import substitution inherited from the preindependence period at the cost of severely hampering the economy's productive efficiency. Consumer imports, foreign capital flows, and, in particular, outflows of private capital have been greatly restricted. A large and generally inefficient public enterprise sector includes many loss-making firms, which contribute significantly to the overall CNFPS deficit. Private investment is constrained by state licensing and controls that raise the cost of doing business in Zimbabwe and by an uncertain policy and property-rights framework.

Private consumption and investment have been crowded out by a public sector able to finance part of its deficit by relying on nonmarket mechanisms to generate and make use of a significant private sector surplus. When the public sector requires more resources from the domestic private sector to finance its deficit, it restricts allocation of foreign exchange to the private sector, constraining aggregate private investment and consumption. During the past decade high saving and low investment were reflected in large private surpluses. By exerting strict controls on private capital flight and placing more government debt in domestic banks and nonbanking financial institutions, the government ensures that the private surplus is channeled toward domestic financial markets. This allows the government to capture these private resources at typically negative real interest rates.

Crowding-out of private investment and declining public investment, combined with a low quality of investment projects as a result of the distorted incentive structure, have substantially affected Zimbabwe's growth potential and performance. In addition, to the extent that imported intermediate and capital goods are not perfect substi-

Table 10.1. Macroeconomic Indicators, Zimbabwe, 1980–92

| Indicator | 1980 | 1981 | 1982 | 1983 | 1984 | 1985 | 1986 | 1987 | 1988 | 1989 | 1990 | 1991 | 1992a |
|---|---|---|---|---|---|---|---|---|---|---|---|---|---|
| *Aggregate indicators* | | | | | | | | | | | | | |
| GDP growth (percent) | 10.6 | 12.5 | 2.6 | 1.6 | −1.9 | 6.8 | 2.6 | −1.5 | 9.6 | 4.6 | 2.1 | 4.9 | −8.3 |
| Capacity utilization (percent) | 79.7 | 89.3 | 88.0 | 85.4 | 80.5 | 84.3 | 85.8 | 83.8 | 88.8 | — | — | — | — |
| CPI inflation (percent) | 5.5 | 13.1 | 10.7 | 23.1 | 20.2 | 8.5 | 14.3 | 12.5 | 7.4 | 12.9 | 17.4 | 24.3 | 46.2 |
| Real exchange rate (1980 = 100)b | 100.0 | 104.6 | 110.3 | 123.4 | 131.9 | 163.1 | 150.2 | 138.2 | 145.1 | 158.0 | 164.4 | 192.9 | 202.4 |
| Nominal interest rate on CNFPS domestic debt (percent) | 4.4 | 5.9 | 7.8 | 7.7 | 8.0 | 10.4 | 12.5 | 13.0 | 13.3 | 13.3 | — | — | — |
| Nominal interest rate on three-month bank deposits (percent) | 3.4 | 8.9 | 8.6 | 11.9 | 10.1 | 9.3 | 9.6 | 9.2 | 9.4 | 9.1 | 9.8 | 14.2 | 25.0 |
| *Composition of output (percentage of GDP)* | | | | | | | | | | | | | |
| Resource balance | −3.0 | −7.3 | −5.9 | −3.2 | 0.6 | 1.2 | 4.4 | 4.1 | 5.2 | 2.4 | 0.3 | −3.9 | −12.6 |
| Exports | 30.3 | 25.2 | 22.0 | 21.3 | 26.7 | 29.9 | 30.9 | 31.2 | 30.0 | 30.2 | 30.1 | 32.7 | 33.8 |
| Imports | 33.3 | 32.5 | 27.9 | 24.5 | 26.1 | 28.7 | 26.5 | 27.1 | 24.8 | 27.8 | 29.9 | 35.6 | 44.6 |
| Total consumption | 84.2 | 84.2 | 84.8 | 84.5 | 83.7 | 85.4 | 81.9 | 76.8 | 73.3 | 76.1 | 78.6 | 82.9 | 90.6 |
| Private | 64.5 | 67.0 | 65.0 | 66.1 | 62.4 | 63.2 | 60.1 | 52.7 | 47.0 | — | 55.7 | 59.2 | — |
| Public | 19.7 | 17.2 | 19.8 | 18.4 | 21.3 | 22.2 | 21.8 | 24.1 | 26.3 | — | 22.9 | 23.7 | — |
| Gross domestic investment | 18.8 | 23.0 | 21.1 | 15.9 | 18.9 | 21.0 | 19.4 | 19.1 | 21.5 | 21.5 | 21.1 | 21.0 | 22.0 |
| Gross fixed capital investment | 15.3 | 18.6 | 19.6 | 19.6 | 18.5 | 16.1 | 15.8 | 15.5 | 17.9 | 17.4 | — | — | — |
| Private | 10.6 | 13.3 | 10.0 | 8.2 | 10.6 | 7.9 | 8.4 | 7.8 | 9.0 | — | — | — | — |
| Public | 4.7 | 5.3 | 9.9 | 11.4 | 7.9 | 8.2 | 7.4 | 7.7 | 8.9 | — | — | — | — |
| Change in stocks | 3.5 | 4.4 | 1.2 | −3.7 | 0.4 | 4.9 | 3.6 | 3.6 | 3.6 | 4.1 | — | — | — |

*Government deficit and debt (percentage of GDP)*

| | | | | | | | | | | | | | |
|---|---|---|---|---|---|---|---|---|---|---|---|---|---|
| **CNFFS (fiscal year data)** | | | | | | | | | | | | | |
| Deficit | 9.1 | 13.5 | 13.1 | 14.4 | 12.7 | 14.3 | 14.4 | 10.9 | 10.4 | — | — | — | — |
| Foreign debt | 12.0 | 17.6 | 23.3 | 27.0 | 33.3 | 42.2 | 40.6 | 41.1 | 38.0 | — | — | — | — |
| Domestic debt | 43.4 | 37.2 | 33.7 | 31.3 | 35.7 | 35.5 | 36.6 | 41.7 | 42.9 | — | — | — | — |
| **Central government (calendar year data)** | | | | | | | | | | | | | |
| deficit (including net lending) | — | — | — | — | — | — | — | — | 10.0 | 9.9 | 9.8 | 11.3 | 16.0 |
| **Monetary aggregates (percentage of GDP)** | | | | | | | | | | | | | |
| Base money | 6.9 | 7.1 | 7.3 | 6.2 | 6.7 | 7.5 | 7.2 | 7.0 | 7.4 | 7.6 | 7.9 | 7.9 | — |
| $M_1$ | 18.4 | 15.3 | 15.9 | 11.9 | 13.5 | 14.3 | 13.3 | 13.7 | 14.0 | 14.1 | 14.8 | 13.8 | — |
| Quasi money | 16.8 | 16.3 | 17.7 | 14.9 | 16.4 | 13.7 | 13.7 | 18.1 | 16.3 | 17.7 | 15.3 | 12.6 | — |
| **Current account and foreign debt (percentage of GDP)** | | | | | | | | | | | | | |
| Current account deficit[c] | 5.6 | 11.5 | 11.1 | 8.4 | 3.7 | 3.5 | 1.0 | 0 | -1.2 | 1.2 | 4.4 | 11.0 | 20.7 |
| Total foreign debt[d] | 14.7 | 19.6 | 27.3 | 35.1 | 42.3 | 55.5 | 52.7 | 53.1 | 42.1 | 43.7 | 48.6 | 54.6 | 68.7 |

— Not available.

*Note:* CNFFS, consolidated nonfinancial public sector.

a. Estimated.

b. Defined as the ratio between the product of the Z$/US$ nominal exchange rate and by the U.S. CPI, divided by the Zimbabwe CPI.

c. Excludes official capital grants.

d. End-of-year total (public and private) foreign debt outstanding and disbursed as a ratio of annual GDP.

*Source:* Reserve Bank of Zimbabwe; Zimbabwe Ministry of Finance; Schmidt-Hebbel 1990; and World Bank data.

tutes for domestic goods, import compression and exchange controls have lowered the utilization of existing capacity.

An adjustment program was initiated in 1991 to tackle both macroeconomic disequilibria and structural distortions. More stringent fiscal and monetary policies were put in place, supported by a devaluation of the nominal exchange rate. Progress has been made in the areas of domestic price liberalization, private investment deregulation, and trade liberalization. However, much more progress is required in liberalization of domestic goods and financial markets, as well as in external trade, restructuring and privatization of state enterprises, and deregulation of private sector production and investment. These measures should complement further fiscal adjustment at the general-government level. Only when these conditions are met—and after allowing for the significant period required to establish the credibility of the new policy framework—can Zimbabwe start on a path of sustained growth and poverty reduction.

This chapter attempts to disentangle the macroeconomic implications of Zimbabwe's public deficits and policy framework, following the methodology laid out in Easterly, Rodríguez, and Schmidt-Hebbel (1989).[1] Most of the empirical analysis covers the 1965–89 period, with a strong emphasis on the last decade. As a first step, the next section identifies the main macroeconomic and fiscal policy variables that have contributed to CNFPS deficits, focusing on the sensitivity of the budget to its main determinants. The results show the overriding influence of fiscal policy variables as determinants of deficits, in comparison with domestic or external variables beyond the direct control of fiscal policymakers. A subsequent deficit-sustainability analysis suggests that current public deficits are unsustainable in the sense that they imply exploding public debt profiles.

We then assess the impact of public deficits on monetary and financial markets. The empirical evidence from simulations based on a dynamic portfolio framework suggests that excessive debt financing only postpones inflation or—in line with Sargent and Wallace's (1981) "unpleasant arithmetic"—the current combination of high public deficits and low-to-moderate inflation rates is unsustainable in the longer term. This confirms the simple steady-state results of the preceding section.

We next go a step further by analyzing the impact of the public sector on private consumption and investment. Clear evidence of crowding-out is presented, especially for the post-1980 period. Crowding-out is not only indirect, through higher interest rates, but is also a direct result of semicompulsory placement of public debt in financial institutions. Again the issue is one of longer-run sustainability, as permanently high fiscal deficits inhibit growth, leading to a further deterioration of fiscal stance.

We then deal with the effect of the public deficit and its financing on external accounts, in particular the trade deficit and the real exchange rate. The Rodríguez model in chapter 2 in this volume is modified to take into account the foreign exchange allocation mechanism and the binding constraints on capital movements. These features make the levels of public sector deficits and public sector spending more relevant than deficit financing for the behavior of the trade balance and the real exchange rate in Zimbabwe.

If public deficits crowd out private investment, potential output and growth prospects are affected. In the final sections we tackle this issue, as well as the effects of other policy distortions, and present our conclusions.

## Determination and Sustainability of Public Sector Deficits

A first step in studying the macroeconomics of public sector deficits is to identify the main macroeconomic and fiscal policy variables that have contributed to those deficits and to assess their sustainability under alternative macroeconomic scenarios. These exercises are carried out in this section, which begins with a discussion of public deficits and liabilities in the 1980s.

### Consolidated Public Sector Deficits and Balance Sheets

Zimbabwe's nonfinancial public sector comprises the central government, local authorities, and public enterprises.[2] The financial public sector consists mainly of the Reserve Bank of Zimbabwe and the Post Office and Savings Bank. These financial institutions do not carry out quasi-fiscal operations and do not show significant deficits or surpluses. Hence most of the analysis will be restricted to the public sector deficit of the CNFPS. Consolidated balance sheets were constructed, however, for both the nonfinancial public sector and the total public sector (TPS), the latter being defined as the nonfinancial public sector and the two public financial institutions mentioned above. The decomposition of the public deficit is referred to the CNFPS; the subsequent sustainability analysis is carried out for the asset and liability holdings of the TPS.

Tables 10.2 and 10.3 report data on the deficits and net liabilities of the CNFPS and the TPS. Figure 10.1 shows the evolution of the primary and total nominal deficits during the 1980s, while figure 10.2 does the same for Zimbabwe's domestic and foreign debt ratios. After independence, public sector deficits grew from less than 10 percent of GDP to 13–14 percent of GDP and stayed at that level for six years, until 1986/87. The increase in the primary deficit initially took place mainly in the public enterprise sector, but a rise in the central government

**Table 10.2. Consolidated Public Sector Deficit, Zimbabwe, Fiscal 1980/81–1988/89**
(ratio to fiscal year GDP)

| Item | 1980/81 | 1981/82 | 1982/83 | 1983/84 | 1984/85 | 1985/86 | 1986/87 | 1987/88 | 1988/89 |
|---|---|---|---|---|---|---|---|---|---|
| CNFPS deficit | 0.091 | 0.135 | 0.131 | 0.144 | 0.127 | 0.143 | 0.144 | 0.109 | 0.100 |
| 1. Primary deficit | 0.067 | 0.100 | 0.087 | 0.090 | 0.063 | 0.071 | 0.069 | 0.033 | 0.022 |
| Central government | 0.054 | 0.024 | 0.019 | 0.046 | 0.037 | 0.035 | 0.055 | 0.018 | 0.009 |
| PLA | 0.014 | 0.075 | 0.069 | 0.043 | 0.026 | 0.036 | 0.014 | 0.015 | 0.013 |
| 2. Net interest payments | 0.024 | 0.036 | 0.043 | 0.055 | 0.064 | 0.072 | 0.075 | 0.077 | 0.078 |
| Foreign debt | 0.007 | 0.011 | 0.018 | 0.032 | 0.035 | 0.033 | 0.029 | 0.027 | 0.025 |
| Domestic debt | 0.017 | 0.025 | 0.025 | 0.022 | 0.029 | 0.039 | 0.046 | 0.050 | 0.053 |
| Consolidated TPS deficit | 0.088 | 0.131 | 0.126 | 0.143 | 0.126 | 0.138 | 0.137 | 0.103 | 0.093 |
| 1. Primary deficit | 0.067 | 0.100 | 0.087 | 0.090 | 0.063 | 0.071 | 0.069 | 0.033 | 0.022 |
| Central government | 0.054 | 0.024 | 0.019 | 0.046 | 0.037 | 0.035 | 0.055 | 0.018 | 0.009 |
| PLA | 0.014 | 0.075 | 0.069 | 0.043 | 0.026 | 0.036 | 0.014 | 0.015 | 0.013 |
| 2. Net interest payments | 0.021 | 0.032 | 0.039 | 0.053 | 0.064 | 0.067 | 0.068 | 0.070 | 0.071 |
| Foreign debt | 0.004 | 0.011 | 0.018 | 0.034 | 0.038 | 0.033 | 0.029 | 0.026 | 0.024 |
| Domestic debt | 0.016 | 0.021 | 0.020 | 0.019 | 0.025 | 0.034 | 0.040 | 0.045 | 0.047 |

Note: PLA, public enterprises and local authorities; TPS, total public sector.
Source: Schmidt-Hebbel 1990.

**Table 10.3. Public Sector Liabilities, Zimbabwe, 1980–88**
(ratio to calendar year GDP)

| Item | June 1980 | June 1981 | June 1982 | June 1983 | June 1984 | June 1985 | June 1986 | June 1987 | June 1988 |
|---|---|---|---|---|---|---|---|---|---|
| CNFPS liabilities | 0.541 | 0.544 | 0.568 | 0.617 | 0.746 | 0.813 | 0.793 | 0.864 | 0.830 |
| 1. Cash | 0.013 | −0.004 | −0.004 | −0.041 | −0.065 | −0.045 | −0.030 | −0.044 | −0.030 |
| 2. Net foreign debt | 0.120 | 0.176 | 0.233 | 0.270 | 0.333 | 0.422 | 0.406 | 0.411 | 0.380 |
| 3. Net domestic debt | 0.434 | 0.372 | 0.337 | 0.313 | 0.357 | 0.355 | 0.366 | 0.417 | 0.429 |
| 4. Equity | 0.000 | 0.007 | 0.007 | 0.007 | 0.009 | 0.009 | 0.009 | 0.008 | 0.008 |
| Consolidated TPS liabilities | 0.541 | 0.544 | 0.567 | 0.617 | 0.746 | 0.813 | 0.793 | 0.864 | 0.830 |
| 1. Base money | 0.058 | 0.061 | 0.064 | 0.057 | 0.059 | 0.058 | 0.060 | 0.064 | 0.064 |
| 2. Net foreign debt | 0.074 | 0.169 | 0.237 | 0.289 | 0.366 | 0.419 | 0.393 | 0.393 | 0.365 |
| 3. Net domestic debt | 0.423 | 0.318 | 0.270 | 0.260 | 0.309 | 0.312 | 0.319 | 0.375 | 0.380 |
| 4. Other liabilities | −0.014 | −0.004 | −0.003 | 0.010 | 0.012 | 0.024 | 0.021 | 0.031 | 0.021 |

*Source:* Schmidt-Hebbel 1990.

**Figure 10.1. Consolidated Fiscal Sector Deficit, Zimbabwe, Fiscal 1980/81 to 1988/89**

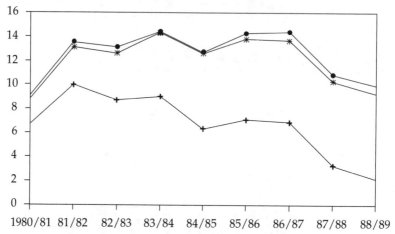

Percentage of GDP

—•— CNFPS deficit
—*— Primary deficit
—+— Consolidated TPS deficit

*Note:* CNFPS, consolidated nonfinancial public sector; TPS, total public sector.
*Source:* Table 10.2.

primary deficit soon followed. Increasing nominal interest rates on domestic debt and rising debt-output ratios during 1980–85 explain the increasing net interest payments throughout the 1980s and up to the present. However, beginning in 1987/88 a partial fiscal adjustment in the central government reduced the deficit by 3.5 percentage points in that fiscal year and by 1 additional percentage point in 1988/89.

The financing requirements of high public sector deficits have contributed to a steady and massive rise in public (CNFPS or TPS) liabilities, from 54.1 percent of GDP in June 1980 to 86.4 percent in June 1988. It is interesting to note how the composition of public debt has been changing. In the early 1980s public deficits relied massively on foreign financing, raising TPS foreign debt from 7.4 percent of GDP in 1980 to a peak of 41.9 percent in 1985. (A similar change was observed in CNFPS debt ratios.) This made possible a reduction in domestic debt from 42.3 percent of GDP in 1980 to 26 percent in 1983. A strong reversal of the composition of debt financing occurred afterward, when the foreign debt ratio fell by a couple of percentage points while the domestic debt ratio increased to reach a level in 1988 only slightly

**Figure 10.2. Consolidated Public Sector Liabilities, Zimbabwe, 1980–88**

Percentage of GDP

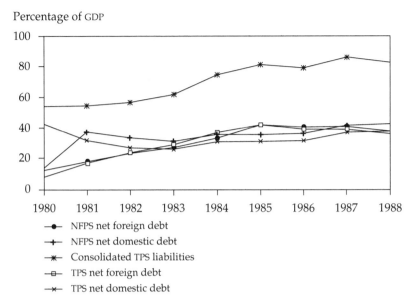

— NFPS net foreign debt
— NFPS net domestic debt
— Consolidated TPS liabilities
— TPS net foreign debt
— TPS net domestic debt

*Note:* NFPS, nonfinancial public sector; TPS total public sector
*Source:* Table 10.3.

below that of 1980. Monetary financing of TPS deficits was relatively small and stable throughout the 1980s. Base money, after increasing slightly in the early 1980s, has remained stable at about 6.4 percent of GDP.

## *Economic and Policy Determinants of Public Sector Deficits*

The following decomposition of public deficits in Zimbabwe during the 1980s compares the role of macroeconomic domestic and foreign variables with that of fiscal policies in generating the initial expansionary phase and the subsequent partial fiscal adjustment. The methodology, which is based on the framework developed by Marshall and Schmidt-Hebbel (1989), is presented in the appendix to Morandé and Schmidt-Hebbel (1991). The main budgetary items of the CNFPS deficit are first identified. By making use of estimated tax revenue functions,[3] the Fisher equation for domestic interest rates, and simple variable transformations, it is possible to identify the effects of the main macroeconomic and policy variables on the deficit. This permits measurement of the sensitivity of Zimbabwe's public

**Table 10.4. Decomposition of Changes in CNFPS Deficits with Changes in Economic and Policy Determinants, Zimbabwe, Fiscal 1981/82–1988/89**

| Item | 1981/82 | 1982/83 | 1983/84 | 1984/85 | 1985/86 | 1986/87 | 1987/88 | 1988/89 |
|---|---|---|---|---|---|---|---|---|
| *Changes attributable to domestic variables* | −0.029 | −0.004 | −0.001 | −0.016 | −0.021 | 0.007 | 0.001 | −0.024 |
| Real GDP *rc* (denominator effects) | −0.009 | −0.003 | 0.000 | −0.004 | −0.007 | −0.001 | −0.004 | −0.009 |
| Real GDP *rc* (economic effects) | −0.016 | −0.004 | 0.000 | −0.005 | −0.010 | −0.001 | −0.005 | −0.012 |
| Real imports *rc* | −0.012 | 0.004 | 0.007 | 0.000 | −0.004 | 0.002 | 0.008 | 0.000 |
| Domestic real interest rate *c* | 0.003 | −0.004 | 0.014 | 0.029 | −0.007 | −0.009 | 0.006 | −0.005 |
| Domestic inflation *c* | 0.006 | 0.007 | −0.013 | −0.020 | 0.015 | 0.010 | −0.005 | 0.005 |
| Real exchange rate *rc* | −0.001 | −0.003 | −0.009 | −0.018 | −0.008 | 0.005 | 0.001 | −0.003 |
| *Changes attributable to foreign variables* | | | | | | | | |
| Foreign nominal interest rate *c* | 0.000 | 0.003 | 0.007 | −0.003 | −0.008 | −0.004 | −0.001 | 0.000 |
| *Changes attributable to policy variables* | | | | | | | | |
| Foreign real debt *rc* | 0.067 | −0.002 | −0.016 | −0.007 | 0.047 | 0.017 | −0.016 | −0.010 |
| Domestic real debt *rc* | 0.004 | 0.003 | 0.001 | −0.001 | 0.006 | 0.003 | 0.000 | −0.002 |
| Wage bill *rc* | −0.001 | −0.002 | 0.000 | 0.003 | 0.000 | 0.004 | 0.005 | 0.005 |
| Goods and services expenditure *rc* | 0.010 | 0.001 | 0.000 | 0.006 | 0.011 | 0.016 | 0.036 | 0.004 |
| Transfers and subsidies *rc* | 0.008 | 0.002 | 0.006 | −0.002 | 0.007 | 0.013 | 0.000 | 0.006 |
| 1980 political regime *c* | 0.008 | 0.011 | 0.003 | −0.005 | 0.012 | −0.010 | −0.051 | 0.001 |
| 1988 direct tax regime *c* | −0.014 | 0.000 | 0.000 | 0.000 | 0.000 | 0.000 | 0.000 | 0.000 |
| 1981 indirect tax regime *c* | 0.000 | 0.000 | 0.000 | 0.000 | 0.000 | 0.000 | −0.024 | 0.000 |
| 1982 customs duties regime *c* | −0.028 | 0.000 | 0.000 | 0.000 | 0.000 | 0.000 | 0.000 | 0.000 |
| 1983 customs duties regime *c* | 0.000 | −0.022 | 0.022 | 0.000 | 0.000 | 0.000 | 0.000 | 0.034 |
| 1988 customs duties regime *c* | 0.000 | 0.000 | −0.037 | 0.000 | 0.000 | 0.000 | 0.000 | −0.052 |
| PLA primary deficit *rc* | 0.015 | 0.006 | −0.014 | 0.001 | −0.011 | −0.002 | 0.012 | 0.001 |
| NFPS investment *rc* | 0.064 | 0.000 | 0.003 | −0.009 | 0.022 | −0.007 | 0.007 | −0.007 |
| Explained sum of changes | 0.038 | −0.003 | −0.010 | −0.026 | 0.018 | 0.020 | −0.016 | −0.034 |
| Changes attributable to other variables | 0.006 | −0.002 | 0.023 | 0.008 | −0.001 | −0.020 | −0.028 | 0.026 |
| Actual change in CNFPS deficits | 0.044 | −0.005 | 0.013 | −0.017 | 0.017 | 0.000 | −0.045 | −0.008 |

*Note:* The figures reflect the annual changes in CNFPS deficits (as ratios to GDP) caused by proportional changes (*rc* is rate of change) or absolute changes (*c*) in the corresponding variables.

*Source:* Authors' calculations based on data from Zimbabwe Ministry of Finance.

budget structure to changes in macroeconomic and policy determinants.

Table 10.4 reports the changes in the CNFPS deficit according to their underlying macroeconomic and policy causes. To illustrate the usefulness of this approach, the principal variables behind the partial fiscal adjustment of 1987/88 to 1988/89 are identified next.

GDP growth was the main macroeconomic variable contributing to deficit reduction during 1987/88 and 1988/89. Its positive effect on tax bases (the "economic effect" in table 10.4) reduced the deficit by 0.5–1.2 percentage points of GDP, in addition to the 0.4–0.9 percentage point reduction resulting from the simple fact that the deficit and every budgetary item are expressed as ratios to GDP (the "denominator effect"). Other macroeconomic variables (apart from imports, whose decline in 1987–89 increased the deficit) tended to cause only minor changes.

Among fiscal variables a stabilization effort was reflected by significantly lower transfers and subsidies in 1987/88 and increased revenue from customs duties in 1988/89. However, other variables under the control of policymakers contributed to an increase in the deficit: the budgetary wage bill expanded significantly, and, to a lesser extent, higher expenditure on goods and services and a higher deficit of public enterprises and local authorities (PLA) raised the CNFPS deficit. In addition, the secular rise in domestic debt increased domestic interest payments.

What was the role of fiscal policies in comparison with the influence of exogenous variables during the 1980s in Zimbabwe? Computation (from table 10.4) of the average contribution of different variables to the explained variation in the deficit over the 1981/82 to 1988/89 period yields the following results: domestic variables, 7 percent; external variables, −11 percent; and fiscal policy variables, 110 percent.[4] These figures confirm the massive predominance of fiscal policy changes in both the cyclical variation and the trend change of CNFPS deficits in Zimbabwe. Both domestic and external variables play a secondary role in shaping deficits—external variables even contribute negatively to deficit changes. This contrasts with the large contribution of fiscal policy variables, which actually offset the influence of foreign interest effects. We conclude that policymakers are responsible for both the fiscal deterioration in the early 1980s and the partial fiscal adjustment since 1987/88.

The decomposition performed above also permits identification of the structural sensitivity of Zimbabwe's CNFPS deficit to its main determinants. Table 10.5 reports measures of the responsiveness (or semielasticities) of the deficit to changes in underlying macroeconomic and policy variables, computed as absolute changes in the CNFPS deficit in response to a 1 percent (or 1 percentage point) change

**Table 10.5. Sensitivity of CNFPS Deficits to Changes in Economic and Policy Determinants**
(percentage points of GDP)

| Change in determinant | Change in CNFPS deficit | |
|---|---|---|
| *Domestic variable* | | |
| 1 percent growth of real GDP | | |
| Denominator effect | −0.16 | |
| Economic effect | −0.21 | |
| 1 percent growth of real imports | −0.10 | |
| 1 percentage point increase in domestic real interest rate | 0.40 | (0.30) |
| 1 percentage point increase in domestic inflation | 0.31 | |
| 1 percent increase in real exchange rate | −0.06 | (−0.04) |
| *Foreign variable* | | |
| 1 percentage point increase in foreign nominal interest rate | 0.25 | (0.18) |
| *Policy variable* | | |
| 1 percent growth in foreign real debt | 0.03 | (0.01) |
| 1 percent growth in domestic real debt | 0.05 | (0.02) |
| 1 percent growth in real wage bill | 0.14 | (0.10) |
| 1 percent growth in expenditure on goods and services | 0.07 | |
| 1 percent growth in transfers and subsidies | 0.10 | |
| Change in political regime, 1980 | −1.4 | |
| Change in direct tax regime, 1988 | −2.4 | |
| Change in indirect tax regime, 1981 | −2.8 | |
| Change in customs duties regime, 1982 | −2.2 | |
| Change in customs duties regime, 1983 | −3.5 | |
| Change in customs duties regime, 1988 | −5.2 | |
| 1 percent growth in NFPS investment | 0.10 | (0.05) |

*Note:* The changes in CNFPS deficits were obtained by dividing the 1987/88–1988/89 change in the deficit caused by the corresponding economic or policy determinant by the change in the corresponding determinant. Values for 1981/82–1982/83 that differ from the 1987/88–1988/89 levels are given in parentheses. The changes in political and tax regimes are measured by the changes in the corresponding dummies estimated by the tax revenue functions.

*Source:* Authors' calculations based on data from Zimbabwe Ministry of Finance.

in the corresponding determinants during the recent past.[5] These elasticities reflect the share in the budget of the respective budgetary variable (which may change over time) and, in the case of the behavioral tax revenue functions, the size of the corresponding coefficients (estimated as time-invariant parameters).

The deficit appears to be quite sensitive to changes in macroeconomic variables. Its semielasticity with respect to GDP (−0.37, the sum of the denominator elasticity of −0.16 and the economic elasticity of −0.21) is surpassed only by that of the domestic real interest rate (0.40). The responsiveness with respect to inflation is also relatively high, at 0.31, but is lower than the real interest rate semielasticity because the negative effect of inflation on the deficit through higher direct taxes partly offsets its positive effect through higher domestic nominal interest payments. Slightly lower is the semielasticity of the deficit with respect to foreign nominal interest rates (0.25). Finally, the deficit is only weakly responsive to the real exchange rate. A 1 percent real depreciation reduces the deficit by 0.06 percentage point of GDP; in other words, the strong effect on the deficit through higher interest payments on foreign debt is almost offset by the higher tax revenue, as both direct and indirect tax payments are boosted by a depreciation.

Among policy variables, changes in tax regime that lead to higher tax burdens or impose stricter controls on evasion tend to affect the CNFPS deficit significantly. Expenditure variables (by decreasing weight) include the wage bill, public investment, transfers and subsidies, and expenditure on other goods and services. Finally, although it has been omitted from the table, a change in the PLA primary noninterest current deficit is obviously of enormous importance, as it and the CNFPS deficit change one-for-one.

The preceding discussion has shed light on the sensitivity of Zimbabwe's public finances to the major macroeconomic and fiscal policy determinants of the deficit. Future fiscal programming and stabilization efforts could be based on this kind of quantitative framework, which complements the usual policy considerations with a clear determination of the effectiveness of policy instruments.

### Sustainable Deficits

We now turn to the derivation of bounds for sustainable deficits of the public sector. The sustainability concept applied here refers to the time path of public liabilities for given demands for these liabilities by the domestic private and foreign sectors.[6]

The analysis starts with the standard budget constraint of the consolidated TPS in current prices. By simple manipulation, the primary (that is, noninterest) deficit of the TPS as a share of GDP ($pd$) can be financed by the following sources:

$$(10.1) \qquad pd = \dot{h} + \dot{b} + \dot{f} + \dot{ol} + h(\hat{P} + \hat{y}) + b(\hat{y} - r)$$
$$+ f(\hat{y} - r^* - \hat{e}) + ol(\hat{P} + \hat{y})$$

where $h$ is the ratio of total base money to GDP, $b$ is the ratio of domestic public debt to GDP, $f$ is the ratio of foreign public debt to GDP, $ol$ is the ratio of other public liabilities to GDP, $P$ is the GDP deflator, $y$ is real GDP, $r$ is the domestic real interest rate, $r^*$ is the foreign real interest rate, and $e$ is the real exchange rate. Dots over variables denote absolute (not relative) rates of change per time unit, and hats denote percentage rates of change.[7]

On the basis of the steady-state notion of the ratios of fixed public liability to GDP, a sustainable deficit is defined as a level consistent with the maintenance of constant holdings of public liabilities, in proportion to GDP, by domestic private and foreign creditors. Under this condition ($\dot{h} = 0 = \dot{b} = \dot{f} = \dot{o}l$), a primary deficit level that can be sustained over time has to be financed by a combination of the four sources identified in equation 10.1: inflation tax and seigniorage from GDP growth; the excess of domestic growth over the domestic real interest rate; the excess of domestic growth over the foreign real interest rate (adjusted for real exchange rate depreciations); and inflation tax cum seigniorage on other liabilities.[8]

Table 10.6 presents simulation results for sustainable public sector deficits in Zimbabwe, consistent with the structure of its public finances and with its recent evolution of macroeconomic variables. The first part of the table summarizes the recent evolution of macro-economic variables required for applying equation 10.1. This helps to identify relevant values for the base. The ratios to GDP of the four main liabilities of the consolidated TPS for June 1988 (obtained from table 10.3) are presented in the second part of table 10.6. They are used as the relevant (constant) liability ratios for the simulations reported in the third part of the table.

Three scenarios are considered. The base scenario assumes GDP growth, real interest rates, and a real exchange rate constancy broadly consistent with Zimbabwe's performance during the late 1980s. Under the high scenario, growth exceeds the corresponding base-case value by 1 percentage point, and the domestic real interest rate falls short of the base case by 1 percentage point. The low scenario has lower GDP growth, higher real interest rates, and a real exchange rate depreciation of 7 percent. This combination could reflect a reform scenario featuring both financial reform (and higher real interest rates on public debt) and trade liberalization (which could require a more depreciated real exchange rate); transitional adjustment costs would be reflected in lower short-term growth.

Changes in growth and interest rates have the strongest effects on the sustainable deficit level simply because stocks of domestic and foreign debt are large in relation to base money and other public liabilities. In addition, capital losses on foreign debt from the real exchange rate devaluation can severely limit sustainable deficits, as shown in the low scenario.

Under the base case the sustainable primary deficit is estimated at 1.7 percent of GDP. The deficit increases slightly, to 2.9 percent of GDP, under the high case and declines significantly, to −4.2 percent of GDP, under the low scenario. Sustainable nominal deficits vary accordingly.

**Table 10.6. Estimation of Sustainable Public Deficits, Zimbabwe**

| Variable | 1986/87 | 1987/88 | 1988/89 |
|---|---|---|---|
| *Macroeconomic variable* | | | |
| GDP growth | −0.038 | −0.014 | 0.044 |
| Domestic inflation | 0.172 | 0.148 | 0.137 |
| Domestic nominal interest rate | 0.130 | 0.130 | 0.135 |
| Domestic real interest rate | 0.042 | 0.018 | 0.002 |
| Foreign nominal interest rate | 0.075 | 0.072 | 0.072 |
| Foreign inflation | 0.022 | 0.042 | 0.045 |
| Foreign real interest rate | 0.013 | 0.030 | 0.028 |
| Domestic devaluation | 0.015 | 0.044 | 0.129 |
| Real exchange rate depreciation | −0.114 | −0.053 | 0.037 |

| *Sustainable nonfinancial public sector deficit* | Base | High | Low |
|---|---|---|---|
| GDP growth | 0.040 | 0.050 | 0.020 |
| Domestic inflation | 0.110 | 0.110 | 0.110 |
| Domestic nominal interest rate | 0.140 | 0.130 | 0.170 |
| Domestic real interest rate | 0.030 | 0.020 | 0.060 |
| Foreign nominal interest rate | 0.080 | 0.080 | 0.090 |
| Foreign inflation | 0.040 | 0.040 | 0.040 |
| Foreign real interest rate | 0.040 | 0.040 | 0.050 |
| Real exchange rate depreciation | 0.000 | 0.000 | 0.070 |
| Inflation tax | 0.007 | 0.007 | 0.007 |
| Seigniorage from growth | 0.003 | 0.003 | 0.001 |
| Domestic debt effect | 0.004 | 0.011 | −0.015 |
| Foreign debt effect | 0.000 | 0.004 | −0.011 |
| Foreign debt capital gain | 0.000 | 0.000 | −0.027 |
| Effect of other liabilities | 0.003 | 0.003 | 0.003 |
| Sustainable primary deficit | 0.017 | 0.029 | −0.042 |
| Interest payments on foreign debt | 0.029 | 0.029 | 0.033 |
| Interest payments on domestic debt | 0.053 | 0.049 | 0.065 |
| Sustainable nominal deficit | 0.099 | 0.107 | 0.056 |

*Note:* The underlying ratios of public sector liabilities to GDP are those of the total public sector for June 1988, reported in table 10.3.

*Source:* Authors' estimations based on data from Reserve Bank of Zimbabwe, Zimbabwe Ministry of Finance, and World Bank.

The actual 1988/89 primary deficit of 2.2 percent of GDP (see table 10.2) is at the midpoint of the base and high scenarios and exceeds the low-case deficit by a large amount (6.4 percentage points of GDP). The nominal deficits of the base and unfavorable cases are similar to the actual TPS and CNFPS nominal deficits, but again, the low scenario shows a sustainable nominal deficit that is almost 5 percentage points below the latter measures.

We may conclude that while public sector deficits in the range observed during the late 1980s in Zimbabwe may be sustainable from the limited perspective of the ratio of constant liability to GDP and under macroeconomic conditions ranging from normal to favorable, they are clearly unsustainable under adverse macroeconomic shocks or when significant devaluations are required in response to policy changes. This is precisely what occurred in 1990–92, when the combination of low growth and real devaluations, aggravated by the drought-induced increase in the deficit, led to a deterioration of the fiscal stance. It is not surprising that total public debt has been rising massively since 1990.

## Deficit Financing and Financial Markets

Having identified the determinants of public deficits and their sustainability, we now invert the focus of causality by looking at the effects of public deficits on Zimbabwe's macroeconomic performance. This section discusses the macroeconomic impact of public sector deficits on financial markets, deriving and estimating a version of Easterly's (1989) model. The model emphasizes the determination of the real interest and inflation rates, with money demand as its main behavioral component. The framework used here incorporates additional features distinctive to the Zimbabwean economy, reflecting three dimensions of financial control exercised by the government to capture private surpluses for financing public deficits: controls on foreign exchange and capital flows, compulsory placement of public debt in financial institutions, and partial interest rate controls. (For a similar view see Chhibber and others 1989.)

Foreign exchange rationing effectively constrains private demands for consumer, capital, and intermediate goods, leading to a private sector aggregate surplus of investment over saving. The government exerts strict control on capital outflows to prevent the private sector from shifting its surplus abroad.

Domestic banking and financial markets have rather smoothly intermediated the private sector surplus to the public sector as a result of several regulations that force financial intermediaries to purchase public debt. In fact, the financial system is exceptionally deep for a country at Zimbabwe's stage of development.[9] A number of institutions comprise the monetary sector (the Reserve Bank of Zimbabwe, two discount houses, five commercial banks, and accepting houses), and monetary assets exceed 30 percent of GDP. The nonmonetary sector (building societies, finance houses, the Post Office and Savings Bank, insurance companies, and pension funds) is significantly larger than the monetary sector, in part because institutional investors (insurance companies and pension funds) have been capturing most private long-term savings. Financial institutions (in particular, institutional investors and the Post Office and Savings Bank) are required to hold significant shares of their portfolio in the form of public sector liabilities.

Nominal interest rates have been kept low by the combination of controls on external and domestic financial flows that has already been described. However, partial interest rate controls have also contributed to the low cost of private funds. Some nominal interest rates are free but tend to follow controlled rates. Both categories of nominal rates have been relatively stable, with the exception of large temporary bursts in the early 1980s and early 1990s, in the aftermath of inflationary shocks. In addition, it is argued that the monetary authority manages required reserve ratios to help maintain stable and low interest rates whenever net private credit demand is inconsistent with stability (Chhibber and others 1989). As a result of such nominal stability, real interest rates have fluctuated much more, in tandem with inflation. On average, real interest rates paid on public debt and banking deposits have been negative.

In sum, the public sector has a major captive source of private funds in the institutional investors, which have financed a significant share of the central government deficit. What is left of private sector saving is distributed among deposits at monetary institutions, the Post Office and Savings Bank, and building societies. Other nonmonetary institutions, such as finance houses, are of less importance. A small share of private saving is directly channeled to treasury bills and government stocks and bonds, all of which are issued by the central government.

## Model Structure and Estimation Results

The model starts by relating base money creation to the consolidated government budget constraint. Taking into account the main features of Zimbabwe's financial system and its public sector budget structure and financing, such a dependence is summarized as:

(10.2) $$h_t = g'_t - b_{rt} - l_{st} + (1 + \pi_t)^{-1}(z_{t-1})$$

where all variables are in real terms (that is, deflated by the domestic CPI), $b_r$ is public sector bonds in private sector hands, $l_s$ is government stock and bonds plus treasury bills in the banking system, $g'$ is the government's primary deficit net of the part financed by foreign debt and the placement of domestic debt with institutional investors, $h$ is base money, and $i_g$ is the common interest rate paid by all types of public debt. Additionally, $z_{t-1} = h_{t-1} + (1 + i_{gt-1})(b_{rt-1} + l_{st-1})$, and $\pi_t$ is the inflation rate. (See Morandé and Schmidt-Hebbel 1991 for a detailed discussion.) Here and below, time periods are denoted by subscripts $t-1$, $t$, and $t+1$.

The nonfinancial private sector voluntarily holds three broad assets: money, interest-earning deposits in the banking system, and public sector bonds. Domestic residents are allowed to hold neither foreign assets nor foreign liabilities, a prohibition that seems to be binding. Private bank loans to the private sector are netted out of the demand for interest-earning deposits. These three asset demands follow a standard portfolio setup. However, interest-bearing deposits can be seen as an indirect way for the nonfinancial private sector to hold public debt ($l_s$). The substitution of public debt and money from the banking system's balance sheet for interest-bearing deposits yields a portfolio of two financial assets: public debt and money. The demand for the latter is:

$$(10.3) \qquad m_t^d = m(r_{ct}, \pi_{t+1}^e, nfa_t)$$
$$\qquad\qquad\qquad (-)\ (-)\ (+)$$

where $r_c$ is the real interest rate paid on interest-earning deposits of the banking system, $\pi_{t+1}^e$ is the inflation expected to prevail between periods $t$ and $t + 1$, and $nfa$ is the real value of the private sector's net financial asset holdings or net wealth (inclusive of compulsory savings in the pension funds and insurance companies). Expected signs of the corresponding partial derivatives are noted in parentheses below each variable. The real interest rate is defined as the nominal interest rate minus expected inflation.

With two financial markets, one for money and the other for public sector bonds, equilibrium in one determines either the real interest rate or expected inflation. We opt for focusing on the determination of the real interest rate while assuming that expected inflation is linked to actual inflation, which in turn is linked to the goods market equilibrium. A reduced-form equilibrium condition for the latter implies the following stochastic equation for inflation:

$$(10.4) \qquad \pi_t = \pi(dH_{t-1}, dE_{t-1}, dW_{t-1}) + \epsilon_t$$
$$\qquad\qquad\qquad (-)\quad (+)\quad (-)$$

where $H$ is base money, $E$ is the nominal exchange rate (domestic currency per unit of foreign currency), $W$ is average wages, $\epsilon$ is a zero-mean constant-variance stochastic term, and $d$ denotes percentage change.

In addition, inflationary expectations are assumed to be rational such that:

$$(10.5) \qquad \pi_{t+1}^e = EX_t[\pi_{t+1}/\text{all information available at time } t]$$

where $EX$ denotes expected value.

The model is made up of three equations: 10.4 and 10.5 determine actual and expected inflation, and a monetary equilibrium condition consistent with money demand in 10.3 determines the real interest rate. To this system we could add equation 10.2, the government budget constraint, in order to endogenize the money supply. This permits determination of the effects that changes in the government deficit and the corresponding financing decisions have on the real interest rate and inflation.[10]

The following log-linear implicit adjustment cost version of money demand equation 10.3 is used:

(10.6)   $\ln m_t^d = b_0 + b_1 r_t + b_2 \pi_{t+1}^e + b_3 \ln nfa_t + b_4 \ln m_{t-1} + v_t$

where $b_1$, $b_2 < 0$, $b_3$, $b_4 > 0$, and $v_t$ is a zero-mean constant-variance stochastic term.

Estimation of equation 10.4 for inflation permits the derivation of values for expected inflation according to the rational-expectations hypothesis. We estimate a linear version of equation 10.4 in which inflation and the percentage variation of $H$, $W$, and $E$ are measured in annual terms, for consistency with the estimated money demand.[11] The empirical results are reported in table 10.7. The short-run semi-elasticities of money demand with respect to the real interest rate and inflation are significantly different from zero and are similar to each other, as expected: $-1.41$ and $-1.13$, respectively. The long-run values of the semielasticities are $-4.55$ for the real interest rate and $-3.65$ for expected inflation.[12] The elasticity with respect to private net financial assets, which is also significantly different from zero, is 0.26 in the short run and 0.84 in the long run. Similar results are found in Elbadawi and Schmidt-Hebbel (1991a), using a general equilibrium setup.

The results reported for inflation include as independent variables the one-quarter-lagged percentage variations of base money and the exchange rate (in addition to the lagged dependent variable).[13] Strong price inertia is evidenced by the large role of the one-quarter-lagged inflation rate.

## Simulation Results for Alternative Deficit Financing Forms

On the basis of the results reported in table 10.7, we perform simulations of government policies related to the size and financing of the public deficit. For convenience the estimated money demand is inverted, and the inflation equation is restated as follows:

(10.7)         $r_t = -0.71 \ln m_t + 0.18 \ln nfa_t - 0.80 \pi_{t+1}^e$

$+ 0.49 \ln m_{t-1} - 0.71 \epsilon_t$

(10.8)      $\pi_t = 0.128 dH_{t-1} + 0.092 dE_{t-1} + 0.723 \pi_{t-1} + v_t.$

The policy simulations are based on equations 10.7 and 10.8, the government budget constraint (equation 10.2), and the assumption of

**Table 10.7. Estimation Results for Money Demand and Inflation, Zimbabwe, 1980–88**

| Variable | Coefficient |
|---|---|
| *Money demand (dependent variable, m)* | |
| Constant | −0.18 |
| | −(0.56) |
| ln $r$ | −1.41 |
| | (−2.76) |
| $\pi^e_{+1}$ | −1.13 |
| | (−2.33) |
| ln *nfa* | 0.26 |
| | (2.37) |
| $m_{-1}$ | 0.69 |
| | (7.27) |
| Adjusted $R^2$ | 0.85 |
| *F*-statistic | 47.7 |
| $Q$ | 19.4 |
| *Inflation rate (dependent variable, $\pi$)* | |
| Constant | −0.003 |
| | (−0.18) |
| $dH_{-1}$ | 0.128 |
| | (4.0) |
| $dE_{-1}$ | 0.092 |
| | (3.1) |
| $\pi_{-1}$ | 0.723 |
| | (7.4) |
| Adjusted $R^2$ | 0.73 |
| *F*-statistic | 35.3 |
| $Q$ | 23.5 |

*Note:* The estimation method is ordinary least squares. Numbers in parentheses are *t*-statistics. $Q$ is the Ljung-Box statistic; marginal significance levels for the $Q$ values are 0.63 (for $Q = 19.4$) and 0.56 (for $Q = 23.5$).

rational expectations. They consider both direct and indirect dynamic feedback effects. For instance, the sensitivity of the real interest rate with respect to changes in real money is such that a 1 percent increase in the latter variable at time $t$ causes a contemporaneous reduction in the real interest rate of 0.7 percentage point and then an increase of 0.49 percentage point in period $t + 1$. However, if changes in $m$ cause changes in base money, there are several other indirect mechanisms of transmission, such as the effect of the change in base money on inflation in $t + 1$ (equation 10.8) and changes in the real value of private net financial assets, that will modify the effect starting in period $t + 1$.

The following exercises simulate a temporary primary deficit, financed alternatively by issuing base money and domestic debt. The simulations consider a one-time 10 percent increase in $G'$ (the nominal adjusted primary deficit) during the first period of a sixteen-quarter simulation horizon, from 1990:1 to 1993:4. At time 0 (1989:4), before the change in $G'$ occurs, the system is assumed to be at a steady-state position, with all variable changes set at zero. In addition, initial real interest and inflation rates are equal to zero.

A. *A 10 percent increase in the primary deficit, financed by issuing base money ($dH_t > 0$).* According to the relative magnitudes of $G'$ and $H$, a 10 percent increase in $G'_t$ requires a 2.83 percent increase in $H_t$. Since there are no further increments in $G'$, $dH_{t+s} = 0$, for $s > 0$. The nominal interest rate remains constant, which is not an unrealistic assumption.

The simulation results are reported in the first part of table 10.8. Figure 10.3 shows the path of the real interest and inflation rates. On impact, the real interest rate declines significantly. Higher base money raises real money balances, since inflation is not affected until the second quarter (1990:2). By 1990:2 the positive lagged effect of real balances on the real interest rate brings the latter back to a level close to its initial value. This effect is offset, however, by the rise in inflation during that quarter. Afterward, the persistence of a positive inflation rate dominates the determination of the real interest rate, even though the same positive inflation rate lowers real balances and net financial assets, thus raising $r$. In the end the price level rises by a cumulative 1.31 percent, about half of the initial increase in base money. The final effect on the real interest rate is a reduction of 1.3 percentage points.

B. *A 10 percent increase in the primary deficit, financed by domestic debt, with monetization of interest payments and debt repayment.* In this simulation the government resorts to issuing new debt to finance the increased deficit in 1990:1 and, beginning in the second quarter, issues base money to pay the interest on the new debt. Issuing debt has a strong positive effect on the real interest rate but no effect on inflation, since no change in base money has occurred. However, the money-financed interest payments (starting in 1990:2) gradually reduce the real interest rate because of the forces that were discussed in the previous simulation. Higher base money also leads to inflation from 1990:3 onward. The debt issued in 1990:1 matures fifteen quarters later, in 1993:3, and is repaid by issuing base money. This portfolio shift causes a large decline in the real interest rate in 1993:3 and an inflationary burst in the subsequent quarter. Although government debt could stay at its increased level for a long time, with interest

**Table 10.8.  Simulation Results for Different Budget Financing Policies, Zimbabwe**

| Policy and quarter | dG' | dH | dB | π | dm | dnfa | dr | r | db(d) |
|---|---|---|---|---|---|---|---|---|---|
| A. 10 percent increase in the primary deficit (G') financed by issuing base money | | | | | | | | | |
| 1990:1 | 10 | 2.83 | 0 | 0.0000 | 2.8300 | 0.8094 | -1.8636 | -1.8636 | -0.6524 |
| 1990:2 | 0 | 0 | 0 | 0.3679 | -0.3679 | -0.1052 | 1.3346 | -0.5290 | 0.4479 |
| 1990:3 | 0 | 0 | 0 | 0.2649 | -0.2649 | -0.0758 | -0.2177 | -0.7467 | -0.0731 |
| 1990:4 | 0 | 0 | 0 | 0.1907 | -0.1907 | -0.0545 | -0.1568 | -0.9035 | -0.0526 |
| 1993:1 | 0 | 0 | 0 | 0.0099 | -0.0099 | -0.0028 | -0.0082 | -1.2857 | -0.0027 |
| 1993:2 | 0 | 0 | 0 | 0.0071 | -0.0071 | -0.0020 | -0.0059 | -1.2915 | -0.0020 |
| 1993:3 | 0 | 0 | 0 | 0.0051 | -0.0051 | -0.0015 | -0.0042 | -1.2958 | -0.0014 |
| 1993:4 | 0 | 0 | 0 | 0.0037 | -0.0037 | -0.0011 | -0.0030 | -1.2968 | -0.0010 |
| B. 10 percent increase in the primary deficit (G') financed by domestic debt, with monetization of interest payments and debt repayment | | | | | | | | | |
| 1990:1 | 10 | 0 | 1.3 | 0.0000 | 0.0000 | 0.6637 | 0.6221 | 0.6221 | 0.2088 |
| 1990:2 | 0 | 0.2819 | 0 | 0.0000 | 0.2819 | 0.0806 | -0.1856 | 0.4365 | -0.0623 |
| 1990:3 | 0 | 0.2819 | 0 | 0.0366 | 0.2453 | 0.0701 | -0.0527 | 0.3838 | -0.0177 |
| 1990:4 | 0 | 0.2819 | 0 | 0.0630 | 0.2189 | 0.0626 | -0.0744 | 0.3094 | -0.0250 |
| 1993:1 | 0 | 0.2819 | 0 | 0.1274 | 0.1545 | 0.0442 | -0.1273 | -0.7261 | -0.0427 |
| 1993:2 | 0 | 0.2819 | 0 | 0.1283 | 0.1536 | 0.0439 | -0.1281 | -0.8541 | -0.0430 |
| 1993:3 | 0 | 3.1119 | -1.3 | 0.1291 | 2.9828 | 0.8531 | -1.9923 | -2.8464 | -0.6686 |
| 1993:4 | 0 | 0 | 0 | 0.4975 | -0.4975 | -0.1423 | 1.3912 | -1.4552 | 0.4669 |

C. 10 percent increase in the primary deficit (G') financed by domestic debt with interest payments financed by further debt until 1993:3, when out-standing debt is monetized

| | | | | | | | | | |
|---|---|---|---|---|---|---|---|---|---|
| 1990:1 | 10 | 0 | 1.30 | 0 | 0 | 0.6637 | 0.6221 | 0.6221 | 0.2088 |
| 1990:2 | 0 | 0 | 0.13 | 0 | 0 | 0.0664 | 0.0622 | 0.6843 | 0.0209 |
| 1990:3 | 0 | 0 | 0.13 | 0 | 0 | 0.0664 | 0.0622 | 0.7465 | 0.0209 |
| 1990:4 | 0 | 0 | 0.13 | 0 | 0 | 0.0664 | 0.0622 | 0.8087 | 0.0209 |
| 1993:1 | 0 | 0 | 0.13 | 0 | 0 | 0.0664 | 0.0622 | 1.3686 | 0.0209 |
| 1993:2 | 0 | 0 | 0.13 | 0 | 0 | 0.0664 | 0.0622 | 1.4308 | 0.0209 |
| 1993:3 | 0 | 6.495 | -2.99 | 0 | 6.4948 | 0.3311 | -4.5517 | -3.1208 | -1.5275 |
| 1993:4 | 0 | 0 | 0 | 0.8443 | -0.8443 | -0.2415 | 3.063 | -0.0579 | 1.0279 |

*Note:* Variables are defined as follows: $dG'$, percentage change in $G'$; $dH$, percentage change in $H$; $dB$, percentage change in $B$; $\pi$, inflation rate; $dm$, percentage change in real base money; $h$; $dnfa$, percentage change in real private net financial assets; $nfa$; $dr$, change in real interest rate; $r$; $db(d)$, percentage change in demand for government bonds. Simulation results for 1991:1–1992:4 are omitted for brevity.

payments financed through the creation of base money, the situation tends toward an unsustainable path of positive, slightly increasing inflation rates. At some point it will be convenient to pay back the debt with a higher but declining inflation rate. In the end the debt-financing strategy ends up being more inflationary after sixteen quarters (with a 1.88 percent cumulative inflation) than the alternative of resorting directly to monetization from the beginning.

C. *A 10 percent increase in the primary deficit, financed by domestic debt, with interest payments financed by further debt until 1993:3, when total debt is monetized.* While the debt stock is increasing, the real interest rate also rises, but the price level is not affected since no change in base money has taken place (figure 10.4). This changes drastically when the government debt is paid back in 1993:3; the real interest rate falls significantly and inflation rises. After that both variables follow the path of the first simulation, but the size of the effects is different. The increase in base money has to be sufficiently large to pay back the accumulated debt, implying a drastic inflationary surge: cumulative inflation in just five more quarters (up to 1994:4) exceeds significantly cumulative inflation in 1990–94.

**Figure 10.3. Effects of Primary Deficit on Inflation and the Real Interest Rate in Zimbabwe: Simulation A**

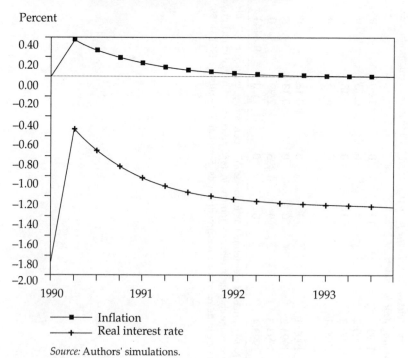

Percent

*Source:* Authors' simulations.

**Figure 10.4. Effects of Primary Deficit on Inflation and the Real Interest Rate in Zimbabwe: Simulation C**

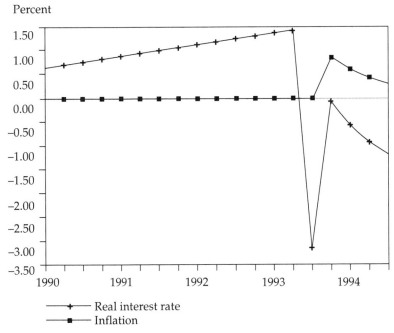

*Source:* Authors' simulations.

Despite their simplicity, the reported simulations show the dynamic sensitivity of the main endogenous variables—inflation and the real interest rate—to government policies for financing a budget deficit. Inflation ensues in all cases, but cumulative inflation never matches the increase in base money. It is not the nonneutrality of the simulation system that is behind this result but the simulation assumption of a constant nominal exchange rate. Indeed, the empirical estimates of the coefficients of the inflation equation add up to almost 1. Moreover, when equation 10.4 is estimated with the homogeneity-of-degree-one restriction imposed on its coefficients, results do not change much. To account for nominal exchange rate devaluations, however, a simulation was performed for the first case, adding a devaluation of 2.83 percent in 1990:1 and distinguishing between an almost neutral case (the one presented above) and a fully neutral case by imposing the restriction that the estimated coefficients of equation 10.4 add up to 1. The qualitative results do not differ much from those reported above.

Real interest rates follow very different patterns that depend on how the government finances its deficit and on the assumption of fixed nominal interest rates. The simulations illustrate Sargent and

**Table 10.9. Public and Private Sector Saving and Investment, Zimbabwe, Fiscal 1980/81–1988/89**

| Item | 1980/81 | 1981/82 | 1982/83 | 1983/84 | 1984/85 | 1985/86 | 1986/87 | 1987/88 | 1988/89 |
|---|---|---|---|---|---|---|---|---|---|
| *Current price investment and saving (millions of Zimbabwe dollars)* | | | | | | | | | |
| Foreign saving | 270.8 | 451.8 | 440.8 | 293.1 | 166.2 | 91.1 | 1.7 | −45.8 | −30.8 |
| Gross national saving | 544.6 | 611.1 | 610.0 | 810.4 | 1,173.8 | 1,453.5 | 1,658.1 | 2,026.5 | 2,544.1 |
| Central government | −186.4 | −89.9 | −55.1 | −279.8 | −306.6 | −386.1 | −575.6 | −155.6 | −170.4 |
| PLA | 30.6 | −45.0 | −92.8 | 51.3 | 105.3 | 152.3 | 242.1 | 157.5 | 132.0 |
| Nonfinancial public sector | −155.8 | −134.9 | −147.9 | −228.5 | −201.3 | −233.8 | −333.5 | 1.9 | −38.4 |
| Private sector | 700.4 | 746.0 | 757.9 | 1,038.9 | 1,375.1 | 1,687.3 | 1,991.6 | 2,024.6 | 2,582.5 |
| Gross domestic investment | 815.4 | 1062.8 | 1,050.9 | 1,103.5 | 1,340.0 | 1,544.5 | 1,659.8 | 1,980.7 | 2,513.3 |
| Central government | 65.1 | 122.2 | 191.9 | 208.7 | 203.2 | 221.2 | 293.1 | 485.0 | 523.0 |
| PLA | 136.2 | 394.2 | 412.0 | 479.6 | 447.2 | 643.5 | 615.3 | 587.4 | 600.0 |
| Nonfinancial public sector | 201.3 | 516.4 | 603.9 | 688.3 | 650.4 | 864.7 | 908.4 | 1,072.4 | 1,123.0 |
| Private sector | 614.1 | 546.4 | 447.0 | 415.2 | 689.6 | 679.8 | 751.4 | 908.3 | 1,390.3 |
| Nonfinancial public sector deficit | 357.1 | 651.3 | 751.8 | 916.8 | 851.7 | 1,098.5 | 1,241.9 | 1,070.5 | 1,161.4 |
| Private sector deficit | −86.3 | −199.5 | −311.0 | −623.7 | −685.5 | −1,007.4 | −1,240.2 | −1,116.3 | −1,192.2 |
| *Ratios of investment and saving ratios to GDP* | | | | | | | | | |
| Foreign saving | 0.069 | 0.094 | 0.077 | 0.046 | 0.025 | 0.012 | 0.000 | −0.005 | −0.003 |
| Gross national saving | 0.139 | 0.127 | 0.106 | 0.128 | 0.175 | 0.190 | 0.193 | 0.207 | 0.219 |
| Central government | −0.048 | −0.019 | −0.010 | −0.044 | −0.046 | −0.050 | −0.067 | −0.016 | −0.015 |
| PLA | 0.008 | −0.009 | −0.016 | 0.008 | 0.016 | 0.020 | 0.028 | 0.016 | 0.011 |
| Nonfinancial public sector | −0.040 | −0.028 | −0.026 | −0.036 | −0.030 | −0.031 | −0.039 | 0.000 | −0.003 |
| Private sector | 0.179 | 0.155 | 0.132 | 0.163 | 0.205 | 0.220 | 0.231 | 0.207 | 0.223 |
| Gross domestic investment | 0.209 | 0.221 | 0.183 | 0.174 | 0.200 | 0.202 | 0.193 | 0.202 | 0.217 |
| Central government | 0.017 | 0.025 | 0.033 | 0.033 | 0.030 | 0.029 | 0.034 | 0.050 | 0.045 |
| PLA | 0.035 | 0.082 | 0.072 | 0.075 | 0.067 | 0.084 | 0.071 | 0.060 | 0.052 |
| Nonfinancial public sector | 0.052 | 0.107 | 0.105 | 0.108 | 0.097 | 0.113 | 0.105 | 0.110 | 0.097 |
| Private sector | 0.157 | 0.113 | 0.078 | 0.065 | 0.103 | 0.089 | 0.087 | 0.093 | 0.120 |
| Nonfinancial public sector deficit | 0.091 | 0.135 | 0.131 | 0.144 | 0.127 | 0.143 | 0.144 | 0.109 | 0.100 |
| Private sector deficit | −0.022 | −0.041 | −0.054 | −0.098 | −0.102 | −0.132 | −0.144 | −0.114 | −0.103 |

*Source:* Reserve Bank of Zimbabwe, various years; Schmidt-Hebbel 1990.

Wallace's (1981) "unpleasant monetarist arithmetic"; debt financing of government deficits in Zimbabwe would only postpone inflation to the future. Of course, this is true as long as government debt cannot be increased beyond some point. The issue, then, is when this point is achieved; the answer will depend on both macroeconomic conditions (that is, the extent to which private consumption and private investment can continue to be restricted) and, to a lesser degree, on conditions in financial markets. We have found that changes in deficit-financing decisions have some effects on real interest rates that could destabilize financial markets, especially when nominal interest rates are fixed for long periods of time, as assumed here.

Finally, the implications of a purely monetary policy that is effected through, say, the required reserve ratio, should be equivalent to those of an increase in base money. Both impinge on the government budget constraint sooner or later.

## Crowding-Out of Private Consumption and Private Investment

This section goes a step further in analyzing the macroeconomic implications of public sector deficits by assessing the impact of the public sector on private sector spending. The focus is on the sensitivity to fiscal variables of private consumption and investment, in addition to the indirect effects of fiscal variables through interest rates, inflation, and private disposable income. How private saving and capital formation are affected by fiscal policies has significant implications for both short-term stabilization and long-run growth prospects.

Table 10.9 presents data on Zimbabwe's saving and investment record, by sectors, from 1980/81 to 1988/89.[14] Between 1981/82 and 1987/88 a major external adjustment took place that was reflected in a reduction of the current account deficit by 10 percentage points of GDP. This improvement came exclusively from the private sector. Whereas until 1986/87 the CNFPS deficit hovered around 14 percent of GDP, the private sector surplus rose from 4.1 percent of GDP in 1981/82 to 14.4 percent of GDP in 1986/87. In fact, in 1986/87, when the public deficit matched its previous record, 100 percent of that deficit was financed by the private sector. As mentioned above, a partial public sector adjustment was initiated in 1987/88, bringing about a reduction of 3.5 percentage points in the deficit in that year and an additional 0.9 percentage point decline in 1988/89. The private sector benefited directly from this decline, with a similar reduction in its required surplus.

To generate a surplus that financed 100 percent or more of the public deficit after 1986/87, the private sector significantly raised its saving rate: since 1984/85 the rate has exceeded 20 percent of GDP, which is larger than the economy's gross domestic investment. This private saving rate is extremely high for a developing economy—a counterpart of very low private consumption rates that barely exceeded 50 percent of GDP. High private saving, channeled through Zimbabwe's developed financial system to the public sector, is a result of restrictions on private consumption (particularly on imported consumer durables) and on formal or illegal capital outflows, coupled with a perception by the private sector that the domestic financial system is stable.

Aggregate or domestic gross investment did not show a strong downward trend during the 1980s. However, when in 1986/87 the public deficit returned to its 1983/84 peak, the domestic investment rate was a couple of percentage points lower than in the period 1980/81 to 1981/82, when the high deficits started. Conversely, when fiscal adjustment took place after 1987/88–1988/89, the domestic investment rate recovered by 2.4 percentage points of GDP. The sector composition of investment changed significantly with the fiscal expansion of the early 1980s. In fact, the deficit increased approximately one-to-one with public investment, while private investment fell. With fiscal adjustment after 1986/87, both the absolute level and the share of private investment in domestic capital formation recovered, with a rise of more than 3 percentage points in the private investment rate, while public investment did not suffer significantly.

The fact that both total investment and the share of private investment recovered under fiscal adjustment is an encouraging sign. Growth—which was rather modest throughout the 1980s—depends on the quantity and quality of investment, and a higher share of private investment is likely to improve quality. Hence additional investment gains, particularly in the private sector, could be positively influenced by continued fiscal adjustment. Fiscal adjustment should rely on additional gains in public saving, over and above the increase of the public saving rate from −3.9 percent in 1986/87 to −0.3 percent in 1988/89.

### Private Consumption

This section, based on a framework developed by Corbo and Schmidt-Hebbel (1991), addresses the effects of public policies on consumption in Zimbabwe. Private consumption, expressed as a ratio to private disposable income, depends on neoclassical determinants (permanent disposable income, interest rates, and relative prices), Keynesian variables and borrowing constraints (current income, consumer credit, money, and foreign saving), public saving, inflation, and public spending on private goods. The presence of (permanent) public saving reflects two very different hypotheses. The first, Ricardian equivalence, states that private consumption increases one-to-one with permanent public saving.[15] The second, direct crowding-out, asserts that under an institutional arrangement whereby the public sector captures private saving either directly or through domestic financial markets, current private saving is crowded out one-to-one by current public saving. In the case of Zimbabwe the second interpretation is probably more relevant than the assumption of rational, forward-looking private consumers who internalize the public sector's intertemporal budget constraint.

The following specification for the ratio of private consumption to private disposable income reflects these variables.[16] In addition it allows for testing the simple Keynesian, permanent income, and Ricardian or direct-crowding-out hypotheses.[17]

$$(10.9) \quad \frac{C_{pt}}{DY_{pt}} = \beta_0 + \beta_1 \frac{PDY_{pt}}{DY_{pt}} + \beta_2 \frac{PS_{gt}}{DY_{pt}} + \beta_3 \, r_{ct} + \beta_4 \pi_{ct} + \beta_5 \frac{P_{cmt}}{P_{cnt}}$$

$$+ \beta_6 \frac{CPTR_t}{DY_{pt}} + \beta_7 \frac{H_t}{DY_{pt}} + \beta_8 \frac{FS_t}{DY_{pt}} + \beta_9 \frac{CC_t}{DY_{pt}} + \eta_{1t}$$

where $C_p$ is private consumption expenditure, $DY_p$ is current private disposable income,[18] $PDY_p$ is permanent private disposable income, $PS_g$ is permanent public saving, $r_c$ is the consumption-based real interest rate, $\pi c$ is the expected rate of change of the private consumption deflator, $P_{cm}$ and $P_{cn}$ are the deflators for imported and national private consumption goods, respectively, CPTR is the sum of public expenditure on privately appropriated services and direct transfers to consumers,[19] $H$ is base money, $FS$ is foreign saving, $CC$ is banking sector credit to consumers, and $\eta_1$ is a stochastic error term.[20] Expected signs of the coefficients are $\beta_0, \beta_1, \beta_2, \beta_7, \beta_8, \beta_9 > 0$; $\beta_6 < 0$; $\beta_3, \beta_4, \beta_5 \gtrless 0$.

Table 10.10 reports the main estimation results for equation 10.9, using annual data for the 1965–88 period. The complete specification yields not very satisfactory results for both expectational alternatives. Most variables are not statistically significant, and two borrowing constraints (consumer credit and base money), have opposite signs from those expected a priori, although the coefficients are not significant. Less surprising is the low significance of the inflation and interest rates, which have ambiguous a priori signs. As in many other developing countries (see, for instance, the cross-country studies by Corbo and Schmidt-Hebbel 1991, Giovannini 1985, and Schmidt-Hebbel, Webb, and Corsetti 1992), interest rates are not significant. This may be either because substitution, income, and wealth effects offset each other or because direct crowding-out

**Table 10.10. Estimation Results for Private Consumption, Zimbabwe, 1965–88**

| Equation | Constant | $PDY_p/DY_p$ | $PS_g/DY_p$ | $r_c$ | $\pi_c$ | $P_{cnt}/P_{cnt}$ | $CC/DY_p$ | $H/DY_p$ | $FS/DY_p$ | $D74$ | $D6586$ | $\rho$ | $R^2A$ | Durbin-Watson |
|---|---|---|---|---|---|---|---|---|---|---|---|---|---|---|
| *Static expectations* | | | | | | | | | | | | | | |
| 1.1 OLS | 1.10 | 0.01 | 0.21 | −0.53 | −0.30 | −0.15 | −4.64 | −0.86 | 0.08 | | | | 0.62 | 1.91 |
| | (8.4) | (0.1) | (0.4) | (−0.8) | (−0.4) | (−2.3) | (−1.3) | (−1.1) | (0.4) | | | | | |
| 1.2 ML | 0.61 | 0.12 | 0.67 | | | | | | | −0.06 | 0.06 | 0.72 | 0.50 | 1.61 |
| | (7.6) | (1.7) | (3.3) | | | | | | | (−2.0) | (2.4) | (5.0) | | |
| *Partial perfect foresight* | | | | | | | | | | | | | | |
| 2.1 OLS | 1.06 | 0.02 | 0.08 | −0.39 | −0.07 | −0.16 | −3.9 | −0.93 | 0.06 | | | | 0.62 | 1.97 |
| | (8.0) | (0.1) | (0.1) | (−0.6) | (−0.1) | (−2.1) | (−0.7) | (−0.8) | (0.3) | | | | | |
| 2.2 ML | 0.66 | 0.09 | 0.05 | | | | | | | −0.03 | 0.03 | 0.13 | 0.16 | 1.70 |
| | (10.6) | (1.7) | (0.2) | | | | | | | (−0.8) | (1.8) | (0.6) | | |

*Note:* The dependent variable is the ratio of private consumption to private disposable income ($C_p/DY_p$). Numbers in parentheses are *t*-statistics. Estimation methods are ordinary least squares (OLS) and maximum likelihood with first-order autoregressive residuals (ML); $\rho$ is the first-order residual correlation coefficient, and $R^2A$ is adjusted $R^2$. The results that include CPTR are not reported here because of the high positive value of its coefficient, which seriously affects the signs and significance levels of other variables. A blank denotes the omission of the specific variable from the regression.

of public savings dominates the role of intertemporal substitution in consumption.

We next followed a different approach, concentrating on the Keynesian (current income), permanent income, and Ricardian direct-crowding-out (public saving) determinants. When dummies for the 1987/88 structural decline in private consumption and the 1984 outlier are added to these variables, the results reported in rows 1.2 and 2.2 in table 10.10 are obtained. Both the overall fit and the separate significance of the contributing variables are more acceptable under the static-expectations alternative for permanent income and permanent public saving.

The magnitude of current income is surprisingly high compared with permanent income—a feature that is even more extreme under the partial perfect foresight specification. In fact, the relative magnitudes of current and permanent income (0.61 and 0.12) are higher than the 0.60 and 0.24 values obtained for thirteen developing countries with a similar methodology. (See the panel data results reported in Corbo and Schmidt-Hebbel 1991.) This suggests that current income is a stringent borrowing constraint that effectively limits intertemporal consumption smoothing.

By contrast, public saving strongly affects private consumption in Zimbabwe under the static-expectations alternative. The fact that the current public saving alternative (the measure for permanent public saving under static expectations) is significant whereas the three-year moving average of current and future public saving is not (under partial perfect foresight) confirms the initial presumption that it is direct crowding-out of private saving by public saving and not Ricardian anticipation of future taxes that is behind this high value.

The main conclusion of our results points toward the overwhelming dominance of the direct effects on private consumption of public sector deficits (or dissaving) over other indirect effects of deficit financing (via interest or inflation rates). An increase in the deficit of 1 Zimbabwe dollar (Z$), caused by a corresponding rise in public consumption, reduces private consumption by Z$0.67, without significant additional effects through interest and inflation rates.[21]

*Private Investment*

Private investment is specified to depend on neoclassical profit and cost variables, borrowing constraints, and risk determinants. To avoid spurious correlation, nonstationary variables are scaled to GDP, yielding the following equation for the ratio of private investment to GDP:[22]

**Table 10.11. Estimation Results for Private Investment, Zimbabwe, 1965–88**

| Equation | Constant | $P_f/P$ | RIL | $K_{p-1}/Y$ | $K_{g-1}/Y$ | PCOT/Y | PRO/Y | FS/Y | H/Y |
|---|---|---|---|---|---|---|---|---|---|
| 1 OLS | -0.09 (-0.8) | -0.01 (-0.1) | -0.27 (-1.5) | 0.22 (1.1) | -0.27 (-1.1) | 1.68 (2.4) | 0.28 (2.0) | 0.16 (1.7) | -0.13 (-0.2) |
| 2 OLS | 0.11 (1.2) | -0.13 (-3.7) | -0.42 (-2.6) | -0.18 (-1.3) | 0.23 (1.2) | | 0.47 (3.5) | 0.33 (3.9) | 1.07 (3.5) |
| 3 ML | 0.16 (2.4) | -0.16 (-6.5) | -0.41 (-3.3) | -0.21 (-2.4) | 0.25 (2.0) | | 0.40 (4.1) | 0.34 (5.1) | 1.29 (6.0) |
| 4 OLS | 0.09 (2.2) | -0.12 (-6.5) | -0.45 (-5.7) | -0.12 (-2.8) | 0.14 (2.1) | | 0.49 (4.8) | 0.31 (5.6) | 1.08 (6.3) |
| 5 ML | 0.14 (2.2) | -0.14 (-6.8) | -0.49 (-5.7) | -0.14 (-2.8) | 0.15 (2.1) | | 0.44 (4.8) | 0.30 (5.6) | 1.31 (6.3) |

| Equation | FC/Y | VPIP | VRIL | VY | D7375 | D84 | $\rho$ | $R^2A$ | Durbin-Watson |
|---|---|---|---|---|---|---|---|---|---|
| 1 OLS | 0.18 (1.7) | 0.66 (0.9) | 1.68 (0.9) | -0.0001 (-1.9) | 0.01 (1.5) | 0.02 (1.0) | | 0.95 | 2.26 |
| 2 OLS | | -0.37 (-0.5) | 1.12 (0.5) | | 0.02 (2.3) | 0.03 (1.4) | | 0.93 | 2.76 |
| 3 ML | | -0.62 (-1.1) | -0.86 (-0.4) | | 0.02 (4.1) | 0.06 (2.3) | -0.70 (-3.5) | 0.99 | 2.40 |
| 4 OLS | | | | | 0.02 (2.3) | 0.05 (4.1) | | 0.94 | 2.74 |
| 5 ML | | | | | 0.02 (4.0) | 0.06 (5.2) | -0.66 (-3.2) | 0.98 | 2.32 |

*Note:* The dependent variable is the ratio of private investment to GDP ($I_p/Y$). Numbers in parentheses are *t*-statistics; $\rho$ is the first-order residual correlation coefficient, and $R^2A$ is adjusted $R^2$. Estimation methods are ordinary least squares (OLS) and maximum likelihood (ML). A blank denotes the omission of the specific variable from the regression.

490

$$(10.10) \quad \frac{I_{pt}}{Y_t} = \frac{I}{Y}\left( \underset{(-)}{\text{PUCK}_t}, \ \underset{(+)}{\text{PMPK}_t}, \ \underset{(?)}{\frac{P_{ipmt}}{P_{ipnt}}}, \ \underset{(-)}{\frac{\text{PCOT}_t}{Y_t}}, \ \underset{(+)}{\frac{K_{gt-1}}{Y_T}}, \ \underset{(+)}{\frac{\text{PRO}_T}{Y_Y}}, \ \underset{(+)}{\frac{\text{FC}_T}{Y_t}}, \right.$$

$$\left. \underset{(+)}{\frac{H_t}{Y_t}}, \ \underset{(+)}{\frac{FS_t}{Y_t}}, \ \underset{(-)}{\text{VUCK}_t}, \ \underset{(-)}{VY_t} \right)$$

where $I_p$ is private fixed-capital investment, $Y$ is GDP, UCK is the user cost of capital,[23] PUCK is estimated permanent UCK, MPK is the marginal product of capital (approximated by the average product of capital equal to the ratio of current-period GDP to lagged end-of-period private sector capital stock), PMPK is estimated permanent MPK, $P_{ipm}/P_{ipn}$ is the price ratio of imported and national private investment components, COT is corporate tax revenue, PCOT is estimated permanent COT, $K_{gt-1}$ is lagged end-of-period public sector capital stock, PRO is corporate profits, FC is banking credit flows to firms, $H$ is base money, FS is foreign saving, VUCK is the coefficient of variation of UCK, and $VY$ is the coefficient of variation of GDP. Expected signs of the corresponding partial derivatives are shown below each variable. Expected investment inflation is based on an autoregressive structure. All expected permanent variables are specified according to partial perfect foresight or static expectations. The two coefficients of variation, which reflect risk and the incidence of investment, are defined as five-period moving coefficients based on two periods back, the current period, and two periods into the future.

A linear form of equation 10.11 was estimated for Zimbabwe using ratios of annual private investment to GDP for the 1965–88 period (table 10.11). Some differences arise between equation 10.10 and the reported results. First, better results were obtained when the user cost of capital was split into its two components, the relative price of investment goods ($P_I/P$) and the sum of the relevant real interest rate and the depreciation rate (RIL). Second, for RIL, as well as for other variables involving estimates of permanent values (the relative price of investment goods, the marginal product of capital, and corporate tax revenue), only the static-expectations results are reported.

The results are very satisfactory. Most neoclassical, borrowing-constraint, and uncertainty variables present the expected signs and are highly significant. The results for the most general specification are reported in row 1 of table 10.11, although not many degrees of freedom are left over. Of all the variables, only the ratio of corporate tax revenue to output has an opposite sign—with a significant coefficient—from what was expected a priori. This variable, in addition to the ratio of firm credit to output and the coefficient of variation of GDP, is deleted from the next set of results.

The two components of the user cost of capital are highly significant. The magnitudes of their signs differ: private investors in Zimbabwe react three times as strongly to the real interest rate as to the relative price of investment goods. The ratio of private capital stock to output (the inverse of the current average product of capital) presents the correct sign but achieves acceptable significance levels only under the maximum likelihood estimations, correcting for residual first-order correlation. In addition, the magnitude of the ratio is small in relation to that of the real interest rate.

The significant role of the ratio of the public capital stock to output (with a coefficient similar in magnitude and significance to that of the ratio of the private capital stock) suggests a strong complementarity between public and private capital in Zimbabwe. This crowding-in of private investment by public capital is an important result that reflects the importance of the composition of public expenditure for the country's growth prospects.

Two flow variables (firm profits and foreign lending as reflected in the current account deficit) and one stock variable (base money) suggest the significant role of the borrowing constraints faced by private investors. This result is not surprising for a period dominated by interest rate controls, which were partially relaxed only in recent years. Even under complete domestic financial liberalization, one should expect that borrowing constraints will affect private capital formation, in addition to the influence of completely deregulated interest rates.

Finally, there is only weak evidence for the effect of our uncertainty proxies on private capital formation. In the most general specification (row 1 of table 10.11), the coefficient of variation of GDP affects private investment negatively and significantly. In rows 2 and 3 the variation of the relative price of investment goods (VPIP) is negative but weak; its coefficient does not achieve acceptable significance levels.

The results support the notion that public deficits and their structures affect private investment in Zimbabwe through various direct and indirect channels. Real interest rates have a strong negative influence on private investment. Hence domestic debt financing of public deficits, which tends to push up interest rates—as observed during the 1980s in Zimbabwe—has a significant crowding-out effect. Public investment, by contrast, has a significant crowding-in effect. For each 1 percentage point of GDP increase in public investment (which raises the ratio of public capital stock to GDP by a similar amount), private investment rises by 0.15–0.25 percentage point of GDP.

## External Accounts, Real Exchange Rates, and the Public Deficit

The next step in analyzing the macroeconomic implications of public sector deficits is to evaluate their effects on Zimbabwe's external accounts and the real exchange rate. These variables are influenced to a large extent by the foreign exchange allocation system. In deciding how to allocate foreign exchange, the foreign exchange allocation commission first makes a projection of the availability of net foreign capital inflows, subject to the government's foreign debt target. The commission then projects total exports on the basis of different assumptions regarding foreign prices and domestic supply conditions, such as future harvests of the main crops. These projections of total available foreign exchange resources provide the basis for the allocation of foreign exchange to importers of capital, intermediate, and consumer goods, which proceeds according to sectoral and historical criteria.

The commission's efforts to decrease dependence on foreign lending led to a significant reduction of the current account deficit during the 1980s, from 10.3 percent of GDP in 1982 to a small surplus in 1988. As mentioned above, this external adjustment fell entirely on the private sector, since the public deficit did not decline below 10 percent of GDP throughout the 1980s.

The government, however, does not only manage the quantitative mechanism of centrally allocating foreign exchange; it also controls the nominal exchange rate and sets import tariffs and quantitative trade restrictions. The exchange rate policy could be important for the foreign exchange projection for total exports; exports are sensitive to the real exchange rate, which in turn can be affected by the nominal exchange rate policy. Trade taxes, however, seem to have been set more to raise fiscal revenue than to protect national production or limit imports. Despite this, custom duties have increased significantly during the past decade, contributing to reduced imports.

## The Model

In estimating the relationship between external variables, such as the trade balance and the real exchange rate, and fiscal policies, the two-stage framework of Rodríguez (chapter 2 in this volume) is amended in two ways. First, less emphasis is placed on the accumulation of net foreign assets (or debt) as the long-term driving force because of the lack of access by Zimbabwe's private sector to foreign financial markets. Second, the determination of the trade balance involves a two-step procedure. At the start of each year the government projects the trade balance on the basis of the difference between income and absorption, which is equivalent to specifying a behavioral relation. At the same time, the government projects total exports on the basis of estimated values of the appropriate relative prices—the terms of trade and the real exchange rate for exports. Given the projected trade balance and export levels, the government instructs the foreign

**Table 10.12. Estimation Results for Real Exchange Rates, Trade Surplus, and Exports, Zimbabwe, 1965–88**

| Dependent variable | Constant | $\ln(G/Y)$ | $\ln(G_N/G)$ | $\ln(TS_{-1}/Y)$ | $\ln(TT^*)$ | $t_M$ | $\rho$ | $R^2A$ | Durbin-Watson |
|---|---|---|---|---|---|---|---|---|---|
| 1. Export real exchange rate $(ex)$[a] | −0.94 (−4.13) | −0.52 (−3.84) | −0.23 (−1.58) | 0.06 (2.15) | 0.37 (1.77) | −0.26 (−1.15) | 0.58 | 0.61 | 1.78 |
| 2. Import real exchange rate $(em)$[a] | −0.55 (−2.16) | −0.38 (−2.74) | −0.11 (−0.85) | 0.06 (2.02) | −0.60 (−2.93) | 0.15 (0.93) | 0.68 | 0.53 | 1.79 |

| Dependent variable | Constant | $ird$ | $\ln(OD_G/Y)$ | $\ln(TS/Y)_{-1}$ | $\ln(CA/Y)$ | $\ln(B_{-1}/Y)$ | $\pi tax$ | $\rho$ | $R^2A$ | Durbin-Watson |
|---|---|---|---|---|---|---|---|---|---|---|
| 3. Trade surplus as share of GDP $(TS/Y)$ | −4.67 (−0.94) | −3.47 (−0.62) | −0.65 (−2.78) | 0.38 (2.16) | −0.24 (−2.59) | 0.46 (1.09) | | | 0.78 | 1.76 |
| 4. $\ln(TS/Y)$ | −3.62 (−0.82) | −2.3 (−0.44) | −0.58 (−2.36) | 0.35 (1.95) | −0.23 (−2.28) | | −0.56 (−0.40) | | 0.76 | 1.75 |

| Dependent variable | Constant | $\ln(ex)$ | $\ln(TT)$ | $\ln(Y/YP)$ | | | | Rho | $R^2A$ | Durbin-Watson |
|---|---|---|---|---|---|---|---|---|---|---|
| 5. Export share of GDP $(X/Y)$ | −1.25 (−15.6) | 0.54 (3.8) | −0.43 (−1.7) | | | | | 0.87 | 0.70 | 1.78 |
| 6. $\ln(X/Y)$ | −1.31 (−17.3) | 0.25 (1.3) | −0.52 (−1.5) | | | | | 0.78 | 0.55 | 1.68 |

*Note:* Numbers in parentheses are $t$-statistics; $\rho$ is the first-order residual correlation coefficient, and $R^2A$ is adjusted $R^2$. The simulation method is generalized least squares, with a maximum-likelihood procedure to correct for first-order autocorrelation and instrumental variables to correct for possible simultaneity bias. A blank denotes the omission of the specific variable from the regression.

a. The deflator is the average wage index.

exchange allocation commission to allocate projected available foreign exchange to imports. Naturally, projected and actual foreign exchange resources differ as a result of unexpected changes in the exogenous variables driving the trade balance and exports.

The government can also affect the trade balance through its exchange rate policy, at least in the short term. The extent of these effects can be tested in the empirical setup that follows. These amendments result in a model comparable to that of Rodríguez in one important respect: fiscal policies are still reflected by the trade balance and real exchange rate equations. The relative prices of (or relevant exchange rates for) exports and imports are specified as:

(10.11)    $ex \equiv (P_X/P_N) = ex(TT^*, t_M, TS/Y, G/Y, G_N/G)$
$$(+)\ (?)\ (+)\ (-)\ (-)$$

(10.12)    $em \equiv (P_M/P_N) = em(TT^*, t_M, TS/Y, G/Y, G_N/G).$
$$(-)\ (+)\ (+)\ (-)\ (-)$$

The ratio of the trade surplus to GDP is given by:

(10.13)       $TS/Y = ts(ird, OD_G/Y, CA/Y, B_{-1}/Y, \pi tax.$
$$(-)\ (-)\ (-)\ (-)\ (-)$$

Finally, the export function is specified as:

(10.14)    $X/Y = x(TT, Y/YP)$    or    $X/Y = x'(ex, Y/YP)$
$$(-)\ (-) \qquad\qquad (+)\ (-)$$

where *ex* is the real exchange rate for exports, *em* is the real exchange rate for imports, *TS* is the trade surplus, *X* is total exports, *Y* is GDP, $P_X$ is the price index of exports, $P_M$ is the price index of imports, $P_N$ is the price index of nontradable goods, $TT^*$ is foreign terms of trade, *TT* is domestic terms of trade, $t_M$ is the average tariff rate, *G* is government spending (public consumption plus public investment),[24] $G_N$ is government spending on nontradable goods, *i* is the average domestic interest rate, *ird* is the uncovered interest rate differential, $OD_G$ is the operational public sector deficit, *CA* is the capital account surplus, $B_{-1}$ is lagged-end-of-period domestic public sector debt stock, *YP* is potential GDP, and $\pi tax$ is the inflation tax.

*Empirical Results*

Equations 10.11 through 10.14 were estimated in log-linear form, using annual data for 1965–88 (see table 10.12).[25]

Both real exchange rates show significantly negative elasticities with respect to the share of government spending in GDP, which confirms the theoretical prediction. In addition, the share of government spending on nontradable goods negatively affects both real exchange rates, although this effect is not significantly different from zero. Foreign terms of trade also exhibit the expected sign in both equations, but they affect more significantly the real import exchange rate. The implicit tariff rate also shows expected signs but again is not significantly different from zero. (In the real export exchange rate, the theoretical sign is ambiguous.) Finally, the effect of the one-year-lagged trade surplus is small but significant and shows the expected sign.[26]

Two alternative specifications were estimated for the trade surplus, depending on how the government finances its deficit. First the debt-output ratio $(B_{-1}/Y)$ was specified following Rodríguez (chapter 2 in this volume). We then tried inflationary financing by including the inflation tax ($\pi$tax). The overall adjustment is slightly better for the former option, but the coefficient of $B_{-1}/Y$ has the incorrect sign. In terms of other individual variables, all coefficients show the expected signs in both equations. The interest rate differential is not significantly different from zero in both estimated versions of equation 10.13. The results were not improved by using an alternative definition of *ird* based on the actual implicit interest rate paid by Zimbabwe on its foreign debt.

Most interesting are the estimated effects of the operational deficit of the public sector, on the one hand, and the ratio of the capital account surplus to GDP $(CA/Y)$, on the other. The latter variable is used as a flow proxy for net foreign assets (NFA) in the Rodríguez setup, and it seems even more appropriate than NFA in the Zimbabwean context, given the way the government allocates foreign exchange. As one could expect from a theoretical point of view, the effect of $CA/Y$ on the ratio of trade surplus to GDP is negative: the more foreign funds flow in, the more financing is available for imports without resort to increased exports. What the estimated elasticities suggest is that an increase in capital inflows does not bring an equal decline in the trade surplus but a substantially smaller reduction.[27] This result reflects government policy to save some of those capital inflows in the form of foreign reserve accumulation and so reduce net foreign indebtedness.

In the case of the operational deficit of the public sector, the estimations indicate that an increase of 1 percentage point of GDP in this variable lowers the trade surplus by 0.6 percentage point of GDP. This confirms the theoretical presumption: the rise in $OD_G$ increases absorption and thus, for a given income level, reduces $TS$. However, the form by which this deficit is financed does not influence the ratio of trade surplus to GDP to any significant extent. Neither the outstanding stock of public sector debt nor the inflation tax are statistically significant. The reason is clear: during the mid- to late 1980s most of the public deficit was financed by issuing domestic debt, which is held either compulsorily—as is true of the share held by institutional investors—or, in the case of private savers, because alternative portfolio choices are lacking. In addition, private saving has significantly increased since the early 1980s because of restrictions on imports of foreign goods. Private investors have not been so much affected by the public indebtedness process as by the restrictions on the acquisition of foreign capital goods imposed by the foreign exchange allocation system. In the end, it is not surprising that the increase in public debt has not been reflected in a lower trade surplus. The two are, temporarily at least, disconnected from each other. The low significance of the inflation tax is even less surprising, as the public sector has resorted to debt to finance its deficits.

Exports were also specified according to two different versions. The first is more in agreement with the spirit of the Rodríguez model because it uses the domestic terms of trade as the relevant relative price variable, while the second specifies the export real exchange rate. Both versions include the ratio of current to potential GDP as an additional explanatory variable. The reasoning is that the higher this ratio, the lower is the share of exportable goods produced that effectively ends up in foreign markets. Both the sign of this variable and the signs of either measure of relative prices are correct, although $Y/YP$ does not reach conventional significance levels.

The reported empirical results confirm the influence of public deficits and spending in Zimbabwe's external sector. They reflect the particular way in which Zimbabwe's government regulates imports through the foreign exchange allocation commission and restricts capital flows.

## Growth Prospects

This section assesses Zimbabwe's growth prospects in connection with the earlier discussion on the macroeconomic effects of public sector deficits. A behavioral function for the relative output supply (the ratio of actual to potential GDP) is specified, following a standard neoclassical setup in which output depends on relative factor prices and prices of intermediate goods.[28] This sheds light on the effect of domestic investment and the incentive system on this ratio and on future growth prospects. The final discussion of the effects of public deficits and structural distortions focuses on the overall performance of the Zimbabwean economy and its growth prospects.

### *Growth*

A neoclassical output supply is specified for GDP by substituting conditional factor demands into an aggregate production function that

**Table 10.13. Estimation Results for Relative Output Supply, Zimbabwe, 1966–88**

| Equation | $\gamma$ | $\lambda$ | $\alpha$ | $\beta$ | $\delta_1$ | $\delta_2$ | $R^2A$ | Durbin-Watson |
|---|---|---|---|---|---|---|---|---|
| 1. NLLS | -0.10 (-4.7) | 0.45 (4.7) | 0.92 (6.7) | 0.19 (1.2) | -0.13 (-5.8) | -0.25 (-9.2) | 0.92 | 2.26 |
| 2. NLTSLS | -0.11 (-2.9) | 0.44 (2.6) | 0.85 (3.1) | | -0.12 (-3.7) | -0.25 (-5.6) | 0.91 | 2.12 |
| 3. NLLSMS | -0.11 (-5.9) | 0.40 (4.6) | 0.80 (0.14) | | -0.12 (-5.9) | -0.24 (-9.3) | 0.91 | 2.12 |

*Note:* The equation is:

$$\ln(y/yp) = \gamma + \lambda \left[ \alpha \ln(P/We^{-\mu t}) + (1-\alpha)\ln(P/P_{Impint}) \right] + \beta (r_p - 0.05) + \delta_1 D_1 + \delta_2 D_2$$

The first dummy is 1.0 for 1974, 1975, 1976, 1980, 1984 (0 otherwise) and the second dummy, for the stronger recessionary years, is 1 for 1977, 1978, and 1979 (0 otherwise). Estimation methods are nonlinear least squares (NLLS) and nonlinear two-stage least squares (NLTSLS). The NLTSLS estimation uses the following list of instruments: the constant; the lagged values of the logarithms of the productivity-adjusted real wage, the real price of intermediate imports, and the dependent variable; and the contemporaneous values of the two dummies, the real interest rate, and the log of the ratio of public expenditure to potential output. Numbers in parentheses are *t*-statistics.

depends on capital, variable factors (labor and working capital), and intermediate imports. By substituting potential output for capital, aggregate supply can be restated as the ratio of actual and potential output levels, defined as a function of the real wage adjusted for productivity gains, the real exchange rate relevant for intermediate imports, the real interest rate relevant for working capital, and period-specific dummies for Zimbabwe's conflictive preindependence period.

$$(10.15) \qquad \ln\left(\frac{y}{yp}\right) = \gamma + \lambda\left[\alpha \ln\frac{P}{We^{-\mu t}} + (1 - \alpha)\ln\frac{P}{P_{Impint}}\right]$$

$$+ \beta(r_p - 0.05) + \sum_s \delta_s D_s$$

where $y$ is actual GDP, $yp$ is potential GDP, $P$ is the GDP deflator, $W$ is the nominal unit wage, $P_{Impint}$ is the price of intermediate imports, $t$ is time, $D_s$ represents supply-specific dummies, and $r_p$ is the real interest rate relevant for production decisions (that is, taking into account the nominal lending rate and expected inflation of the GDP deflator).[29]

The results of equation 10.15 are reported in table 10.13. The first row shows the results for the complete specification, with a positive but not significant coefficient for the real interest rate, which is dropped from the subsequent estimations. The second row reports nonlinear two-stage least-squares (NTSLS) results to take care of possible simultaneity biases caused by the nonindependence of the real wage and the real price of imports stemming from the interaction of aggregate supply and demand. The results are not very different from the OLS results reported in the third row, in terms of both the excellent overall fit and the individual coefficients.

The price-elasticity of aggregate supply is relatively low; it is 0.44 in the NTSLS equation. This result implies that aggregate demand shocks (for a given aggregate demand elasticity) will have a strong relative price response and a weak output effect. The coefficient $\alpha$ (the share of labor in gross output net of capital value added) is very high and significant, reflecting a strong weight of the real product wage in comparison with the real exchange rate in determining short-run output. Finally, $\delta_1$ and $\delta_2$ reflect the relative intensity of supply disruptions, which coincide mostly with the international oil shocks and the preindependence period of civil war.

### Deficits, Distortions, and Growth

Zimbabwe's economic fragility increased during the 1980s and early 1990s, culminating in the particularly intense drought of 1992. But the country's zero per capita growth in twelve years is a symptom of deeper problems. Major fiscal disequilibria, a large and inefficient state enterprise sector, and a distorted incentive structure have led to a weak performance in private investment and exports. How are these problems reflected in the empirical results reported shown in the preceding sections?

A sharp contraction in economic activity occurred in the early 1970s as oil shocks and domestic turmoil hit the economy. After independence in 1980 a hesitant recovery began, initially financed by external debt. When, after 1983, foreign capital inflows fell, a major reduction in current account deficits was achieved during 1982–86 and was preserved until 1989. To finance the large deficit, the government turned to the private sector; both high private saving and low private investment contributed to the large private surplus required to finance public deficits in the range of 10–14 percent of GDP. Since 1990, however, the public sector has again been turning to foreign financing as private surpluses dry up.

Because the decline in private investment after the early 1980s was not matched by higher public investment, total investment in fixed capital has been decreasing as a share of GDP. It is in this sense that public deficit financing has been detrimental to Zimbabwe's growth prospects. But private investment in Zimbabwe is not only crowded out by public deficits; it is also affected by an environment not supportive enough of private business. The scarcity of financial resources available for private investment, restrictions on private business, a heavy tax burden, and limited access to intermediate and capital goods imports as a result of foreign exchange quotas and uncertain property rights inhibit private capital formation. These interventions affect both the level and the productivity of investment. Indeed, they generate a distorted relative price and overall incentive structure that induces investment to flow to sectors in which social returns are low. Therefore both fiscal stabilization and structural reforms are required to attain a higher quantity and quality of investment and improve Zimbabwe's growth prospects.

## Conclusions and Policy Implications

The Zimbabwean case is most interesting because of the coexistence of persistent large public deficits with a private surplus that is sizable in comparison with those in comparable developing countries. Private investment has been squeezed to less than 10 percent of GDP and private consumption to about 60 percent of GDP through a combination of foreign exchange and capital flow restrictions with compulsory public debt placement in a relatively developed—albeit managed—financial market. This has been possible at moderate inflation rates, reflecting a generally conservative monetary policy and low, typically negative, real interest rates.

This chapter has identified various determinants and consequences of public deficits in Zimbabwe. The results point to ten implications related to deficit financing and the role of macroeconomic and fiscal policy variables in determining deficits.

• The large financing requirements of public deficits have led to a steady and massive buildup of total public liabilities, from 54 percent of GDP in 1980 to 81 percent in 1987, and well beyond 100 percent in 1992. Issuing money has traditionally been a minor source of deficit financing, with the exception of the recent past, when the inflation tax rose substantially.

• Among macroeconomic variables (and in decreasing order), real GDP growth, real import growth, and a real exchange rate devaluation have a negative impact on the public sector deficit. Also in declining order, higher domestic real interest rates, domestic inflation, and foreign nominal interest rates boost the deficit. Among central government policy variables, cuts in the wage bill, transfers and subsidies, public investment, and expenditure on goods and services affect the deficit. Further policy measures on the revenue side, such as tax reforms and reductions in the public enterprise deficit, have major and immediate effects on public finance. Reductions in outstanding domestic and foreign debt stocks affect deficits with a lag by lowering interest payments.

• Measuring the role of fiscal policy variables in relation to variables beyond the direct control of policymakers, we conclude that fiscal policy had an absolutely predominant role in shaping the size of deficits throughout the 1980s. Therefore policymakers—not domestic or foreign shocks—are to blame for the fiscal deterioration in the early 1980s and are to be praised for fiscal adjustment, such as the partial deficit reduction in the late 1980s.

Bounds for sustainable public deficits were derived by relating primary deficits and interest payments to financing sources: money, domestic debt, and foreign debt. Sustainability was defined in the sense of holding constant the 1988 ratio of total public sector liabilities to GDP. The main conclusions follow.

• Under a base scenario showing a macroeconomic environment similar to that of the recent past, the sustainable primary deficit is estimated at 1.7 percent of GDP. It increases to 2.9 percent of GDP under a high case of higher growth and lower real domestic interest rates. The corresponding nominal (primary plus interest payments) deficits are 9.9 and 10.7 percent of GDP, slightly lower than the actual 10 to 11 percent of GDP for the nominal deficit observed in 1987–91.

• Under a low-case scenario of low growth, high real domestic and foreign interest rates, and a real exchange rate depreciation of 7 percent per year, the sign of the sustainable deficit is reversed; a primary surplus of 4.2 percent of GDP (or a nominal deficit of 5.6 percent of GDP) is required to avoid unbounded growth of public sector liabilities. Since 1990, Zimbabwe has been experiencing such an adverse scenario. Low growth, real exchange rate depreciations, and rising drought-induced deficits are leading to a rapid buildup of total public sector debt.

Next we reversed the focus by looking at the macroeconomic effects of public deficits. On the financial market implications of deficit financing, we concluded the following.

• Since 1983 the government has intervened in financial markets in order to generate a private sector surplus that can be used to finance its large deficits. Reliance on domestic debt financing seems to be an optimal choice because it avoids inflationary finance. However, our simulations indicate that this solution is not sustainable in the long term; larger inflation should be anticipated eventually, when domestic financial markets become less willing to absorb additional public liabilities. In spite of financial sector regulation and the large policy-induced private sector surplus, a moderate upward trend of real interest rates was observed during the 1980s as a result of increasing public debt (see also Elbadawi and Schmidt-Hebbel 1991b).

On the question of how the large public deficits and their composition have affected private saving and investment, Zimbabwe's recent experience suggests the following conclusions.

• Between 1981/82 and 1987/88 a major external improvement was achieved, turning a current account deficit of 9.4 percent of GDP into a balanced account. This improvement relied exclusively on the private sector, as the public sector deficit hovered around 10–14 percent of GDP. Both an increase in private saving and a decline in private investment were behind the rise in the private surplus. Private saving exceeded 20 percent of GDP and exceeded the economy's domestic investment between 1984/85 and 1988/89. Low private consumption was made possible through a combination of consumer import repression, strict controls on capital outflows, and a perception that the financial system was stable. The empirical results show that private consumption is strongly influenced by current private disposable income and by public saving. However, after 1989 the private sector surplus shrank, and in 1992 there was actually a deficit.
• The declining private investment up to 1986/87 implies lower aggregate capital formation and, probably, a lower efficiency of domestic investment, which contributed to Zimbabwe's meager

growth. The effect on private investment of the partial fiscal adjustment in 1987–89 is encouraging because it allowed a recovery of 2.4
percentage points of the gross domestic investment rate. Our results
indicate that real interest rates have a strong negative influence on
private investment. Hence domestic debt financing of public sector
deficits, which (as observed in the 1980s) tends to push up interest
rates, has a significant indirect crowding-out effect. This is in part
offset by crowding-in of private investment as a result of higher public investment.

To what degree does the fiscal deficit spill over abroad?

• Public deficits and public spending strongly affect the trade surplus and relative export and import prices in Zimbabwe. An increase
of 5 percentage points of GDP in the public deficit lowers the trade
surplus by 3 percentage points of GDP, causing an appreciation of the
real exchange rate. The particular way in which Zimbabwe's government administers imports through the foreign exchange allocation
commission and the binding restraints placed on capital movements
are the central factors in this outcome.

Finally, on growth prospects and required policy reforms, our findings are as follows.

• Crowding-out of private investment and declining public investment, combined with the low quality of investment projects resulting
from the distorted incentive structure, have substantially affected
Zimbabwe's growth performance and potential. In addition, to the
extent that imported intermediate and capital goods are not perfect
substitutes for domestic goods, import compression and exchange
controls have reduced utilization of existing capacity. The 1991 adjustment program has started to deal with macroeconomic disequilibria
and structural distortions. The main conclusion of this study is that
this adjustment program has to be deepened substantially along two
lines.

First, much more fiscal adjustment is needed to achieve macroeconomic and financial stability and to foster higher growth. This
point is strongly supported by a recent study on macroeconomic
adjustment in Zimbabwe (Elbadawi and Schmidt-Hebbel 1991b). The
simulation results in this study, which uses a macroeconomic general
equilibrium model, show that fiscal reform would help Zimbabwe
achieve a sustainable debt path, a decline in interest rates paid on
public debt, and a recovery of private consumption and investment.

Second, much more progress is required in liberalizing domestic
goods and financial markets as well as foreign trade, restructuring
and privatizing state enterprises, and deregulating private sector production and investment. Only when these conditions are met—and

after allowing for a significant lapse of time required to raise the credibility of the new policy framework—will Zimbabwe be able to start on a path of sustained growth and poverty reduction.

## Notes

The authors thank Rob Davies and Lloyd McKay for illuminating discussions in Harare. They are also grateful to Bela Balassa, Ibrahim Elbadawi, Sarshar Khan, Steve O'Connell, Jorn Rattso, Ragnar Torvik, and Michael Walton for their comments on previous drafts. Efficient research assistance was provided by Maria Cristina Almero-Siochi, Rodney Chun, and Heinz Rudolph. Responsibility for any remaining errors is the authors'.

1. Among recent papers on Zimbabwe's macroeconomic situation and prospects are Chhibber and others (1989); Dailami and Walton (1989); Khadr and others (1989); Davies and Rattso (1990); Elbadawi and Schmidt-Hebbel (1991a, 1991b); Davies, Rattso, and Torvik (1993); and Mehlum and Rattso (1993).

2. The 1980–89 data presented in this section are based on Schmidt-Hebbel (1990), the most comprehensive attempt to date to build consistent consolidated and stock-flow data for nonfinancial public sector deficits and for nonfinancial and financial public sector balance sheets. For a detailed discussion, see the above reference and Morandé and Schmidt-Hebbel (1991). A first application to Zimbabwe of a framework for macroeconomic consistency in current and constant prices for a six-sector disaggregation (for 1981 and 1987) can be found in Khadr and Schmidt-Hebbel (1989a, 1989b). An application of the RMSM-X macroeconomic consistency model for a five-sector disaggregation to Zimbabwe, covering the 1985–87 historical period and the 1988–95 projection period, was done by Khadr and others (1989). A significant extension of the former, in terms of behavioral specification, sector disaggregation, and period coverage, is the macroeconomic general equilibrium model for Zimbabwe by Elbadawi and Schmidt-Hebbel (1991a, 1991b), with base year 1988 and simulations covering 1988–95.

3. Tax revenue functions were estimated separately for direct taxes, indirect taxes, and customs duties. The results in Morandé and Schmidt-Hebbel (1991) show that tax revenues depend on relevant tax bases (GDP and imports) and tax reforms. An interesting finding is that inflation boosts income tax payments as a result of income bracket creep, without any evidence of negative Olivera-Tanzi effects of inflation on any tax revenue category.

4. The average relative contribution of each group of deficit determinants to the explained variation of the deficit is computed as

$$\left[ \sum_{i=81/82}^{88/89} dv_i(\text{sign} d_i) \right] \bigg/ \sum_{i=81/82}^{88/89} |d_i|$$

where $d_i$ is the explained change of the deficit in period $i$ (the third-from-last line in table 10.4) and $dv_i$ is the change in the deficit caused by variable category $v$ (the total variations resulting from domestic, foreign, and fiscal policy variables in table 10.4).

5. The semielasticities were computed for 1987/88 to 1988/89. If the semi-elasticities changed over the 1980s, the values for the early 1980s (1981/82 to 1982/83) were added in parentheses after the 1987/88 to 1988/89 values.

6. This follows work on fiscal sustainability developed by Buiter (1983, 1985) and van Wijnbergen (1989), with applications such as that by van Wijnbergen, Rocha, and Anand (1988) to Turkey.

7. The real exchange rate, denoted by $e$, is defined as $(E\ P^*/P)$, where $E$ is the nominal exchange rate and $P^*$ is the foreign price index. The domestic and foreign real interest rates are defined according to the Fisher equations in their simplified linear form.

8. Note that this equation determines the size of the primary deficit from the effects of inflation, output growth, the real exchange rate, and domestic and external interest rates on interest payments and deficit financing sources. Hence the direct effects of the first three variables on the primary deficit, considered in the preceding section, are omitted here. The same holds for the simulations carried out below. For instance, whereas a higher real devaluation worsens the sustainable deficit calculation in table 10.6 because of higher external real interest payments (for a given primary deficit), the net total effect—through both higher interest payments and a lower primary deficit—was shown in table 10.5 to reduce the deficit, as a result of the dominating effect on the primary deficit.

9. The pattern and depth of the financial system were inherited from the preindependence period and have remained intact, preventing the development of informal credit markets. This has been the result of a combination of factors: a strict regulatory framework, relatively conservative monetary and exchange rate policies, and high confidence in public debt as a result of strict debt servicing.

10. As an alternative to the money market equilibrium condition for determining the real interest rate, one can use the equilibrium condition in the public sector bonds market, which also depends on the demand for money. That is, $b = b^d = nfa' - (1/s)m(r,\ \pi^e_{+1},\ nfa)$, where $s$ is the money multiplier. This equation, in combination with the government budget constraint in equation 10.2, determines the amount of monetary financing of the deficit.

11. Some variable measurement and data features merit a discussion. Money is seasonally adjusted $M_1$; the nominal interest rate is a weighted average of public sector stock and bonds, annual interest rates, and deposit rates at commercial banks (also on an annual basis); and net financial assets is the sum of private sector deposits in the financial system, including compulsory savings in pension funds and insurance companies. All series are deflated by the consumer price index (CPI) of the rich, which is less affected by (at times pervasive) price controls than the CPI of the poor during the sample period. The same CPI of the rich is also used for actual and expected inflation. The data frequency is quarterly, and the sample period, determined by data availability, is 1979:1 to 1988:3 for most estimations.

12. In the reported results we do not restrict $b_1$ to be equal to $b_2$, although this was also tried. The estimates were similar but were not so close as to reject the hypothesis that they are significantly different from each other.

13. Coefficients of other variables in equation 10.4, such as wages, were not significantly different from zero.

14. The fiscal year macroeconomic aggregates of table 10.9 (foreign saving, national saving, gross domestic investment, and GDP) are consistent with calendar year data from national accounts.

15. Note that this refers to the separate effect of public saving (taxes minus current public expenditure) on consumption, in addition to the effect of disposable income (gross income minus taxes).

16. All nonstationary variables are scaled to current private disposable income to reduce the incidence of spurious correlation. An alternative procedure, combining cointegration tests and dynamic error-correction models, is not feasible because of the short time series.

17. Three simple null hypotheses are tested with this specification: (a) Keynesian: $\beta_0 > 0$, $\beta_1 = \beta_2 = 0$; (b) permanent income hypothesis without Ricardian equivalence: $\beta_1 > 0$, $\beta_0 = \beta_2 = 0$; and (c) Ricardian equivalence or direct-crowding-out hypotheses: $\beta_0 = 0$, $\beta_1 = \beta_2 > 0$.

18. For a precise definition of disposable income and public saving, see Corbo and Schmidt-Hebbel (1991).

19. Privately appropriated services paid by the government are measured as the sum of public expenditure on education and health. These, plus direct transfers to consumers, could reduce private consumption (if they are substitutes for the latter) or increase it (if they strongly crowd in complementary private expenditure categories).

20. We specify two alternatives for the expected permanent values of public saving in equation 10.9 and other variables in the investment equation (10.10) that follows. The first alternative is partial perfect foresight, defined as the simple average of the variables for the current period and two periods into the future. The second alternative is the static-expectations specification, which assigns a 100 percent weight to the current value.

Similar assumptions are made with respect to expected consumption inflation (and expected investment inflation, below). A first alternative takes actual inflation between today and tomorrow as the relevant proxy for rationally expected inflation. The second alternative is adaptive expectations, specifying the expected price change either from an autoregressive moving average (ARMA) backward-looking process or by assigning 100 percent weight to the actual price change between yesterday and today, consistent with static expectations. Permanent private disposable income is defined as the ratio of permanent to current GDP (see the last section of this chapter) multiplied by current private disposable income.

21. Similar results can be found in Elbadawi and Schmidt-Hebbel (1991a) in an estimated general equilibrium model for Zimbabwe. Our approach allows for more freedom in choosing lags and variables, but it is less explicit on potential feedbacks among variables.

22. This setup is similar to a private investment model implemented for Morocco by Schmidt-Hebbel and Müller (1992).

23. UCK is defined as the product of the investment deflator ratio to the GDP deflator $(P_I/P)$ and the sum of the real interest rate relevant for investment and the depreciation rate (RIL).

24. Government spending has a negative effect on both real exchange rates when more than half of G is made up of nontradable goods.

25. The application of logarithms to variables with negative values necessitated adding a constant to them. In estimating equations 10.11 and 10.12, nontradable prices—the deflator in the definition of both *ex* and *em*—were proxied alternatively by the average wage index and the domestic price level. Results were clearly better when using the former, on which the results reported in table 10.13 are based. Prices of exports and imports were proxied by the corresponding national accounts deflators. $G$, $G_N$, TS, and $Y$ are measured at current prices. $G_N$, government spending on health, housing, and education, stands for spending on nontradable goods. TT* was constructed as the ratio of export and import deflators, adjusted by the average tariff rate implicit in custom duty revenues. This implicit average tariff rate is also present in the regression as $t_M$. The expected rate of devaluation was assumed to be equal to the actual rate, a perfect-foresight approximation of the rational-expectations hypothesis. The foreign interest rate is LIBOR, and the domestic interest rate is a weighted average of active rates in the financial sector.

26. The use of the lagged trade surplus rules out a potential simultaneity bias. We also used the current account deficit as an alternative to the lagged trade surplus, without success.

27. Since we are considering percentage rates of change, there could be some differences between 1 percent of $CA/Y$ and 1 percent of $TS/Y$. On average, however, these differences are not large.

28. This section draws heavily on Elbadawi and Schmidt-Hebbel (1991a).

29. Note that equation 10.15 is not a reduced-form market equilibrium equation but a structural aggregate supply form. Hence no demand variables are included. The specification is homogeneous of degree zero in absolute prices. The real wage is adjusted for Harrod-neutral productivity increases at an annual rate of $\mu = 0.008$, its 1965–72 trend growth rate, which is deemed to be representative for a normal period of productivity-related wage increases when the economy is operating at levels close to full employment. From 1972 to 1979 real wages stagnated; after 1979 they grew strongly, probably reflecting both the partial recovery of output and the change in political regime.

## References

Balassa, Bela. 1989. "The Effects of Interest Rates on Saving in Developing Countries." Policy Research Working Paper 56. World Bank, Office of the Vice President, Development Economics, Washington, D.C.

Barro, Robert F. 1974. "Are Government Bonds Net Wealth?" *Journal of Political Economy* 82(6): 1095–1117.

Bernheim, B. Douglas. 1987. "Ricardian Equivalence: An Evaluation of Theory and Evidence." In *NBER Macroeconomics Annual*. Cambridge, Mass.: Massachusetts Institute of Technology Press.

Buiter, Willem H. 1983. "Measurement of the Public Sector Deficit and Its Implications for Policy Evaluation and Design." *International Monetary Fund Staff Papers* 30 (2): 307–49.

———. 1985. "A Guide to Public Sector Debt and Deficits." *Economic Policy* 1 (November): 13–79.

Chhibber, Ajay, Joaquin Cottani, Resa Firuzabadi, and Michael Walton. 1989. "Inflation, Exchange Rates, and Fiscal Adjustment: The Case of Zimbabwe. Policy Research Working Paper 192. World Bank, Country Economics Department, Washington, D.C.

Corbo, Vittorio, and Jaime de Melo. 1989. "External Shocks and Policy Reforms in the Southern Cone: A Reassessment." In Guillermo A. Calvo, ed., *Debt, Stabilization and Development: Essays in Memory of Carlos Diaz Alejandro*. Oxford, U.K.: Basil Blackwell.

Corbo, Vittorio, and Klaus Schmidt-Hebbel. 1991. "Public Policies and Saving in Developing Countries." *Journal of Developing Economics* 36 (1): 89–115.

Dailami, Mansoor, and Michael Walton. 1989. "Private Investment, Government Policy, and Foreign Capital in Zimbabwe." Policy Research Working Paper 248. World Bank, Country Economics Department and Southern Africa Department, Washington, D.C.

Davies, Rob, and Jorn Rattso. 1990. "Macroeconomic Policies for Medium Term Development: The Zimbabwe Case Study." University of Trondheim, Department of Economics, Trondheim, Norway.

Davies, Rob, Jorn Rattso, and Ragnar Torvik. 1993. "The Macroeconomics of Zimbabwe in the Eighties—A CGE-Model Analysis." University of Trondheim, Norway, Department of Economics.

Dornbusch, Rudiger. 1983. "Real Interest Rates, Home Goods, and Optimal External Borrowing." *Journal of Political Economy* 91 (1): 141–53.

———. 1985. "Overborrowing: Three Case Studies." In Gordon W. Smith and John T. Cuddington, eds., *International Debt and the Developing Countries*. A World Bank Symposium. Washington, D.C.

———. 1989. "Capital Flight: Theory, Measurement and Policy Issues." Massachusetts Institute of Technology, Department of Economics, Cambridge, Mass.

Easterly, William. 1989. "Fiscal Adjustment and Deficit Financing during the Cost Crisis." In Ishrat Husain and Ishac Diwan, eds., *Dealing with the Debt Crisis*. Washington, D.C.: World Bank.

Easterly, William, Carlos A. Rodríguez, and Klaus Schmidt-Hebbel. 1989. "Research Proposal: The Macroeconomics of the Public Sector Deficit." World Bank, Country Economics Department, Washington, D.C.

Elbadawi, I. A., and Klaus Schmidt-Hebbel. 1991a. "Macroeconomic Structure and Policy in Zimbabwe: An Analysis with an Empirical Model (1965–1988)." World Bank, Policy and Research Department, Washington, D.C.

———. 1991b. "Macroeconomic Adjustment in Zimbabwe: 1988–95 Simulations with a RMSM-XX Model." World Bank, Policy and Research Department, Washington, D.C.

Fry, Maxwell J. 1988. *Money, Interest and Banking in Economic Development*. Baltimore, Md.: Johns Hopkins University Press.

Giovannini, Alberto. 1985. "Saving and the Real Interest Rate in LDCs." *Journal of Development Economics* 18 (August): 197–217.

Hall, R. E. 1978. "Stochastic Implications of the Life-Cycle Permanent Income Hypothesis: Theory and Evidence." *Journal of Political Economy* 86(6): 971–87.

Haque, Nadeem Ul, and Peter J. Montiel. 1989. "Consumption in Developing Countries: Tests for Liquidity Constraints and Finite Horizons." *Review of Economics and Statistics* 71(3): 408–15.

Hayashi, Fumio. 1982. "The Permanent Income Hypothesis: Estimation and Testing by Instrumental Variables." *Journal of Political Economy* 90(5): 895–916.

———. 1985. "Tests for Liquidity Constraints: A Critical Survey." NBER Working Paper 1729. National Bureau of Economic Research, Cambridge, Mass.

Hubbard, R. G., and K. L. Judd. 1986. "Liquidity Constraints, Fiscal Policy, and Consumption." *Brookings Papers on Economic Activity* 1: 1–50.

IMF (International Monetary Fund). 1990. "Zimbabwe: Recent Economic Developments." Policy Framework Paper. Washington, D.C.

Khadr, Ali, and Klaus Schmidt-Hebbel. 1989a. "A Method for Macroeconomic Consistency in Current and Constant Prices." Policy Research Working Paper 306. World Bank, Country Economics Department, Washington, D.C.

———. 1989b. "A Framework for Macroeconomic Consistency for Zimbabwe." Policy Research Working Paper 310. World Bank, Country Economics Department, Washington, D.C.

Khadr, Ali, L. McKay, Klaus Schmidt-Hebbel, and J. Ventura. 1989. "A RMSM-X Model for Zimbabwe." World Bank, Country Economics Department, Washington, D.C.

Kiguel, Miguel A., and Nissan Liviatan. 1989. "The Old and the New in Heterodox Stabilization Plans: Lessons from the 1960s and 1980s." Policy Research Working Paper 323. World Bank, Country Economics Department, Washington, D.C.

Leiderman, Leonardo, and Mario I. Blejer. 1988. "Modeling and Testing Ricardian Equivalence: A Survey." *International Monetary Fund Staff Papers* 35 (1): 1–35.

Marshall, Jorge, and Klaus Schmidt-Hebbel. 1989. "Economic and Policy Determinants of Public Sector Deficits." Policy Research Working Paper 321. World Bank, Country Economics Department, Washington, D.C.

Mehlum, H., and Jorn Rattso. 1993. "Import Compression and Growth Restrictions in Zimbabwe." University of Trondheim, Norway, Department of Economics.

Morandé, Felipe, and Klaus Schmidt-Hebbel. 1991. "Macroeconomics of Public Sector Deficits: The Case of Zimbabwe." Policy Research Working Paper 688. World Bank, Country Economics Department, Washington, D.C.

Reserve Bank of Zimbabwe. Various years. *Quarterly Economic and Statistical Review*. Harare.

Rossi, Nicola. 1988. "Government Spending, the Real Interest Rate, and the Behavior of Liquidity-Constrained Consumers in Developing Countries." *International Monetary Fund Staff Papers* 35 (1): 104–40

Sargent, Thomas J., and Neil Wallace. 1981. "Some Unpleasant Monetarist Arithmetic." *Federal Reserve Bank of Minneapolis Quarterly Review* 9: 15–31.

Schmidt-Hebbel, Klaus. 1987. "Terms of Trade and the Current Account under Uncertainty." *Análisis Económico* 2 (1): 67–89.

Schmidt-Hebbel, Klaus. 1990. "Zimbabwe: The Need for Fiscal Adjustment." World Bank, Country Economics Department, Washington, D.C.

Schmidt-Hebbel, Klaus, and Tobias Müller. 1992. "Private Investment under Macroeconomic Adjustment in Morocco." In Ajay Chhibber, Mansoor Dailami, and Nemat Shafik, eds., *Reviving Private Investment in Developing Countries*. Amsterdam: North-Holland.

Schmidt-Hebbel, Klaus, Steven B. Webb, and Giancarlo Corsetti. 1992. "Household Saving in Developing Countries: First Cross-Country Evidence." *World Bank Economic Review* 6 (3): 529–47.

Summers, Lawrence H. 1985. "Issues in National Savings Policy." NBER Working Paper 1710. National Bureau of Economic Research, Cambridge, Mass.

Tanzi, Vito, Mario I. Blejer, and Mario O. Teijeiro. 1987. "Inflation and the Measurement of Fiscal Deficits." *International Monetary Fund Staff Papers* 34 (December): 711–38.

van Wijnbergen, Sweder. 1989. "External Debt, Inflation and the Public Sector: Toward Fiscal Policy for Sustainable Growth." *World Bank Economic Review* 3 (3): 297–320.

van Wijnbergen, Sweder, Robert Rocha, and Ritu Anand. 1988. "Inflation, External Debt, and Financial Sector Reform: A Quantitative Approach to Consistent Fiscal Policy." Policy Research Working Paper 261. World Bank, Latin America and the Caribbean Country Department II, Washington, D.C.

World Bank. 1990. *Adjustment Lending Policies for Sustainable Growth*. Policy and Research Series 14. Washington, D.C.

———. Various years. *World Development Report*. New York: Oxford University Press.

Zimbabwe, Central Statistical Office. Various issues. *National Accounts*. Harare.

Zimbabwe, Minstry of Finance. Various issues. *Financial Statements*. Harare.

# Afterword

# The Political Economy of Fiscal Deficit Reduction

*Vito Tanzi*

The project that resulted in the case studies presented in this volume has generated a wealth of interesting conclusions and analyses. As one who has been deeply interested in fiscal policy for many years, I can appreciate the importance of the results. The managers of the project and the authors of the papers should be congratulated on the outcome. This project is as close to a scientific experiment as, perhaps, is possible in economics. I hope that similar "experiments" will be carried out in the future by the World Bank on other important policy issues.

In these remarks, I shall focus not on the technical aspects of the macroeconomics of public sector deficits—since these have been exhaustively studied in the case studies—but on the political and institutional reasons why promoting policies aimed at reducing large fiscal deficits has proved so difficult in many countries. These reasons deserve much more attention from economists than they have received in the past.

By now, many economists would agree that, especially in developing countries, the public sector, far from being the "balancing factor" advocated by Keynes, has often been an accomplice—if not the main culprit—in generating major macroeconomic imbalances. This reality must be kept in mind when the government is called on to pursue "stabilizing" fiscal policies.

## Macroeconomic Imbalances and Government Response

A question to be addressed first is the following: if, in fact, the government has been part of the macroeconomic problem, what is the likelihood that it will become part of the solution? Two views are possible: an optimistic and a pessimistic one. The optimistic view is that policymakers are fast learners and have enough political flexibility to be willing *and* able to change their minds *and* their policies when the situation requires these changes. For some countries—it is to be hoped that the number is increasing—this is a realistic view. These are the countries that manage to stay out of major economic difficulties or that manage to make quick policy corrections when required.

The pessimistic view is that policymakers are not fast learners and that governments are not very flexible: policymakers do not learn from other countries' mistakes or even from the mistakes made by their predecessors. They learn mostly from their own mistakes, and they learn slowly and sometimes not very well. Furthermore, by the time they have learned from their own mistakes, they may have convinced themselves that the political situation does not allow them any room for policy changes.[1]

If the pessimistic view represents the reality of a significant number of countries, the policymakers of these countries face a credibility problem. This problem comes about because what may be considered mistaken policies today are often the policies on which the government ran in past elections or which it espoused when it took power. It is difficult for governments to run away from or renege on their past policies and still be credible. Such behavior might bring into question the government's claim to be the one capable of running the country. For this reason governments tend to deny, for a long time and even to themselves, that they have made and are still making mistakes. They naturally tend to rationalize the situation by blaming undesirable developments on the policies of their predecessors or, more often, on world or other exogenous developments. And, in fact, governments do often inherit bad situations.

A recurrent theme is that what is happening to the economy is the result of unfavorable international economic developments: recessions in industrial countries; high international interest rates caused by large fiscal deficits in industrial countries; marketing practices of industrial countries that restrict access to those markets for particular products; and so on. Natural events are also frequently a convenient part of the rationalization. While these factors may contribute to economic difficulties and should thus not be ignored, they are rarely the whole or even the main explanation for poor economic performance.[2]

Policymakers set on achieving the "social good" can escape the credibility problem by claiming that their previous choices were correct but that new developments require a change of course. In this case, the policymakers can argue that the goals they are pursuing are the same as their earlier ones but that achieving these goals requires different policies. These policymakers will retain their credibility if they (a) explain clearly the reason for the change in policies, (b) are seen to be sincere, and (c) are seen to be competent and capable of choosing and implementing the appropriate policies.

The extent to which a government can convincingly blame external events for the failure of its own economic policy choices depends crucially on the existence of asymmetric information between the policymakers and private agents; the government's easier access to data and information and its greater expertise in interpreting them may allow it to mislead people by overstating the importance of adverse external events. Therefore, knowledgeable and independent institutions with technical expertise and data availability comparable to the government's can play an independent role. The more developed and democratic is the country, the more likely it is that such institutions exist, and thus the more difficult it will be for the government to mislead the people about the reasons for its poor economic performance.

Governments unwilling to take serious corrective actions occasionally resort to the use of jargon as a substitute for policies. For example, they may announce that they will "rationalize public expenditure" or "improve tax administration." These "commitments" may occasionally be included in formal agreements with international organizations. The inevitable question is: if public spending was in need of being "rationalized" (whatever that means), or if tax administration needed to be improved, why wasn't this done earlier? Periods of crisis, when real wages may have been reduced and civil servants may be demoralized, are rarely the best time to make these changes. One is led to suspect that at times the promises may not reflect real, specific commitments. For example, after making strong statements about improving tax administration, the government may not even bother to discuss this intention with the ranking tax administrators, who, as the ones who will implement the changes, will determine whether tax administration really improves.

Such an example relates again to the issue of observability of economic policy actions. In order to be credible, a commitment should be clearly defined and should promise an action or a result that can be readily observed so that the private sector can easily see when the government deviates from the promised course of action. The government will then know that eventual misconduct will be spotted and "punished" at the next elections or the next opinion poll. Vague and general promises will be discounted by the people because of the much greater difficulty in verifying whether they are carried out. A specific commitment is easier to monitor and is therefore more credible.

## Macroeconomic Policy, Credibility, and Biases in Official Pronouncements

Biases in official pronouncements or forecasts have important implications for economic policy. For example, in official statements about future growth, inflation, fiscal deficits, and so on, the future often looks rosier than the present, and the solution to current macroeconomic difficulties, regardless of how big they are, always seems to be just over the horizon. This problem affects developing countries in particular but is not limited to them. Consider, for example, the projections about the federal fiscal deficit in the United States made throughout the 1980s: each year the federal budget was going to be balanced three or four years in the future. Another (admittedly extreme) example was a public speech in August 1989 in which the president of a Latin American country forecast that the rate of inflation for that country would be *negative* by the end of that year. In December 1989 the consumer price index rose at a *monthly* rate of 40 percent. In January 1990 it rose again, at a *monthly* rate of 79 percent. A survey of the official pronouncements or even the official forecasts of many countries would indicate that an optimistic bias is common. Furthermore, the size of the bias is generally not insignificant.

Official optimism in economic forecasts is not without important policy consequences. A legitimate question is whether the policymakers really believe in their own official pronouncements or make them for political reasons. If they believe in them, it is difficult to see how the same individuals who make the projections can pursue the policies necessary for remedying the macroeconomic imbalances—policies that are both correct and adequate to the situation. If they do not believe in their projections, that raises questions about the government's credibility and reputation and about the expectations that individuals will form about the future. When forecasts are systematically biased, either the policymakers are not competent or they are not sincere. The second interpretation appears more plausible: governments try to gain popularity by making sanguine forecasts that underestimate the seriousness of the problems and overestimate policymakers' ability to deal with them. This behavior is bound to create significant credibility problems.

Optimistic forecasts lead to insufficient policies and vice versa, and, of course, if people come to believe that the forecasts are biased, the government's credibility and individuals' faith in the effectiveness of its policies will be weakened.[3] The result will be a magnified negative impact on the effectiveness of policies. Governments have to be consistent; they cannot forecast a rosier future and at the same time demand "blood, sweat, and tears" from their citizens.[4] We find here a common cause of inadequate macroeconomic policies. Careful research on the connection between rosy forecasts and insufficient policies would be useful. I am not aware of any study that has specifically analyzed this connection.

Good economic policies, however, require more than unbiased forecasts. They require competence, honesty, willingness to acknowledge past mistakes, and precise, down-to-earth explanations of the problems and the proposed solutions. It would be refreshing and salutary if a president or prime minister were to face the citizens and state clearly: ''Ladies and gentlemen, we goofed; we goofed badly. We take full blame, and we will take corrective action. The medicine will hurt a lot, but it is necessary. We will make sure that the burden of adjustment is shared equitably. I will explain now, and I promise to explain in the future, exactly what we are going to do and why.''[5] People will understand that sacrifices are necessary if policymakers can explain to them clearly and simply why the sacrifices must be made and can convince them that the burden of adjustment will be fairly shared.[6] Popular support will be necessary to override the opposition of powerful groups that will try to shift onto others the burden of the adjustment.

The problem of credibility may persist even after such an explanation. A government that in the past has not been sincere about the real macroeconomic situation and has made major mistakes will not have a good reputation. As we learn from Aesop's fable, the boy who cried wolf too many times when there was no wolf was not believed when he finally told the truth. Governments may learn from past mistakes, but if they are slow learners, by the time they have learned they may have lost much of their credibility. As a consequence, policies that require credibility to be effective will lose their effectiveness. It will thus be necessary to rely on alternative policies that require less credibility to be effective. These alternative policies may be less than optimal, raising the cost of the adjustment.

For example, it may be necessary to shift up front the policy package contained in the adjustment program rather than follow an optimal sequencing path. It may also be necessary to push through some measures before they have been fully prepared. This may be interpreted as a signal of firm commitment. When private agents are uncertain as to whether the government is really committed to the reform, the government can try to signal its intention by paying immediately a cost that only a government with good intentions would be willing to bear. As a consequence of these changes, the cost of adjustment will rise.[7]

The above discussion may explain why the ability of a new government to solve the country's problems is likely to decline with the passing of time, even though the government acquires more information and experience with time.[8] Beginning with the more costly measures can be an effective signal that might increase credibility, have a favorable effect on expectations, and thus raise the probability of success. But those policies must be seen as durable, and major technical mistakes must be avoided. A government that waits too long to face the macroeconomic situation realistically will have less chance of success when it finally decides to take action.

It might be argued that a good reputation is not sufficient to generate credibility. At least two other conditions are required: (a) the policy proposed should be internally consistent—that is, it should pursue nonconflicting goals with adequate instruments, and (b) to avoid the "time consistency problem," future incentives and constraints should be so designed that the government will not find it in its interest to deviate from the preannounced course of action. Moreover, in particular cases—as, for example, when a new government takes office—a good reputation may not even be a necessary condition for credibility. Because private agents have not yet had any experience with the government on which to base their assessments, the characteristics of the policy itself, or the way it is announced, will be crucial.

## The Control Issue

One reason why governments occasionally pursue economic policies that are unsustainable over the longer run is that in too many instances they come to power with little experience and with unrealistic commitments. Often, this course of action leads to misguided policies or makes the new policymakers abstain from quickly introducing policies that would solve the existing problems. But let us assume that the government has credibility or has somehow regained it. In this situation the problems encountered in the pursuit of sound macroeconomic policies may be of a different nature. Here I shall focus on the extent to which the policymakers who are responsible for economic policy control (a) the decisionmaking process and (b) the instruments of economic policy.

The issue of control over the decisionmaking process and the instruments of economic policy has attracted little attention on the part of mainstream economists. Perhaps the formal or mathematical approach that dominates modern macroeconomics has distracted us from what is, admittedly, a messy area. Formal relationships that simply establish a stable and, often, single-valued relationship between a policy instrument and a policy objective ignore the issues I wish to discuss here.[9] An example is an expression such as $R = f(t)$ that implies a direct and single-valued relationship between tax revenue, $R$, and a statutory tax rate, $t$.

In this necessarily brief discussion the focus will be on fiscal instruments, but similar issues may arise in connection with other instruments. The discussion will be divided into two parts: control over policymaking and control over policy instruments.

### Control over Policymaking

It is convenient to start by focusing on the nerve centers where the basic decisions in macroeconomic policy are or should be made. In most countries these are the office of the minister of finance or economics and the office of the governor of the central bank. In countries where the central bank is not independent, the former plays the leading role.

The minister of finance and the governor of the central bank should have in mind the public interest (however defined) when they formulate their economic policies.[10] This, of course, raises questions about the individuals in charge—questions that relate to their economic sophistication, their biases, their honesty, their ability to withstand pressures from organized interest groups, their political ambitions (which may influence their short-run actions), and so forth.

In discussions of economic policy, economists take for granted that the policymakers have the technical background to understand the finer points of economics and the wisdom and independence to apply that sophistication to the solution of current economic problems. However, some of these individuals have no background, or only a limited one, in economics. Thus their ability to discriminate between good and bad economic advice may not be as great as is generally assumed.[11] This helps explain why at times the people in charge of economic policy seem to pay undue attention to advice that good economists would consider obviously poor.[12] The technical ability of the minister of finance and his immediate advisers can determine the extent to which *technical* errors are made.

But assume, for the sake of argument, that the minister of finance is competent enough to choose good economic policy, or at least that he has able advisers who help him sort out the good policies.[13] Fiscal decisions (budgetary cuts and, to a lesser extent, tax increases) must often be sold to the rest of the cabinet before they become official government policy. Here we face another common difficulty. While the minister of finance has (or at least should have) the *public* interest in mind, the other ministers will usually have more parochial or *particular* interests which it is, essentially, their job to promote or represent. It is therefore natural for them to assess the proposed policies from their own angle: will the policies help the interests of the particular groups they represent? For these ministers, fiscal retrenchment is a negative-sum game. To the extent possible, each will try to shift the burden of deficit reduction onto some other part of the public sector, even when they agree that a reduction is desirable.[14] The minister of defense will be interested in protecting defense spending; the minister of planning, in protecting public investment; the minister of education, in protecting educational expenditure; and so forth.

A minister who is forced by his responsibilities to play Scrooge among ministers who want to play Santa Claus is not going to be very popular. And unless he gets strong support from his superiors, he will not have much political power. At times his recommendations will be distorted and emasculated before they become government policy. At other times he will be voted down, or his budgetary instructions may be ignored. This is one reason why budgetary overruns on the part of some ministries are so common.[15] In many countries some of the other ministers are politically more powerful than the minister of finance and thus have the clout to neutralize or even ignore his decisions. (The same reasoning points to the importance of a politically independent central bank.)

To get leverage over other ministers, the minister of finance will have to sell his proposed policies to the people above him—the president or the prime minister; indeed, he will need their full support. But presidents and prime ministers, who are rarely economists, often take a political rather than an economic view of policymaking. They worry more about the next six months than about the medium run, and in the short run political and economic objectives may appear to conflict. Furthermore, the voice of the minister of finance will be just one among many. Other ministers will try to get the ear of their superiors in order to obtain backing for their own positions.

Unlike decisions concerning the exchange rate or monetary developments, which usually do not require legislative approval, many fiscal decisions must be approved by the legislature. A legislature may simply refuse to go along with the policies that the government proposes.[16] Its members represent special interests or at least certain geographic areas, and they often see their role as protecting those interests, even though in the abstract they may share broad governmental objectives such as reduction of the fiscal deficit. When the proposed policies conflict with their particular interests, they will try to prevent the policies from being enacted or will try to water them down. This explains why it is so difficult to enact major tax reforms in countries with powerful and independent legislatures. Unfortunately, fiscal decisions are always decomposable into many subdivisions related to tax rates, kinds of tax, specific spending cuts, and so forth, and it is therefore impossible to avoid friction among groups. Even when there is no legislature to deal with, the difficulties outlined above are still significant; they just take different forms.

### Control over Policy Instruments

Up to now I have discussed problems encountered in the *formulation* of policy, which exist in all countries, industrial and developing. This section deals with issues of policy *implementation* that are particularly relevant to developing countries. Economists often suffer from what could be called the "rich country syndrome." To use an analogy, they generally assume that if an architect draws a sketch of a house, that house can be built to specifications and without particular problems as long as the financing is there. In other words, they take for granted that all the required skills and materials exist and are available and that the architectural directions will be followed faithfully. Applying the analogy to policy change, the assumption is that when an economist sketches a proposed tax reform and the proposal is accepted by the government and becomes law, the actual or effective tax system (as distinct from the statutory one) will be reformed accordingly. However, this assumption is often unrealistic. Between conception of a policy—the decision to enact a particular policy—and its effective implementation there are many steps, and each is a potential trap that can prevent the policy from having its full effect or can even change its effect. The final outcome is often somewhat different from the original intention. As a consequence, it becomes difficult to establish a formal relationship between a change in the policy instrument (for example, in a tax rate) and a change in some objective (say, increased tax revenue).[17]

In the implementation stage of policy reform, policy decisions must often be decomposed and must then be carried out by lower-level government employees. For decisions to be implemented at the lower levels as decided at the top, at least three conditions must be met.

First, the signals that the policymakers send downward must be very clear. The subordinates must have no doubts about exactly what the policymakers want. Making a decision is different from conveying that decision to those who have to implement it. When the signals sent are confused or conflicting or are seen as timid, they are not going to raise to action those in charge of making the policy operational. The policymakers must put their full prestige and authority on the line when they convey the policy decisions.

Second, the policymakers must have the power to force those immediately in charge of the various operational departments to push through the decisions. There have been cases in which the minister of finance, intent on reducing the fiscal deficit, was unable to get the cooperation of the director of taxation or the director of the budget and could not remove those individuals from their important jobs.

Third, the incentive structure for those who must implement the policy decisions must be such that they do not have a strong interest in sabotaging the decisions. Powerful, entrenched bureaucracies often have enough power to determine the success or failure of the policies chosen by the political leaders. For example, freezes on wages may be circumvented by faster promotions, and reductions in spending may be defeated by accumulation of arrears.

Tax reform provides examples of some of these problems. It is frustrating to see simple tax proposals become distorted during the drafting stage to such an extent as to be unrecognizable. Once the laws are drafted and approved, extensive new regulations are needed; here and during the approval process, further pitfalls can be encountered. Finally, the new laws must be administered and enforced, and how they are implemented depends on the incentive structure of the tax administrators. A new law that cannot be administered is worth not much more than the paper on which it is written.

At times, the tax administration will not have the information or the means to administer the new laws. For example, in some countries lack of gasoline may prevent tax inspectors from visiting taxpayers' premises. Often the tax reform will change the power structure within the tax administration, and the losers will try to ambush the changes. And of course, the more complex the tax laws become, the more widespread evasion or avoidance will be, occasionally with the help of corrupt tax inspectors (who may become even more corrupt if their wages are too low). In some instances, politically powerful taxpayers will see to it that the new laws are not fully implemented, or the tax administration will be reluctant to go after powerful taxpayers.

What the government really controls (with the cooperation of the legislature) is the *statutory tax rates* and some *statutory definition of tax bases*. However, the *effective tax rates* and the *effective tax bases* may change in ways that are not predictable.[18] Thus an equation indicating that tax revenue, $R$, is a function of the tax rate, $t$, may not be very meaningful since, depending on the circumstances, a given change in the rate might generate a whole range of revenue outcomes. As a result, forecasting the effect of changes in tax rates on tax revenue is very difficult. Similar problems occur with respect to policies aimed at changing public spending. Because of such problems, governments find it easier to change the look of the fiscal situation than its substance.

## Conclusion

I have tried here to call attention to issues that do not lend themselves to easy formal treatment but that nonetheless are very important in determining the success or failure of macroeconomic policy and that help explain why some countries appear to have major difficulties in reducing fiscal imbalances. These issues have not so far received the attention that they deserve, and they could benefit from more formal study.

## Notes

This chapter is an edited and expanded version of remarks delivered at the roundtable session of the World Bank Conference on the Macroeconomics of the Public Sector Deficit, held June 20–21, 1991. The views expressed are strictly personal and are not official views of the International Monetary

Fund. Comments received from Marco Annunziata and Karim Nashashibi were much appreciated.

1. It is a common experience to hear policymakers admit that a situation requires policy changes (for example, a reduction in the size of the fiscal deficit) but that political realities would make it impossible for them to implement such changes. In recent years discussions of policies have often been replaced by discussions about implementation of those policies.

2. International developments should be similar to random shocks: they should help in some periods and hurt in others. Policymakers, however, rarely attribute improvements in their country's macroeconomic situation to positive external developments.

3. For example, investors will not believe that these policies will improve the situation and thus will not invest; individuals who have taken their money out of the country will not repatriate it; and so forth.

4. Of course, they could promise a rosier future *as a result of* drastic policy changes. But this would imply that the forecast is conditional on those policies.

5. Interestingly enough, after the rosy inflation prediction mentioned above proved completely wrong, the president started considering drastic policy changes and talking about "surgery without anesthesia." As a consequence, the credibility of the government improved, and so did the economic situation.

6. Governments have a tendency to try to hide or minimize the required sacrifices or, worse, to blame them on international institutions that require "demand adjustment."

7. This shift up front of the adjustment effort will mean that less efficient but more quickly implemented policies will be preferred. It may also mean that temporary policies may be chosen over more permanent policies. Thus, a strong adjustment effort based on temporary (that is, nondurable) policies may not solve the credibility problem. (On this subject, see Tanzi 1989, 1990.)

8. It is often stated that the first one hundred days are crucial for a new government.

9. That formal approach was pioneered by Jan Tinbergen for industrial countries and has been extended, without the necessary qualifications, to developing countries. Econometric models depend heavily on it. Even in industrial countries it is questionable whether the simple relationships between instruments and objectives ever exist. The recent literature on models of voting and lobbying is probably a first step in the right direction. However, the issues I raise here extend beyond issues of voting and lobbying.

10. Here is another crucial problem that raises serious doubt about the validity of many theoretical analyses of macroeconomic policy: how can we define the public interest? Society is made up of different groups of individuals with different and often conflicting interests. To reconcile these particular interests within some definition of public interest (or social welfare function), we need to make arbitrary value judgments about the relative importance of different groups. This is what the traditional approach to macroeconomics often does, failing to analyze how the strategic interaction between various groups of individuals influences the decisionmaking process in economic policy. We must recognize that in particular circumstances what

we call poor policy may simply be a policy that has given excessive weight to the interests of some groups. Thus, in some sense, it may still be a "rational" policy.

11. Of course, to the extent that the minister and the governor surround themselves with competent economists and rely on the advice of these advisers, their own lack of economic sophistication becomes less relevant.

12. It would be interesting to study whether the technical competence of policymakers has made a difference for macroeconomic adjustment. My impression is that it has. Of course, political considerations explain some of the choices made.

13. This is itself a strong assumption: economists of equivalent quality and reputation may disagree as to the appropriate policies in particular circumstances.

14. As a U.S. senator once put it: "Don't tax him, don't tax me, tax the fellow behind the tree." The argument presented here is, of course, highly relevant to legislators' reactions to proposed cuts. A current example is the way U.S. legislators who had argued over many years for cuts in defense spending reacted to proposed closings of military bases in their own districts ("When Defense Cuts Hit Home: Lawmakers Look for Creative Reasons to Keep Bases Open," *Washington Post*, March 22, 1993, p. 1).

15. The existence of effective institutions that permit the minister of finance to monitor most expenditures on a timely basis is particularly important. These institutions are often nonexistent or highly inadequate.

16. The relationship between the government in power and the majority in parliament will play an important role.

17. The greater the number of policy changes taking place, the more complex the relationship becomes.

18. They also change as a consequence of the impact of macroeconomic policies on tax bases (see Tanzi 1991, ch.8).

## References

Tanzi, Vito. 1989. "Fiscal Policy, Growth, and the Design of Stabilization Programs." In Mario I. Blejer and Ke-young Chu, eds., *Fiscal Policy, Stabilization, and Growth in Developing Countries*. Washington, D.C.: International Monetary Fund.

———. 1990. "Fiscal Issues in Adjustment Programs." *Ricerche Economiche* 2–3: 173–94.

———. 1991. *Public Finance in Developing Countries*. Aldershot, U.K.: Edward Elgar.

# Appendix
# and Indexes

# Statistical Appendix

*William Easterly and Klaus Schmidt-Hebbel*

This appendix presents annual data on public sector deficits, the monetary sector, and the financial sector for a large and varying sample of member countries of the Organization for Economic Coopera-tion and Development (OECD) and developing countries for periods of varying length, from 1965–90 to 1980–89. A brief description of the data methodology for each table follows.

*Public sector surplus or deficit (table A.1)*. Coverage of the public sector in OECD countries is for the general government. Coverage for devel-oping countries may be for the central government, general govern-ment, or the consolidated nonfinancial public sector. Sources: For OECD countries, OECD, *OECD Economic Outlook* (various issues). For developing countries, IMF, *International Financial Statistics* (various issues), ECLAC data, and World Bank data, except for the following: Brazil 1979–86, from Polak (1989); Colombia 1967–83, from García and Guterman (1988); Dominican Republic 1970–79, from Greene and Roe (1989); Indonesia 1980–84, from Reisen and Trotsenburg (1988); Korea 1976–79, from Tanzi (1985); Korea 1980–84, from Reisen and Trotsen-burg (1988); Mexico 1979–86, from Polak (1989); Peru 1979–86, from Polak (1989).

*Seigniorage (table A.2)*. Annual seigniorage is calculated as the sum of monthly increases in the constant-price monetary base as a ratio to constant-price annual GDP. The monthly increase in constant-price monetary base is defined as the monthly change in monetary base divided by the monthly consumer price index (CPI). Hence annual seigniorage ($S_T$) is:

$$S_T = \left[ \sum_{t=1}^{12} (M_t - M_{t-1})/CPI_t \right] \bigg/ GDP_T$$

where $M_t$ is current-price money base at the end of month $t$, $CPI_t$ is the consumer price index in month $t$, and $GDP_T$ is constant-price gross domestic product in year $T$. Source: IMF, *International Financial Statis-tics* (various issues).

*Inflation (table A.3)*. Inflation is defined as the annual percentage rate of change between annual average CPIs for the preceding and the current year. Source: IMF, *International Financial Statistics* (various issues).

*Real interest rate (table A.4)*. The annualized real interest rate ($r_t$) is calculated as the geometric mean of ex-post monthly real interest rates and is defined as:

$$r_t = [(1 + i_t)/(1 + \pi_t)] - 1$$

where $i_t$ is the annualized monthly nominal interest rate and $\pi_t$ is the annualized monthly CPI inflation rate. The nominal interest rate is the average deposit (or passive) rate in the banking system for maturities that vary between 30 days and 18 months. Sources: For OECD countries, IMF, *International Financial Statistics* (various issues), except Turkey 1966–85, from World Bank data. For developing countries, sources are as follows: for Argentina, Brazil, Chile, Côte d'Ivoire, Ecuador, India, Mexico, Pakistan, Peru, Zaire, and Zambia for the years 1965–84, World Bank data; for Jamaica, Malaysia, Morocco, Philippines, Sri Lanka, and Venezuela for the years 1965–85, for Nigeria for the years 1966–84, for Korea, Malawi, Sierra Leone, and Thailand for the years 1966–85, for Indonesia for the years 1969–84, and for Colombia for the years 1970–88, World Bank data; for Argentina 1985–88, Brazil 1985–86, Chile 1987–87, Indonesia 1985–89, and Morocco 1986–88, Easterly (1989); for Zaire 1985–87 and India 1985, de Melo (1988); for Zimbabwe 1970–88, Central Bank of Zimbabwe; for Ghana 1967–88, Mexico 1985–89, and Zimbabwe 1979–88, country case studies listed in the references to chapter 1; for all other figures, IMF data.

*Taxes from financial repression in ten countries (table A.5)*. This table presents estimates from various sources for the implicit tax revenue from financial repression, using methodologies specified in the tables. Sources: OECD data and IMF, *International Financial Statistics* (various issues).

## References

Chamley, Christophe, and Patrick Honohan. 1990. "Taxation of Financial Intermediation." Policy Research Working Paper 421. World Bank, Country Economics Department, Washington, D.C.

De Melo, Jaime. 1988. "Computable General Equilibrium Models for Trade Policy Analysis in Developing Countries: A Survey." World Bank, Country Economics Department, Washington, D.C.

Easterly, William. 1989. "Fiscal Adjustment and Deficit Financing during the Debt Crisis." In Ishrat Husain and Ishac Diwan, eds., *Dealing with the Debt Crisis*. A World Bank Symposium. Washington, D.C.

García, Jorge, and Lia Guterman. 1988. "Medición del Déficit del Sector Público Colombiano y su Financiación: 1950–1986." *Ensayos de Política Económica* 14 (December).

Green, Duty D., and Terry L. Roe. 1989. *Trade, Exchange Rate, and Agricultural Pricing Policies in the Dominican Republic.* World Bank Comparative Studies. Washington, D.C.

IMF (International Monetary Fund). Various issues. *International Financial Statistics.* Washington, D.C.

OECD (Organization for Economic Cooperation and Development). Various issues. *OECD Economic Outlook.* Paris.

Polak, Jacques J. 1989. *Financial Policies and Development.* Development Centre Studies. Organization for Economic Cooperation and Development, Paris.

Reisen, Helmut, and Axel van Trotsenburg. 1988. *Developing Country Debt: The Budgetary and Transfer Problem.* Development Centre Studies. Organization for Economic Cooperation and Development, Paris.

Tanzi, Vito. 1985. "Fiscal Management and External Debt Problems." In Hassanali Mehran, ed., *External Debt Management.* Washington D.C.: International Monetary Fund.

**Table A.1. Consolidated Public Sector Surplus (+) or Deficit (−), 1970–90**

(percentage of GDP, except as specified)

| Country | 1970 | 1971 | 1972 | 1973 | 1974 | 1975 | 1976 | 1977 | 1978 | 1979 | 1980 | 1981 |
|---|---|---|---|---|---|---|---|---|---|---|---|---|
| *OECD countries* | | | | | | | | | | | | |
| Australia | 2.8 | 2.3 | 2.1 | -0.2 | 2.3 | -0.6 | -2.9 | -0.8 | -2.8 | -2.6 | -1.9 | -1.2 |
| Austria | 1.2 | 1.5 | 2.0 | 1.3 | 1.3 | -2.5 | -3.7 | -2.4 | -2.8 | -2.4 | -1.7 | -1.8 |
| Belgium | -2.1 | -3.0 | -4.2 | -3.8 | -2.9 | -5.3 | -6.0 | -6.3 | -6.7 | -7.5 | -9.2 | -13.1 |
| Canada | 0.8 | 0.0 | 0.0 | 0.9 | 1.9 | -2.5 | -1.8 | -2.5 | -3.1 | -2.0 | -2.8 | -1.5 |
| Denmark | 3.2 | 3.9 | 3.9 | 5.2 | 3.1 | -1.4 | -0.3 | -0.6 | -0.4 | -1.7 | -3.3 | -6.9 |
| Finland | 4.3 | 4.5 | 3.9 | 5.7 | 4.6 | 2.7 | 4.9 | 3.2 | 1.4 | 0.4 | 0.3 | 1.2 |
| France | 1.1 | 0.8 | 0.8 | 0.8 | 0.1 | -2.2 | -0.6 | -0.8 | -2.1 | -0.8 | 0.0 | -1.9 |
| Germany, Fed. Rep. | 0.2 | -0.2 | -0.5 | 1.2 | -1.3 | -5.6 | -3.4 | -2.4 | -2.4 | -2.6 | -2.9 | -3.7 |
| Greece | -0.1 | -0.9 | -0.3 | -1.4 | -2.2 | -3.4 | -2.6 | -2.1 | -1.7 | -2.5 | -2.9 | -10.9 |
| Ireland | -3.6 | -3.5 | -3.2 | -3.8 | -6.9 | -11.1 | -7.4 | -6.6 | -8.6 | -11.0 | -12.2 | -13.3 |
| Italy | -4.0 | -5.9 | -8.6 | -7.9 | -7.8 | -12.9 | -9.8 | -8.6 | -10.4 | -10.2 | -8.6 | -11.6 |
| Japan[a] | 1.7 | 1.2 | -0.1 | 0.5 | 0.4 | -2.8 | -3.7 | -3.8 | -5.5 | -4.7 | -4.4 | -3.8 |
| Netherlands | -1.1 | -1.0 | -0.4 | 0.6 | -0.5 | -3.0 | -2.7 | -1.8 | -2.8 | -3.7 | -4.1 | -5.5 |
| Norway | 3.2 | 4.3 | 4.5 | 5.7 | 4.7 | 3.3 | 2.5 | 1.2 | -0.1 | 1.3 | 5.7 | 4.7 |
| Spain | 0.0 | -1.0 | -0.1 | 0.8 | -0.4 | -0.5 | -1.1 | -1.4 | -2.4 | -2.2 | -2.6 | -3.9 |
| Sweden | 4.6 | 5.3 | 4.4 | 4.1 | 2.0 | 2.8 | 4.7 | 1.7 | -0.5 | -3.0 | -4.0 | -5.3 |
| United Kingdom | 2.9 | 1.3 | -1.3 | -2.7 | -3.9 | -4.6 | -5.0 | -3.4 | -4.4 | -3.3 | -3.4 | -2.6 |
| United States | -1.1 | -1.8 | -0.3 | 0.5 | -0.3 | -4.1 | -2.2 | -1.0 | 0.0 | 0.5 | -1.3 | -1.0 |
| Average OECD | 0.8 | 0.4 | 0.1 | 0.4 | -0.3 | -3.0 | -2.3 | -2.1 | -3.1 | -3.2 | -3.3 | -4.6 |
| *Developing countries* | | | | | | | | | | | | |
| Argentina | -1.9 | -4.4 | -5.8 | -7.5 | 8.1 | -15.1 | -11.7 | -5.1 | -6.8 | -6.1 | -7.5 | -13.3 |
| Bangladesh | — | — | — | — | — | — | — | — | — | — | -14.5 | -11.8 |
| Bolivia | — | — | — | — | — | — | — | — | — | -8.4 | -9.0 | -7.8 |
| Brazil | — | — | — | — | — | — | — | — | — | — | -1.2 | -3.4 |
| Bulgaria | — | — | — | — | — | — | — | — | — | — | — | — |
| Burkina Faso | — | — | — | — | — | — | — | — | — | — | — | — |
| Chile | -2.1 | -7.5 | -8.1 | -19.0 | -5.5 | -2.1 | 4.0 | 0.4 | 1.4 | 4.6 | 5.4 | 0.4 |
| Colombia | -6.9 | -7.4 | -6.5 | -7.1 | -0.9 | 0.9 | -1.9 | -2.7 | -1.2 | -4.0 | -5.8 | -6.8 |

| | | | | | | | | | | | | |
|---|---|---|---|---|---|---|---|---|---|---|---|---|
| Côte d'Ivoire | −3.4 | −4.5 | −1.3 | 0.0 | −0.9 | −2.3 | −12.4 | −3.6 | −8.4 | −10.3 | −12.2 | −11.8 |
| Dominica | — | — | — | — | — | — | — | — | — | — | — | — |
| Dominican Rep. | −0.3 | 0.8 | 1.4 | −0.4 | −2.6 | 1.9 | −0.7 | −1.3 | −5.4 | −5.1 | −6.5 | −5.8 |
| Ecuador | −4.4 | −4.3 | −1.9 | 3.1 | 0.8 | −2.2 | −3.3 | −8.3 | −6.2 | −2.0 | −4.6 | −5.6 |
| Ghana | 1.3 | −2.7 | −3.5 | −3.2 | −5.1 | −13.2 | −9.2 | −7.5 | −7.1 | −4.0 | −6.0 | −7.4 |
| Honduras | −8.5 | −8.7 | −8.5 | −6.1 | −6.9 | −8.3 | −8.3 | −5.6 | −6.5 | −7.5 | −8.5 | −8.5 |
| India | −3.4 | −3.7 | −4.6 | −2.9 | −3.5 | −4.6 | −5.6 | −5.4 | −3.7 | −5.0 | −6.4 | −7.3 |
| Indonesia | −3.0 | −2.5 | −2.4 | −2.3 | −1.5 | −3.5 | −4.2 | 1.0 | 0.4 | 1.3 | 2.6 | 0.1 |
| Jamaica | −2.5 | −3.7 | −4.1 | −5.3 | −7.8 | −7.9 | −15.5 | −14.5 | −16.7 | −15.3 | −20.8 | −16.6 |
| Jordan | — | — | — | — | — | — | — | — | — | — | — | — |
| Kenya | 0.0 | 0.0 | −3.8 | −5.1 | −2.8 | −4.8 | −5.9 | −4.1 | −3.8 | −7.0 | −6.3 | −9.9 |
| Korea, Rep. of | −0.8 | −0.3 | −3.8 | −0.5 | −2.2 | −2.0 | −1.4 | −1.7 | −1.3 | −1.4 | −3.2 | −4.6 |
| Malawi | — | −6.5 | −7.7 | −6.7 | −4.8 | −6.7 | −5.5 | −3.6 | −11.6 | −13.6 | −16.4 | −12.1 |
| Malaysia | — | — | −9.7 | −5.9 | −5.6 | −8.3 | −7.1 | −9.3 | −6.6 | −3.2 | −11.9 | −17.0 |
| Mexico | −3.5 | −2.2 | −4.4 | −5.4 | −5.7 | −8.4 | −7.2 | −4.9 | −5.1 | −6.7 | −8.3 | −11.6 |
| Morocco | −3.0 | −3.0 | −4.0 | −2.0 | −3.9 | −9.5 | −18.1 | −15.8 | −11.3 | −10.1 | −9.0 | −13.6 |
| Nigeria | — | — | −0.7 | −1.5 | 6.4 | −6.5 | −6.5 | −1.8 | −0.9 | −1.1 | −3.6 | −9.1 |
| Oman | — | — | — | — | — | — | — | — | — | — | — | — |
| Pakistan | — | — | −3.7 | −5.2 | −9.3 | −8.8 | −8.3 | −7.4 | −8.3 | −5.8 | −4.8 | −4.9 |
| Paraguay | −1.1 | −2.2 | −2.9 | −0.6 | 0.2 | −5.0 | −3.3 | −0.8 | 0.0 | 1.2 | 0.1 | −2.9 |
| Peru | — | −1.6 | −0.9 | −2.9 | −1.9 | −2.7 | −3.7 | −2.8 | −4.5 | −1.5 | −5.4 | −6.7 |
| Philippines | — | — | — | — | — | — | — | — | — | — | — | −7.5 |
| Poland | — | — | — | — | — | — | — | — | — | −1.9 | −10.7 | −2.5 |
| Sierra Leone | — | — | — | — | −4.4 | −10.5 | −7.9 | −6.9 | −9.3 | −11.6 | −12.8 | −9.3 |
| Sri Lanka | −6.2 | −7.1 | — | −5.0 | −3.1 | −6.3 | −8.2 | −4.4 | −12.1 | −11.8 | −18.3 | −12.4 |
| Thailand | −5.2 | −5.4 | −5.3 | −3.1 | 0.7 | −2.2 | −5.0 | −4.5 | −5.3 | −5.3 | −6.3 | −7.1 |
| Trinidad and Tobago | — | — | — | — | — | — | — | — | — | — | — | — |
| Turkey | −3.5 | −4.9 | −3.1 | −2.6 | −2.6 | −1.9 | −2.8 | −8.7 | −10.6 | −9.0 | −11.9 | −6.1 |
| Venezuela | −1.2 | 0.4 | −0.2 | 1.3 | 3.9 | 1.4 | −2.5 | −1.6 | −3.3 | 3.8 | 4.4 | 3.6 |
| Zaire | −2.2 | −7.3 | −7.0 | −9.6 | −1.8 | −11.3 | −21.9 | −11.4 | −10.9 | −5.2 | −1.9 | −7.2 |
| Zambia | — | — | −13.1 | −16.8 | 3.4 | −21.7 | −14.0 | −14.4 | −10.3 | −12.2 | −16.4 | −15.6 |
| Zimbabwe | −7.1 | −6.5 | −6.9 | −8.1 | −4.3 | −7.8 | −4.1 | −4.8 | −3.2 | −1.3 | −9.1 | −13.5 |

*(Table continues on the following page with the year 1982.)*

**Table A.1** (continued)

| Country | 1982 | 1983 | 1984 | 1985 | 1986 | 1987 | 1988 | 1989 | 1990 | Average, (1970–90, except as specified) |
|---|---|---|---|---|---|---|---|---|---|---|
| *OECD countries* | | | | | | | | | | |
| Australia | -1.0 | -4.5 | -4.0 | -3.3 | -3.0 | -1.1 | 0.7 | 1.2 | 2.3 | -0.8 |
| Austria | -3.4 | -4.0 | -2.6 | -2.5 | -3.7 | -4.3 | -3.1 | -2.7 | -1.1 | -1.8 |
| Belgium | -11.2 | -11.4 | -9.3 | -8.7 | -8.8 | -7.2 | -6.8 | -6.5 | -6.1 | -7.0 |
| Canada | -5.9 | -6.9 | -6.5 | -6.8 | -5.5 | -4.4 | -2.6 | -3.4 | -3.0 | -2.7 |
| Denmark | -9.1 | -7.2 | -4.1 | -2.0 | 3.4 | 2.5 | 0.3 | -0.4 | -0.5 | -0.6 |
| Finland | -0.6 | -1.7 | 0.4 | 0.1 | 0.8 | -1.2 | 1.4 | 2.7 | 2.5 | 2.0 |
| France | -2.8 | -3.1 | -2.8 | -2.9 | -2.7 | -1.9 | -1.8 | -1.4 | -1.2 | -1.2 |
| Germany, Fed. Rep. | -3.3 | -2.5 | -1.9 | -1.1 | -1.3 | -1.8 | -2.1 | 0.2 | -0.8 | -1.8 |
| Greece | -7.6 | -8.6 | -10.2 | -14.0 | -12.7 | -12.0 | -14.5 | -17.8 | -17.2 | -6.9 |
| Ireland | -14.1 | -12.0 | -10.1 | -11.8 | -11.6 | -9.2 | -2.6 | -2.8 | -1.1 | -7.9 |
| Italy | -11.3 | -10.7 | -11.6 | -12.5 | -11.7 | -11.1 | -10.9 | -10.2 | -10.2 | -9.8 |
| Japan[a] | -3.6 | -3.7 | -2.1 | -0.8 | -0.9 | 0.7 | 2.1 | 2.7 | 3.1 | -1.3 |
| Netherlands | -7.1 | -6.4 | -6.3 | -4.8 | -6.0 | -6.5 | -5.0 | -5.1 | -5.1 | -3.7 |
| Norway | 4.4 | 4.2 | 7.5 | 10.4 | 5.9 | 4.8 | 3.1 | 1.0 | 1.2 | 4.0 |
| Spain | -5.6 | -4.8 | -5.5 | -7.0 | -6.1 | -3.2 | -3.1 | -2.1 | -2.0 | -2.6 |
| Sweden | -7.0 | -5.0 | -2.9 | -3.9 | -1.3 | 4.2 | 3.4 | 5.3 | 4.6 | 0.7 |
| United Kingdom | -2.4 | -3.3 | -3.9 | -2.7 | -2.4 | -1.2 | 1.1 | 1.3 | 0.7 | -2.1 |
| United States | -3.5 | -3.8 | -2.8 | -3.3 | -3.4 | -2.4 | -2.0 | -2.0 | -1.3 | -1.7 |
| Average OECD | -5.3 | -5.3 | -4.4 | -4.3 | -3.9 | -3.1 | -2.4 | -2.2 | -2.0 | -2.5 |
| *Developing countries* | | | | | | | | | | |
| Argentina | -15.1 | -15.2 | -11.9 | -6.0 | -4.7 | -5.5 | -7.4 | -0.5 | 1.6 | -7.5 (1980–90) |
| Bangladesh | -13.6 | -13.0 | -10.4 | -10.5 | -9.9 | -8.4 | -7.1 | -7.2 | -8.1 | -10.4 (1979–90) |
| Bolivia | -14.7 | -19.1 | -27.4 | -9.1 | -3.4 | -7.8 | -6.7 | -4.3 | -3.3 | -9.4 (1980–90) |
| Brazil | -4.8 | -7.0 | -3.5 | -2.7 | -4.3 | -3.7 | -5.5 | -4.3 | 1.2 | -4.1 (1980–90) |
| Bulgaria | — | — | — | -0.8 | -2.6 | 1.2 | -0.9 | -0.6 | — | -0.7 (1985–89) |
| Burkina Faso | — | — | -6.2 | -3.1 | -6.9 | -5.5 | -6.1 | 4.1 | — | -4.0 (1984–89) |
| Chile | -3.9 | -3.5 | -4.6 | -2.9 | -2.0 | -0.2 | 3.6 | 3.8 | 0.5 | -1.8 |
| Colombia | -8.9 | -8.5 | -6.3 | -3.5 | -0.3 | -2.0 | -2.1 | -1.8 | -2.2 | -4.1 |

| Country | | | | | | | | | | Average | Period [a] |
|---|---|---|---|---|---|---|---|---|---|---|---|---|
| Côte d'Ivoire | -15.9 | -11.4 | -1.7 | 2.0 | -2.4 | -7.3 | -13.5 | -14.4 | — | -6.8 | (1970-89) |
| Dominica | — | — | -3.9 | 1.4 | 3.2 | 3.5 | 2.3 | -8.8 | — | -0.4 | (1984-89) |
| Dominican Rep. | -7.0 | -5.3 | -6.7 | -0.9 | -5.6 | -3.8 | -5.6 | — | — | -3.1 | (1970-88) |
| Ecuador | -6.7 | 0.0 | -0.6 | 1.9 | -5.1 | -9.6 | -5.1 | -1.6 | 0.9 | -3.6 | |
| Ghana | -4.1 | -2.3 | -0.5 | -1.9 | 2.2 | 3.4 | 2.2 | — | -6.3 | -8.1 | (1970-88) |
| Honduras | -12.5 | -12.7 | -11.4 | -8.2 | -6.3 | -6.1 | -6.7 | -7.2 | -7.0 | -6.2 | |
| India | -7.0 | -7.5 | -7.7 | -9.1 | -8.6 | -8.6 | -8.9 | -9.3 | -0.5 | -1.6 | |
| Indonesia | -4.1 | -2.6 | 0.9 | -2.9 | -4.8 | -2.0 | -0.7 | -2.7 | -3.5 | -10.5 | |
| Jamaica | -15.3 | -13.6 | -15.1 | -13.2 | -5.6 | -5.4 | -13.4 | -4.3 | -14.2 | -9.6 | (1983-90) |
| Jordan | — | -4.7 | -8.5 | -8.7 | -4.7 | -14.3 | -15.3 | -6.6 | -5.5 | -4.8 | |
| Kenya | -6.9 | -3.9 | -4.4 | -4.9 | -5.3 | -6.6 | -4.3 | -4.7 | -0.7 | -1.4 | |
| Korea, Rep. of | -4.3 | -1.6 | -1.4 | -1.2 | -0.1 | 0.9 | -1.6 | 0.2 | — | -7.0 | (1971-89) |
| Malawi | -8.7 | -7.0 | -4.9 | -9.6 | -6.6 | 1.2 | -0.4 | -1.8 | -6.1 | -8.6 | (1972-90) |
| Malaysia | -17.9 | -14.8 | -11.1 | -5.9 | -10.3 | -5.6 | -2.7 | -4.2 | -3.9 | -6.4 | |
| Mexico | -15.4 | -6.0 | -2.9 | -3.3 | -7.0 | -5.4 | -10.7 | -5.9 | — | -8.3 | (1970-88) |
| Morocco | -9.2 | -11.5 | -8.1 | -8.6 | -5.7 | -6.1 | -5.5 | — | -6.1 | -4.4 | (1972-90) |
| Nigeria | -8.4 | -10.1 | -4.2 | -1.8 | -2.8 | -8.7 | -10.8 | -8.1 | — | -3.9 | (1983-89) |
| Oman | — | -0.2 | -3.2 | -2.1 | -16.8 | 1.9 | -7.4 | 0.2 | -6.1 | -6.8 | (1972-90) |
| Pakistan | -6.4 | -5.5 | -7.1 | -7.1 | -8.3 | -7.4 | -7.8 | -6.4 | 1.2 | -2.1 | |
| Paraguay | -2.7 | -5.3 | -6.4 | -2.6 | -1.8 | -2.9 | -3.1 | 0.6 | -8.1 | -5.3 | |
| Peru | -7.2 | -9.5 | -6.8 | -3.7 | -6.2 | -7.5 | -8.1 | -13.6 | -3.5 | -4.3 | (1972-90) |
| Philippines | -6.5 | -4.6 | -3.3 | -2.1 | -4.8 | -2.6 | -3.0 | -4.3 | — | -2.8 | (1981-89) |
| Poland | -0.9 | -0.5 | 0.0 | -0.3 | -0.8 | -0.9 | -9.0 | — | -2.9 | -8.4 | (1980-89) |
| Sierra Leone | -10.5 | -13.7 | -7.1 | -10.9 | -10.8 | -3.8 | -4.0 | -5.9 | -7.7 | -9.3 | (1974-90) |
| Sri Lanka | -14.0 | -10.6 | -6.8 | -9.7 | -10.1 | -8.7 | -12.6 | -10.8 | 1.2 | -4.3 | |
| Thailand | -7.6 | -5.7 | -7.3 | -8.6 | -6.6 | -3.2 | 0.2 | 0.8 | -4.8 | -7.4 | |
| Trinidad and Tobago | — | — | — | — | -7.4 | -10.9 | -7.4 | -6.4 | -3.7 | -5.3 | (1985-89) |
| Turkey | -6.1 | -6.6 | -9.2 | -5.6 | -4.3 | -4.4 | -3.9 | -3.8 | 0.2 | -3.8 | |
| Venezuela | -5.6 | 0.4 | 8.6 | 5.0 | -7.5 | -5.4 | -9.4 | -1.3 | — | 0.6 | |
| Zaire | -10.5 | -4.1 | -0.7 | 1.2 | -1.7 | -3.1 | -4.6 | 1.2 | -8.2 | -6.0 | (1970-89) |
| Zambia | -16.8 | -8.9 | -9.3 | -19.8 | -28.5 | -10.8 | -12.1 | -9.4 | -9.8 | -13.4 | (1972-90) |
| Zimbabwe | -13.1 | -14.4 | -12.7 | -14.4 | -14.4 | -10.9 | -10.4 | -9.9 | — | -8.9 | (1972-90) |

— Not available.

a. Percentage of GNP.

**Table A.2. Seigniorage, 1965–89**
(percentage of GDP)

| Country | 1965 | 1966 | 1967 | 1968 | 1969 | 1970 | 1971 | 1972 | 1973 | 1974 | 1975 | 1976 | 1977 |
|---|---|---|---|---|---|---|---|---|---|---|---|---|---|
| *OECD countries* | | | | | | | | | | | | | |
| Austria | 0.3 | 1.0 | 1.2 | 0.2 | 0.5 | 0.7 | 1.4 | 1.7 | 0.2 | 1.0 | 1.8 | 0.5 | 0.5 |
| Belgium | 1.1 | 0.5 | 0.3 | 0.4 | 0.0 | 0.4 | 1.0 | 1.8 | 1.6 | 0.5 | 0.7 | 0.7 | 0.9 |
| Canada | 0.6 | 0.5 | 0.3 | 0.5 | 0.3 | 0.3 | 0.8 | 0.9 | 0.9 | 0.7 | 0.8 | 0.5 | 0.6 |
| Denmark | — | — | 0.4 | 1.7 | 0.4 | -0.6 | 0.2 | 0.2 | 0.1 | 0.1 | 1.6 | 0.2 | -0.6 |
| Finland | 0.0 | 0.4 | 0.4 | 0.3 | -0.1 | 0.1 | 0.3 | 0.4 | 0.3 | 0.4 | 0.4 | 0.0 | 0.2 |
| France | 0.9 | 0.6 | 1.0 | 0.9 | -0.1 | 0.5 | 1.0 | 3.0 | 0.8 | 1.4 | -2.3 | 0.5 | 0.6 |
| Germany, Fed. Rep. | 0.6 | 0.8 | -0.5 | 0.9 | 0.3 | 1.8 | 1.4 | 2.5 | 0.8 | -0.1 | 0.3 | 1.0 | 0.8 |
| Greece | 1.5 | 2.6 | 1.8 | 1.9 | 0.6 | 1.0 | 2.9 | 3.0 | 2.8 | 3.1 | 3.1 | 2.9 | 3.2 |
| Italy | 2.2 | 1.6 | 2.0 | 1.4 | 2.0 | 1.7 | 3.0 | 3.1 | 5.2 | 3.1 | 3.1 | 2.9 | 2.4 |
| Japan | 0.7 | 0.9 | 1.3 | 1.3 | 1.3 | 1.1 | 1.1 | 2.2 | 2.6 | 1.3 | 0.3 | 0.7 | 0.7 |
| Netherlands | 1.1 | 0.8 | 0.3 | 0.2 | 0.6 | 0.5 | 0.4 | 0.7 | 0.4 | 0.5 | 0.8 | 0.6 | 0.6 |
| Norway | 0.9 | 0.8 | 0.4 | 0.7 | 0.7 | 0.3 | 1.0 | 0.4 | 0.5 | 1.1 | 1.3 | 1.0 | 0.9 |
| Spain | 1.5 | 1.4 | 1.2 | 1.1 | 1.2 | 1.3 | 2.6 | 1.8 | 2.1 | 2.3 | 1.9 | 1.5 | 2.0 |
| Sweden | 0.3 | 0.4 | 0.5 | 0.5 | 0.5 | 0.2 | 0.8 | 0.6 | 0.6 | 2.3 | -0.1 | 0.7 | 0.6 |
| Turkey | — | — | — | — | 2.0 | 2.3 | 3.4 | 4.1 | 3.2 | 3.1 | 3.5 | 2.5 | 5.0 |
| United States | 0.5 | 0.5 | 0.4 | 0.6 | 0.3 | 0.5 | 0.8 | 0.1 | 0.7 | 0.5 | 0.4 | 0.4 | 0.4 |
| *Developing countries* | | | | | | | | | | | | | |
| Argentina | 2.6 | 4.1 | 0.4 | 3.5 | 2.7 | 1.9 | 5.6 | 2.9 | 5.0 | 6.5 | 13.2 | 4.9 | 1.1 |
| Bangladesh | — | — | — | — | — | — | — | — | — | — | 0.0 | 0.1 | 1.4 |
| Bolivia | 1.7 | 2.3 | 0.5 | 0.6 | 1.1 | 1.2 | 1.9 | 1.6 | 2.9 | 2.0 | 1.4 | 3.5 | 2.4 |
| Brazil | 4.7 | 2.1 | 1.9 | 3.1 | 2.0 | 1.4 | 2.3 | 0.5 | 2.2 | 1.9 | 1.8 | 2.0 | 2.1 |
| Burkina Faso | 0.3 | 0.4 | 0.0 | 0.3 | 1.1 | 0.2 | -0.1 | 0.5 | 1.1 | 1.1 | 2.1 | 1.0 | 1.3 |
| Chile | — | — | — | — | — | 3.3 | 10.5 | — | — | — | — | — | — |
| Colombia | — | — | — | — | — | 1.4 | 0.9 | 1.7 | 2.2 | 1.5 | 1.7 | 2.7 | 2.9 |

| | | | | | | | | | | | | | |
|---|---|---|---|---|---|---|---|---|---|---|---|---|---|
| Côte d'Ivoire | 0.2 | 1.4 | 0.8 | 0.7 | 1.2 | 2.0 | 1.2 | 1.4 | 1.0 | 3.3 | 1.4 | 1.8 | 3.3 |
| Dominican Rep. | 2.7 | −0.9 | 0.0 | 1.8 | 1.7 | 1.3 | 1.6 | 0.9 | 2.4 | 4.5 | −0.9 | −0.1 | 2.3 |
| Ecuador | 0.3 | 1.1 | 0.5 | 1.8 | 1.7 | 3.1 | 1.2 | 3.1 | 4.1 | 3.9 | −0.1 | 2.0 | — |
| Ghana | — | — | — | 0.3 | 1.6 | 1.3 | 0.2 | 4.1 | 2.4 | 3.3 | 5.0 | 5.6 | 6.7 |
| Honduras | 0.7 | 0.2 | 0.5 | 0.7 | 1.0 | 0.8 | 0.6 | 0.2 | 1.3 | −0.2 | 0.8 | 2.3 | 1.2 |
| India | 1.0 | 0.8 | 0.7 | 0.6 | 1.2 | 1.1 | 1.2 | 1.0 | 2.2 | 0.5 | 0.5 | 1.6 | 2.1 |
| Indonesia | — | — | — | — | — | 1.3 | 1.4 | 3.1 | 2.3 | 2.9 | 2.3 | 1.6 | 1.8 |
| Jamaica | — | — | — | — | — | 1.0 | 1.7 | 0.8 | 1.6 | 1.2 | 1.2 | 1.1 | 1.3 |
| Jordan | — | — | — | — | — | — | — | — | — | — | — | 8.0 | 6.2 |
| Kenya | — | — | — | 1.7 | 3.2 | 1.1 | −0.9 | 0.8 | 1.0 | 1.9 | −0.9 | 1.3 | 3.6 |
| Korea, Rep. of | — | — | — | — | — | 3.1 | −0.4 | 3.3 | 3.6 | 1.9 | 2.8 | 2.5 | 3.4 |
| Malawi | — | — | — | — | — | — | — | — | — | — | — | — | — |
| Malaysia | 0.8 | 1.1 | −1.4 | 0.6 | 1.6 | 1.0 | 0.8 | 3.1 | 3.5 | 1.6 | 0.3 | 2.0 | 1.7 |
| Mexico | 0.3 | 0.6 | 0.8 | 1.0 | 0.7 | 0.4 | 0.8 | 4.2 | 2.4 | 3.2 | 3.2 | −0.8 | 8.5 |
| Morocco | 1.5 | −0.1 | 1.5 | 1.6 | 1.6 | 1.0 | 1.3 | 2.3 | 1.9 | 2.3 | 1.9 | 2.8 | 2.0 |
| Nigeria | 0.1 | 0.6 | −0.4 | −0.7 | 1.8 | 2.0 | 0.3 | 0.4 | 0.9 | 4.4 | 3.6 | 2.2 | 1.9 |
| Pakistan | 2.3 | 2.3 | 0.1 | 0.9 | 1.7 | 2.0 | 2.1 | 3.4 | 2.1 | 0.0 | 1.2 | 2.3 | 2.8 |
| Paraguay | 1.1 | 0.3 | 3.1 | −0.9 | 1.1 | 1.7 | 1.3 | 2.2 | 2.6 | 1.5 | — | 2.1 | 3.1 |
| Peru | 1.4 | 0.5 | 0.9 | 0.5 | 0.7 | 3.3 | 0.5 | 2.7 | 0.6 | 3.2 | 0.5 | 2.9 | 1.6 |
| Philippines | — | — | — | — | 1.0 | 0.5 | 0.6 | 1.6 | 0.6 | 1.2 | 0.7 | 0.6 | 1.2 |
| Sri Lanka | 0.9 | −0.8 | 1.5 | 1.0 | 0.2 | 0.3 | 1.4 | 1.8 | 1.7 | 0.7 | −0.6 | 1.9 | 3.5 |
| Thailand | 1.2 | 1.3 | 0.8 | 0.9 | 0.6 | 0.9 | 1.1 | 1.7 | 1.5 | 1.2 | 1.0 | 1.1 | 0.8 |
| Trinidad and Tobago | −0.4 | 0.9 | 0.0 | 1.5 | −0.3 | 0.6 | 1.6 | 0.6 | 0.8 | 3.4 | 3.2 | 2.7 | 0.2 |
| Venezuela | — | — | 0.7 | 0.9 | 0.4 | 0.2 | 1.5 | 0.9 | 1.7 | 2.4 | 3.3 | 2.0 | 2.3 |
| Zaire | — | — | — | 4.4 | 0.8 | 2.0 | −1.7 | 2.8 | 1.7 | 2.3 | 4.8 | 8.6 | 4.2 |
| Zambia | — | — | — | 1.1 | 2.3 | 3.7 | −4.1 | 1.2 | 2.0 | −0.5 | 3.4 | 2.3 | −0.2 |
| Zimbabwe | — | — | — | — | — | — | — | — | — | — | — | — | — |

(*Table continues on the following page with the year 1978.*)

**Table A.2** (continued)

| Country | 1978 | 1979 | 1980 | 1981 | 1982 | 1983 | 1984 | 1985 | 1986 | 1987 | 1988 | 1989 |
|---|---|---|---|---|---|---|---|---|---|---|---|---|
| *OECD countries* | | | | | | | | | | | | |
| Austria | 1.6 | 0.2 | 0.7 | 0.7 | 0.4 | 0.7 | 0.2 | 0.1 | 0.7 | -0.3 | 0.1 | 1.2 |
| Belgium | 0.8 | 0.3 | 0.1 | 0.2 | 0.0 | 0.3 | 0.1 | -0.1 | 0.4 | 0.2 | 0.1 | 0.2 |
| Canada | 0.6 | 0.4 | 0.5 | 0.1 | 0.2 | 0.1 | 0.1 | 0.2 | 0.3 | 0.3 | 0.2 | 0.1 |
| Denmark | 0.3 | 0.5 | 0.1 | 0.3 | 0.1 | 0.2 | 0.3 | 4.6 | -2.4 | 0.7 | 0.2 | -0.2 |
| Finland | 0.5 | 1.4 | 1.1 | -0.1 | 0.4 | 0.9 | 1.5 | 0.6 | -0.1 | 2.2 | 1.2 | 1.5 |
| France | 0.8 | 0.4 | 0.9 | 0.3 | 1.0 | 0.3 | 0.6 | 1.0 | 0.1 | 0.7 | -0.1 | -0.0 |
| Germany, Fed. Rep. | 1.3 | 0.5 | -0.4 | -0.2 | 0.5 | 0.5 | 0.3 | 0.3 | 0.6 | 0.8 | 1.0 | 0.6 |
| Greece | 3.0 | 1.5 | 3.5 | 6.4 | 4.6 | 1.6 | 4.8 | 1.1 | 2.8 | 3.7 | — | — |
| Italy | 3.4 | 1.7 | 1.7 | 1.6 | 2.0 | 2.1 | 1.9 | 2.4 | 1.1 | 1.2 | 1.3 | 1.8 |
| Japan | 1.2 | 0.6 | 0.5 | 0.2 | 0.5 | 0.5 | 0.7 | 0.3 | 0.7 | 0.8 | 1.2 | 1.3 |
| Netherlands | 0.5 | 0.5 | 0.5 | 0.1 | 0.5 | 0.8 | 0.4 | 0.3 | 0.3 | 0.7 | 0.9 | 1.3 |
| Norway | 1.0 | 0.7 | 0.6 | -0.1 | 0.4 | 0.2 | 0.9 | 0.4 | -0.1 | 0.4 | -0.1 | 0.3 |
| Spain | 2.4 | 3.6 | 1.8 | 1.5 | 2.5 | 11.7 | 0.7 | 0.8 | 1.5 | 4.9 | 1.4 | 3.1 |
| Sweden | 0.9 | 2.4 | -0.9 | 0.8 | 0.5 | 0.1 | 0.3 | 0.1 | 1.4 | 0.3 | 1.1 | 0.7 |
| Turkey | 4.6 | 4.9 | 3.4 | 3.9 | 3.2 | 3.1 | 3.6 | 2.9 | 1.9 | 2.7 | 4.2 | — |
| United States | 0.7 | 0.4 | 0.3 | 0.2 | 0.3 | 0.3 | 0.3 | 0.5 | 0.8 | 0.3 | 0.4 | 0.2 |
| *Developing countries* | | | | | | | | | | | | |
| Argentina | 3.3 | 1.4 | 6.5 | 4.4 | 7.6 | 6.2 | 5.8 | 1.5 | 3.2 | 4.9 | — | — |
| Bangladesh | 1.2 | 0.7 | 0.9 | 0.5 | 0.5 | 1.7 | 1.9 | 0.8 | 0.5 | 2.6 | 0.8 | 0.9 |
| Bolivia | 1.2 | 1.0 | 3.2 | 1.4 | 10.4 | 8.3 | 9.9 | — | — | — | — | — |
| Brazil | 1.8 | 2.9 | 1.7 | 1.7 | 1.8 | 1.8 | 2.4 | 2.2 | 4.4 | — | — | — |
| Burkina Faso | -0.3 | 1.9 | 0.6 | 1.5 | 0.7 | 3.5 | 4.0 | -1.2 | — | — | — | — |
| Chile | — | 2.4 | 2.2 | -0.6 | -1.8 | 0.5 | 0.9 | — | — | — | — | — |
| Colombia | 4.4 | 2.6 | 2.6 | 2.1 | 1.5 | — | 1.8 | — | — | — | 1.2 | — |

| | | | | | | | | | | | | |
|---|---|---|---|---|---|---|---|---|---|---|---|---|
| Côte d'Ivoire | 3.1 | 0.4 | 0.1 | 0.9 | -0.6 | 0.7 | 1.8 | 2.5 | 0.5 | — | — | — |
| Dominican Rep. | 0.8 | 0.4 | -0.4 | 1.9 | 0.2 | 1.4 | 2.1 | -0.2 | 6.5 | -0.6 | 7.2 | 2.6 |
| Ecuador | 1.5 | 2.6 | 1.7 | 0.6 | 0.6 | 1.5 | 2.2 | 1.3 | 1.9 | 2.2 | 2.9 | 2.1 |
| Ghana | 7.8 | 2.4 | 2.9 | 4.2 | 1.2 | 2.9 | 2.8 | 2.2 | 3.3 | 2.5 | 3.4 | — |
| Honduras | 1.4 | 1.0 | 0.1 | 0.4 | — | — | 0.0 | 0.4 | 0.9 | 0.6 | 1.0 | 1.9 |
| India | 2.1 | 2.4 | 1.7 | 1.2 | 2.0 | 1.4 | 1.9 | 2.7 | 2.0 | 2.4 | 2.1 | — |
| Indonesia | 0.7 | 1.8 | 2.0 | 0.9 | 0.3 | 1.3 | 0.6 | 1.1 | 1.4 | 0.7 | -0.4 | 1.4 |
| Jamaica | 0.6 | 1.3 | 3.0 | -0.5 | -0.9 | 4.3 | 7.3 | 4.3 | 2.6 | 2.8 | — | 3.6 |
| Jordan | 5.8 | 8.9 | 7.3 | 4.6 | 3.9 | 3.9 | 0.7 | 1.4 | 2.3 | 1.6 | 8.4 | 6.7 |
| Kenya | 0.6 | 1.8 | 0.1 | 0.3 | 1.6 | -0.2 | 0.7 | 1.0 | 2.4 | 1.5 | 0.1 | 1.2 |
| Korea, Rep. of | 2.9 | 2.0 | -1.1 | -1.0 | 1.8 | 0.4 | 0.2 | 0.1 | 0.7 | 2.2 | 1.7 | 2.1 |
| Malawi | — | — | 1.5 | 1.7 | 1.2 | 0.7 | 4.1 | 0.4 | 5.3 | 5.6 | 1.1 | -1.2 |
| Malaysia | 1.7 | 1.6 | 1.9 | 1.2 | 1.9 | 0.5 | 0.4 | 0.9 | 0.6 | 0.7 | 1.3 | 2.8 |
| Mexico | 3.5 | 4.2 | 4.7 | 5.4 | 11.0 | 6.2 | 5.6 | 1.5 | 3.0 | 2.7 | — | — |
| Morocco | 2.1 | 2.3 | 0.7 | 1.7 | 0.9 | 1.7 | 1.5 | 1.0 | 2.7 | 0.7 | 2.0 | 2.5 |
| Nigeria | -0.4 | 1.3 | 5.1 | -0.6 | 0.8 | 0.2 | 0.3 | 0.6 | 0.5 | 1.3 | 2.7 | 2.6 |
| Pakistan | 2.2 | 3.7 | 2.4 | 1.2 | 2.3 | 1.9 | 2.2 | 1.2 | 2.7 | 2.9 | 1.7 | 2.5 |
| Paraguay | 3.6 | 2.5 | 2.7 | 1.8 | 0.1 | 3.2 | 1.9 | 1.8 | 2.8 | 3.4 | 1.9 | — |
| Peru | 3.0 | 5.3 | 6.3 | 3.5 | 3.6 | 6.0 | 7.3 | 10.3 | 3.4 | 5.5 | 9.6 | 6.9 |
| Philippines | 1.3 | 1.2 | 0.6 | 0.5 | 0.3 | — | — | — | — | 1.0 | 1.1 | 2.7 |
| Sri Lanka | 1.2 | 1.9 | 1.7 | 1.4 | 1.9 | 2.1 | 1.3 | 2.3 | 0.3 | 1.0 | 2.4 | 0.6 |
| Thailand | 1.3 | 1.2 | 1.1 | 0.5 | 0.9 | 0.8 | 0.4 | 0.7 | 0.8 | 1.7 | 1.1 | 1.2 |
| Trinidad and Tobago | 0.4 | 4.6 | 0.6 | 1.8 | 5.2 | -0.8 | -1.1 | 0.9 | -2.4 | -2.6 | — | — |
| Venezuela | 1.0 | 0.9 | 0.5 | 1.4 | 1.7 | 3.3 | 2.7 | 0.4 | -2.3 | 1.8 | 0.3 | 2.6 |
| Zaire | 7.5 | 0.1 | 6.4 | 5.2 | 9.3 | 5.4 | 5.7 | 3.1 | 5.7 | 10.2 | — | — |
| Zambia | 0.5 | 0.4 | 1.0 | 1.3 | 1.4 | 1.2 | 1.5 | 1.4 | 8.0 | 4.0 | 10.1 | — |
| Zimbabwe | — | — | 1.9 | 1.8 | 1.2 | 0.2 | 0.6 | 1.3 | 1.0 | — | — | — |

a. For developing countries, 1970–89.

## Table A.3. Inflation Rates, 1965–90

(percent)

| Country | 1965 | 1966 | 1967 | 1968 | 1969 | 1970 | 1971 | 1972 | 1973 | 1974 | 1975 | 1976 | 1977 |
|---|---|---|---|---|---|---|---|---|---|---|---|---|---|
| *OECD countries* | | | | | | | | | | | | | |
| Australia | 4.0 | 3.0 | 3.2 | 2.7 | 2.9 | 3.9 | 6.1 | 5.9 | 9.5 | 15.1 | 15.1 | 13.5 | 12.3 |
| Austria | 4.9 | 2.2 | 4.0 | 2.8 | 3.1 | 4.4 | 4.7 | 6.4 | 7.5 | 9.5 | 8.4 | 7.3 | 5.5 |
| Belgium | 4.1 | 4.2 | 2.9 | 2.7 | 3.7 | 3.9 | 4.3 | 5.4 | 7.0 | 12.7 | 12.8 | 9.2 | 7.1 |
| Canada | 2.5 | 3.7 | 3.6 | 4.1 | 4.5 | 3.4 | 2.8 | 4.8 | 7.6 | 10.9 | 10.8 | 7.5 | 8.0 |
| Denmark | 5.5 | 7.1 | 8.2 | 8.0 | 3.5 | 6.5 | 5.9 | 6.6 | 9.3 | 15.3 | 9.6 | 9.0 | 11.1 |
| Finland | 4.8 | 3.9 | 5.6 | 9.2 | 2.2 | 2.7 | 6.5 | 7.1 | 11.0 | 16.7 | 17.8 | 14.4 | 12.7 |
| France | 2.7 | 2.6 | 2.8 | 4.6 | 6.1 | 5.9 | 5.5 | 6.2 | 7.3 | 13.7 | 11.8 | 9.6 | 9.4 |
| Germany, Fed. Rep. | 3.2 | 3.6 | 1.6 | 1.6 | 1.9 | 3.4 | 5.2 | 5.5 | 7.0 | 7.0 | 5.9 | 4.3 | 3.7 |
| Greece | 3.1 | 5.0 | 1.7 | 0.3 | 2.5 | 2.9 | 3.0 | 4.3 | 15.5 | 26.9 | 13.4 | 13.3 | 12.2 |
| Ireland | 5.0 | 3.1 | 3.2 | 4.7 | 7.4 | 8.2 | 9.0 | 8.6 | 11.4 | 17.0 | 20.9 | 18.0 | 13.6 |
| Italy | 4.2 | 2.4 | 3.9 | 1.5 | 2.2 | 5.1 | 4.9 | 6.0 | 10.6 | 19.2 | 17.1 | 16.6 | 17.0 |
| Japan | 6.6 | 5.1 | 4.0 | 5.4 | 5.2 | 7.7 | 6.3 | 4.9 | 11.6 | 23.2 | 11.8 | 9.4 | 8.2 |
| Netherlands | 5.8 | 5.8 | 3.5 | 3.7 | 7.4 | 3.7 | 7.5 | 7.8 | 8.0 | 9.6 | 10.2 | 9.1 | 6.5 |
| Norway | 4.3 | 3.3 | 4.4 | 3.5 | 3.11 | 0.6 | 6.3 | 7.2 | 7.4 | 9.4 | 11.7 | 9.2 | 9.0 |
| Portugal | 3.5 | 5.0 | 6.5 | 6.1 | 7.4 | 4.5 | 7.5 | 8.9 | 10.4 | 28.0 | 20.4 | 18.2 | 27.1 |
| Spain | 13.1 | 6.2 | 6.5 | 4.9 | 2.2 | 5.7 | 8.2 | 8.3 | 11.4 | 15.7 | 17.0 | 15.0 | 24.5 |
| Sweden | 5.0 | 6.4 | 4.3 | 1.9 | 2.7 | 7.0 | 7.4 | 6.0 | 6.7 | 9.9 | 9.8 | 10.3 | 11.5 |
| Turkey | 5.9 | 4.4 | 6.8 | 0.4 | 7.9 | 6.9 | 15.7 | 11.7 | 15.4 | 15.8 | 19.2 | 17.4 | 27.1 |
| United Kingdom | 4.8 | 3.9 | 2.4 | 4.7 | 5.5 | 6.4 | 9.4 | 7.1 | 9.2 | 15.9 | 24.2 | 16.5 | 15.9 |
| United States | 1.7 | 3.0 | 2.8 | 4.2 | 5.4 | 5.9 | 4.3 | 3.3 | 6.2 | 11.0 | 9.1 | 5.7 | 6.5 |
| *Developing countries* | | | | | | | | | | | | | |
| Argentina | 28.6 | 31.9 | 29.2 | 16.2 | 7.6 | 13.6 | 34.7 | 58.4 | 61.2 | 23.5 | 182.9 | 444.0 | 176.0 |
| Bangladesh | 7.9 | 8.0 | 5.4 | 2.7 | 5.0 | 3.0 | -0.5 | 40.7 | 49.0 | 54.8 | 21.9 | 2.3 | 4.8 |
| Bolivia | 2.9 | 7.0 | 11.2 | 5.5 | 2.2 | 3.8 | 3.7 | 6.5 | 31.5 | 62.8 | 8.0 | 4.5 | 8.1 |
| Botswana | — | — | — | — | — | — | — | — | — | — | 12.0 | 11.7 | 13.2 |
| Brazil | 65.7 | 41.3 | 30.5 | 22.0 | 22.7 | 22.4 | 20.1 | 16.6 | 12.7 | 27.6 | 29.0 | 42.0 | 43.7 |
| Burkina Faso | -0.7 | 2.4 | -4.3 | -0.3 | 9.7 | 1.8 | 2.1 | -2.9 | 7.6 | 8.7 | 18.8 | -8.4 | 30.0 |
| Burundi | — | 4.4 | -1.1 | 6.1 | 4.0 | -0.2 | 3.9 | 3.8 | 6.0 | 15.7 | 15.7 | 6.9 | 6.8 |
| Cameroon | — | — | — | — | -1.1 | 5.9 | 4.0 | 8.1 | 10.4 | 17.2 | 13.6 | 9.9 | 14.7 |

| Country | | | | | | | | | | | | | |
|---|---|---|---|---|---|---|---|---|---|---|---|---|---|
| Central African Rep. | — | — | — | — | — | — | — | — | — | — | — | — | — |
| Chad | 28.8 | 23.1 | 18.8 | 26.3 | 30.4 | 32.5 | 20.0 | 74.8 | 361.5 | 504.7 | 374.7 | 211.8 | 91.9 |
| Chile | 3.5 | 19.9 | 8.2 | 5.8 | 10.1 | 6.8 | 9.1 | 13.4 | 20.8 | 24.3 | 22.9 | 20.2 | 2.5 |
| China | 4.1 | 5.3 | 3.5 | 3.8 | -0.7 | 1.5 | 4.1 | 9.8 | 3.5 | 5.4 | 17.4 | 7.2 | 33.1 |
| Colombia | -0.7 | 0.2 | 1.2 | 4.1 | 2.6 | 4.7 | 3.1 | 4.6 | 15.2 | 30.1 | 17.4 | 3.5 | 14.0 |
| Congo | 1.3 | 5.5 | 2.3 | 5.3 | 4.5 | 9.4 | -1.5 | 0.3 | 11.1 | 17.4 | 11.4 | 12.1 | 4.2 |
| Costa Rica | | | | | | | -0.3 | 3.7 | | 0.7 | | 27.4 | 27.4 |
| Côte d'Ivoire | | | | | | | -0.4 | 8.6 | | | | 0.8 | 1.3 |
| Czechoslovakia | 2.5 | 2.5 | 1.0 | 5.4 | 4.2 | 12.4 | 3.6 | 3.7 | 12.1 | 34.4 | 19.9 | 10.9 | 86.5 |
| Dominica | -1.9 | 0.3 | 1.2 | 0.0 | 1.0 | 3.8 | 3.6 | 8.6 | 15.1 | 13.1 | 14.5 | 7.8 | 12.9 |
| Dominican Rep. | 3.1 | 5.5 | 3.8 | 6.3 | 6.3 | 5.1 | 8.4 | 7.9 | 13.0 | 23.3 | 15.4 | 10.7 | 13.0 |
| Ecuador | 14.8 | 9.0 | 0.7 | 3.4 | 3.4 | 3.8 | 3.1 | 2.1 | 5.1 | 10.0 | 9.7 | 10.3 | 12.7 |
| Egypt | 0.5 | -1.2 | 1.5 | -0.2 | -0.2 | 2.8 | 0.4 | 1.5 | 6.4 | 16.9 | 19.1 | 7.0 | 11.8 |
| El Salvador | | -1.4 | 0.8 | 1.4 | 1.4 | 10.1 | 0.5 | -6.1 | 8.9 | 8.6 | 6.6 | 28.5 | 16.7 |
| Ethiopia | 1.2 | 0.2 | 1.4 | 5.0 | 5.0 | 4.1 | 9.1 | 22.0 | 11.1 | 14.5 | 13.1 | 11.4 | 7.0 |
| Fiji | 26.4 | 13.2 | -8.4 | 7.3 | 7.3 | -2.0 | 3.1 | 8.7 | 6.9 | 9.2 | 25.9 | 17.0 | 12.4 |
| Gambia, The | -0.8 | 0.7 | 0.5 | 2.1 | 2.1 | 3.0 | 9.6 | 10.1 | 17.7 | 18.1 | 29.8 | 56.1 | 116.5 |
| Ghana | 2.3 | 8.3 | -2.9 | 1.4 | 1.4 | 2.3 | -0.5 | 0.5 | 13.8 | 16.5 | 13.2 | 10.7 | 12.3 |
| Guatemala | 3.2 | 1.8 | 2.1 | 1.3 | 1.4 | 1.4 | 9.6 | 3.2 | 22.7 | 15.0 | 16.8 | 7.0 | 6.5 |
| Haiti | | | | 1.9 | 1.3 | 2.9 | 2.2 | 3.6 | 5.2 | 12.8 | 8.4 | 4.9 | 8.4 |
| Honduras | | | | | | | | | 3.4 | 1.8 | 3.8 | 5.2 | 3.9 |
| Hungary | 9.5 | 10.8 | 13.1 | 3.0 | 0.6 | 5.1 | 3.1 | 6.5 | 16.9 | 28.6 | 5.7 | -7.6 | 8.3 |
| India | 305.5 | 14.5 | 106.0 | 128.8 | 15.5 | 12.3 | 4.4 | 6.5 | 31.0 | 40.6 | 19.1 | 19.9 | 11.0 |
| Indonesia | 2.2 | -0.4 | 1.6 | 0.7 | 3.6 | 1.7 | 4.2 | 6.4 | 9.8 | 14.2 | 12.9 | 11.3 | 27.3 |
| Iran | 7.7 | 7.9 | 1.7 | 2.1 | 2.4 | 6.1 | 12.0 | 12.9 | 20.0 | 39.7 | 39.3 | 31.4 | 34.6 |
| Israel | 2.6 | 1.9 | 3.0 | 5.9 | 6.3 | 14.7 | 5.3 | 5.4 | 17.7 | 27.2 | 17.4 | 9.8 | 11.2 |
| Jamaica | | | | | | 5.9 | 4.8 | 7.7 | 11.1 | 19.4 | 12.0 | 11.5 | 14.6 |
| Jordan | 3.6 | 5.0 | 1.8 | 0.4 | -0.2 | 2.2 | 3.8 | 5.8 | 9.3 | 17.8 | 19.1 | 11.4 | 14.8 |
| Kenya | | | | 10.9 | 12.5 | 16.1 | 13.4 | 11.7 | 3.2 | 24.3 | 25.3 | 15.3 | 10.2 |
| Korea, Rep. of | | | 10.6 | | | | | | | 13.4 | 14.2 | 11.4 | 16.7 |
| Lesotho | 4.2 | 3.2 | 0.8 | 1.0 | 3.8 | 2.9 | 5.4 | 5.6 | 6.1 | 22.1 | 8.2 | 5.0 | — |
| Madagascar | -0.1 | 1.0 | 4.6 | -0.2 | -0.4 | 1.8 | 1.6 | 3.2 | 10.6 | 17.3 | 4.5 | 2.6 | 3.1 |
| Malaysia | 1.6 | 0.5 | 0.7 | 2.0 | 2.3 | 3.7 | 2.3 | 3.4 | 7.7 | 7.3 | 8.8 | 0.6 | 4.8 |
| Malta | | | | | | | | | | | | | 10.0 |

(Table continues on the following page.)

539

**Table A.3** *(continued)*

| Country | 1965 | 1966 | 1967 | 1968 | 1969 | 1970 | 1971 | 1972 | 1973 | 1974 | 1975 | 1976 | 1977 |
|---|---|---|---|---|---|---|---|---|---|---|---|---|---|
| *Developing countries (continued)* | | | | | | | | | | | | | |
| Mauritius | 1.7 | 2.5 | 1.9 | 6.9 | 2.3 | 1.6 | 0.3 | 5.4 | 13.5 | 29.1 | 14.8 | 12.9 | 9.2 |
| Mexico | 3.6 | 4.2 | 3.0 | 2.3 | 3.4 | 5.2 | 5.3 | 5.0 | 12.0 | 23.8 | 15.2 | 15.8 | 29.0 |
| Morocco | 3.5 | -1.0 | -0.7 | 0.4 | 2.9 | 1.3 | 4.2 | 3.8 | 4.1 | 17.6 | 7.9 | 8.5 | 12.6 |
| Nicaragua | — | — | — | — | — | — | — | — | 27.0 | 13.3 | 7.5 | 2.8 | 11.4 |
| Niger | 4.4 | 10.6 | 0.4 | -2.9 | 10.6 | 1.1 | 4.2 | 9.8 | 11.8 | 3.4 | 9.1 | 23.5 | 23.3 |
| Nigeria | 4.1 | 9.7 | -3.7 | -0.5 | 10.2 | 13.8 | 16.0 | 3.5 | 5.4 | 12.7 | 33.9 | 24.3 | 13.8 |
| Pakistan | 5.6 | 7.2 | 6.8 | 0.2 | 3.2 | 5.3 | 4.7 | 5.2 | 23.1 | 26.7 | 20.9 | 7.2 | 10.1 |
| Panama | 0.5 | 0.2 | 1.4 | 1.6 | 1.8 | 3.1 | 1.9 | 5.4 | 6.9 | 16.3 | 5.9 | 4.0 | 4.6 |
| Papua New Guinea | — | — | — | — | — | — | — | 6.1 | 8.3 | 23.2 | 10.5 | 7.7 | 4.5 |
| Paraguay | 3.8 | 2.9 | 1.4 | 0.7 | 2.1 | -0.7 | 4.8 | 9.5 | 12.5 | 25.2 | 6.8 | 4.6 | 9.3 |
| Peru | 16.4 | 8.8 | 9.8 | 19.1 | 6.2 | 5.0 | 6.8 | 7.2 | 9.5 | 16.9 | 23.6 | 33.5 | 38.1 |
| Philippines | 3.1 | 4.9 | 5.7 | 2.0 | 1.3 | 15.3 | 21.4 | 8.2 | 16.6 | 34.2 | 6.8 | 9.2 | 9.9 |
| Poland | — | — | — | — | — | — | 1.1 | -0.1 | 2.5 | 7.1 | 2.3 | 4.4 | 4.9 |
| Rwanda | — | — | 1.5 | 3.1 | 0.5 | 0.5 | 0.5 | 3.1 | 9.4 | 31.1 | 30.2 | 7.2 | 13.7 |
| Senegal | — | — | — | 0.1 | 4.0 | 2.8 | 3.9 | 6.2 | 11.3 | 16.6 | 31.7 | 1.1 | 11.3 |
| Sierra Leone | 4.7 | 4.3 | 4.9 | 1.6 | 3.1 | 6.4 | -1.3 | 5.5 | 5.7 | 14.4 | 19.9 | 17.2 | 8.3 |
| Somalia | 13.0 | -3.3 | -0.3 | 3.4 | 6.3 | 0.9 | -0.6 | -3.0 | 6.5 | 18.2 | 19.4 | 14.2 | 10.6 |
| South Africa | 3.9 | 3.6 | 3.4 | 2.1 | 3.2 | 4.1 | 5.7 | 6.5 | 9.5 | 11.6 | 13.5 | 11.2 | 11.1 |
| Sri Lanka | 0.2 | -0.2 | 2.2 | 5.9 | 7.5 | 5.9 | 2.7 | 6.3 | 9.6 | 12.3 | 6.6 | 1.3 | 1.2 |
| Sudan | -2.4 | 1.7 | 11.0 | -10.0 | 12.6 | 4.0 | 1.3 | 13.6 | 15.3 | 26.2 | 24.0 | 1.7 | 17.1 |
| Syria | -3.9 | 3.8 | 6.5 | 2.8 | -2.0 | 4.6 | 5.6 | 2.1 | 20.4 | 15.5 | 11.5 | 11.4 | 12.0 |
| Tanzania | — | 9.8 | 12.2 | 15.6 | 16.4 | 3.5 | 4.7 | 7.6 | 10.5 | 19.2 | 26.5 | 6.9 | 11.6 |
| Thailand | 0.2 | 4.0 | 4.3 | 1.8 | 2.5 | -0.1 | 0.5 | 4.8 | 15.5 | 24.3 | 5.3 | 4.1 | 7.6 |
| Togo | — | — | -2.3 | 0.3 | 6.0 | 4.5 | 6.5 | 7.7 | 3.6 | 12.8 | 18.0 | 11.6 | 22.5 |
| Trinidad and Tobago | 1.8 | 4.1 | 2.1 | 8.2 | 2.4 | 2.5 | 3.5 | 9.3 | 14.8 | 22.0 | 17.0 | 10.7 | 11.7 |
| Tunisia | 6.6 | 4.0 | 3.0 | 2.5 | 4.1 | 1.1 | 5.7 | 2.1 | 4.6 | 4.2 | 9.6 | 5.3 | 6.7 |
| Uruguay | 56.6 | 73.5 | 89.3 | 125.3 | 21.0 | 16.3 | 24.0 | 76.5 | 97.0 | 77.2 | 81.4 | 50.6 | 58.2 |
| Venezuela | 1.7 | 1.8 | 0.0 | 1.3 | 2.4 | 2.5 | 3.2 | 2.8 | 4.1 | 8.3 | 10.3 | 7.6 | 7.8 |
| Yugoslavia | 33.3 | 25.5 | 6.6 | 5.1 | 9.1 | 9.5 | 15.7 | 15.9 | 19.5 | 22.0 | 23.5 | 11.2 | 14.7 |
| Zaire | -2.7 | 15.8 | 36.9 | 53.3 | 6.2 | 8.0 | 5.8 | 15.8 | 15.6 | 29.5 | 28.7 | 80.4 | 68.9 |
| Zambia | 8.1 | 10.2 | 5.0 | 10.8 | 2.4 | 2.7 | 6.0 | 5.1 | 6.5 | 8.1 | 10.1 | 18.8 | 19.8 |
| Zimbabwe | 2.5 | 3.1 | 2.4 | 1.4 | 0.4 | 2.1 | 3.0 | 2.8 | 3.1 | 6.6 | 10.0 | 11.0 | 10.3 |

| Country | 1978 | 1979 | 1980 | 1981 | 1982 | 1983 | 1984 | 1985 | 1986 | 1987 | 1988 | 1989 | 1990 |
|---|---|---|---|---|---|---|---|---|---|---|---|---|---|
| *OECD countries* | | | | | | | | | | | | | |
| Australia | 7.9 | 9.1 | 10.1 | 9.7 | 11.1 | 10.1 | 4.0 | 6.7 | 9.1 | 8.5 | 7.2 | 7.6 | 7.3 |
| Austria | 3.6 | 3.7 | 6.3 | 6.8 | 5.4 | 3.3 | 5.7 | 3.2 | 1.7 | 1.4 | 1.9 | 2.6 | 3.3 |
| Belgium | 4.5 | 4.5 | 6.7 | 7.6 | 8.7 | 7.7 | 6.3 | 4.9 | 1.3 | 1.6 | 1.2 | 3.1 | 3.5 |
| Canada | 8.9 | 9.1 | 10.2 | 12.5 | 10.8 | 5.8 | 4.3 | 4.0 | 4.2 | 4.4 | 4.0 | 5.0 | 4.8 |
| Denmark | 10.0 | 9.6 | 12.3 | 11.7 | 10.1 | 6.9 | 6.3 | 4.7 | 3.7 | 4.0 | 4.6 | 4.8 | 2.6 |
| Finland | 7.8 | 7.5 | 11.6 | 12.0 | 9.6 | 8.4 | 7.1 | 5.9 | 2.9 | 4.1 | 5.1 | 6.6 | 6.1 |
| France | 9.1 | 10.8 | 13.3 | 13.4 | 11.8 | 9.6 | 7.4 | 5.8 | 2.5 | 3.3 | 2.7 | 3.5 | 3.4 |
| Germany, Fed. Rep. | 2.7 | 4.1 | 5.4 | 6.3 | 5.3 | 3.3 | 2.4 | 2.2 | -0.1 | 0.2 | 1.3 | 2.8 | 2.7 |
| Greece | 12.5 | 19.0 | 24.9 | 24.5 | 21.0 | 20.2 | 18.4 | 19.3 | 23.0 | 16.4 | 13.5 | 13.7 | 20.4 |
| Ireland | 7.6 | 13.2 | 18.2 | 20.4 | 17.1 | 10.5 | 8.6 | 5.4 | 3.8 | 3.1 | 2.2 | 4.0 | 3.4 |
| Italy | 12.2 | 14.8 | 21.0 | 17.9 | 16.5 | 14.7 | 10.8 | 9.2 | 5.9 | 4.7 | 5.0 | 6.3 | 6.4 |
| Japan | 4.2 | 3.7 | 7.7 | 4.9 | 2.7 | 1.9 | 2.3 | 2.0 | 0.6 | 0.0 | 0.7 | 2.3 | 3.1 |
| Netherlands | 4.1 | 4.2 | 6.5 | 6.7 | 5.9 | 2.8 | 3.3 | 2.2 | 0.1 | -0.7 | 0.7 | 1.1 | 2.5 |
| Norway | 8.2 | 4.8 | 10.9 | 13.6 | 11.4 | 8.4 | 6.3 | 5.7 | 7.2 | 8.7 | 6.7 | 4.6 | 4.1 |
| Portugal | 22.7 | 23.6 | 16.6 | 20.0 | 22.7 | 25.1 | 28.9 | 19.3 | 11.7 | 9.4 | 9.6 | 12.6 | 13.4 |
| Spain | 19.8 | 15.7 | 15.6 | 14.5 | 14.4 | 12.2 | 11.3 | 8.8 | 8.8 | 5.3 | 4.8 | 6.8 | 6.7 |
| Sweden | 9.9 | 7.2 | 13.7 | 12.1 | 8.6 | 8.9 | 8.0 | 7.4 | 4.2 | 4.2 | 5.8 | 6.4 | 10.5 |
| Turkey | 45.3 | 58.7 | 110.2 | 36.6 | 30.8 | 31.4 | 48.4 | 45.0 | 34.6 | 38.8 | 75.4 | 69.6 | 60.3 |
| United Kingdom | 8.2 | 13.5 | 18.0 | 11.9 | 8.6 | 4.6 | 5.0 | 6.1 | 3.4 | 4.1 | 4.9 | 7.8 | 9.5 |
| United States | 7.6 | 11.3 | 13.5 | 10.3 | 6.2 | 3.2 | 4.3 | 3.6 | 1.9 | 3.7 | 4.0 | 4.8 | 5.4 |
| *Developing countries* | | | | | | | | | | | | | |
| Argentina | 175.5 | 159.5 | 100.8 | 104.5 | 164.8 | 343.8 | 626.7 | 672.1 | 90.1 | 131.3 | 343.0 | 3,080.0 | 2,314.0 |
| Bangladesh | 5.3 | 14.7 | 13.4 | 16.2 | 12.5 | 9.4 | 10.5 | 10.7 | 11.0 | 9.5 | 9.3 | 10.0 | 8.1 |
| Bolivia | 10.4 | 19.7 | 47.2 | 28.6 | 133.3 | 269.0 | 1,281.0 | 11,750.0 | 276.3 | 14.6 | 16.0 | 15.2 | 17.1 |
| Botswana | 9.0 | 11.7 | 13.6 | 16.4 | 11.1 | 10.5 | 8.6 | 8.1 | 10.0 | 9.8 | 8.4 | 11.6 | 11.4 |
| Brazil | 38.7 | 52.7 | 82.8 | 105.6 | 97.8 | 142.1 | 197.0 | 226.9 | 145.2 | 229.7 | 682.3 | 1,287.0 | 2,938.0 |
| Burkina Faso | 8.3 | 15.0 | 12.2 | 7.6 | 12.1 | 8.3 | 4.8 | 6.9 | -2.6 | -2.8 | 4.4 | -0.5 | -0.5 |
| Burundi | 23.9 | 36.5 | 2.5 | 12.0 | 5.8 | 8.4 | 14.4 | 3.6 | 1.9 | 7.1 | 4.5 | 11.6 | 7.0 |
| Cameroon | 12.5 | 6.6 | 9.6 | 10.7 | 13.3 | 16.6 | 11.4 | 1.3 | 7.7 | 6.0 | 8.6 | 0.0 | 10.7 |

*(Table continues on the following page.)*

**Table A.3** (continued)

| Country | 1978 | 1979 | 1980 | 1981 | 1982 | 1983 | 1984 | 1985 | 1986 | 1987 | 1988 | 1989 | 1990 |
|---|---|---|---|---|---|---|---|---|---|---|---|---|---|
| *Developing countries (continued)* | | | | | | | | | | | | | |
| Central African Rep. | — | — | — | — | 13.3 | 14.6 | 2.5 | 10.4 | 2.2 | -7.0 | -4.0 | 0.7 | 0.0 |
| Chad | 40.1 | — | — | — | — | — | 20.3 | 5.2 | -13.1 | -6.0 | 15.5 | -4.9 | 0.6 |
| Chile | 0.9 | 33.4 | 35.1 | 19.7 | 9.9 | 27.3 | 19.9 | 30.7 | 19.5 | 19.9 | 14.7 | 17.0 | 26.0 |
| China | 17.8 | 2.0 | 7.4 | 2.5 | 2.0 | 1.9 | 2.7 | 11.9 | 7.0 | 8.8 | 20.7 | 16.3 | 1.4 |
| Colombia | 10.5 | 24.7 | 26.5 | 27.5 | 24.5 | 19.8 | 16.1 | 24.0 | 18.9 | 23.3 | 28.1 | 25.8 | 29.1 |
| Congo | 6.0 | 8.1 | 7.3 | 17.0 | 12.8 | 7.8 | 12.7 | 6.1 | 2.5 | 2.3 | 3.7 | 4.1 | -4.8 |
| Costa Rica | 13.0 | 9.2 | 18.1 | 37.1 | 90.1 | 32.6 | 12.0 | 15.1 | 11.8 | 16.8 | 20.8 | 16.5 | 19.0 |
| Côte d'Ivoire | 1.6 | 16.6 | 14.7 | 8.8 | 7.3 | 5.9 | 4.3 | 1.8 | 7.3 | 0.4 | 7.0 | 1.0 | -0.8 |
| Czechoslovakia | -36.7 | 3.9 | 2.9 | 0.8 | 5.1 | 0.9 | 0.9 | 2.3 | 0.5 | 0.1 | 0.1 | 1.4 | 10.0 |
| Dominica | 3.5 | 25.2 | 25.2 | 13.3 | 4.4 | 4.1 | 2.2 | 2.1 | 3.0 | 5.3 | 1.8 | 6.8 | 1.4 |
| Dominican Rep. | 11.7 | 9.2 | 16.8 | 7.5 | 7.6 | 4.8 | 27.0 | 37.5 | 9.7 | 15.9 | 44.4 | 45.4 | 59.4 |
| Ecuador | 11.1 | 10.3 | 13.0 | 16.4 | 16.3 | 48.4 | 31.2 | 28.0 | 23.0 | 29.5 | 58.2 | 75.6 | 48.5 |
| Egypt | 13.3 | 9.9 | 20.7 | 10.3 | 14.8 | 16.1 | 17.0 | 12.1 | 23.9 | 19.7 | 17.7 | 21.3 | 16.8 |
| El Salvador | 14.3 | 14.6 | 17.4 | 14.8 | 11.7 | 13.3 | 11.5 | 22.3 | 31.9 | 24.9 | 19.8 | 17.6 | 24.0 |
| Ethiopia | 6.1 | 16.0 | 4.5 | 6.1 | 5.9 | -0.7 | 8.4 | 19.1 | -9.8 | -2.4 | 7.1 | 7.8 | 5.2 |
| Fiji | 8.9 | 7.8 | 14.5 | 11.2 | 7.0 | 6.7 | 5.3 | 4.4 | 1.8 | 5.7 | 11.8 | 6.2 | 8.2 |
| Gambia, The | 73.1 | 6.1 | 6.8 | 5.9 | 10.9 | 10.6 | 22.1 | 18.3 | 56.6 | 23.5 | 11.7 | 8.3 | 12.2 |
| Ghana | 8.3 | 54.4 | 50.1 | 116.5 | 22.3 | 122.9 | 39.7 | 10.3 | 24.6 | 39.8 | 31.4 | 25.2 | 37.3 |
| Guatemala | -2.7 | 11.3 | 10.8 | 11.4 | 0.3 | 4.5 | 3.4 | 18.7 | 36.9 | 12.3 | 10.8 | 11.4 | 41.2 |
| Haiti | 5.7 | 13.1 | 17.8 | 10.9 | 7.4 | 10.2 | 6.4 | 10.6 | 3.3 | -11.4 | 4.1 | 6.9 | 21.5 |
| Honduras | 4.7 | 12.1 | 18.1 | 9.4 | 9.0 | 8.3 | 4.7 | 3.4 | 4.4 | 2.5 | 4.5 | 9.9 | 23.3 |
| Hungary | 2.5 | 9.0 | 9.3 | 4.5 | 7.0 | 6.4 | 8.7 | 7.0 | 5.3 | 8.2 | 16.3 | 16.9 | 29.0 |
| India | 8.1 | 6.3 | 11.4 | 13.1 | 7.9 | 11.9 | 8.3 | 5.6 | 8.7 | 8.8 | 9.4 | 6.2 | 9.0 |
| Indonesia | 11.7 | 18.3 | 18.0 | 12.2 | 9.5 | 11.8 | 10.5 | 4.7 | 5.8 | 9.3 | 8.0 | 6.4 | 12.5 |
| Iran | 50.6 | 10.5 | 20.6 | 24.2 | 18.7 | 19.7 | 12.5 | 4.4 | 18.4 | 28.6 | 28.7 | 22.3 | 7.6 |
| Israel | 34.9 | 78.3 | 131.0 | 116.8 | 120.4 | 145.6 | 373.8 | 304.6 | 48.1 | 19.8 | 16.3 | 20.2 | 17.2 |
| Jamaica | 6.9 | 29.1 | 27.3 | 12.7 | 6.5 | 11.6 | 27.8 | 25.7 | 15.1 | 6.7 | 8.3 | 14.4 | 22.0 |
| Jordan | 16.9 | 14.2 | 11.1 | 7.7 | 7.4 | 5.0 | 3.8 | 3.0 | 0.0 | -0.2 | 6.6 | 30.0 | 16.2 |
| Kenya | 14.5 | 8.0 | 13.9 | 11.8 | 20.4 | 11.5 | 10.2 | 13.1 | 3.9 | 5.2 | 8.3 | 12.9 | 15.6 |
| Korea, Rep. of | 12.5 | 18.3 | 28.7 | 21.3 | 7.2 | 3.4 | 2.3 | 2.5 | 2.8 | 3.0 | 7.1 | 5.7 | 8.6 |
| Lesotho | 6.5 | 16.0 | 15.7 | 12.4 | 12.1 | 17.5 | 11.0 | 13.3 | 18.0 | 11.8 | 11.5 | 14.7 | 11.6 |
| Madagascar | 4.9 | 14.1 | 18.2 | 30.5 | 31.8 | 19.3 | 9.9 | 10.6 | 14.5 | 15.0 | 26.8 | 9.0 | 11.8 |
| Malaysia | 3.7 | 3.7 | 6.7 | 9.7 | 5.8 | 3.7 | 3.9 | 0.3 | 0.7 | 0.9 | 2.0 | 2.8 | 2.6 |
| Malta | 4.7 | 7.1 | 15.7 | 11.5 | 5.8 | -0.9 | -0.4 | -0.2 | 2.0 | 0.4 | 0.9 | 0.8 | 3.0 |

| | | | | | | | | | | | | | |
|---|---|---|---|---|---|---|---|---|---|---|---|---|---|
| Mauritius | 8.5 | 14.5 | 42.0 | 14.5 | 11.4 | 5.6 | 7.4 | 6.7 | 1.6 | 0.5 | 9.2 | 12.7 | 13.5 |
| Mexico | 17.5 | 18.2 | 26.4 | 27.9 | 58.9 | 101.8 | 65.5 | 57.7 | 86.2 | 131.8 | 114.2 | 20.0 | 26.7 |
| Morocco | 9.7 | 8.3 | 9.4 | 12.5 | 10.5 | 6.2 | 12.4 | 7.7 | 8.7 | 2.7 | 2.4 | 3.1 | 6.8 |
| Nicaragua | 4.6 | 48.2 | 35.3 | 23.9 | 24.8 | 31.1 | 35.4 | 219.5 | 681.4 | 911.9 | 10,205.0 | 48.7 | 75.8 |
| Niger | 10.1 | 7.3 | 10.3 | 22.9 | 11.6 | -2.5 | 8.4 | -0.9 | -3.2 | -6.7 | -1.4 | -2.8 | -0.8 |
| Nigeria | 21.7 | 11.7 | 10.0 | 20.8 | 7.7 | 23.2 | 39.6 | 5.5 | 5.4 | 10.2 | 38.3 | 50.4 | 7.4 |
| Pakistan | 6.1 | 8.3 | 11.9 | 11.9 | 5.9 | 6.4 | 6.1 | 5.6 | 3.5 | 4.7 | 8.8 | 7.8 | 9.1 |
| Panama | 4.2 | 8.0 | 13.8 | 7.3 | 4.3 | 2.1 | 1.6 | 1.0 | -0.1 | 1.0 | 0.3 | 0.0 | 0.6 |
| Papua New Guinea | 5.8 | 5.8 | 12.1 | 8.1 | 5.5 | 7.9 | 7.4 | 3.7 | 5.5 | 3.3 | 5.4 | 4.5 | 6.9 |
| Paraguay | 10.6 | 28.3 | 22.4 | 14.0 | 6.8 | 13.4 | 20.3 | 25.2 | 31.7 | 21.8 | 22.8 | 26.4 | 38.2 |
| Peru | 57.8 | 66.7 | 59.1 | 75.4 | 64.4 | 111.2 | 110.2 | 163.4 | 77.9 | 85.8 | 667.0 | 3,399.0 | 7,482.0 |
| Philippines | 7.3 | 17.5 | 18.2 | 13.1 | 10.2 | 10.0 | 50.3 | 23.1 | 0.8 | 3.8 | 8.8 | 12.2 | 13.4 |
| Poland | 8.1 | 7.0 | 9.4 | 21.2 | 100.8 | 22.1 | 15.0 | 15.1 | 17.7 | 25.2 | 60.2 | 251.1 | 555.4 |
| Rwanda | 13.3 | 15.7 | 7.2 | 6.5 | 12.6 | 6.6 | 5.4 | 1.7 | -1.1 | 4.1 | 3.0 | 1.0 | 4.2 |
| Senegal | 3.4 | 9.7 | 8.7 | 5.9 | 17.4 | 11.6 | 11.8 | 13.0 | 6.2 | -4.1 | -1.8 | 0.4 | 0.3 |
| Sierra Leone | 10.9 | 21.3 | 12.9 | 23.4 | 26.9 | 68.5 | 66.6 | 76.6 | 80.9 | 178.7 | 34.3 | 60.8 | 110.9 |
| Somalia | 10.0 | 24.3 | 58.8 | 44.4 | 23.6 | 36.0 | 91.2 | 37.8 | 35.8 | 28.2 | 81.9 | — | — |
| South Africa | 10.2 | 13.1 | 13.8 | 15.2 | 14.7 | 12.3 | 11.5 | 16.3 | 18.6 | 16.1 | 12.8 | 14.7 | 14.4 |
| Sri Lanka | 12.1 | 10.7 | 26.1 | 18.0 | 10.8 | 14.0 | 16.6 | 1.5 | 8.0 | 7.7 | 14.0 | 11.6 | 21.5 |
| Sudan | 19.2 | 31.1 | 25.4 | 24.6 | 25.7 | 30.6 | 34.1 | 45.4 | 24.5 | 20.6 | 64.7 | 63.5 | 68.5 |
| Syria | 4.8 | 4.6 | 19.3 | 18.4 | 14.3 | 6.1 | 9.2 | 17.3 | 36.1 | 59.5 | 34.6 | 11.4 | 19.4 |
| Tanzania | 11.3 | 13.8 | 30.3 | 25.6 | 28.9 | 27.1 | 35.3 | 34.1 | 32.4 | 29.9 | 31.2 | 25.8 | 19.7 |
| Thailand | 7.9 | 9.9 | 19.7 | 12.7 | 5.3 | 3.7 | 0.9 | 2.4 | 1.8 | 2.5 | 3.9 | 5.4 | 5.9 |
| Togo | 0.4 | 7.5 | 12.3 | 19.7 | 11.1 | 9.4 | -3.5 | -1.8 | 4.1 | 0.1 | -0.1 | -0.8 | 1.0 |
| Trinidad and Tobago | 10.3 | 14.7 | 17.5 | 14.3 | 11.6 | 15.2 | 13.3 | 7.6 | 7.7 | 10.8 | 7.8 | 11.4 | 11.0 |
| Tunisia | 5.3 | 7.8 | 10.0 | 8.9 | 13.7 | 8.9 | 8.4 | 8.0 | 5.8 | 7.2 | 6.4 | 7.4 | 6.5 |
| Uruguay | 44.5 | 66.8 | 63.5 | 34.0 | 19.0 | 49.2 | 55.3 | 72.2 | 76.4 | 63.6 | 62.2 | 80.4 | 112.5 |
| Venezuela | 7.1 | 12.4 | 21.5 | 16.2 | 9.6 | 6.3 | 12.2 | 11.4 | 11.5 | 28.1 | 29.5 | 84.2 | 40.8 |
| Yugoslavia | 14.1 | 20.7 | 30.9 | 39.8 | 31.5 | 40.2 | 54.7 | 72.3 | 89.8 | 120.8 | 194.1 | 1,240.0 | 583.1 |
| Zaire | 48.8 | 101.1 | 46.6 | 35.4 | 36.7 | 76.5 | 52.2 | 23.8 | 46.7 | 90.4 | 82.7 | 104.1 | 81.3 |
| Zambia | 16.3 | 9.7 | 11.6 | 13.0 | 13.6 | 19.6 | 20.0 | 37.3 | 51.8 | 43.0 | 55.6 | 127.9 | 117.5 |
| Zimbabwe | 5.7 | 18.2 | 5.4 | 13.2 | 10.6 | 23.1 | 20.2 | 8.5 | 14.3 | 12.5 | 7.4 | 12.9 | 17.4 |

— Not available.

543

## Table A.4. Real Interest Rates, 1965–89
(percent)

| Country | 1965 | 1966 | 1967 | 1968 | 1969 | 1970 | 1971 | 1972 | 1973 | 1974 | 1975 | 1976 | 1977 |
|---|---|---|---|---|---|---|---|---|---|---|---|---|---|
| *OECD countries* | | | | | | | | | | | | | |
| Australia | — | — | — | — | — | 0.1 | -1.8 | 0.5 | -6.8 | -7.1 | -4.4 | -5.0 | -0.2 |
| Austria | — | — | — | — | — | — | — | — | — | — | — | — | — |
| Belgium | -0.6 | 0.5 | 0.2 | 0.6 | 0.2 | 2.7 | -0.9 | -3.3 | -2.8 | -7.7 | -5.1 | -1.8 | -0.8 |
| Canada | — | — | — | — | — | — | — | — | — | — | -1.6 | 3.3 | -1.8 |
| Denmark | — | — | — | — | — | — | — | — | — | — | — | — | — |
| Finland | — | — | — | — | — | — | — | — | — | — | — | — | — |
| France | — | — | — | — | — | -1.4 | -1.8 | -2.9 | -4.3 | -9.0 | -3.5 | -2.1 | -1.8 |
| Germany, Fed. Rep. | — | — | — | — | — | — | — | — | — | — | — | — | — |
| Greece | 0.3 | 0.6 | 7.2 | 3.2 | 3.5 | 2.0 | 2.8 | -0.7 | -19.0 | -3.2 | -5.3 | -2.5 | -3.8 |
| Ireland | -0.7 | -0.9 | 0.5 | -1.3 | -3.0 | -5.0 | -4.5 | -4.5 | -5.0 | -8.7 | -7.8 | -10.8 | -4.2 |
| Italy | — | — | — | — | — | — | — | — | — | — | — | — | — |
| Japan | -2.4 | -0.4 | -1.6 | 0.2 | -2.2 | -3.9 | -0.7 | -1.8 | -12.0 | -13.0 | -2.3 | -5.6 | -1.1 |
| Netherlands | — | — | — | — | — | — | — | — | — | — | — | — | — |
| Norway | — | — | — | — | — | — | — | — | — | — | — | — | — |
| Spain | — | — | — | — | — | — | — | — | — | — | — | — | — |
| Sweden | — | 1.3 | 2.5 | 3.8 | 2.9 | -0.6 | -0.6 | -0.7 | -1.9 | -4.4 | -1.7 | -2.6 | -3.7 |
| Turkey | — | -5.6 | -9.9 | -3.0 | -8.4 | -5.6 | -9.9 | -3.0 | -8.4 | -8.3 | -8.9 | -6.8 | -24.6 |
| United Kingdom | -0.2 | 0.9 | 1.6 | -0.4 | 1.1 | -2.5 | -4.5 | -3.3 | -2.4 | -7.9 | -14.3 | -6.6 | -6.5 |
| United States | — | — | — | — | — | — | — | — | — | — | — | — | — |
| Average OECD | -0.7 | -0.5 | 0.1 | 0.4 | -0.8 | -1.6 | -2.4 | -2.2 | -7.0 | -7.7 | -5.5 | -4.1 | -4.9 |
| *Developing countries* | | | | | | | | | | | | | |
| Argentina | -20.4 | -15.4 | -13.4 | 0.4 | 3.0 | -9.8 | -18.8 | -27.2 | -17.0 | -16.3 | -72.4 | -65.1 | -19.2 |
| Bangladesh | — | — | — | — | — | — | — | — | — | — | — | -7.6 | 1.6 |
| Bolivia | — | — | — | — | — | — | — | — | — | — | — | — | — |
| Brazil | -4.3 | -8.6 | 5.5 | 5.2 | 3.5 | 8.1 | 8.3 | 9.9 | 7.4 | -5.0 | -2.9 | -9.4 | -1.5 |
| Burkina Faso | — | — | — | — | — | 2.0 | -4.2 | 15.7 | -8.8 | 5.2 | -8.3 | 5.3 | -15.5 |
| Chile | -10.0 | -0.9 | -5.0 | -9.2 | -7.9 | -11.1 | -3.1 | -52.9 | -78.8 | -52.6 | -18.5 | 3.9 | 21.7 |
| Colombia | -8.8 | -7.9 | -1.6 | -0.6 | -6.5 | 5.0 | -5.0 | -2.6 | -6.9 | -5.9 | 2.2 | -3.9 | -3.8 |
| Côte d'Ivoire | 1.0 | -0.5 | 0.0 | -6.3 | 2.7 | -3.3 | 3.9 | 1.2 | -8.8 | -12.0 | -4.6 | -7.2 | -13.2 |
| Czechoslovakia | — | — | — | — | — | 2.7 | 3.1 | 3.0 | 2.4 | 2.1 | 2.0 | 1.9 | 1.4 |

| | | | | | | | | | | | | |
|---|---|---|---|---|---|---|---|---|---|---|---|---|
| Dominica | — | — | — | — | — | — | — | — | — | — | — | — | — |
| Ecuador | -1.1 | 2.0 | -1.5 | 1.4 | -1.3 | -3.1 | 1.1 | 1.0 | -10.0 | -10.5 | -4.2 | -4.1 | -1.1 |
| Ghana | — | — | 13.6 | -3.6 | -3.1 | 0.9 | -2.8 | -3.2 | -10.6 | -10.9 | -17.0 | -31.0 | -50.2 |
| Honduras | — | — | — | — | — | — | — | — | — | — | — | — | — |
| India | -2.9 | -7.6 | -0.5 | 7.0 | 0.2 | 2.6 | 0.5 | -4.8 | -16.7 | -7.5 | 21.2 | -1.0 | 3.0 |
| Indonesia | 1.7 | 1.0 | 0.9 | -0.7 | 46.2 | 16.5 | 23.7 | -4.9 | -8.9 | -10.3 | -3.0 | 1.6 | 0.8 |
| Jamaica | — | — | 1.5 | 3.0 | -4.0 | 3.5 | 2.5 | -1.1 | -14.5 | -6.6 | -3.2 | 3.5 | -2.1 |
| Kenya | — | — | — | — | 3.6 | 1.9 | -3.7 | 0.5 | -10.8 | -9.4 | -12.6 | -2.3 | -13.1 |
| Korea, Rep. of | 5.0 | 12.9 | 14.0 | 13.7 | 11.6 | 8.9 | 7.2 | 8.4 | 3.2 | -9.1 | -8.3 | 4.0 | 4.7 |
| Malawi | 3.3 | 1.8 | 1.1 | -5.1 | 0.4 | -4.9 | -1.7 | 2.6 | -2.7 | -8.5 | -8.6 | 6.1 | -1.4 |
| Malaysia | — | 3.7 | -0.6 | 9.1 | 3.7 | 5.4 | 3.1 | 2.0 | -10.1 | -2.7 | 6.8 | 3.6 | 1.1 |
| Mexico | — | 2.6 | 3.3 | 4.7 | 3.8 | -0.5 | 1.5 | 0.8 | -11.7 | -7.9 | 0.2 | -12.9 | -6.4 |
| Morocco | — | — | 4.5 | 1.2 | 2.5 | 0.5 | -1.6 | 0.4 | -6.6 | -8.7 | -0.6 | -7.0 | -3.2 |
| Nigeria | — | 3.0 | 7.3 | -4.6 | -9.3 | -9.7 | -6.7 | 4.7 | -8.3 | -18.6 | -19.8 | -9.8 | -13.2 |
| Oman | -3.0 | -4.3 | -4.2 | — | — | — | — | — | — | — | — | — | — |
| Pakistan | -6.9 | -0.6 | -9.9 | 3.5 | 4.1 | -2.0 | 1.4 | 0.6 | -9.4 | -20.1 | -11.7 | 2.8 | 0.1 |
| Peru | 5.7 | -0.9 | -2.1 | -2.6 | 1.2 | 1.3 | -0.6 | 2.6 | -5.9 | -10.2 | -13.7 | -21.9 | -12.4 |
| Philippines | — | — | — | 5.0 | 2.6 | -8.3 | -4.2 | 3.4 | -15.2 | -15.7 | 7.7 | 5.0 | 2.1 |
| Poland | — | — | — | — | — | — | — | — | — | — | — | — | — |
| Sierra Leone | 2.3 | -1.6 | -2.0 | 3.7 | 0.1 | -4.1 | 8.5 | -0.7 | -1.5 | -12.1 | -6.2 | -13.5 | -2.4 |
| Sri Lanka | — | 3.2 | -2.5 | -3.1 | -2.0 | 1.2 | -1.1 | 0.7 | -8.4 | -5.4 | 6.3 | 4.9 | 13.3 |
| Thailand | — | 2.1 | 3.9 | 5.7 | 4.0 | 8.3 | 5.7 | -1.8 | -10.9 | -8.3 | 3.4 | 4.4 | -0.8 |
| Trinidad and Tobago | — | — | — | — | — | — | — | — | — | — | — | — | — |
| Venezuela | 4.9 | 6.6 | 7.5 | 4.9 | 5.1 | 3.7 | 4.8 | 4.5 | 1.7 | -3.7 | -0.4 | 0.5 | -0.1 |
| Zaire | -2.9 | -10.3 | -41.7 | -8.1 | -8.1 | -3.7 | 3.7 | 1.0 | -7.8 | -10.6 | -14.9 | -35.7 | -26.6 |
| Zambia | -4.9 | -3.1 | -0.3 | -5.9 | 3.0 | 0.2 | -0.3 | -0.6 | -2.9 | -2.1 | -6.5 | -14.4 | -7.6 |
| Zimbabwe | — | — | — | — | — | 1.1 | 0.2 | 0.4 | 0.1 | -3.1 | -6.1 | -6.9 | -6.6 |

*(Table continues on the following page with the year 1978.)*

# Table A.4 (continued)

| Country | 1978 | 1979 | 1980 | 1981 | 1982 | 1983 | 1984 | 1985 | 1986 | 1987 | 1988 | 1989 |
|---|---|---|---|---|---|---|---|---|---|---|---|---|
| *OECD countries* | | | | | | | | | | | | |
| Australia | 0.7 | -1.6 | -0.6 | -0.8 | 1.3 | 1.9 | 7.0 | 2.0 | 3.8 | 6.3 | 3.9 | 7.0 |
| Austria | 1.3 | 0.3 | -1.6 | -1.3 | 0.3 | 0.4 | -1.0 | 1.1 | 2.5 | 1.4 | 0.9 | 0.1 |
| Belgium | 0.6 | 0.3 | 0.1 | -0.6 | -0.6 | -0.5 | 2.0 | 2.6 | 4.7 | 3.5 | 2.6 | 1.5 |
| Canada | 0.4 | 2.1 | 1.5 | 5.4 | 4.1 | 3.2 | 6.1 | 3.9 | 3.9 | 3.4 | 5.3 | 6.6 |
| Denmark | — | -0.9 | -0.1 | 0.7 | 3.6 | 3.7 | 3.2 | 4.5 | 2.2 | 2.9 | 3.1 | 3.3 |
| Finland | — | — | — | -0.8 | -0.3 | 0.2 | 2.9 | 3.6 | 3.8 | 3.3 | 1.8 | -1.3 |
| France | -2.7 | -5.3 | -6.4 | -5.4 | -0.4 | -0.9 | 0.9 | 2.0 | 3.0 | 2.1 | 1.9 | 2.3 |
| Germany, Fed. Rep. | 0.6 | -0.3 | 2.3 | 2.8 | 2.8 | 1.9 | 2.8 | 2.7 | 4.8 | 2.2 | 1.5 | 2.4 |
| Greece | -1.4 | -10.3 | -9.3 | -6.5 | -3.8 | -4.5 | -2.3 | -7.6 | -1.2 | -0.3 | 2.9 | 2.1 |
| Ireland | -1.5 | -4.3 | -5.3 | -9.7 | 0.3 | -0.9 | 1.1 | 2.0 | 3.2 | 3.1 | 0.9 | -0.1 |
| Italy | -0.7 | -6.9 | -7.1 | -3.2 | -0.7 | 0.3 | 1.3 | -0.7 | 4.5 | 1.7 | 1.2 | 0.6 |
| Japan | -1.1 | -2.3 | -1.3 | -0.2 | 1.7 | 2.0 | 0.8 | 2.1 | 2.2 | 1.0 | 0.8 | -0.3 |
| Netherlands | 1.1 | 0.7 | -0.7 | -1.1 | 1.5 | 1.0 | 1.3 | 2.4 | 4.0 | 4.3 | 2.2 | 2.2 |
| Norway | — | 0.4 | -7.6 | -6.1 | -5.9 | -1.7 | -0.6 | 4.2 | 1.9 | 4.3 | 5.5 | 5.2 |
| Spain | — | -5.1 | -1.9 | -2.6 | -1.5 | 0.1 | 3.0 | 2.1 | 0.7 | 4.2 | 3.0 | 2.5 |
| Sweden | 0.0 | -1.8 | -2.5 | 3.0 | 1.6 | 0.6 | 1.9 | 5.9 | 6.1 | 3.6 | 2.7 | — |
| Turkey | -18.0 | -33.7 | -35.6 | 9.0 | 10.2 | 5.8 | -3.1 | 3.1 | 8.6 | -12.7 | -14.9 | -9.1 |
| United Kingdom | -2.1 | -4.7 | -0.9 | -1.2 | 3.2 | 1.1 | 1.7 | 3.1 | 3.1 | 1.3 | -2.1 | -1.8 |
| United States | — | — | — | — | — | — | — | — | — | — | — | — |
| Average OECD | -1.6 | -4.3 | -4.5 | -1.0 | 1.0 | 0.8 | 1.6 | 2.2 | 3.4 | 2.0 | 1.3 | 1.4 |
| *Developing countries* | | | | | | | | | | | | |
| Argentina | -14.7 | -9.5 | -4.4 | 9.4 | -27.7 | -30.2 | -39.0 | -21.9 | -12.0 | -14.8 | 9.2 | — |
| Bangladesh | 0.9 | -9.4 | -3.9 | -6.4 | 5.5 | 1.1 | 1.5 | -0.8 | 1.3 | 1.7 | 4.1 | 3.2 |
| Bolivia | — | -20.2 | -4.8 | 2.6 | -67.1 | -67.3 | -90.8 | — | — | — | — | — |
| Brazil | 3.2 | -15.8 | -19.1 | -4.5 | 4.6 | -10.0 | 6.9 | -0.8 | -1.8 | -5.9 | -13.3 | 223.9 |
| Burkina Faso | -13.1 | 5.2 | -3.9 | 8.0 | 1.2 | 9.3 | 1.7 | 4.1 | 1.4 | 3.1 | — | 7.7 |
| Chile | 27.0 | 5.0 | 5.2 | 28.0 | 23.3 | 5.3 | 8.6 | 5.6 | 8.6 | 5.5 | — | — |
| Colombia | 3.1 | -4.6 | 0.7 | 2.1 | 3.2 | 9.9 | 5.2 | — | — | — | 4.8 | — |
| Côte d'Ivoire | -11.4 | -7.6 | -4.4 | -0.3 | 3.6 | 1.1 | — | — | — | — | — | — |
| Czechoslovakia | 1.1 | -1.1 | -0.2 | 1.8 | -2.3 | 1.7 | 1.7 | 0.3 | 2.1 | 2.6 | 2.5 | 0.9 |

| | | | | | | | | | | | | |
|---|---|---|---|---|---|---|---|---|---|---|---|---|
| Dominica | -5.0 | -22.4 | -14.3 | -2.9 | 0.9 | 2.3 | 2.9 | 1.4 | 1.8 | 2.0 | — | 0.3 |
| Ecuador | -2.9 | -0.4 | -2.0 | -6.2 | -9.2 | -24.9 | -6.1 | -2.7 | -4.7 | -5.4 | -27.8 | -9.1 |
| Ghana | -35.3 | -27.5 | -25.4 | -45.3 | -10.5 | -49.3 | -16.4 | 6.8 | -3.8 | -12.2 | -9.0 | — |
| Honduras | — | — | — | — | 1.7 | 3.2 | 6.4 | 5.4 | 6.3 | 6.5 | 1.8 | — |
| India | 2.5 | -4.8 | -4.5 | -0.7 | -1.7 | -2.8 | 2.8 | 0.1 | 5.0 | 7.8 | 11.5 | 11.8 |
| Indonesia | 2.5 | -13.0 | -6.6 | 1.9 | -0.6 | 3.6 | 8.1 | 12.7 | 7.5 | 8.4 | 8.6 | 1.6 |
| Jamaica | -25.0 | -7.9 | -14.3 | 5.0 | 1.8 | -2.0 | -12.9 | -1.8 | 7.0 | 4.4 | 0.3 | 2.1 |
| Kenya | -7.5 | -3.7 | -6.5 | -8.8 | -1.0 | 3.0 | 0.8 | 0.8 | 8.6 | 3.8 | 2.6 | 4.7 |
| Korea, Rep. of | 1.9 | -2.1 | -7.8 | 7.0 | 3.0 | 5.9 | 6.4 | 6.6 | -1.0 | -15.7 | -10.2 | 13.9 |
| Malawi | 2.3 | -10.6 | -2.6 | 1.6 | -1.6 | -4.9 | 0.7 | 1.5 | 5.7 | 2.6 | 3.6 | — |
| Malaysia | 2.5 | 2.1 | 0.8 | 0.7 | 5.0 | 5.2 | 7.7 | 7.9 | -7.9 | -21.7 | 6.9 | 22.1 |
| Mexico | -2.7 | -5.9 | -6.6 | 0.2 | -25.3 | -11.5 | -6.2 | -0.5 | 3.9 | 6.0 | 7.6 | — |
| Morocco | -2.5 | -1.8 | -1.1 | -4.2 | 3.1 | -2.3 | 2.3 | 1.4 | -3.8 | 3.1 | 2.1 | — |
| Nigeria | -7.7 | -0.1 | -13.4 | -4.1 | -5.0 | -21.4 | -8.7 | 8.0 | 8.3 | 7.5 | — | 8.7 |
| Oman | — | — | — | — | — | — | 9.0 | 9.0 | — | — | — | — |
| Pakistan | 2.1 | 2.6 | -2.3 | -1.8 | 2.9 | 3.8 | 3.2 | — | 11.6 | 0.7 | — | 0.1 |
| Peru | -23.1 | -20.1 | -17.6 | -11.4 | -10.4 | -31.1 | -24.3 | 15.8 | -9.8 | -18.4 | -27.6 | — |
| Philippines | 3.7 | -8.8 | -3.2 | 3.0 | 5.4 | -9.7 | -23.4 | -7.0 | -45.5 | -48.0 | -15.9 | -35.3 |
| Poland | — | — | -6.8 | -20.5 | -45.1 | -11.4 | -6.9 | -45.5 | 2.9 | 1.2 | -1.6 | 1.2 |
| Sierra Leone | -2.7 | -11.4 | -0.9 | -11.6 | -11.6 | -30.5 | -37.1 | 20.2 | 7.9 | 5.6 | 6.2 | 3.2 |
| Sri Lanka | -1.7 | 0.3 | -3.7 | 1.5 | 15.8 | 0.5 | 13.3 | 9.4 | -3.4 | -2.1 | — | -2.8 |
| Thailand | 0.2 | -5.2 | -3.8 | 0.6 | 10.2 | 8.8 | 13.4 | -1.3 | -3.4 | -22.3 | — | -28.6 |
| Trinidad and Tobago | — | — | — | — | -4.1 | -6.7 | -6.4 | 3.9 | -6.6 | -51.6 | -19.6 | — |
| Venezuela | 0.9 | -8.8 | -6.4 | -4.5 | 7.6 | 6.4 | -5.0 | 3.9 | -6.6 | -24.8 | — | — |
| Zaire | -40.4 | -33.0 | -1.2 | 2.7 | -7.8 | -35.3 | 13.7 | -3.7 | — | -1.4 | -32.1 | — |
| Zambia | -4.7 | -2.5 | -2.8 | -15.0 | -5.2 | -10.3 | -7.6 | -27.2 | -12.5 | — | — | — |
| Zimbabwe | -2.5 | -13.7 | -1.9 | -5.0 | -5.1 | -18.4 | -3.0 | 0.8 | -6.5 | — | 0.1 | -6.4 |

— Not available.

**Table A.5. Taxes from Financial Repression in Ten Countries**
(percentage of GDP)

| Study and country | 1980 | 1981 | 1982 | 1983 | 1984 | 1985 | 1986 | 1987 | 1988 | 1989 | Average |
|---|---|---|---|---|---|---|---|---|---|---|---|
| *Tax on financial intermediation (excluding inflation tax)* | | | | | | | | | | | |
| Easterly (1989)[a] | | | | | | | | | | | |
| Argentina | 0.1 | 0.1 | 0.7 | 1.7 | 0.3 | -2.9 | 0.1 | — | — | — | 0.0 |
| Chile | 0.2 | 1.8 | 1.4 | 0.0 | 0.0 | 0.1 | — | — | — | — | 0.4 |
| Colombia | -0.2 | -0.2 | -0.2 | -0.8 | -0.9 | -0.6 | — | — | — | — | -0.3 |
| Mexico | 0.7 | -0.2 | 5.5 | 3.9 | 1.7 | 0.1 | 5.1 | — | — | — | 2.0 |
| Morocco | 0.6 | 1.0 | 0.0 | 0.8 | 0.2 | 0.3 | -0.6 | — | — | — | 0.4 |
| Thailand | 0.3 | -0.0 | -0.8 | -0.7 | -1.2 | -1.0 | -0.8 | — | — | — | -0.4 |
| Giovannini and de Melo (1990)[b] | | | | | | | | | | | |
| Colombia | 0.2 | 0.2 | 0.2 | 0.3 | 0.4 | — | — | — | — | — | 0.3 |
| Mexico | — | — | — | — | 0.5 | 0.8 | 10.8 | 11.0 | — | — | 5.8 |
| Morocco | 1.1 | 5.5 | 2.9 | 3.7 | 4.7 | 3.3 | — | — | — | — | 2.3 |
| Pakistan | — | — | 3.6 | 2.9 | — | — | — | — | — | — | 3.3 |
| Thailand | 1.2 | 0.9 | 0.2 | 0.5 | 1.7 | -0.9 | — | — | — | — | 0.4 |
| Zimbabwe | 5.8 | 4.6 | 9.1 | 6.7 | 7.4 | -0.5 | — | — | — | — | 5.5 |
| Chamley and Honohan (1990)[c] | | | | | | | | | | | |
| Côte d'Ivoire | — | 0.0 | 0.0 | 0.0 | 0.0 | 0.0 | 0.0 | 0.0 | 0.0 | — | 0.0 |
| Ghana | — | 0.0 | 2.4 | 0.0 | 0.1 | 0.4 | 0.5 | 0.2 | 0.2 | — | 0.5 |

*Tax on time deposits: estimations from this study[a]*

| | | | | | | | | | | |
|---|---|---|---|---|---|---|---|---|---|---|
| Argentina | 0.8 | -1.4 | 3.4 | 3.5 | 4.0 | 2.2 | 1.6 | 2.1 | -1.2 | 1.7 |
| Chile | -0.4 | -4.7 | -3.8 | -0.5 | -0.1 | -0.4 | -0.1 | -0.4 | — | -1.2 |
| Côte d'Ivoire | 0.4 | 0.1 | -0.2 | 0.0 | -0.4 | — | — | — | — | 0.0 |
| Colombia | 0.0 | -0.1 | -0.1 | -0.5 | -0.4 | -0.3 | -0.4 | -0.2 | -0.2 | -0.2 |
| Ghana | 0.9 | 1.3 | 0.4 | 0.9 | 0.3 | -0.1 | 0.1 | 0.4 | 0.3 | 0.5 |
| Morocco | 0.1 | 0.4 | -0.2 | 0.3 | -0.2 | -0.0 | -0.4 | -0.7 | -0.8 | -0.2 |
| Mexico | 1.3 | 0.2 | 3.9 | 1.9 | 1.2 | 0.2 | 1.2 | 2.9 | — | 1.6 |
| Pakistan | 0.4 | 0.3 | -0.3 | -0.4 | -0.3 | — | — | — | — | -0.1 |
| Thailand | 1.2 | 0.1 | -3.2 | -3.1 | -5.7 | -4.2 | -3.6 | -2.5 | -2.7 | -2.5 |
| Zimbabwe | 0.5 | 0.6 | 1.0 | 2.6 | 0.6 | 0.0 | 1.0 | 0.4 | — | 0.8 |

— Not available.

a. The tax is calculated as the negative of the real interest rate multiplied by government domestic debt outside the central bank as a percentage of GDP. (The implicit long-run real interest rate is assumed to be zero.)

b. The tax is calculated as the difference between the foreign and the domestic interest rate multiplied by government domestic debt outside the central bank as a percentage of GDP.

c. The tax is calculated as 1 minus the domestic real interest rate multiplied by government debt outside the central bank as a percentage of GDP.

d. The tax is calculated as the OECD average real interest rate (0.9 percent) minus domestic real interest rate multiplied by time deposits as a percentage of GDP. The data used are from IMF, *International Financial Statistics*, various years.

# Author Index

# Subject Index

Agricultural marketing boards, reforms of, 35, 38, 313

Argentina: Austral plan in, 53, 128, 142; Banco Hipotecario (state mortgage bank) of, 117; bank deposits in, 122, 136–39, 139–45; BONEX plan in, 101, 126; Bunge and Born plan in, 53, 126, 142; currency devaluations in, 6–7, 106; currency overvaluation in, 102; debt meltdowns in, 53, 106–07; deficit financing in, 3, 9, 116–22; demand for real cash balances in, 128, 130–32; dollar-denominated instruments (LEDOL and BONEX) in, 101, 142–45; ''dollarization'' in, 6–7, 107; domestic debt of, 9, 124; economic growth in, 9–10, 102, 104, 106; Erman plan in, 53, 142; external debt of, 9, 82, 153; financial markets in, 135–36; fiscal policies in, 34, 35; foreign shocks in, 26; indexation in, 106; inflation and hyperinflation in, 3, 6, 46, 101–02, 104, 122; inflation tax in, 117–19; interest rates in, 122, 136–39; measuring CNFPS of, 105–06, 108–09, 114; money creation (seigniorage) in, 43, 114, 117–19, 120–22; money demand in, 126, 128, 130–32; Primavera plan in, 53, 124, 126, 128, 142; private consumption in, 158–62; private credit as share of GDP in, 53; private investment and saving in, 4, 5, 162–64; public sector primary deficits in, 104, 106–07, 143, 155–56, 157; public sector primary surpluses in, 102; public spending (consumption) in, 102, 104, 108, 162–64; quasi-fiscal deficits and expenditures in, 6, 19, 114–16, 118; real exchange rate in, 149–54; *Rodrigazo* inflation in, 122; saving in, 5; stabilization plans in, 53, 101–02, 124, 126, 128, 142, 143, 164 n1; terms of trade for, 151, 152, 153, 155; trade balance in, 9, 154–56. *See also* Central Bank (Argentina)

Asset portfolio model, 183–85, 213–15

Bank deposits: in Colombia, 230, 244–45; interest elasticity of demand in Argentina and, 139–45; interest rates in Argentina and, 136–39; Islamic noninterest basis for, 446; *plazos fijos,* 122. *See also* Interest rates

Banque Centrale des Etats de l'Afrique de l'Ouest (BCEAO), 273

Barro-Ricardian hypothesis on consumption, 64–65. *See also* Ricardian equivalence hypothesis

Black market: fiscal balances and, 24, 336–37, 339–46; in Ghana, 316–17, 336–46; methodology for analyzing, 337–46

Bolivia, 46

Budget deficit. *See* Public sector deficit